Electric Vehicles: Prospects and Challenges

Electric Vehicles: Prospects and Challenges

Edited by

Tariq Muneer
Edinburgh Napier University, Edinburgh, Scotland, United Kingdom

Mohan Lal Kolhe
University of Agder, Kristiansand, Norway

Aisling Doyle
Edinburgh Napier University, Edinburgh, Scotland, United Kingdom

ELSEVIER

Elsevier
Radarweg 29, PO Box 211, 1000 AE Amsterdam, Netherlands
The Boulevard, Langford Lane, Kidlington, Oxford OX5 1GB, United Kingdom
50 Hampshire Street, 5th Floor, Cambridge, MA 02139, United States

Notices
Knowledge and best practice in this field are constantly changing. As new research and experience broaden our understanding, changes in research methods, professional practices, or medical treatment may become necessary.

Practitioners and researchers must always rely on their own experience and knowledge in evaluating and using any information, methods, compounds, or experiments described herein. In using such information or methods they should be mindful of their own safety and the safety of others, including parties for whom they have a professional responsibility.

To the fullest extent of the law, neither the Publisher nor the authors, contributors, or editors, assume any liability for any injury and/or damage to persons or property as a matter of products liability, negligence or otherwise, or from any use or operation of any methods, products, instructions, or ideas contained in the material herein.

Library of Congress Cataloging-in-Publication Data
A catalog record for this book is available from the Library of Congress

British Library Cataloguing-in-Publication Data
A catalogue record for this book is available from the British Library

ISBN: 978-0-12-803021-9

For information on all Elsevier publications visit our
website at https://www.elsevier.com/books-and-journals

  Working together
to grow libraries in
developing countries

www.elsevier.com • www.bookaid.org

Publisher: Joe Hayton
Acquisition Editor: Lisa Reading
Editorial Project Manager: Peter Jardim
Production Project Manager: Sruthi Satheesh
Cover Designer: Mark Rogers

Typeset by SPi Global, India

Contents

List of contributors

Girard Aymeric Adolfo Ibáñez University, Viña del Mar, Chile

T.P. Chathuri Madusha University of Agder, Kristiansand, Norway

Aisling Doyle Edinburgh Napier University, Edinburgh, Scotland, United Kingdom

Simon François University of Granada, Granada, Spain

Eulalia Jadraque Gago University of Granada, Granada, Spain

Irene Illescas García Edinburgh Napier University, Edinburgh, Scotland, United Kingdom

Michael Jeffrey Edinburgh Napier University, Edinburgh, Scotland, United Kingdom

Matjaž Knez University of Maribor, Celje, Slovenia

Mohan Lal Kolhe University of Agder, Kristiansand, Norway

Yash Kotak Heriot Watt University, Riccarton, Edinburgh, United Kingdom

Ross Milligan Edinburgh College, Dalkeith, United Kingdom

Parimita Mohanty UN Environment, Asia Pacific office

Tariq Muneer Edinburgh Napier University, Edinburgh, Scotland, United Kingdom

Koki Ogura Kawasaki Heavy Industries, Ltd., Kobe, Japan

Foreword

Cities, businesses, and national governments around the world have recognized electric vehicles (EVs) as an essential part of a smarter and more sustainable future. The multiple environmental, economic, and energy system benefits offered by EVs have established broad consensus on why this transformation is needed. The challenge that we face today is therefore finding ways to make this happen as quickly as possible.

Since 2009, I have travelled the world collecting good ideas, crafting strategies, and developing projects to accelerate EV adoption. I have found that many of the challenges that we face today are common to different countries. I have also experienced a growing confidence in how advances in technologies, business models, and infrastructure deployment will deliver new solutions to these challenges.

As director of E-cosse, a partnership to accelerate widespread adoption of EVs in Scotland, I have also had the opportunity to work with the Scottish Government to put some of these solutions into practice. This has been guided by a comprehensive roadmap that sets out the ambition to free Scotland's towns, cities, and communities from the damaging emissions of fossil-fueled vehicles.

My work in Scotland has greatly benefited from the expertise and enthusiasm of the authors of this book from Edinburgh Napier University. Professor Muneer and his team have been active champions for EVs in Scotland and in promoting a wider understanding of the linked opportunities for innovation in transport and energy systems.

This book presents critical insights on the technologies and systems that will enable widespread transport electrification. It builds on real-world data and practical experiences from countries including Scotland, Norway, Spain, Chile, and India. Pulling together these insights represents a considerable achievement and makes an important contribution that will inform and inspire further research and positive actions around the world.

David Beeton
Urban Foresight Ltd.

Preface

The aims of this monograph are threefold: (a) present a case for electric vehicles, (b) discuss the relevant science and performance issues, and (c) highlight case studies for a number of countries across the world. Following the year 2015 disclosure of automobile emission-cheating scandal of a number of global manufacturers and the clearly identified link between diesel engines and pollution-related deaths, a number of countries and cities across the world have now committed to phase out the latter mobile power plants. With hydrogen-based vehicle development as far away as it ever was and the rather limited overall energetic efficiency of production, storage, and reuse of hydrogen, the only logical alternative has to be electric propulsion. However, there are a few challenges that need to be addressed: the range anxiety, the limitations of the electrical grid, the cost of electric batteries, the sourcing of lithium for Li-ion batteries, and the extra cost of cabin heating as opposed to the free energy that is available for the latter purpose for fossil-fuel burning vehicles. The latter follows under the dictum of the second law of thermodynamics—for any given heat engine, heat rejection is inevitable. However, for electric vehicles, no such free energy is available, and in countries of high latitude, the cost of vehicle cabin heating can be as much as 50% of the total journey cost. An attempt has presently been made to critically discuss the above-highlighted issues, and solutions are offered.

Hopefully, this book will meet the needs of energy and transport policy planners and graduate students and researchers. The book attempts to fill the present knowledge gap by providing the necessary fundamentals, explaining the practical aspects. It also covers topics such as traction modelling, energy use for heating and cooling of cabin space, electric vehicle battery technologies, their modelling, and next-generation battery-driven light rail vehicles.

Between them, the authors have over 6 year' experience of driving three different makes of electric vehicles (EVs) and have managed a programme of research and development related to fast- and rapid-charging network for EVs. They have also gained over 25 years' experience related to solar photovoltaic electricity generation and have dabbled in the art of hydrogen production through electrolysis and fuel-cell evaluation. The authors maintain regular contact with electric vehicle users—private, corporate, and public. They also work with government agencies to promote the use of electric vehicles.

The authors are indebted to their respective academic institutions—Edinburgh Napier University (ENU), Scotland, and University of Agder, Norway, for moral and research funding support without which this work would not have seen the light of the day. Steve Paterson of ENU designed the cover for our book, and indeed, we appreciate his help.

We are also grateful to our publisher Elsevier's editorial project manager Mr. Peter Jardim for his ceaseless support and gentle nudging to meet our deadlines.

The authors welcome suggestions for additions and improvements.

Tariq Muneer and Aisling Doyle
Edinburgh

Mohan Lal Kolhe
Agder

The automobile

Tariq Muneer, Irene Illescas García
Edinburgh Napier University, Edinburgh, Scotland, United Kingdom

1.1 Introduction

In the year 2000, a leading magazine carried out a survey in which the readers were asked to name a single word that would encompass the development of the previous century. There were a large number of quite potent competitors—computer, phones, and Internet; however, the word that won the competition was 'automobile'. Indeed, the automobile is one of the greatest inventions of modern times. While the Internet enables transfer of data over large distances with speed, the automobile transports people with speed, comfort, safety, and economy!

Currently, there are over 1.2 billion vehicles on this planet. That figure is expected to rise to 2 billion by 2035. The runaway success of the automobile is due to its impressive performance—travels more than 300 mi without refuelling the fuel tank and can attain speeds of over 100 mph. Fig. 1.1.1 shows the historical rise of global automobile fleet.

Automobiles are a liberating technology for people around the world. The personal automobile allows people to live, work, and play in ways that were unimaginable a century ago. Automobiles provide access to markets, to doctors, and to jobs. Nearly every car trip ends with either an economic transaction or some other benefits to our quality of life. Table 1.1.1 presents the merits and demerits of an automobile.

Due to economic and other reasons, people around the world are passionate with respect to automobiles. 'Violent protests and looting, prompted by a big hike in petrol prices have erupted across Mexico' reports the 'i' newspaper, London in its 6 Jan. 2017 edition. The change takes the price of a gallon (5 l) of petrol to £3 that equates to the minimum daily wage ('i' newspaper, London, 6 Jan. 2017, p. 24).

Having acquired the taste of freedom of movement, the population at large would simply be unwilling to give up their automobile. Drawn on the theme presented in the famous 'Thousand and One Nights' tale from Baghdad, Fig. 1.1.2 depicts the above dilemma.

The global auto industry is a key sector of the economy for every major country in the world. The industry continues to grow, registering a 30% increase over the past decade (1995–2005). Building over 66 million vehicles a year (year 2005 statistics) requires the employment of about 9 million people directly in making the vehicles. That is over 5% of the world's total manufacturing employment. It is estimated that each direct automobile job supports at least five indirect jobs in the community, resulting in more than 50 million jobs. Automobiles are built using the goods of many industries, including steel, iron, aluminium, glass, plastics, glass, carpeting, textiles,

Electric Vehicles: Prospects and Challenges. http://dx.doi.org/10.1016/B978-0-12-803021-9.00001-X

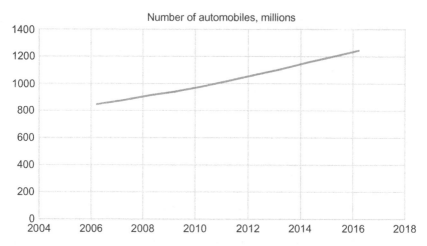

Fig. 1.1.1 Global insatiable demand for automobiles.
http://www.oica.net/category/sales-statistics/.

Table 1.1.1 **Why own an automobile?**

Advantages	Disadvantages
Freedom of movement	Air pollution, climate change
Personal security	Road congestion
Large employment sector	Fossil fuel exhaustion
Personal or work space	Loss of building material
Ability to travel long distances	Space consumption on roads
Increased speed	Society stratification
Enjoining of communities	Accidents

computer chips, rubber, and more. This level of output is equivalent to a global turn-over (gross revenue) of over €2 trillion. It is said that, if auto manufacturing were a country, it would be the sixth largest economy.

The automobile industry is also a major innovator, investing over €84 billion in research, development, and production. The auto industry plays a key role in the technology level of other industries and of society. Vehicle manufacturing and use are also major contributors to government revenues around the world, contributing well over €400 billion. Box 1.1.1 summarizes the efforts of auto industry to improve the efficiency of the modern fleet.

However, the savings in emissions per unit is negated by the rapid increase of auto fleet. The only true sustainable solution is to introduce zero tailpipe emission via electric vehicles that are charged with renewable electricity. Stopgap solutions that offer reduced emissions will only shift, and not arrest, the time of occurrence of runaway greenhouse effect. In this respect, the following discussion is offered with respect to introduction of higher efficiency Chinese coal-fired power plants.

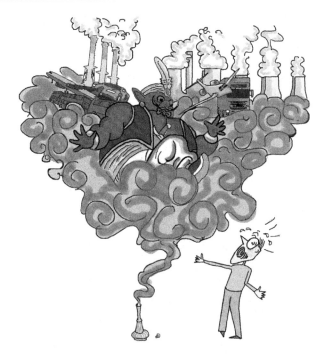

Fig. 1.1.2 The Genie of freedom to travel has been released making it difficult to reverse that process.

Box 1.1.1 Technological developments that led to an increase of thermal efficiency of automobile

Direct injection

Port fuel-injection systems long ago replaced carburettors in cars because of their efficiency and lower maintenance requirements. But now, automakers are moving towards the even more efficient direct injection. With port fuel injection, gasoline is sprayed into the intake manifold, where it mixes with air, and then is sucked down into the cylinders. Direct injection places an injector on each cylinder, spraying gasoline into the cylinder itself. Direct injection leads to more efficient engines, as the gasoline-air mixture burns more completely. The individual injectors also ensure that each cylinder gets the same amount of fuel, and the spray can be more precisely timed.

Turbocharger

Turbocharging is usually associated with performance, but now, it is being used for fuel economy. A turbocharger is a turbine that uses an engine's own exhaust

Continued

Box 1.1.1 Technological developments that led to an increase of thermal efficiency of automobile—cont'd

gases for its power. At the other end is a compressor that forces air into an engine's intake manifold. That pressurized air running into the cylinders creates more power. During steady-speed cruising, the turbocharger is idle, and the engine gets higher fuel economy.

Deceleration fuel shut-off

Engine designers explore every part of the drive cycle to obtain efficiency gains. A recent innovation is shutting off the fuel flow to the cylinder when the car is decelerating.

Cylinder deactivation

Cylinder deactivation is based on the idea that, when cruising at a steady speed, the driver does not need full power output. The technology has been improved with the advent of fuel injection and computer control of engines

Idle stop

Idle stop saves fuel by shutting down the engine when the car comes to a halt and is being adopted for standard gasoline engines. The engine is switched off once the foot brake is pressed at a stop. The engine consequentially shuts off. The engine starts up again once the brake is released.

Brake regeneration

This technology borrows heavily from hybrid cars. When decelerating or braking, the car captures the kinetic energy and converts it to electricity. This electricity is used to charge up the car's battery. One car manufacturer claims that this technology can increase fuel economy by 3%.

Lock-up clutch for automatic transmissions

Historically, manual transmissions have offered better fuel efficiency than automatics, as the latter use a torque converter, which wastes engine energy when shifting gears. A lockup clutch in an automatic transmission helps keep gear shifts efficient by maintaining a fixed connection between the transmission and drive shaft whenever possible.

Dual-clutch transmission

Derived from race car technology, a dual-clutch transmission combines the efficiency of a manual transmission with the convenience of an automatic. Although primarily used on performance cars, dual-clutch transmissions are starting to be used for fuel efficiency.

Continuously variable transmission

The continuously variable transmission (CVT) obviates fixed gears in favour of a pulley system, which can constantly adjust the drive ratio to best match the optimum engine speed. Letting the engine run at an optimum speed for as much driving time as possible leads to greater fuel efficiency.

Box 1.1.1 Technological developments that led to an increase of thermal efficiency of automobile—cont'd

Electric power steering

Past power-steering systems used hydraulics, relying on the engine to build pressure in the system. The hydraulic pump created a drag on the engine that caused it to burn more fuel. Using an electric motor for power-steering boost eliminates this drag. Almost all automakers are moving to electric power-steering systems with the view to improve fuel efficiency.

https://www.cnet.com/roadshow/pictures/11-gas-saving-technologies-photos/

The pollution generated by combustion engine (CE) vehicles though will eventually be its undoing. The UK industry executives believe that the diesel technology will be a thing of the past. They have also observed that a greater investment will be needed in battery technology over the next 5 years. That was the consensus, according to 93% of motoring executives ('i' newspaper, London, 9 Jan. 2017, p. 3).

For example, some of the most polluted roads in the UK capital, London, that include Brixton Road, Oxford Street, King's Road, and Putney High Street quite regularly exceed the 200 $\mu g/m^3$ hourly legal limit for nitrogen dioxide in their respective atmosphere. The report also claims that a direct result of the prevalent air pollution problem is that it reduces the life expectancy by 17 months, thus causing 9000/annum premature deaths. The problem is not just confined to London alone. Some 40% of local authorities in the United Kingdom that include Glasgow, Leeds, and Birmingham regularly exceed the above legal limit for nitrogen dioxide ('i' newspaper, London, 9 Jan. 2017, p. 20).

The world CO_2 annual share of transport sector is over 2.7 billion tonnes that is 23% of the total emissions. Most of the latter emissions result from private road vehicles. The year 2016 was the hottest ever on record, it being 0.2°C warmer than the year 2015 that has set record for warmest year. It was also reported that 2016 recorded 20°C, like with like, higher than normal temperatures in the Arctic. At 14.8°C, the 2016 global temperature was 1.5°C higher than those recorded at the beginning of the industrial revolution ('i' newspaper, London, 6 Jan. 2017, p. 23). Furthermore, across the world, automobile-related road accidents claim 1.25 million lives each year, and the average urban speeds hardly ever exceed 10 mph.

With the view to reduce the carbon loading of the atmosphere, the United Kingdom and other governments around the world have started a subsidy scheme to introduce electric vehicles. Within the past 5 years, this has resulted in the introduction of over 70,000 electricity-propelled automobiles that now populate British roads, for example.

Recent research has uncovered the scale of the problem, with 3 million premature deaths a year attributed to nitrogen dioxide, particulate matter, and other pollutants.

The city of Barcelona has taken a strong step in its policy of abatement of greenhouse gas emission by announcing that it will ban one million cars from its city centre by year 2020. These measures follow a monitored increase of 11% in annual average nitrogen dioxide levels. Partial bans on heavily polluting vehicles will begin in Apr. 2017 with a permanent inner-city prohibition introduced 4 years later. Petrol-powered cars registered between years 2000 and 2007 that represent 50% of the city's car fleet will be affected by the above ban ('i' newspaper, London, 23 Nov. 2016.)

The mayors of all major European cities are grappling with how to tackle pollution due to diesel emission. Therefore, four of the world's biggest cities are to ban diesel vehicles from their centres within the next decade, as a means of tackling air pollution, with campaigners urging other city leaders to follow suit. The mayors of Paris, Madrid, Athens, and Mexico City announced plans on Friday to take diesel cars and vans off their roads by 2025.

Note that automobile manufacturers have used technical fixes to suppress the true emission reports. Box 1.1.2 shows the level and spread of such activity.

Today's automobiles, in general, are large and heavy objects that occupy a big area of road space while they are parked (space pollution), emit poisonous gases on the kerbside (local pollution) in addition to carbon dioxide emission (global pollution). Additionally, due to their weight and powerful engines, the vehicles produce sound pollution.

Box 1.1.2 Technical fixes adopted by automobile manufacturers for suppressing true emission reports

The Volkswagen's emissions-cheating software was discovered in September of 2015. (http://www.roadandtrack.com/new-cars/car-technology/a29293/vehicle-emissions-testing-scandal-cheating/). In the ensuing months, nearly every other major automaker worldwide has come under increased scrutiny regarding real-world vehicle emissions and fuel economy. However, Volkswagen was only one of many manufacturers who were engaged in similar activities. The following is a synopsis reported by the above-cited website.

Opel

In May 2016, 'Der Spiegel' and 'ARD' television program 'Monitor', and the environmentalist group Deutsche Umwelthilfe discovered software on diesel-powered Zafira minivans and Insignia sedans that turn off emissions controls during real-world driving. However, while the software does indeed turn off the affected vehicles' emissions controls in most real-world driving scenarios, it is found to be 100% compliant with European Union laws. That is because the EU law allows automakers to program their emissions controls to deactivate when necessary to protect the engine from harm. And it allows the automakers to define for themselves what counts as a protective shutdown. So Opel's affected diesel vehicles shut off all emissions controls at ambient temperatures below 20°C, or above 30°C, or at speeds over 145 km/h, or engine speeds more than 2400 RPM, or at elevations higher than 850 m. These are quite a few exemptions!

Box 1.1.2 Technical fixes adopted by automobile manufacturers for suppressing true emission reports—cont'd

Chevrolet/GMC/Buick

It was discovered that the year 2016 models produced by the US automaker 'Chevy Traverse', 'GMC Acadia', and 'Buick Enclave' were all sold with window stickers that indicate fuel economy ratings a full two MPG better than the US Environmental Protection Agency's official figures.

Daimler

In same the year, i.e. 2016, a group of US-based Mercedes owners filed a class action lawsuit, alleging that the automaker's BlueTEC diesel-powered vehicles shut off their emissions controls in real-world driving. A second class action suit filed in the same year characterized the software as a 'defeat device' akin to Volkswagen's TDI. Both suits allege that Mercedes BlueTEC diesels, which use a costly and complex injection system to combat diesel NOx emissions, shut down at ambient temperatures below 10°C.

Fiat Chrysler

Fiat recently found itself in an emissions-related problem over the European-market diesel-powered 500X. The problem is similar to the Opel situation. In Apr. 2016, German news magazine 'Bild am Sonntag' reported that Fiat's 2.0 l diesel-powered 500X almost entirely shuts off its emissions control devices after 22 min of driving. German environmental activist group DUH says a diesel 500X it tested put out between 11 and 22 times the legal limit of NOx emissions when tested with a warm engine. Note that European emissions test starts with a cold engine and lasts 20 min.

Mitsubishi

In Apr. 2016, Nissan revealed that Mitsubishi had been artificially boosting its official fuel economy ratings by up to 10% by overinflating tires during testing. At first, the act was thought to only affect about 600,000 Japanese-market kei cars, with 660 cc engines. Around 470,000 of these were built by Mitsubishi. Within a week, Mitsubishi admitted that the fuel economy deception reached all the way back to 1991, a systematic effort affecting millions of vehicles.

PSA and Renault

In 2016, the manufacturer recalled nearly 16,000 European-market diesel-powered SUVs and offered a 'voluntary' software fix to reduce the NOx emissions of nearly 700,000 diesel-powered vehicles.

Independent investigations in Europe have found that nearly every diesel-powered vehicle emits far more in real-world driving than in government-designed tests. The result has been increased scrutiny of nearly every major automaker and not just for diesels. Environmental groups and governments will continue to probe the problems related to vehicle emissions and fuel economy.

Note that the ratio of driver's to gross weight of the vehicle is only around 5%. Presently, there are three competing energy sources for driving automobiles—electricity, fossil fuels (compressed or liquefied natural gas, diesel, or petrol), or hydrogen. Hydrogen is not a naturally occurring gas and hence has to be derived from energy intensive processes. Analysis presented in two independent researchers show that for 1 MWh of electricity sourced the hydrogen-powered vehicle, involving a two-stage conversion: electrolysis and fuel cell, would deliver 1790 km journey length. However, the battery electric vehicle (BEV) would deliver 5525 km.

Traffic-related road congestion ought to be avoided as it leads to wastage of time, lost productivity, high levels of pollution from the current auto fleet, and risks to society such as delays to the flow of emergency vehicles such as ambulance, fire engines, and police. Note that cities account for only 2% of the global land mass but 75% of energy consumption. Furthermore, the UN has a predicted that 80% of the global population will reside in urban areas by 2030. Hence, traffic congestion will only become far worse with time. Fig. 1.1.3 presents the pattern of the likely population shift from rural to urban areas (Source: Mitchell, Borroni-Bird, & Burns, 2010, p. 158). Fig. 1.1.4 shows the relationship between car ownership and affluence (Source: Mitchell et al., 2010, p. 160). Note that electric vehicles lend themselves easily towards 'self-driven' technology with several potential advantages.

Congested roads impact negatively in a number of ways—longer journey times, increased fuel consumption due to start-stop traffic that results in idling and longer running cabin comfort and entertainment systems, increased kerbside and global pollution, and noise. Fig. 1.1.5 presents the relationship between fuel efficiency and average vehicular speed. City-wide congestion is bound to further increase with time as depicted in Fig. 1.1.6. Note, however, that the level of congestion stress will be of

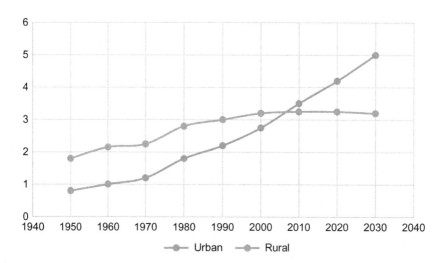

Fig. 1.1.3 Shifting urban and rural human population demographic. Note: units for Y-axis is billion.

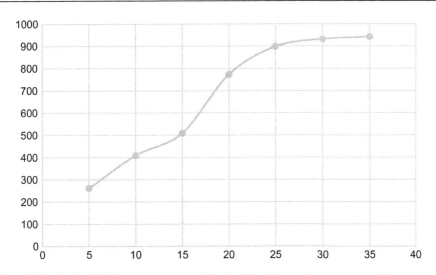

Fig. 1.1.4 Relationship between car ownership (vehicles/thousand persons: *y*-axis) and affluence (annual income, thousand USD/annum: *x*-axis).

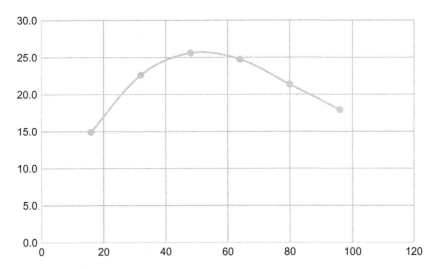

Fig. 1.1.5 Relationship between fuel efficiency (km/l of fuel: *y*-axis) and average vehicular speed (kmph: *x*-axis).

different order of magnitude for the most densely populated cities such as Delhi as shown in Table 1.1.2.

A research study carried out for UK Department for Transport found that converting all cars on UK roads to self-driving mode would cut delays due to congestion by 40%. This would be achieved due to driverless cars can change lanes more efficiently, safely drive close to the car in front and travel at consistent speed without repeatedly

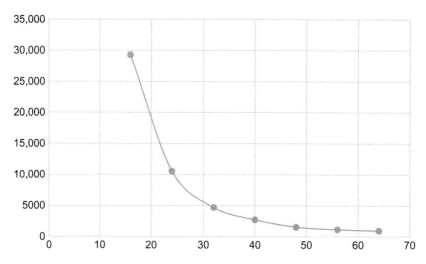

Fig. 1.1.6 Relation between average driving speed (kmph: x-axis) and population density (humans/km^2: y-axis).

Table 1.1.2 **Comparison of road congestion stress for mega-cities**

City	Population/km^2	Vehicles, million
Karachi	24,000	1.7
Delhi	19,698	7.5
Cairo	19,376	0.6
Jakarta	16,992	5.5
Manila	14,500	2.5
Mexico City	14,276	3.2
Mumbai	12,500	2.5
Sao Paolo	7420	3.4
Los Angeles	2734	7.3

Note: Delhi and Los Angeles have almost equal number of vehicles and yet Delhi's human population density is nearly nine times that of latter.

braking and accelerating. The study also found that the benefits of driverless cars would be much more felt in built-up areas where bunching cars together would cut delays. The Times, London (7 Jan. 2017, p. 12).

Any given vehicle will deliver higher efficiency if it was travelling at a constant speed, though the lower that speed is the higher efficiency. Efficiency suffers badly in a stop-start mode—for older petrol or diesel vehicle, the deceleration energy is simply lost and even for a hybrid or electric vehicles the reversing of charge in the regenerative mode still falls short of 100% efficiency. The road congestion is therefore detrimental to the vehicle's efficiency.

City and country planners have to bear in mind that short of a totalitarian edict it is not possible to wean away the masses from car ownership. Once the freedom of movement has been tasted, people are simply unwilling to give up or alter their travel habits. The solution is to introduce a fleet that is truly sustainable on the counts of energy, environment, and space pollution.

The solution to the congestion and space-pollution problem posed by autos may be sought by a fundamental review of the most populous automobiles' design (length and breadth of passenger cars). In this respect, some good solutions have been provided by Mitchell et al. (2010). Fig. 1.1.7 presents those conceptual designs. The design concept for an ultrasmall vehicle (USV) presented by the latter team has a length of 100 in (2.54 m), weigh less than 1000 lbs (454 kg), is propelled by 2×5 kW motors, has a 4 kWh lithium-ion battery and have an energy consumption of 200 mi/mile of US gallon of gasoline (>85 km/l). Note that the average modern car weighs as much as 20 times the weight of its driver, occupies 10 m^2 of space and is parked for 90% of the time, thus occupying road or garage space that causes 'space pollution'. Space in the modern world is at a premium and ought to be included in the overall economics of an automobile. Fig. 1.1.8 presents information related to the average daily distance covered by automobiles for the United States (Source Mitchell p. 165).

Let us take the case of Great Britain, as an example, to ascertain the growing problem of automobile population. At the end of 2015, there were 36.5 million vehicles licensed for use on the roads in Great Britain, of which 30.3 million were cars. The total number of licensed vehicles has increased every year since the end of World War II except 1991. Between 1996 and 2007, the annual growth in licensed vehicles averaged 2.5% a year, although from the mid-2000s, it had already begun to slow down. Following the recession of 2008–09 it slowed further, but did not stop, averaging 0.3% a year between 2008 and 2011. Between 2012 and 2013, the total vehicle stock increased by 1.5%, the first substantial year-on-year increase since 2007. Since 1994, the number of licensed cars in Great Britain has increased by 37%. Over the same period, the numbers of vans (light good vehicles/LGVs) and motorcycles have increased by 57 and 69%, respectively. However, the average carbon dioxide emissions of newly registered cars fell by 3.4% in 2013 compared with 2012 to 128 g/km (vehicles.stats@dft.gsi.gov.uk, accessed on 30 Aug. 2016).

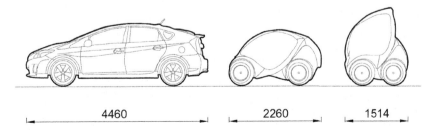

Fig. 1.1.7 Conceptualization of an ultra-compact car which offers solution with respect to energy efficiency and space pollution.

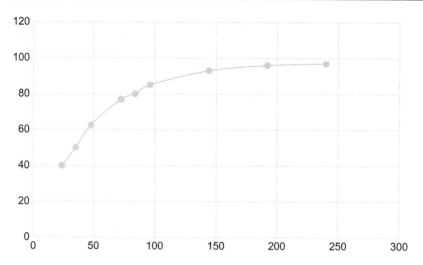

Fig. 1.1.8 Average daily driving distance (km) driven in the United States. While 90% of journeys undertaken in the United States are of 140 km length those in the United Kingdom, in contrast, are only 20 km long.

From the very onset, it is important to understand the mechanism that will bring a true revolution in sustainable transport. The means by which electricity is sourced to charge the electric vehicles (e-vehicles) is the key. It will be shown in this chapter and the book that fossil-fuel-based electricity actually makes the e-vehicles more unsustainable. The route to sustainability will only be achieved through renewable electricity. It is true that, historically, grid electricity carbon intensity within the EU member states, the United States, Japan, China, and India, has been reducing almost continuously. This is due to the closure of a significant number of coal-fired plants and their replacement with gas-, wind-, and solar-powered generation. However, true sustainability will only be achieved when all electric vehicles are charged with renewable electricity.

With the cost of wind and solar electricity dropping at an exponential rate, we are truly witnessing a revolution in plentiful and sustainable electric generation. Thus, it is conceivable that within the next 15–20 years one of the two legs, i.e. the decrease of wind- and solar-generated electricity, would truly deliver sustainable transport solution, the other leg being road congestion.

In 1995, the United Kingdom signed up to international air quality standards. However, in the past 20 years, there has been very little improvement in air quality. Research carried out by Messrs Chatterton and Parkhurst from the University of West of England has shown that currently the poor air quality is contributing to more than 50,000 premature deaths per year in the United Kingdom ('i' newspaper, 30 Aug. 2016).

Tailpipe emissions from automobiles may be eliminated via introduction of electric vehicles and as a start there are three points to consider. First, plug-in hybrids should not be confused with electric cars. Clever engineering produces cars that are significantly heavier and can run solely on battery power for maybe 30 mi. For urban driving, they are brilliant and reduce air pollution; for longer journeys, they offer no advantage at all. Second, fully electric cars can also be excellent for short journeys but cannot replace conventional cars until charging or battery exchange points are as ubiquitous as petrol stations and are comparably quick and easy to use. Third, all this is immaterial unless a much higher proportion of electricity is from nuclear or renewable resources. Until then, and bearing in mind that adding more wind and solar farms without an effective way of storing their output is simply ineffective; emissions are simply shifted from exhaust pipes to power stations. A 1.4 t electric vehicle driven at an average speed of 20 mph draws 3.3 kW of traction power and consumes 0.35 kWh of energy. It, however, regenerates and returns 0.1 kWh of energy to the battery while braking during city driving, the regenerative power at that speed being 2 kW. Table 1.1.3 presents well-to-tank, tank-to-wheel, and well-to-wheel efficiencies and CO2 emissions for a number of vehicles.

Electric vehicles are not a new invention, the first such contraption being built around 1800 by Scottish inventor, Robert Anderson. Increase in battery capacity was achieved due to early work of Plant and Faure beginning in the 1880s. However, it was noted that the batteries were expensive and took a long time to charge. The runaway success of fossil-fuelled automobiles can be attributed to the low cost of oil that was discovered soon after and its energy density. Table 1.1.4 compares the latter property on a volumetric and mass basis (source: Mitchell et al., 2010, p. 92).

Table 1.1.3 Well-tank, tank-wheel, and well-wheel analysis

	Energy (kWh/km)			CO$_2$ emission (g/km)		
Vehicle type	W^a-T^b	T^b-W^c	W^a-W^c	W^a-T^b	T^b-W^c	W^a-W^c
Petrol ICE	0.14	0.60	0.73	31	138	169
Petrol hybrid	0.08	0.32	0.51	22	97	119
Fuel-cell e-vehicle[a]	0.18	0.27	0.45	88	0	88
BEV	0.15	0.14	0.28	56	0	56

[a]Well.
[b]Tank.
[c]Wheel.

Table 1.1.4 Mass and volumetric densities of energy stored in automobile fuels

	Mass density (kWh/kg)	Volumetric density (kWh/l)
Lead-acid	0.04	0.08
Nickel metal hydride	0.09	0.11
Lithium-ion	0.08	0.12
Compressed hydrogen, 350 bar	1.01	0.71
Petrol	6.06	8.81

A typical internal-combustion-engine (ICE)-based automobile has several thousand moving parts, and its energy delivery efficiency is no more than 13% as shown in Fig. 1.1.9. In contrast, owing to the simplicity of design, the greatly reduced total of moving parts and higher grade power source the electric vehicle offers much higher overall efficiency as shown in Fig. 1.1.10. There is another subtle factor that makes an electric vehicle a safer means of transport, i.e. the range anxiety. As a result of continuous display features offered within the e-vehicle including the remainder of the battery charge, the driver is cautious with respect to the vehicle speed—the slower one drives, the more mileage one gets. In this respect, Table 1.1.5 displays the relevant information. A drop in speed from 30–20 mph increases the driving range by 40%,

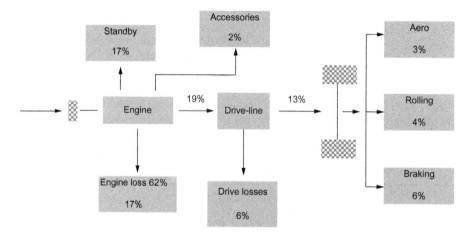

Fig. 1.1.9 Sankey diagram for a Combustion Engine (CE) vehicle.

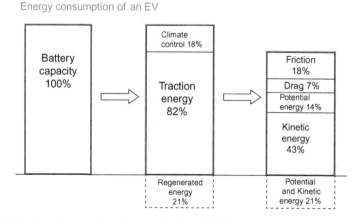

Fig. 1.1.10 Sankey diagram for a Battery Engine Vehicle (BEV).

Table 1.1.5 **The influence of vehicular speed on energy consumption**

City of Edinburgh: Newington to Chesser (4 miles)			
Speed (mph)	$E_{traction}$ (kWh)	$E_{recovered}$ (kWh)	E_{nett} (kWh)
30	1.4	0.4	1.0
25	1.1	0.3	0.8
20	0.8	0.2	0.6

Table 1.1.6 **Three 'e' (energy, environmental, and economic) analysis for Renault Zoe electric vehicle**

Energy used, kWh/km (kWh/mile) 0.164 (0.262) Economics (UK pence/mile): Electricity cost 3.15, Battery cost 0.78, servicing cost 0.04, vehicle depreciation cost 33.16. Total 37.12 CO_2-emissions (g/mile). Charging based on UK grid 142, Solar PV 12, Nuclear, Hydro or Wind 3.3

thus offering higher safety to other road users. Mitchell et al. (2010) have shown that the probability of fatalities to pedestrian in a crash with an automobile drops in the following pattern: from 40 to 30 mph speed the probability drops from 87% to 45%. It then further drops to 4% when the speed is reduced to 20 mph. Note that Figs. 1.1.3–1.1.9 and Tables 1.2.2–1.2.4 have been redrawn and reconstructed based on data presented by Mitchell et al. (2010). In most cases, conversion of that data was carried out to transform that information in SI units.

Table 1.1.6 presents the 'three-e' analysis for an e-vehicle (a 1.5 tonne, Renault Zoe with a 22 kWh battery). The three 'e's being energy, economy, and environment.

1.2 Societal impact

Transport represents a crucial sector in today's economy and society, having a large impact on growth and employment. From a societal point of view, the importance of leisure and its related activities in modern societies makes transport an essential activity for the normal development of human relations.

Means of transport have evolved a great deal over time. The use of cars has become widespread mainly because they are an easy mode of transportation that offers liberty and independence. Table 1.2.1 presents a summary of the drivers that perpetuate the use of automobiles in the society. Most of this section has been based on the work by Jeekel from his book 'The car-dependent society' (Jeekel, 2013).

Among all means of transport, the car is considered the most convenient. With a car, there is no need to rely on public transportation for the daily commutes, providing

Table 1.2.1 Scheme of driving forces for car use

Individual attitudes and motives	Characteristics related to risk society
Convenience and comfort	The urge for flexibility, combining tasks in tight time frames
Avoidance of contact with strangers	Geographical spreading out of activities
Cherishing the feeling of freedom with a car	Fears, feeling vulnerable, seeking protection
Cherishing the idea of status related to the car	Implicit expectations in the social sphere
Habit in choice of transport modes	
Creating possibilities for participating in life's experience	Everything on appointment
Geographical spreading out of activities	Convenience and comfort
The creation of highway location	The wish for 'community light'
The urge of flexibility, combining tasks in tight timeframes	The lack of ethical boundary on mobility
	Cherishing the idea of identity related to the car
Creating possibilities for participating in life's experience	Fears, feeling vulnerable, seeking protection
	Habit in choice of transport modes

Jeekel, H. (2013). The car dependent society (1st ed.). Burlington, VT: Ashgate Pub., pp. 40–63, 236–244.

flexibility to decide when to leave or arrive at a place. Cars have been developed to offer comfort and make life easier by saving time and avoiding the frustration that often comes with public transport.

The use of the car is also very flexible, is fast and agile, and seems to be the only means of transport able to connect quickly with different locations, when offices, shops, hospitals, schools, or sport facilities are far away from homes. A car allows to combine multiple activities and appointments at different locations in a tight time schedule.

In terms of safety, which is a top concern, driving a car makes us feel safer than for example driving a motorcycle, cycling, or walking. A car is a solid and robust object that gives protection to drivers and passengers in car crashes and can withstand impacts. Fortunately, today's modern cars have evolved with the new technologies, offering more occupant protection and avoiding more accidents than old cars.

Owning a car gives a sense of freedom, not having to depend on anyone or on public transport, avoiding for example bad driving by taxi or bus, the bad behaviour of some passengers, the anxiety about arriving on time or the laboriousness of the payment system of some buses as in the United Kingdom.

Due to all the aforementioned benefits, nowadays, the car has become a necessity, and in the past decades, the transport sector has experienced a significant growth in the world economy.

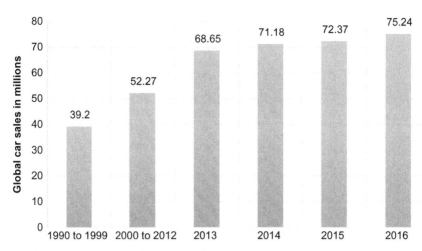

Fig. 1.2.1 Number of cars sold worldwide from 1990 to 2016 (in million units).
Statista (2016a). Number of cars sold worldwide from 1990 to 2016 (in million units) [online]
Available at: http://www.statista.com/graphic/5/200002/international-car-sales-since-1990.jpg
Accessed 18.11.16; Statista. (2016b). Car production: Number of cars produced worldwide
2015| Statista [online]. Available at: https://www.statista.com/statistics/262747/worldwide-
automobile-production-since-2000/ Accessed 08.12.16; Statista. (2016c). Quality of
infrastructure: countries with best infrastructure 2016| Statista [online]. Available at: https://
www.statista.com/statistics/264753/ranking-of-countries-according-to-the-general-quality-
of-infrastructure/ Accessed 11.12.16.

For example, the number of cars sold worldwide from 1990 to 2016 (in million units) is shown in Fig. 1.2.1. The figure shows that from 1990–99 to the forecast for 2016 the number of cars sold in the world has almost doubled, around 75.24 million cars are expected to be sold by 2016, and the sales will continue to grow, as they are expected to exceed 100 million units by 2020 (Statista, 2016a).

In 2015, Europe had the strongest increase in new passenger vehicle registrations. Around million more cars were sold than in the previous year, which accounts for 16% increase, Germany, Britain, and France being the three largest markets. However, China remains the world's largest car market; for the first time, the Chinese car market exceeded 20 million cars, which entails an increase of 1.7 million cars compared with year 2014, nearly 3 million more than in the United States.

Car sales in Japan, for the same year, decreased around 10% (approximately half a million vehicles less) compared with 2014. In India, the new car sales grew by 8% (nearly 2.8 million cars). And in Russia and Brazil, the car sales were down by more than a third and a quarter, respectively (Bekker, 2016).

Figure 1.2.2 shows the vehicles sales from 2007 to 2012 in different countries, China being the country with the fastest-growing vehicle sales over the years.

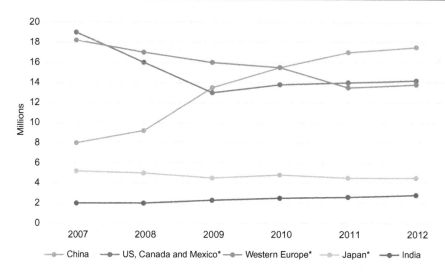

Fig. 1.2.2 Vehicle sales from 2007 to 2012. *Light vehicles only.
This information was resourced from Automotive News, manufactures' associations, 2012 as reported in BBC News. (2012). China's economic miracle [online]. Available at: http://www. bbc.co.uk/news/world-asia-china-20069627 Accessed 18.11.16.

Table 1.2.2 Ownership of cars, by households (%)

Country	Zero car	One car	Two cars	More than two cars
Germany	18	53	23	4
The United Kingdom	24	45	53	7
France	19	45	32	4
Switzerland	18.8	50.6	25.1	5.4
Flanders	18.2	53.6	24.7	3.5
Sweden	25	52	20	3
The Netherlands	20.9	54.9	21.6	2.6

Based on data from Jeekel, H. (2013). The car dependent society (first edition), Ashgate Publishers: Burlington, Vermont, USA.

Table 1.2.2 below shows the car ownership in seven countries, revealing that now there are more households with more than two cars than households with one car or no car at all. According to the rule of thumb for these seven countries, around 21% of the households do not have a car, 49% possess only one car, and a 30% of the households have two or more cars (Jeekel, 2013).

Overall, the passenger car fleet in almost all of the EU member states has grown from 2009 to 2013 and will keep growing; see Table 1.2.3. In 2013, Germany reached the highest number of registered passenger cars with 44 million cars, followed by Italy (37 million passenger cars) and France (32 million cars). The highest growth from

Table 1.2.3 **Registered passenger cars (1000 cars)**

	2009	2010	2011	2012	2013
Belgium	5193	5276	5407	5444	5493
Bulgaria	2502	2602	2695	2807	2910
Czech Republic	4435	4496	4582	4706	4729
Denmark[a]	:	:	:	:	:
Germany	41,738	42,302	42,928	43,431	43,851
Estonia	546	553	574	602	629
Ireland	:	:	1962	1951	1985
Greece[a]	:	:	:	:	:
Spain	21,984	22,148	22,277	22,248	:
France	31,394	31,657	31,754	32,132	32,244
Croatia	1541	1521	1518	1445	1448
Italy	36,372	36,751	37,113	37,078	36,963
Cyprus	461	463	470	475	475
Latvia[b]	904	637	612	618	635
Lithuania	1695	1692	1713	1753	1809
Luxembourg	332	337	346	356	:
Hungary	3014	2984	2968	2986	3041
Malta	234	241	247	250	256
The Netherlands	7622	7736	7859	:	:
Austria	4360	4441	4513	4584	4641
Poland	16,495	17,240	18,125	18,744	19,389
Portugal	:	4692	4712	4259	4327
Romania	4245	4320	4335	4487	4696
Slovenia	1059	1062	1066	1066	1064
Slovakia	1589	1669	1749	1824	1880
Finland[c]	2777	2877	2978	3037	3106
Sweden	4300	4334	4401	4446	4495
The United Kingdom[d]	28,247	28,421	28,467	:	:
Iceland[a]	:	:	:	:	:
Liechtenstein	26	27	27	28	28
Norway	2244	2308	2376	2443	2500
Switzerland[e]	4010	4076	4163	4255	:
Montenegro[a]	:	:	:	:	:
The former Yugoslav Republic of Macedonia	282	310	313	302	:
Serbia	1641	1565	:	:	:
Turkey	7094	7545	8113	8649	9284

[a]Denmark, Greece, Iceland, Montenegro—data not available (:).
[b]Vehicles with no technical inspection for 5 years are excluded.
[c]Including Åland.
[d]Great Britain only.
[e]Estimated values.
Ec.europa.eu (2016a).

2009 to 2013 was recorded in Slovakia and Poland (both 18%), followed by Bulgaria (16%) and Estonia (15%) (Ec.europa.eu (2016a)).

In 2013, cars powered by diesel engines led the European market with 58% of new registrations, 38% belonged to cars powered by petrol engines, and the alternative fuel engines (electricity, liquefied petroleum gas (LPG), natural gas, and other alternative fuels) played a minor role, accounting for just 4% (Ec.europa.eu (2016a)). One of the reasons why customers would prefer diesel cars could be the national taxation systems or special incentive schemes.

Table 1.2.4 shows that in 2013 the highest share of diesel cars was recorded in Latvia and Ireland with 74% and 73% of new registrations. On the other hand, countries such as Cyprus or Malta had the highest share of petrol cars with 80% and 60% of the new registrations.

At the end of 2015, in Great Britain, there were 36.5 million vehicles licensed, of which 30.3 million were cars (Grove, 2016). Fig. 1.2.3 shows how diesel and alternative fuel vehicles have been growing along the years, while petrol cars have been decreasing. For example, during 2015, there were 29,963 new ultralow emission vehicles (ULEVs) registered for the first time compared with 15,833 in 2014, which represents a 89% increase (Department of Transport, 2016).

Note that in year 2016 diesel cars came under increased scrutiny due to their excessive nitrogen dioxide and particulate emissions.

In many countries, there is a greater use of urban transport in daily journeys, especially those related to transportation to the workplace. Table 1.2.5 below shows that around 70 of the journeys by car are made for work purposes, education represents 31, shopping around 55, and leisure 51 from the total average.

The development of mobility could be analysed; the growth of car mobility could take place in three ways (Jeekel, 2013):

1. Through the car mi/km driven—lower car occupancy means that more cars have to be used to drive the same distance
2. Through the increase in travel distance by each member of the population—if the number of inhabitants remains stable, the mobility figures will be higher
3. Through the population growth—if the number of inhabitants increases, the mobility figures will be higher (average distance per person being stable)

Table 1.2.6 shows how the distance travelled per person in the different countries along the year has not changed a lot, the distance travelled per person being between 41 km (Germany) and 25 km (France) around 2008 and the distance per car per year being between 13,000 and 14,300 km. Fig. 1.2.4 also shows the average kilometres travelled by car in different countries of the world in 2010. As illustrated, the distance travelled for each country varies significantly from around 9000 km per year in Japan to close to 19,000 km in the United States. Note that Tables 1.2.1, 1.2.2, 1.2.5, and 1.2.6 are based on information provided by Jeekel (2013).

Table 1.2.4 New passenger cars by type of engine fuel, 2013

	Total	Petrol	Diesel	Alternative fuel
Belgium	490,369	169,665	319,863	841
Bulgaria	199,963	:	:	:
Czech Republic	164,627	:	:	:
Denmark[a]	:	:	:	:
Germany	2,952,431	1,502,784	1,403,113	46,534
Estonia	19,690	12,208	7345	137
Ireland	74,960	19,892	54,428	640
Greece[a]	:	:	:	:
Spain[b]	710,638	224,403	485,043	1192
France	1,756,953	562,849	1,182,129	11,975
Croatia	46,563	15,245	31,046	272
Italy	1,311,334	419,045	707,641	184,648
Cyprus	14,771	11,758	2998	15
Latvia	55,808	14,293	41,502	11
Lithuania[c]	155,855	44,919	110,402	534
Luxembourg[b]	50,398			
Hungary	126,937	52,677	72,343	1917
Malta	13,094	7800	5267	27
The Netherlands[a]	:	:	:	:
Austria	319,035	136,689	181,061	1285
Poland	987,809	449,741	468,097	69,971
Portugal[b]	110,002	:	:	:
Romania	279,740	141,921	136,673	1146
Slovenia	51,968	23,942	28,016	10
Slovakia	113,876	:	:	:
Finland	103,450	64,194	38,697	2000
Sweden	292,162	108,067	176,485	7610
The United Kingdom[d]	1,907,411	924,509	958,536	24,366
Iceland[a]	:	:	:	:
Liechtenstein	1920	1041	858	21
Norway	176,019	67,701	97,464	10,854
Switzerland[b]	334,000	200,600	125,000	8600
Montenegro[a]	:	:	:	:
The former Yugoslav Republic of Macedonia[b]	32,870	:	:	:
Serbia[e]	28,951	18,393	7364	3194
Turkey	654,905	256,506	383,904	14,495

[a]Denmark, Greece, the Netherlands, Iceland, Montenegro—data not available (:).
[b]2012 data.
[c]Including new and reregistered.
[d]Great Britain only, 2011 data.
[e]2010 data.
Ec.europa.eu (2016a)

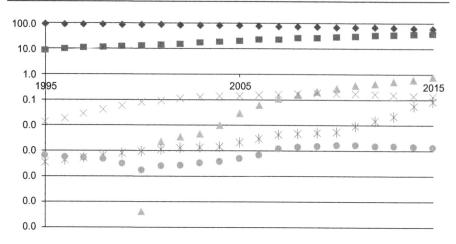

Fig. 1.2.3 Percentage of car licensed by propulsion/fuel type in United Kingdom. Key: *Blue diamond* = Petrol, *Brown square* = Diese, *Grey triangle* = Hybrid, *Yellow cross* = Gas, *Blue asterisk* = Electric, *Green circle* = Other. Note *Y*-axis plotted on loarithmic scale. Gov.uk. (2016a). Cars (VEH02) - GOV.UK [online]. Available at: https://www.gov.uk/government/statistical-data-sets/veh02-licensed-cars Accessed 24.11.16; Gov.uk. (2016b). Plug-in car and van grants - GOV.UK [online]. Available at: https://www.gov.uk/plug-in-car-van-grants/what-youll-get Accessed 23.12.16; Gov.uk. (2016c). Plug-in vehicle charge point grants - GOV.UK [online]. Available at: https://www.gov.uk/government/collections/plug-in-vehicle-chargepoint-grants Accessed 23.12.16.

Table 1.2.5 Journeys by motive made by car for selected countries

	Trips made by car			
Country	Work	Education	Shopping	Leisure
Germany (2008)	70	27	58	49
The United Kingdom (2009)	75	42	63	–
France (2008)	75	38	67	60
Flanders (2008)	70	31	44	48
The Netherlands (2009)	60	16	43	47
Average	70	31	55	51

Jeekel, H. (2013). The car dependent society (1st ed.). Burlington, VT: Ashgate Pub., pp. 40–63, 236–244.

Table 1.2.6 **Distance travelled per person, per car, and transport and population growth**

Country	Distance travelled per person		Distance per car	Transport and population growth 1995–2009
	Around 1998	Around 2008		
Germany	40 km (1998)	41 km (2008)	14.3	7%; 2%
The United Kingdom	34.5 km (2000)	34.1 (2009)	13.2	11%; 5%
France	23 km (only daily mobility, 2000)	25.1 km (2008)	13	12%; 6%
Switzerland	38.1 (2000)	38.2 (2005)	13.9	14%; 5%
The Netherlands	34.8 (2000)	35.1 (2009)	13	12%; 6%

Jeekel, H. (2013). The car dependent society (1st ed.). Burlington, VT: Ashgate Pub., pp. 40–63, 236–244

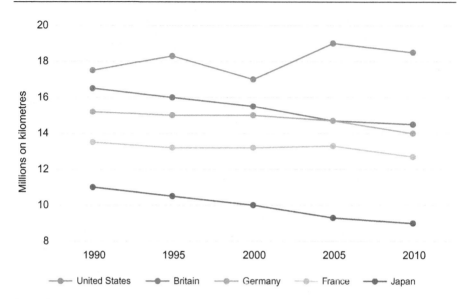

Fig. 1.2.4 Average kilometres travelled per car.
Department of Infrastructure and Transport, Australia; US Department of Transport; The Economist. (2012). Seeing the back of the car [online]. Available at: http://www.economist.com/node/21563280 Accessed 25.11.16.

1.3 Climate change

Climate change is happening in many ways, temperatures and precipitations are increasing, glaciers are melting, and the global mean sea level is rising, which is leading to a wide range of impacts on the environment, the economy, and the society across the world. And the main cause of that has been human activity.

These impacts vary throughout Europe according to geographic, socioeconomic, and climatic conditions. Populations are growing faster and so is energy consumption, the energy demand is continuously increasing along the years, keeping pace with the improvement in quality of life.

Throughout history, the world population has grown fast; in 2013, there were 7162 million inhabitants (Eurostat, 2015), but it was after World War II when the population increased significantly (see Fig. 1.3.1), maybe partly due to the appearance of antibiotics (penicillin—1942, tetracycline—1955, and streptomycin—1943) (Tverberg, 2012).

This growth of population has been linked to the growth of world energy consumption, which has been increasing dramatically over the last 200 years, as Fig. 1.3.2 shows. Global energy consumption in transport accounts for around 27%, and approximately 60% of the energy demand is related to passenger transport (mostly passenger cars) (REN21, 2016) (see Fig. 1.3.3). The advances in technology are one of the reasons why energy consumption is rising so quickly. Sometimes, the quality of life is associated with the amount of energy that a society consumes, which also leads to an increase in the use of fossil fuels.

Population in billion

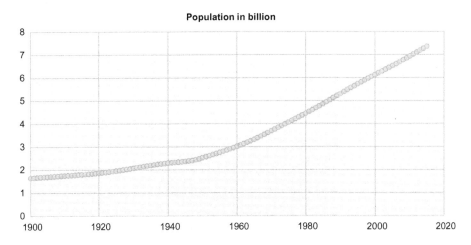

Fig. 1.3.1 World population.
Our World In Data. (2016). World population growth [online]. Available at: https://ourworldindata.org/world-population-growth/ Accessed 15.11.16.

World energy consumption

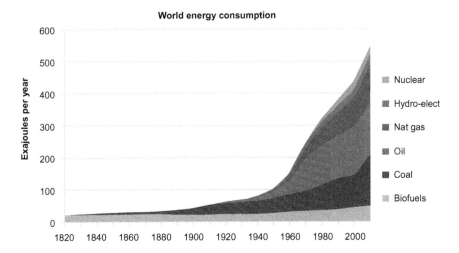

Fig. 1.3.2 World energy consumption.
Tverberg, G. V. (2012). World energy consumption since 1820 in charts [online]. Our finite world. Available at: https://ourfiniteworld.com/2012/03/12/world-energy-consumption-since-1820-in-charts/ Accessed 15.11.16.

According to the International Energy Agency between 2008 and 2035, the global energy-related CO_2 emissions will increase by 21%, from 29.3 to 35.4 Gt (see Fig. 1.3.4) (IEA, 2010). The transport sector is responsible for almost a quarter of Europe's greenhouse gas emissions and is the main cause of air pollution in cities (Ec.europa.eu., 2016a). Despite the fact that transport emissions started to decrease in 2007, they still remain very high compared with 1990 (see Fig. 1.3.4). All sectors

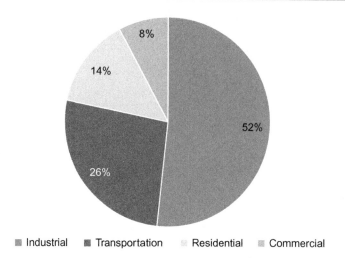

Fig. 1.3.3 World energy consumption by sector.
The International Energy Agency (EIA). (2012). Energy technology perspectives
2012—Pathways to a clean energy system [online]. Available at: https://www.iea.org/
publications/freepublications/publication/ETP2012_free.pdf Accessed 26.12.16.

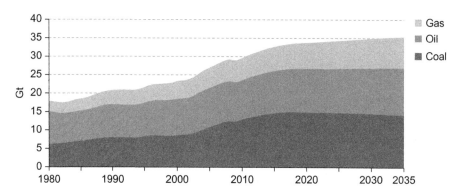

Fig. 1.3.4 World energy-related CO_2 emissions by fuel.
The International Energy Agency (EIA). (2010). World energy outlook, 2010 [online]. France,
pp. 95–96. Available at: http://www.worldenergyoutlook.org/media/weo2010.pdf Accessed
30.06.16.

(industry, domestic, and transport) will contribute to the rise in CO_2 emissions, the
transport sector being one of the most polluting (Fig. 1.3.5).

However, energy use and CO_2 emissions do not increase at the same pace, as the
energy use is growing faster (see Fig. 1.3.6).

The atmosphere contains around 40% more carbon dioxide than before the first
industrial revolution (Climate.nasa.gov., 2016), with global average temperature
increasing continuously, especially in the last 30 years (see Fig. 1.3.7).

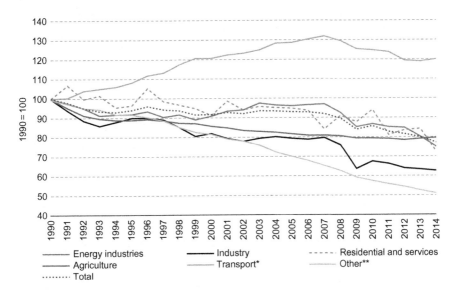

Fig. 1.3.5 Emissions by all sectors.
European Environment Agency (EEA). (2015). Air quality in Europe—2015 report [online].
Luxembourg, pp. 49–53. Available at: http://www.eea.europa.eu/publications/air-quality-in-europe-2015 Accessed 05.12.16.

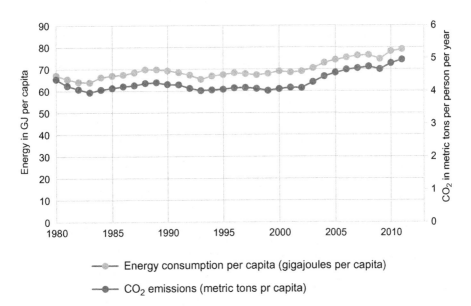

Fig. 1.3.6 World per capita energy consumption and CO_2 emissions.
Data.worldbank.org. (2016). CO2 emissions (metric tons per capita) | Data. [online] Available
at: http://data.worldbank.org/indicator/EN.ATM.CO2E.PC Accessed 18.11.16; The
International Energy Agency (EIA). (2011). Technology roadmap: Electric and plug-in hybrid
electric vehicles [online]. Available at: http://www.iea.org/publications/freepublications/
publication/EV_pHEV_roadmap.pdf Accessed 30.11.16.

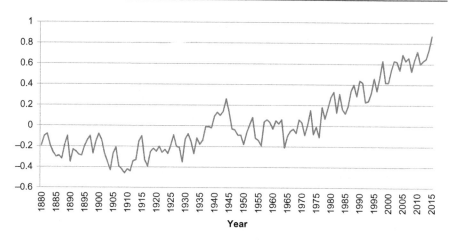

Fig. 1.3.7 World temperature anomaly, degree Celsius.
Climate.nasa.gov. (2016). Scientific consensus: Earth's climate is warming [online]. Available
at: http://climate.nasa.gov/scientific-consensus/ Accessed 10.11.16.

Fig. 1.3.7 illustrates the change in global surface temperature relative to 1951–80 average temperatures. The 10 warmest years in the 134-year record all have occurred since 2000, with the exception of 1998. The year 2015 ranks as the warmest on record (source, NASA/GISS). This research is broadly consistent with similar constructions prepared by the Climatic Research Unit and the National Oceanic and Atmospheric Administration.

In Dec. 2015, delegations from all over the world met in Paris with the aim of reaching a new international agreement on climate change. Unless measures are taken, there is a high risk that the Earth's surface temperatures might increase beyond 2°C, and this global warming could cause irreversible changes in the climate system.

The increase in world population, energy consumption, and transport is seriously affecting climate change. One of the sectors where public interventions are required to reduce CO_2 emissions is transport. For example, the European Union has already ambitious plans to reduce emissions by cutting greenhouse gas emission by 20% in 2020 and 40% by 2030 (Eurostat, 2015).

Therefore, the need for reducing the use of fossil fuels and improving fuel efficiency in the transport sector is crucial to limit the CO_2 burden. The development of innovative vehicle technologies, sustainable biofuels, improved transport infrastructure to avoid traffic congestion, the use of public transport, or the electrification of the transport could help to improve the current situation in Europe and in the whole world.

1.4 Impact on human health

Air pollution remains one of the principal environmental factors linked to preventable illness and premature mortality across the world. It is rising at an alarming rate in world's cities. Outdoor air pollution has grown 8% globally in the past 5 years, Middle East, southeast Asia, and the western Pacific being the most fast-growing areas impacted with pollution levels 5–10 times above World Health Organization (WHO) recommended levels (see Figs. 1.4.1 and 1.4.2), causing more than 3 million

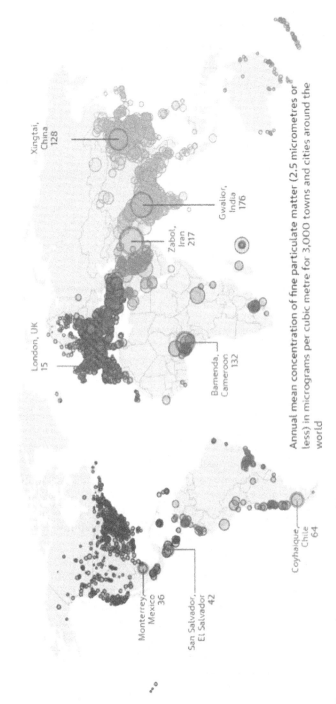

Fig. 1.4.1 Annual mean concentration of fine particulate matter in micrograms per cubic metre for 3000 towns and cities around the world. Vidal, J. (2016). Air pollution rising at an 'alarming rate' in world's cities [online]. The Guardian. Available at: https://www.theguardian.com/ environment/2016/may/12/air-pollution-rising-at-an-alarming-rate-in-worlds-cities Accessed 25.11.16.

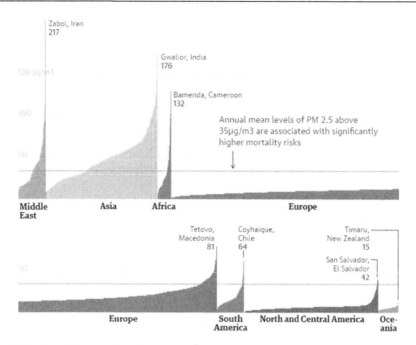

Fig. 1.4.2 Annual mean of PM 2.5 in $\mu g/m^3$ in each of the 3000 cities.
Vidal, J. (2016). Air pollution rising at an 'alarming rate' in world's cities [online]. The Guardian. Available at: https://www.theguardian.com/environment/2016/may/12/air-pollution-rising-at-an-alarming-rate-in-worlds-cities Accessed 25.11.16.

deaths annually (Vidal, 2016), and it is estimated that by 2040, these premature deaths will increase up to 4.5 million (IEA, 2016a).

Fig. 1.4.2 presents the annual mean levels of PM 2.5, the measurement of these fine particles serves as an indicator of air quality. These particles can enter deep into the lungs by breathing them, increasing the risk of heart attack, cardiovascular disease, or cancer. Among all major air pollutants, these particles are the most damaging to human health, and above 35 $\mu g/m^3$ is associated with significantly higher mortality risks (Vidal, 2016). Cars and trucks, coal power plants, agricultural, and industrial emissions are the main factors (Zaragoza, 2015).

From the 3 million premature deaths in 2012 (see Fig. 1.4.3), China was the country with the highest deaths caused by outdoor air pollution from particulate matter, accounting for more than 1 million, followed by India with 620,000 deaths, Russia with 140,000, Indonesia and Pakistan with 60,000 deaths each, Ukraine and Nigeria, both with 50,000, and Egypt and the United States with 40,000 (IEA, 2016a, 2016b, 2016c, 2016d, 2016e). For example, in most of the OECD countries, the number of deaths from lung and heart diseases caused by air pollution is much higher than the number of deaths caused by traffic accidents (Oecd.org., 2014).

As urban life continues to grow more popular, so does the risk of developing various diseases caused by pollutants some of which include respiratory complications and irreversible damage in unborn children. The World Health Organization estimates

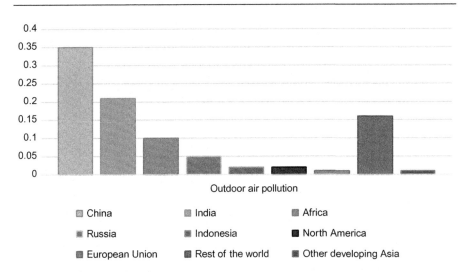

Fig. 1.4.3 Regional proportional deaths attributable to outdoor air pollution, 2012. WHO and IEA analysis (2016). Courtesy of World Health Organization. It is estimated that worldwide 3.3 million premature deaths per year are attributable due to vehicle emitted particulate matter alone.

that 1.25 million people are killed annually due to urban air pollution, a significant portion of which is generated by industry, energy production, and vehicles. While previous studies have linked urban air pollution to type 2 diabetes and increased blood sugar levels in pregnant women, among other ailments, new research suggests that poor air quality might also be responsible for reducing children's IQs, affecting their ability to learn and access information.

Exposure to air pollution and economic hardship lowers children's intelligence. Referred to as a 'first-of-its-kind study', the new report from Columbia University's Mailman School of Public Health concludes that children born to mothers experiencing economic hardship who were exposed to high levels of urban air pollution suffered from reduced IQs. 'The findings are a concern because, as has been shown with lead (poisoning), even a modest decrease in IQ can impact lifetime earnings', researchers wrote.

Unborn children exposed to polycyclic aromatic hydrocarbons, or PAH, scored much lower on IQ tests at age 5 compared with children born to wealthier families who were less exposed to air pollutants, according to the report, which was published in the medical journal neurotoxicology and teratology.

Based on World Health Organization research, Table 1.4.1 presents the breakdown of deaths attributed to specific diseases caused by outdoor air pollution.

Road transport is the main cause of air pollution that has a negative impact on the environment and human health. Vehicles on the road release a number of pollutants including the following chemicals: carbon dioxide (CO_2), carbon monoxide (CO), nitrogen oxides (NO and NO_2), sulphur dioxide (SO_2), hydrocarbons, and particulate matter (PM). All these pollutants can cause severe health problems such as respiratory problems (asthma) or exacerbated cardiovascular problems, which can be fatal for human beings.

Fig. 1.4.4 shows the breakdown of exhaust gas contents of vehicles.

Table 1.4.1 Breakdown by diseases

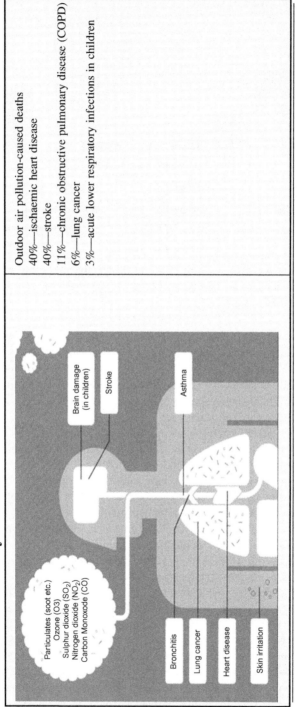

Particulates (soot etc.)
Ozone (O3)
Sulphur dioxide (SO$_2$)
Nitrogen dioxide (NO$_2$)
Carbon Monoxode (CO)

Brain damage (in children)

Stroke

Asthma

Bronchitis

Lung cancer

Heart disease

Skin irritation

Outdoor air pollution-caused deaths
40%—ischaemic heart disease
40%—stroke
11%—chronic obstructive pulmonary disease (COPD)
6%—lung cancer
3%—acute lower respiratory infections in children

World Health Organization. (2016). WHO | 7 million premature deaths annually linked to air pollution [online]. Available at: http://www.who.int/mediacentre/news/releases/2014/air-pollution/en/ Accessed 26.11.16.

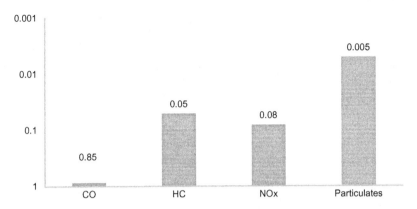

Fig. 1.4.4 Breakdown of harmful exhaust gas that are generated in addition to products of complete combustion, per cent. Note: Y-axis has logarithmic scale. Khaoutiev (2013).

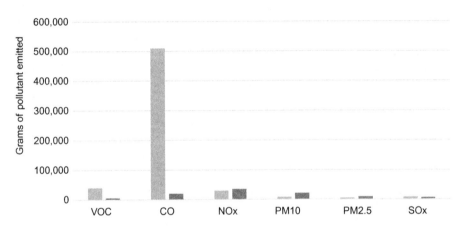

Fig. 1.4.5 Air pollutants emitted by conventional gasoline vehicles and battery electric vehicles. CV = conventional gasoline vehicle, BEV = battery electric vehicle. Note: Blue = ICE, Orange = EV.
Aguirre, K., Eisenhardt, L., Lim, C., Nelson, B., Norring, A., Slowik, P., and Tu, N. (2012). Lifecycle analysis comparison of a battery electric vehicle and a conventional gasoline vehicle, pp. 7–24.

Fig. 1.4.5 below shows a comparative chart between conventional gasoline vehicles and battery electric vehicles emissions in its life cycle. The graph below shows the huge difference between these two kinds of vehicles in terms of emissions, conventional gasoline vehicles being the greatest pollutant.

The reaction of VOCs with NOx in sunlight forms ozone. Ozone in the troposphere is a dangerous pollutant, causing smog and lung diseases in human beings. CO is also responsible for the formation of smog, which reduces visibility, and causes a reduction of the amount of oxygen that reaches the vital organs. NOx, SOx, PM10, and PM2.5 cause mainly respiratory problems.

Most countries have taken measures to reduce pollution from internal combustion engine vehicles. In Europe, for example, transport-specific emissions were reduced due to EU legislation. The target for 2015 was that new cars registered in the EU could not emit more than 130 g CO_2/km and by 2021, 95 g CO_2/km (Ec.europa.eu., 2016b).

Thanks to the strict policies on emissions from vehicles established in some countries such as Canada, Japan, China, etc. air pollution has started to fall (see Fig. 1.4.6). However, the adoption of more strict emission limits for vehicles has been outpaced by the rapid rise in traffic, as nowadays there many more cars on the road.

Also, this downward trend in emissions of pollutants caused by road transport has been off-set by the sale of more diesel than petrol vehicles, which generate more harmful pollutants. That is why some countries such as Switzerland and the United States have established a tax disincentive for the purchase of diesel cars (Oecd.org., 2014). From an environmental perspective, there is no reason to prefer diesel over petrol cars.

As a result, the deaths caused by air pollution have not dropped in line with the overall decline in air emissions in spite of the reduction in CO_2 emissions from vehicles. As an example, between 2005 and 2010, deaths increased by about 5% in China and by 12% in India (Oecd.org., 2014).

Outdoor air pollution is affecting the economies and people's quality of life around the world, causing more deaths every year, and it would be necessary to invest on reducing transport emissions if air quality wants to be improved in order to save millions of lives.

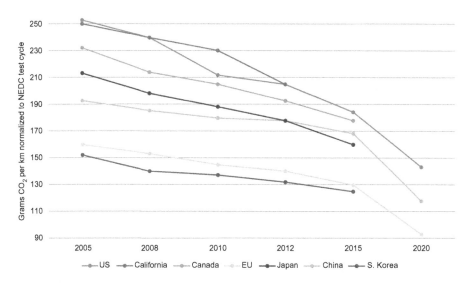

Fig. 1.4.6 Comparison of historical fleet CO_2 emissions performance and the stringency of current or proposed standards for light-duty vehicles, across regions.
This information was resourced from The icct.org, 2011 as reported in Unep.org. (n.d.). Canadian automotive fuel economy policy [online]. Available at: http://www.unep.org/ transport/gfei/autotool/case_studies/northamerica/canada/cs_cn_0.asp Accessed 30.11.16.

1.5   Fauna, flora and heritage

The harmful effects of air pollution are not limited to do with human health. Air pollution has also significant negative effects on vegetation, ecosystems, and, to a lesser extent, natural heritage, all this leading to several important environmental impacts.

The environmental impacts caused by acidification are one of the greatest environmental issues at present in the world, affecting flora, fauna, soils, and waters (The Natural Hazards Partnership, 2015). Acid rain contains harmful amounts of nitrogen oxides and sulphur oxides that are released into the atmosphere when fossil fuels are burned, damaging trees and making the water unsuitable for fishes and other wildlife. Acid rain is also responsible for the degradation of national heritage elements like buildings, sculptures, and statues (Fig. 1.5.1).

According to the European Commission, in 2010, the European area was exposed to eutrophication, including 71% of Natura 2000 ecosystems. Besides, the European Environment Agency estimated that 63% of the total EU'28 ecosystem area was at risk of eutrophication (EEA, 2015). For example, air emissions from vehicles contribute to increase the amount of nitrogen in the atmosphere, and this oversupply of nutrients may lead to changes and losses of plant and animal diversity. For instance, eutrophication can cause a rapid growth of algae placed on the water surface, preventing light from penetrating the water, which leads to the extinction of aquatic plants (EEA, 2015).

Wildlife may also be affected by toxic pollutants in the air, deposited on soil, or water. Animals can also experience health problems like human beings and some studies have shown that air toxics are contributing to reproductive failure, birth defects and disease in animals.

Ozone pollution is responsible for damages in agricultural crops, forests, and plants, as it reduces their growth rates. For example, in 2008, the United Kingdom had £183 million of losses in the yield of crops (The Natural Hazards Partnership, 2015).

Fig. 1.5.2 shows the rural concentration of the O_3 indicator for crops in 2012, the lowest value being in the north of Europe.

Air pollution can also damage heritage sites and historic buildings in a significant and nonreversible way, which leads to the loss of history and culture. Air pollutants emissions will lead to biodegradation, corrosion, and soiling of buildings. In Europe, the costs linked to the damage suffered by buildings were estimated to be approximately 1 billion EUR in 2010 (EC, 2013a).

In 2013, the European Commission (EC) presented a new clean air policy package updating the existing legislation that controls harmful emissions from industry, traffic, energy plants, and agriculture, with the aim of reducing their impact on human health and the environment.

This package will have the following benefits, and it will be fully implemented by 2030 (EEA, 2015):

- Save 123,000 km^2 of ecosystems from nitrogen pollution
- Save 56,000 km^2 of protected Natura 2000 areas from nitrogen pollution
- Save 19,000 km^2 of forest ecosystems from acidification
- Prevent 58,000 premature deaths

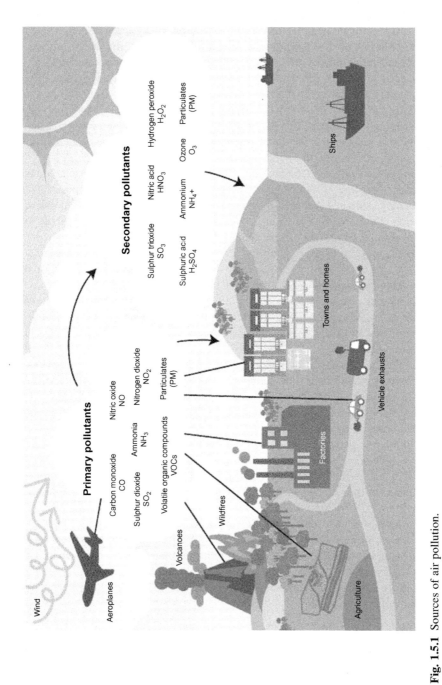

Fig. 1.5.1 Sources of air pollution.
Gov.scot. (2015). Cleaner air for Scotland—The road to a healthier future [online]. Available at: http://www.gov.scot/Publications/2015/11/5671/6 Accessed 30.11.17.

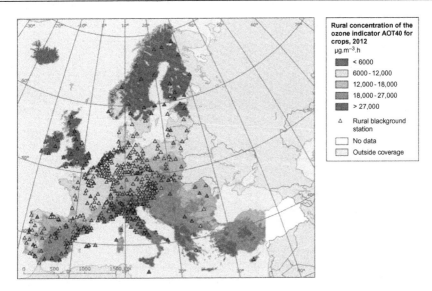

Fig. 1.5.2 Map of the rural concentration of the O3 indicator AOT40 for crops in 2012 (ETC/ACM, 2015).

The package will generate health benefits that will lead to savings of around EUR 40 billion and EUR 140 billion only in reduced damage costs and EUR 3 billion as a result of lower healthcare costs, higher productivity, higher crop yields, and less damage to buildings. This package is expected to have a positive net impact on Europe's economic growth (EC, 2013b).

1.6 Fossil fuels depletion

Since the first industrial revolution, the consumption of fossil fuels and CO_2 emissions from burning fossil fuels have increased substantially. In the future, global oil demand will grow significantly, as the global vehicle fleet is expected to double by 2050 (Gomez et al., 2016). Fig. 1.6.1 below illustrates the global production of fossil energy from 1800 to 2010, showing that the levels of tons of oil equivalents have increased from almost nothing in 1800 up to 10,000 million tons in 2010 (Höök & Tang, 2013).

In 2014, fossil fuels accounted for more than 80% of the global energy consumption, oil representing 31.3%, the most widely consumed fossil fuel (IEA, 2016b) (see Fig. 1.6.2).

Transportation accounts for around 27% of world energy demand. Based on the world oil production, oil-based fuels dominate the transport energy demand accounting for 61.5% (see Fig. 1.6.3), 18 million barrels of oil being consumed every day by cars. Globally according to the International Energy Agency publication World Energy Outlook, 2013, from 2011 to 2035 the diesel demand is expected to increase by 6.4 mb/d and the demand for gasoline by 2.1 mb/d (IEA, 2013a).

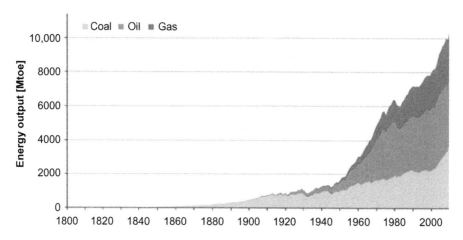

Fig. 1.6.1 Global production of fossil energy from 1800 to 2010.
Höök, M., & Tang, X. (2013). Depletion of fossil fuels and anthropogenic climate change—
A review. *Energy Policy, 52*, 797–809, Elsevier.

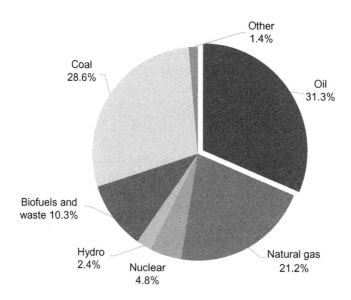

Fig. 1.6.2 Total primary energy supply by fuel in 2014. Total energy use = 13,699 Mtoe.
The International Energy Agency (EIA). (2016b). International energy outlook 2016-
Transportation sector energy consumption—Energy Information Administration [online].
Available at: http://www.eia.gov/outlooks/ieo/transportation.cfm Accessed 01.12.16.

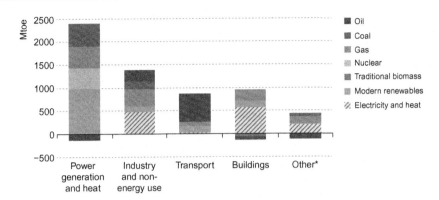

Fig. 1.6.3 Change in energy demand by sector and fuel.
The International Energy Agency (EIA). (2013a). World energy outlook 2013 [online].
France, p. 71 and p. 501. Available at: https://www.iea.org/publications/freepublications/
publication/WEO2013.pdf Accessed 01.12.16.

For example, in 2012, the United States was the world's largest transportation
energy consumer, representing a 25% of global transportation energy demand, which
means 13 million barrels of oil consumed per day. The United States, together with
China and OECD Europe, represented the 55% of the of global transportation energy
demand, see Fig. 1.6.4.

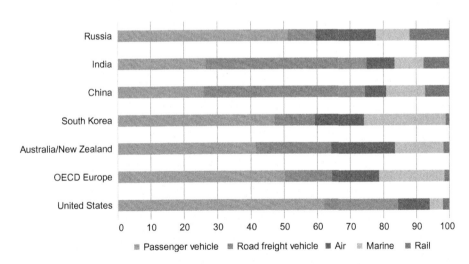

Fig. 1.6.4 Transportation energy consumption by mode in selected countries and regions, 2012.
The International Energy Agency (EIA). (2016c). Key world energy statistics [online]. France:
International Energy Agency, p. 6. Available at: https://www.iea.org/publications/
freepublications/publication/KeyWorld2016.pdf Accessed 02.12.16.

In the future, emerging economies like China and India will play a very important role in terms of oil demand, due to the expected increase in their vehicle density, large populations, and fast economic development.

Fossil fuels dominate the global energy mix, but their depletion is a fact, and some studies have proved that fossil fuels will eventually be depleted in the long term, as these resources are finite and nonrenewable.

According to Hubbert's theory and many other analysts, the oil production will peak, and then, it will decline as quickly as it grew (see Fig. 1.6.5). This happens because the oil taken from larger, more accessible reservoirs with lower costs is produced first, and then, the small reservoirs that are more difficult to find and have more costs involved will be used.

According to World Energy Outlook, 2013, recoverable resources of conventional crude oil stand at some 3300 billion barrels, and 34% of this amount has been already produced, which is equivalent to 1136 billion barrels. On the other hand, the oil reserves are estimated to be close to 1700 billion barrels, which means an equivalent of 54 years of current production (see Fig. 1.6.6).

The continuous ups and downs of the oil price raise serious economic issues. In 2014, oil prices started to drop sharply, thus putting an end to a period of relative price stability that lasted 4 years. In 2015, the oil price dropped by 47%, and in Apr. 2016, the oil price fell to its lowest level, around $43 a barrel as it can be seen in Fig. 1.6.7 (Reed, 2017). This has been the third largest decline in the last 30 years. The last time that oil price experienced such a decline was during the global financial crisis of 2008–09 (see Fig. 1.6.7).

Predicting how the energy consumption will increase in the future is a difficult task; the International Energy Agency in the report 'Energy and Air Pollution' estimated according to the new policies scenario that the global primary energy demand will

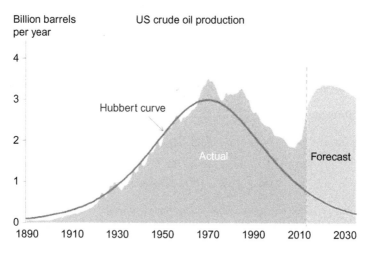

Fig. 1.6.5 Crude oil production.
Based on data from BP Statistical Review (2012).

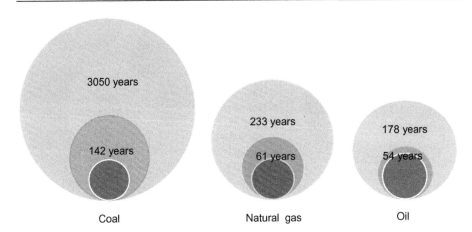

Fig. 1.6.6 Fossil energy resources by type. Outer bubble: total remaining recoverable resources, intermediate bubble: proven reserves, inner bubble: cumulative production to date.
The International Energy Agency (EIA). (2013a). World energy outlook 2013 [online].
France, p. 71 and p. 501. Available at: https://www.iea.org/publications/freepublications/publication/WEO2013.pdf Accessed 01.12.16.

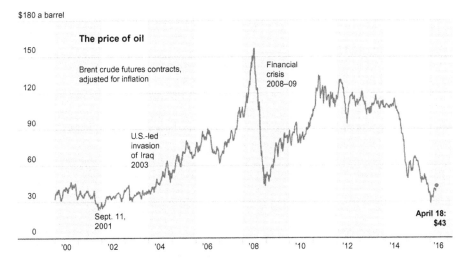

Fig. 1.6.7 The price of oil.
Based on data from Bureau of Labor Statistics (2016).

increase by a third by 2040, see the comparison between the year 2013 and 2040 in global energy demand in million tonnes of oil equivalent in Fig. 1.6.8.

The use of fossil fuels is adversely affecting the environment, causing the degradation of it and threatening human health and quality of life. All this will eventually cause damages in the ecological balance and biological diversity.

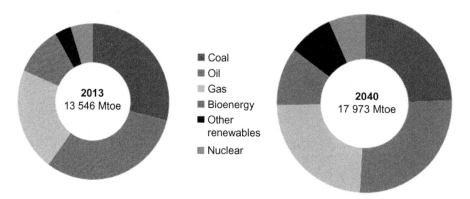

Fig. 1.6.8 World primary energy demand in the new policies scenario.
The International Energy Agency (EIA). (2016c). Key world energy statistics [online]. France: International Energy Agency, p. 6. Available at: https://www.iea.org/publications/ freepublications/publication/KeyWorld2016.pdf Accessed 02.12.16.

What is certain is that energy is crucial for human beings; people will not be willing to give up driving their vehicles, as they have a huge impact on their quality of life. Road transport is the main consumer of crude oil, and this could lead to concerns related to energy security.

For this reason, governments around the world are seriously reconsidering replacing fossil fuels with renewable energy sources such as wind power, hydropower, solar power, biomass, or tidal and wave energy and also turning to new alternatives to transport, which could solve the problem of global climate change and contribute to reducing the dependence on energy imports as fossil fuels production could not keep up with the energy demand throughout the world.

To sum up this section, a profile of six high oil-consuming nations that account for bulk of the year 2012 world's total oil consumption is presented below.

1.6.1   United States

The United States, the world's biggest oil-consuming country, consumed 13 mbd in year 2012, which accounted for nearly 20% of the world's total oil consumption.

1.6.2   China

China's oil consumption stood at 10.3 mbd, accounting for about 11.7% of the world's total oil consumption making it the second biggest oil consumer after the United States. China's oil consumption has more than doubled since 2000, and the consumption in 2012 increased by 5% compared with the previous year.

1.6.3 Japan

Japan consumed 4.7 mbd of oil, becoming the world's third biggest oil consumer, with about 5.3% of the world's total oil consumption.

1.6.4 India

India ranks fourth among the world's biggest oil-consuming countries; its oil consumption in 2012 stood at 3.6 mbd, 5% higher than the previous year, accounting for about 4.2% of the world's average oil consumption. The country's oil consumption has increased about threefold in the last 20 years.

1.6.5 Brazil

Brazil is the world's seventh biggest oil-consuming country and consumed 2.8 mbd of oil, accounting for about 3% of the world's total oil production. The country's oil consumption has steadily increased since 2008 and increased by 2.5% during 2011–12. Brazil is the biggest oil producing country in South America, and its total oil production increased from 1.83 mbd in 2004 to 2.65 mbd in 2012.

1.6.6 Germany

Germany is the eighth biggest oil-consuming country in the world and the second biggest oil consumer in Europe, after Russia. Germany's oil consumption in 2012 stood at 2.4 mbd, accounting for about 2.7% of the world's total oil consumption per day in the year, and has shown a declining trend since 2008. Germany consumed 0.7% less oil in 2012 compared with 2011.

1.7 Employment sector

Transport is a fundamental driver of economic and social development. It is a crucial sector to reduce poverty, boost prosperity, and achieve sustainability. Nowadays, transport and mobility are essential to maintain the quality of life of people, and a competitive, efficient, and reliable transport system is key to job creation and economic growth.

In 2015, the world's automobile production exceeded 68 million cars (see Fig. 1.7.1). From these 68 million cars, 25% were manufactured in the European Union, around 16 million units (Acea.be, 2016).

In 2005, 66 million vehicles including cars, trucks, vans, and buses were manufactured worldwide needing the employment of around 9 million people directly in making the vehicles. Fig. 1.7.2 shows employment in countries with the highest amount of people employed in the transport sector, China leading the market with 1,605,000 employees (Oica.net., 2016).

In the European Union, the transport industry accounts for 4.5% of total employment, employing over 10 million people. An additional 1.5% of employment and 1.7%

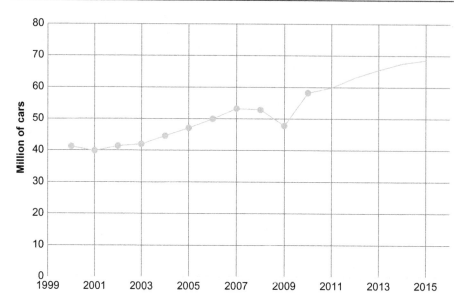

Fig. 1.7.1 Worldwide automobile production from 2000 to 2015 (in million vehicles). Statista. (2016b). Car production: Number of cars produced worldwide 2015| Statista [online]. Available at: https://www.statista.com/statistics/262747/worldwide-automobile-production-since-2000/ Accessed 08.12.16.

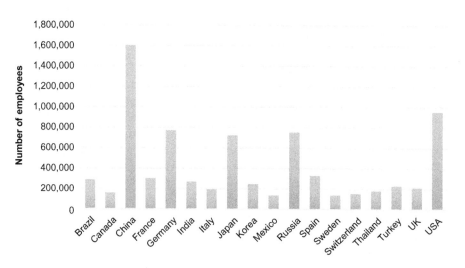

Fig. 1.7.2 Employment in several countries related to the manufacturing of vehicles in 2005. Oica.net. (2016). Auto jobs | OICA [online]. Available at: http://www.oica.net/category/economic-contributions/auto-jobs/ Accessed 07.12.16.

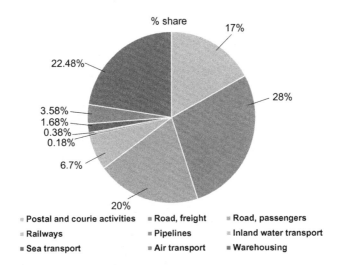

Fig. 1.7.3 Employment in the transport sector in the EU 27 in 2009.
This information was resourced from EC (2012) based on Transport Research and Innovation Portal (TRIP). (2013). Employment in the EU transport sector. Communicating Transport Research and Innovation [online]. pp. 2–24. Available at: http://www.transport-research.info Accessed 08.12.16.

of gross domestic product (GDP) is provided by transport equipment manufacturing (TRIP, 2013). Fig. 1.7.3 shows that road passenger transport accounts for 19.9%, employing more than 2 million people.

After the global economic crisis, employment in the transport sector decreased significantly. Fig. 1.7.4 shows the impact in terms of job losses in vehicle drivers from the crisis period.

Apart from that, the transport sector has always had a negative image for employment due to work schedules, unpredictable shifts, long working hours, overtime, etc. Also, due to demographic changes, which affect consumption patterns, employment in the transport sector will face economic and societal challenges, and this could lead to shortages of skilled employees in the future.

In 2013, the European Council stressed that fighting youth unemployment would be its main and immediate objective, as well as implementing measures to retain older workers and attracting more women to the sector. To achieve an efficient and high-quality transport system, highly skilled people will be fundamental in the sector, but an improvement in the working conditions will be required in the future, creating more satisfying jobs (TRIP, 2013).

In order to retain skilled employees in the transport sector, some strategies should be implemented to ensure the employability of older workers through continuous skills development, updating their knowledge and skills to meet the continuous technological developments in the transport sector.

Besides, the work-life balance for employees should be improved, as stereotypes may be an obstacle to women's employment (Austrian Institute for SME Research, 2012).

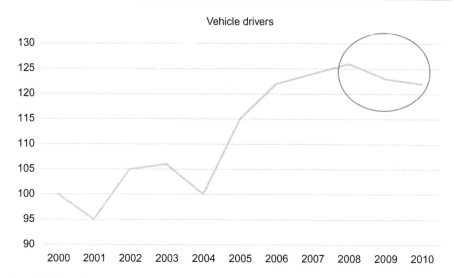

Fig. 1.7.4 Trends in employment in the transport sector (only vehicle drivers) from year 2000 to 2010.
Based on Joint Research Centre of the European Commission. (2014). Analysis of labour supply and demand. Future employment in transport. Luxembourg, pp. 9–38, 94–101.

Fig. 1.7.5 shows the projections for employment in the transport sector by gender. Throughout the entire period shown above, women have accounted for only around 20% of the EU 27 transport workforce.

The Lisbon Strategy for Growth and Jobs wanted to improve this situation with the goal of increasing employment of women by 60%. The strategy applied was successful, as employment increased between 2000 and 2009 from 57.3 to 62.5% (TRIP, 2013).

Also, with the aim of attracting more women to the sector, a project called WISE offered guidelines to create a more friendly work environment for women and also to promote their roles focusing on training and recruitment opportunities and procedures. Finally, in order to catch the attention of young people, different measures have been created to increase their interest in transport, trying to modernize the image of the sector. Schools and universities are trying to attract young people to work in the sector.

Road passenger transport, which includes motor bus, streetcar, tramway, trolley bus, and railway, will employ by 2030 around 2.1 million people that means around more than 200,000 jobs compared with 2010, an increase of 12.1% (JRC, 2014). Table 1.7.1 shows that the number of people employed in 1990 was higher than in 2010, but according to The White Paper projections, in the future employability will recover.

The automobile industry is one of the main investors in research, production, and development. This is a key sector that spends a lot of money, over 84 billion euros, in researching and developing new technologies. Vehicle manufacturing and use

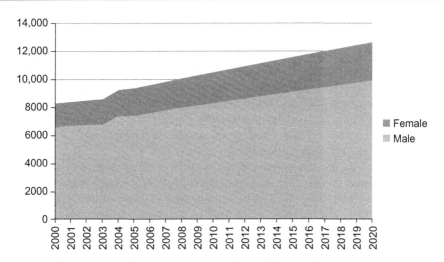

Fig. 1.7.5 Projection of employment in the transport sector by gender.
Based on Joint Research Centre of the European Commission. (2014). Analysis of labour supply and demand. Future employment in transport. Luxembourg, pp. 9–38, 94–101.

Table 1.7.1 Transport sector employment from 1990 to 2030

	1990	2000	2010	2020	2030
Employment (people)	2,288,281	1,962,336	1,913,200	2,060,347	2,145,163

Joint Research Centre of the European Commission. (2014). Analysis of labour supply and demand. Future employment in transport. Luxembourg, pp. 9–38, 94–101.

contribute more than 430 billion euros in 26 countries, being one of the greatest contributors to government revenues worldwide (Oica.net., 2016).

The German automotive industry represents one-third of the total global research and development spending in this sector (VDA, 2016). In 2013, around 24 billion euros was spent only in the automotive sector (vehicles alone), accounting for 37% of all the money invested in R&D (Deuse, 2014). The Volkswagen Company is the world's biggest investor in R&D.

Some of the 10 biggest investors in R&D around the world are automotive companies. As aforementioned, Volkswagen is in the top, having spent $13.5 billion in 2013, with 5.2% of revenues. Their main research area is related to hybrid vehicles and the electrification of vehicles. The Japanese automaker Toyota spent $9.1 billion in hybrid systems like fuel cell and electric vehicles and also safety technologies, achieving a revenue of 3.5%. Google is also investing money in technologies to carry out self-driving cars projects (Casey, 2014).

1.8   Road networks

Road networks are a key element for the economic growth of every country. It is essential to project a strategic and sustained expansion and an adequate maintenance of these networks to guarantee quality connections between the different parts of a geographical territory. They enable the supply of goods and services around the world and connect people to workplaces, schools, hospitals, etc. Road infrastructure improves the effectiveness and efficiency of countries and increases the standard living of people, making their lives easier.

Countries with a good infrastructure system are in a better position to obtain benefits from the trade domestically and internationally, improving their economic conditions as well. Hong Kong and Singapore are a good example of that, being the top two spots in global infrastructure quality. In 2016, the country with the best infrastructure was Switzerland. Fig. 1.8.1 shows the top ten of the best infrastructures in 2016, Switzerland having a value of 6.5 on a scale of 1–7 (Statista, 2016c).

In 2015, the number of cars sold worldwide was 72.37 million. This rapid growth in the number of registered motorized vehicles over the past decade has been accompanied by a huge investment in road infrastructure.

According to the IEA analysis, the infrastructure in the road transport sector needs to increase substantially by 2050, as almost 25 million paved road lane kilometres will be needed in the world. All this investment in building new infrastructures will lead to significant costs.

One of the reasons why global travel has increased along the years has been the support to infrastructural growth. New road projects have been designed or existing

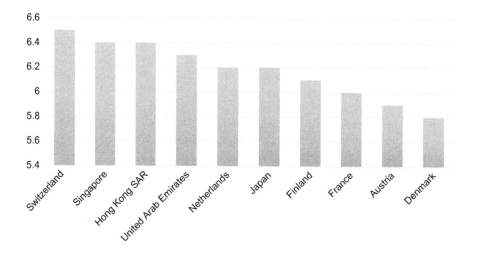

Fig. 1.8.1 Top 10 of the best infrastructures in 2016.
Statista. (2016c). Quality of infrastructure: Countries with best infrastructure 2016| Statista [online]. Available at: https://www.statista.com/statistics/264753/ranking-of-countries-according-to-the-general-quality-of-infrastructure/ Accessed 11.12.16.

roads have been upgraded, improving their safety. Without investment in infrastructure, the economic growth of a country would be difficult, but this growth in road network infrastructures may also lead to some drawbacks, for example, an increase in energy consumption by vehicles, congestion, air pollution, traffic accidents, and environmental costs. However, a poor-quality road network would lead to potential accidents and carelessness.

In non-OECD countries, the projected infrastructure will account for 85% over the next 40 years, leading to a significant growth in passenger and freight travel. According to IEA, due to this travel growth rates in non-OECD countries, by 2030, the expenditures on land transport infrastructure will be higher than in OECD countries, and by 2050, non-OECD countries will have 45% more infrastructure than OECD countries (IEA, 2013b).

According to Fig. 1.8.2, global roadway network length has increased by around 12 million lane kilometres since 2000, which means 35% more in the last 10 years. China and India have had the highest growth in road network, more than half of paved lane-kilometre additions.

According to a new policies scenario, 'The ETP 2012 4°C Scenario', an increase in transport energy consumption and emissions by 40% by 2050 is projected. The ETP 2012 presents detailed scenarios and strategies to 2050 under the four degrees stabilization. Under this scenario, by 2050, it is expected to more than double the annual vehicle kilometres, and for that, a growth around 60% above 2010 infrastructures would be essential. As Fig. 1.8.3 shows, from 2010 to 2050, global roads are expected to increase by around 25 million paved lane kilometres.

The projection for 2050 under the same scenario are that, in OECD countries, the average levels of national infrastructure occupancy, i.e. the average vehicle kilometres per lane kilometre will not increase too much, and in regions like Latin America even the average occupancy levels will decrease, as infrastructure additions keep up with

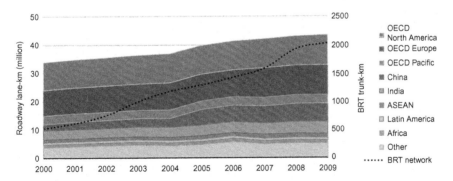

Fig. 1.8.2 Historic paved roadway lane in kilometres.
IRF (International Road Federation). (2012). World road statistics. Geneva: IRF; The International Energy Agency (IEA). (2012). Energy technology perspectives 2012—Pathways to a clean energy system [online]. Available at: https://www.iea.org/publications/freepublications/publication/ETP2012_free.pdf Accessed 26.12.16.

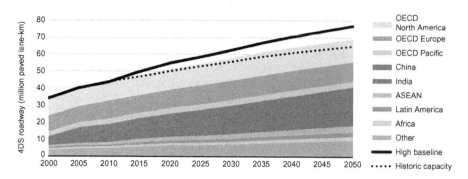

Fig. 1.8.3 Roadway projections for four degrees scenario.
The International Energy Agency (EIA). (2013b). Global land transport infrastructure
requirements. Estimating road and railway infrastructure capacity and costs to 2050 [online].
France, pp. 8–11, 16, 34. Available at: https://www.iea.org/publications/freepublications/
publication/TransportInfrastructureInsights_FINAL_WEB.pdf Accessed 10.12.16.

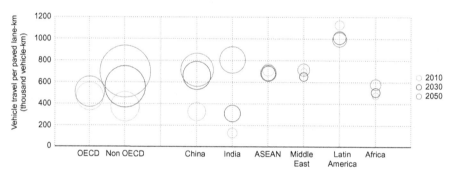

Fig. 1.8.4 Roadway occupancy levels for four degrees scenario.
The International Energy Agency (EIA). (2013b). Global land transport infrastructure
requirements. Estimating road and railway infrastructure capacity and costs to 2050 [online].
France, pp. 8–11, 16, 34. Available at: https://www.iea.org/publications/freepublications/
publication/TransportInfrastructureInsights_FINAL_WEB.pdf Accessed 10.12.16.

travel demand (see Fig. 1.8.4). However, in non-OECD countries, as shown in the
same Fig. 1.8.4, the average national road occupancy will experience a significant
growth as vehicle travel rapidly intensifies. China and India would be the countries
with the highest growth in average road occupancy levels, as it is expected that road
travel will outpace the additions in infrastructure.

The expenses projected for building transport infrastructures, including road transport, high-speed rail (HSR), railway, car parks, and bus rapid transit (BRT) too, can be
seen in Fig. 1.8.5, from 2010 to 2050. Road infrastructure accounts for around 65% of
the total expenditures.

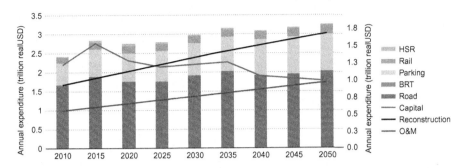

Fig. 1.8.5 Cumulative land transport infrastructure expenditures projected to reach USD 120 trillion by 2050 for four degrees scenario.
The International Energy Agency (EIA), (2013b). Global land transport infrastructure requirements. Estimating road and railway infrastructure capacity and costs to 2050 [online]. France, pp. 8–11, 16, 34. Available at: https://www.iea.org/publications/freepublications/publication/TransportInfrastructureInsights_FINAL_WEB.pdf Accessed 10.12.16.

1.9 Road safety

World Health Organization in its report on road safety for 2015 has highlighted that worldwide around 1.25 million people die every year due to road accidents. The number of deaths has been stable since 2007, but taking into account the growth in worldwide population and the increase in the number of vehicles, it could be said that improvements in road safety have saved a lot of human lives. Fig. 1.9.1 shows the

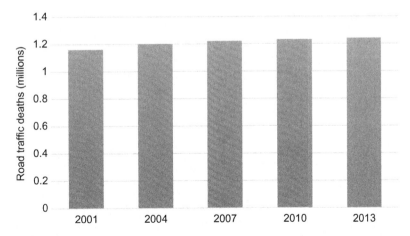

Fig. 1.9.1 Number of road traffic deaths, worldwide.
World Health Organization. (2015). Global status report on road safety 2015. Geneva. Available at: http://www.who.int/violence_injury_prevention/road_safety_status/2015/en/ Accessed 01.12.16.

number of road traffic deaths in the world; from 2010 to 2013, the global population grew by 4%, whereas the number of vehicles increased by 16% (WHO, 2015).

However, this is not enough when it comes to human lives. In Sep. 2015, the heads of state attended the United Nations General Assembly and accepted the proposed sustainable development goals (SDGs), one of their targets being to reduce by half the global number of injuries and deaths caused by road traffic accidents by 2020.

The materialization of this proposal will be beneficial in the future as Fig. 1.9.2 shows. Road traffic injuries are one of the most important causes of death in the world, and the main cause of death for young people between 15 and 29 years old.

An important factor to consider when discussing road traffic related accidents is the relevant location. The fatality rates in low-income countries are higher (more than double) than in high-income countries, and according to the World Health Organization, 90% of these deaths happen in low- and middle-income countries.

The graphics in Fig. 1.9.3 shows the population, the road traffic deaths, and the number of registered vehicles according to the income (low, middle, or high) of the country.

Africa is the region with the highest fatality rates due to road traffic accidents, and in Europe (particularly the countries with high income), the opposite is true; it is the region with the lowest fatality rates (see Fig. 1.9.4). The road traffic fatality rates in Africa are more than double than in Europe, a rate of 26.6 per 100,000 inhabitants compared with 9.3 in Europe. The average in the world is 17.4.

Road traffic incidents are an important subject in terms of human safety. According to the World Health Organization, road safety relies on three pillars: vehicle, road

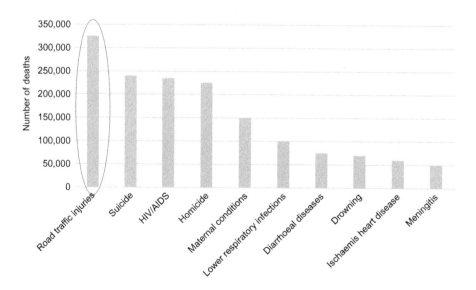

Fig. 1.9.2 Top ten causes of death among people aged 15–29 years in 2010.
World Health Organization. (2014). Global health estimates. Geneva. Available at: http://www.who.int/healthinfo/global_burden_disease/estimates/en/index1.html, http://www.who.int/healthinfo/global_burden_disease/projections/en Accessed 01.12.16.

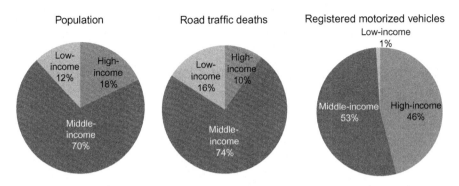

Fig. 1.9.3 Population, road traffic deaths and registered vehicles by country income. World Health Organization. (2015). Global status report on road safety 2015. Geneva. Available at: http://www.who.int/violence_injury_prevention/road_safety_status/2015/en/ Accessed 01.12.16.

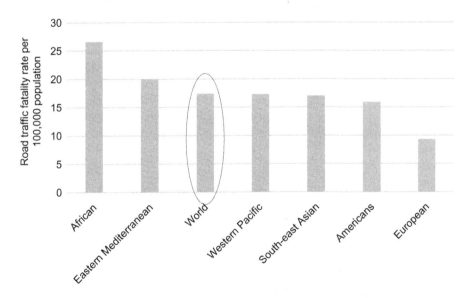

Fig. 1.9.4 Road traffic fatality rates per 100,000 population by region in 2013. World Health Organization. (2015). Global status report on road safety 2015. Geneva. Available at: http://www.who.int/violence_injury_prevention/road_safety_status/2015/en/. Accessed 01.12.16.

user, and infrastructure. Countries should enhance their road safety legislation by improving road user behaviour and education, road infrastructure, vehicle design, road traffic rules, etc. in order to reduce these road traffic injuries. The only way to further improve road safety and to maintain the sustainability of mobility is to combine the efforts of all the parties involved (WHO, 2015).

This section has been based on the report called 'Global status report on road safety 2015' by World Health Organization.

1.10   Impact on other modes of transport

Public transport systems tend to be more efficient than individually owned vehicles. They represent an easy and convenient means of transport in terms of safety and affordability and are less polluting and some of them faster than a private car, whether because of the low amount of energy per passenger that they use, the CO_2 emissions that they release or even the road space. For example, a fully occupied diesel train may have a fuel economy in the order of 9 kWh per 100 passenger kilometres, far better than any car (44–106 kWh/100 km), and with electric trains, this is even better, presenting a fuel economy as low as 1.6 kWh/100 km. In addition, if the materials and the energy used to manufacture the different vehicles are considered, the fact of sharing the means of transport also reduces the ecological footprint and can also reduce congestion as they take up less road space. Just one bus can carry approximately 35 people, equivalent to 9 fully occupied cars.

Table 1.10.1 below presents the fuel life-cycle energy consumption and the CO_2 emissions from different urban means of transport by vehicles and by number of passengers. As the table shows, public modes of transport individually emit much more kilograms of CO_2 per kilometre and use more energy than a car, but as the number of occupants that public transport can have is higher, this amount of CO_2 and energy used would decrease per passenger. These calculations have taken into account all the seats available, assuming that all the seats would be occupied by people.

Table 1.10.1 **Fuel life-cycle energy consumption and CO_2 emissions for urban transport modes, based on UK data**

Mode	Seats	MJ per vehicle km	kg CO_2 per vehicle km	MJ per seat km	kg CO_2 per seat km
Urban electric train	300	117	11.7	0.39	0.039
Urban diesel train	146	74	8.8	0.5	0.06
Light rail	265	47	10.1	0.18	0.038
Metro	555	122	26	0.22	0.046
Bus	49	14.2	1.6	0.29	0.033
Double bus	74	16.2	1.9	0.22	0.026
Minibus	20	7.1	0.8	0.36	0.040
Car	5	3.5	0.39	0.7	0.078

This information was resourced from Carpenter (1994), Potter and Roy (2000), Potter and Yarrow (2002) as reported in Potter, S. (2003). Transport energy and emissions: Urban public transport. In: Hensher, D., & Button, K. (Eds.) *Handbook of transport and the environment*, Vol. 4. Handbooks in transport (pp. 247–262). Amsterdam: Elsevier.

Nevertheless, the results shown in the table referred above may differ from reality according to the time of the day (peak or off-peak hours), as the benefits of public transport could diminish with a decreasing load factor. An off-peak bus with only a handful of passengers can have a carbon footprint per passenger mile as bad as or even worse than that for a car. According to Potter (2000) in Britain, the car occupancy average in peak hours is 1.17 persons, whereas buses and trains at that time remain full loaded. However, in off-peak hours the results are very different. Table 1.10.2 shows some assumptions made by the author based on Potter (2003), about how the CO_2 emissions and the energy used would vary depending on the number of passengers, considering peak, off-peak trips and assuming 50% of the occupancy.

In general, public transport modes release less CO_2 emissions and air pollution per passenger when they are half full. However, when public transport is studied individually, it generates much more pollution than a car. In fact, some studies have shown that bus stops where people are waiting for the bus, there are pollution black spots. Therefore, turning these vehicles into hybrids or electric vehicles could help to tackle this issue.

It could be said that public transport may be incompatible when the density of a city is low and it would be interesting to carry out a feasibility study before investing in public transport infrastructure on whether employment and housing density would be high enough build this infrastructure.

According to Bertaud, there might be a relationship between the spatial structure and the effectiveness of public transport (Bertaud & Malpezzi, 2003), suggesting that there are effectiveness areas for each type of transport. Fig. 1.10.1 below shows that public transport would be a very convenient mean of transport when the population density of the cities is high; however, when the cities have a low density, the only means of transportation possible is the use of the car.

Sometimes, the solution of decrease the pollution by transport sector is not to increase the density or reduce the number of car if not to carry out beneficial and efficient urban structures.

A good example of that is Barcelona, a city highly structured through city planning and a transport system weighted in favour of public transport.

If Barcelona and Atlanta are compared, both have the same demography and income levels. However, in terms of urban transport, Barcelona emits 6 times less CO_2 giving employment and houses to 20% more inhabitants and with an area 11 times smaller than Atlanta, which illustrates the extreme difference in urban footprint (see Fig. 1.10.2).

Due to the high population density in Barcelona, the small travel distances enable people to walk for around 20% of their trips, and the greatest distance between two points of the urban area is only 37 km compared with 137 km in Atlanta (Newman & Kenworthy, 1999), reducing that way the air pollution.

The latest news about Barcelona are that in 2017 the city will offer owners of the most polluting vehicles the opportunity to travel for 3 years free of charge in public transport in exchange for getting rid of their car and turning it into scrap metal, thus contributing to the reduction of air pollution and improving the air quality of the city (www.efe.com, 2016).

Table 1.10.2 Peak and off-peak fuel life-cycle energy consumption and CO_2 emissions for urban transport

Mode	Peak period occupancy (%)	MJ per vehicle km	g CO_2 per vehicle km	Off-peak period occupancy (%)	MJ per vehicle km	g CO_2 per vehicle km	Assuming half occupancy (%)	MJ per vehicle km	g CO_2 per vehicle km
Urban electric train	70	0.56	56	30	1.3	130	50	0.78	78
Urban diesel train	60	0.84	100	25	2.03	241	50	1.01	121
Light rail	70	0.25	54	30	0.59	127	50	0.35	76
Metro	60	0.37	78	25	0.88	187	50	0.44	94
Bus	70	0.41	47	20	1.45	163	50	0.58	65
Double bus	60	0.36	43	20	1.09	128	50	0.44	51
Minibus	50	0.71	80	20	1.78	200	50	0.71	80
Car	30	2.33	260	50	1.4	156	50	1.4	156

Potter, S. (2003). Transport energy and emissions: urban public transport. In: Hensher, D., & Button, K. (Eds.) Handbook of transport and the environment, Vol. 4. Handbooks in transport (pp. 247–262). Amsterdam: Elsevier.

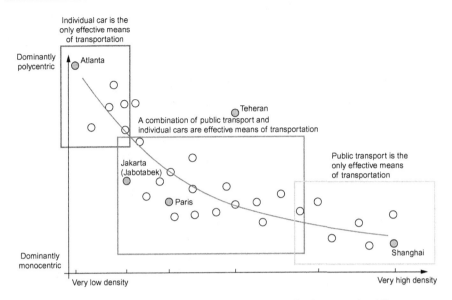

Fig. 1.10.1 Relationship between spatial structure and the effectiveness of public transportation.
Bertaud, A., & Malpezzi, S. (2003). The spatial distribution of population in 48 world cities: Implications for economies in transition. University of Wisconsin

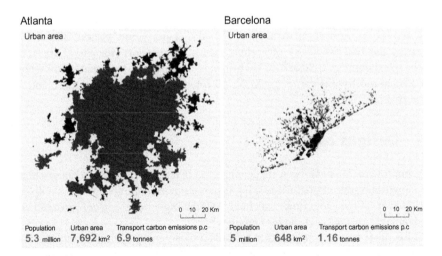

Fig. 1.10.2 Population, urban area and transport carbon emissions in Atlanta and Barcelona. This information was resourced from LSE Cities (2014) as reported in Misra, T. (2015). You won't believe how much sprawl costs America [online]. CityLab. Available at: http://www. citylab.com/housing/2015/03/how-much-sprawl-costs-america/388481/ Accessed 11.12.17.

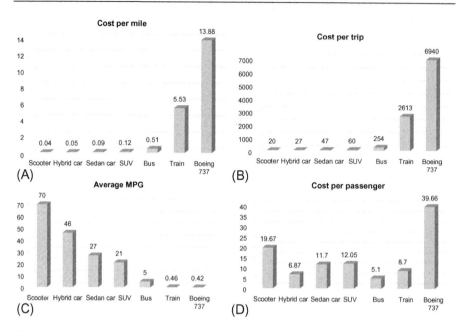

Fig. 1.10.3 (A) Average vehicle miles per gallon, (B) average fuel costs based on 500 mile trip, (C) average fuel cost per mile based on 500 mile trip, (D) average fuel cost per passenger based on 500 mile trip.
Davis, S., Diegel, S., & Boundy, R. (2010). Transportation energy data book (28th ed.) Oak Ridge, TN: Oak Ridge National Laboratory.

To sum up and for comparison purposes, the charts below present an overview comparing the fuel efficiency of various means of transport (see Fig. 1.10.3). The number of passengers assigned to each means of transport has been 1 for scooter, 4 for Sedan and hybrid car, 5 for SUV, 50 for the bus, 300 for the train, and 175 for the Boeing 737.

1.11 Designs aspects

Many improvements have been made since the invention of the first automobile in 1885. Nowadays, the automobile is safer, faster, more reliable, cheaper, less polluting, etc. However, at the same time, cars have become bigger and heavier, counteracting the advantages of the most modern engines.

On the other hand, the revolution towards a clean energy future is moving faster than expected and the electrification of transport is accelerating the transition. Many countries are backing the use of electric vehicles, and thanks to the advances in technology they are becoming a major competitor of the conventional car.

The section of this report has been based mainly in three sources, the report made for the consultory PWC called '2016 auto industry trends. Automakers and suppliers can no longer sit out the industry's transformation' (2016), the article written by Ortego, L. titled 'Size versus consumption' (2013), and the book 'Reinventing the automobile' by Mitchell et al. (2010).

Aerodynamics and weight reduction play a key role in car design; for example, vehicles with an aerodynamic shape consume less fuel, as the air flows easily over them, needing less energy to move them forward. For example, 60%–70% of a vehicle's energy is used to move it through the air at 95 km/h, and only 40% is used at 50 km/h. (SEI, n.d.).

With regard to the automobile weight, many automakers support the use of lighter materials to improve the performance of the car; as an example, in 2014, Ford changed the steel by aluminium, risking that some customers would not trust this change, believing aluminium is less rugged. Nevertheless, this model of truck, the F-150, was the best-selling vehicle in the United States in 2015, having the best mileage among all gasoline pickups (PWC, 2016). According to IDEA, every additional 100 kg represents a respectable increase of 5% in consumption.

The automobiles have changed profoundly over the years, today's automobiles have managed to reduce their consumption by 20%–30% compared with automobiles 30 years ago. However, currently they are much heavier and have a bigger size than automobiles 25 years ago, producing a counterproductive effect.

The pursuit of comfort has widened and lengthened the cars, the passive safety measures and the high technological equipment and comfort of the automobiles have increased their weight around 25%, and this increase in size not only translates into more weight but also gives cars a greater resistance to the wind.

Usually, an automobile transports only one or two passengers, so why this space requirement? If automobiles would not have increased so much in size along the years, the efficiency of the cars would be even better.

On the other hand, automobile companies have been dabbling with new technologies and vehicle concepts that can transform the traditional concept of automobile. The connected car will be the future of the means of transport, as their development and evolution are growing by leaps and bounds. The difference between these cars and the traditional ones is the introduction of an updated technology such as systems equipped with Internet access and a wireless local area network.

Also, the intelligent car is experiencing a fast development, offering the drivers the experience of driving without any control over the vehicle; some of the functions that this vehicle features are self-braking, self-parking, and automatic cruise control based on road conditions or automatic accident-avoidance features, among others (PWC, 2016).

William Mitchell and two industry experts talk about a new form of automobile for the future, an ultrasmall vehicle (USV), introducing these new vehicle concepts and describing a kind of vehicle that is green, smart, connected, and fun to drive. This vehicle allows the drivers to be connected to their social networks, news, or entertainment, making an efficient use of their time, and the best of all is that it will be powered only with renewable energy.

The development of this vehicle will be based on four big ideas. The first one would be to change the well-known internal combustion engine vehicles for electric engine vehicles with wireless communications. The second idea lies in the mobility internet, allowing vehicles to collect, share, and process information to reduce travel times managing the traffic. The third idea would be to integrate these e-vehicles with smart grids that use renewable energy. And the last idea is based on providing real-time control capabilities for urban mobility and energy systems developing electronically managed, dynamically priced markets for electricity, roads, parking, and vehicles thanks to the proposed wireless connectivity and onboard intelligence.

These four ideas together will have a great impact in the society, improving the economic growth and the prosperity and eliminating the negative effects of the current automobile. Table 1.11.1 presents the benefits of this new concept of automobile.

Table 1.11.1 Benefits of the new automotive DNA

Converging ideas		Change in personal mobility		Benefits
New automotive DNA (electric + connected) + Mobility Internet + Clean, smart energy + Dynamically priced markets	=	Zero emissions Renewable energy Crash avoidance Safe social networking while driving Fun driving and autonomous driving Varied designs Shorter, more predictable urban travel times Space- and time-efficient parking Increased roadway throughout Quieter cities Safer pedestrians and bicycles More equitable access Lower cost	=	Enhanced freedom + Sustainable mobility + Sustainable economic growth and prosperity

Mitchell, W., Borroni-Bird, C., & Burns, L. (2010). Reinventing the automobile (1st ed.). Cambridge, MA: Massachusetts Institute of Technology.

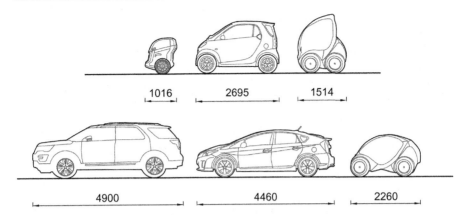

Fig. 1.11.1 Dimensions of the ultra-small vehicle compared to conventional car.
This diagram has been redrawn based on concept presented by Mitchell, W., Borroni-Bird, C.,
and Burns, L. (2010). Reinventing the automobile (1st ed.) Cambridge, MA: Massachusetts
Institute of Technology.

Table 1.11.2 **Current vehicle versus new USV**

Current vehicle	New ultrasmall vehicle
Mechanically driven	Electrically driven
Powered by internal combustion engine	Powered by electric motors
Fuelled by petroleum	Fuelled by electricity
Mechanically controlled	Electronically controlled
Stand-alone operation	Intelligent and interconnected

Mitchell, W., Borroni-Bird, C., & Burns, L. (2010). Reinventing the automobile (1st ed.). Cambridge, MA:
Massachusetts Institute of Technology.

This new kind of vehicle, the ultrasmall vehicle, will reduce energy consumption,
will improve the energy security, and will reduce carbon emissions thanks to its only
2.54 m of length, approximately 450 kg of weight, and a consumption of 58 km/l, thus
improving the quality of urban life (see the design in Fig. 1.11.1). Table 1.11.2 compares the current vehicle with the new USV presented by Mitchell et al., showing the
benefits of the last one.

1.12 Life-cycle

This section will compare the conventional gasoline vehicle (CV), the hybrid vehicle,
and the battery electric vehicle (BEV) in terms of life-cycle energy and CO_2 emissions
based on the article titled 'Life cycle Analysis Comparison of a Battery Electric Vehicle and a Conventional Gasoline Vehicle' by K. Aguirre et al.

According to research conducted by K. Aguirre et al., the weight of the three vehicles used was assumed to be the same in order to obtain the most accurate results (Aguirre et al., 2012).

In this study, the energy inputs of each vehicle in its lifetime have been calculated, considering the main life phases (material, manufacture, transport, use, and end of life). The results in Fig. 1.12.1 show that the conventional gasoline vehicle would require around 858,000 MJ of energy in its lifetime and 95% of that energy will be used in the use phase. This is because of the energy intensity of gasoline itself and because the extraction and processing of gasoline requires huge amounts of energy. The other life phases such as the manufacture of the different parts of the vehicle, engine manufacturing, transportation, and end of life would contribute minimal energy requirements, as the picture shows.

A hybrid vehicle would require much less energy than the conventional gasoline vehicle, around 564,000 MJ; again, the greatest amount of energy is used in the use phase, 89%. The study suggests that the manufacture of a hybrid's battery only represents 4% of the energy inputs along the life cycle.

The battery electric vehicle was the vehicle that required the lowest amount of energy, approximately 507,000 MJ. The battery manufacturing phase played an important role in its life-cycle energy requirements, accounting for a 19%. The use

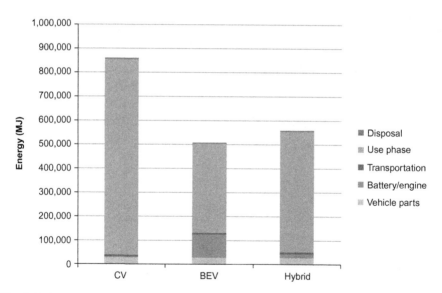

Fig. 1.12.1 Energy inputs lifecycle. CV = conventional gasoline vehicle, BEV = battery electric vehicle.

Aguirre, K., Eisenhardt, L., Lim, C., Nelson, B., Norring, A., Slowik, P., & Tu, N. (2012). Lifecycle analysis comparison of a battery electric vehicle and a conventional gasoline vehicle. pp. 7–24.

phase represents 74% of the total energy required in its lifetime, in this phase; the generation of electricity needed to charge the battery was taken into account. The manufacturing of the vehicle parts, its transportation, and its disposal accounted for a very small amount in the overall energy inputs along the life cycle.

With regard to the amount of CO_2 emissions produced by these vehicles in its entire lifetime, the study points out that the conventional gasoline vehicle would release around 63,000 kg CO_2 equivalents, the hybrid vehicle 41,000 kg and the electric vehicle around 32,000 kg (see Fig. 1.12.2).

To conclude, it could be said that life-cycle energy results and life-cycle emissions results are better for the electric vehicle, followed by the hybrid vehicle and the conventional gasoline vehicle. The electric vehicle is the best option in terms of reducing environmental impacts.

Assuming that the lifetime of the vehicles is 288,000 km, according to Table 1.12.1, a conventional gasoline vehicle would consume 2.98 MJ/km and would emit 0.22 kg CO_2 equivalents, a hybrid vehicle would require 1.96 MJ/mi and would emit 0.14 kg CO_2 equivalents, and an electric vehicle would require 1.76 MJ/mi and would emit 0.11 kg CO_2 equivalents.

When comparing the conventional gasoline vehicle with the electric vehicle, the first one will release 49% more of CO_2 emissions and will require 41% more energy than the electric vehicle.

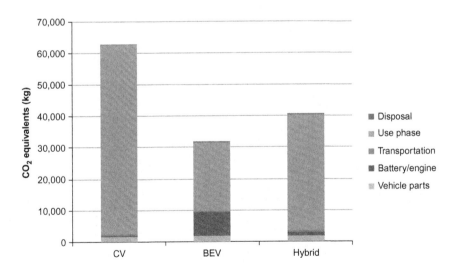

Fig. 1.12.2 Emission of CO_2 equivalents lifecycle.
Aguirre, K., Eisenhardt, L., Lim, C., Nelson, B., Norring, A., Slowik, P., & Tu, N. (2012). Lifecycle analysis comparison of a battery electric vehicle and a conventional gasoline vehicle. pp. 7–24.

Table 1.12.1 **Energy and emissions per kilometre for different types of vehicles**

	Energy (MJ/km)	Emissions (kg CO_2 eq/km)
Gasoline vehicle	2.98	0.22
Hybrid vehicle	1.96	0.14
Electric vehicle	1.76	0.11

Based on Aguirre, K., Eisenhardt, L., Lim, C., Nelson, B., Norring, A., Slowik, P., & Tu, N. (2012). Lifecycle analysis comparison of a battery electric vehicle and a conventional gasoline vehicle. pp. 7–24.

An example is presented below based on future projections of carbon intensity for gasoline and electricity mix in California. California expects that by 2020 gasoline will become dirtier, around 15% more, as they are exploring tar sand technology, and also expects that 33% of its electricity will come from renewable sources by then. As shown in Fig. 1.12.3, if this assumption is true in the future, conventional gasoline

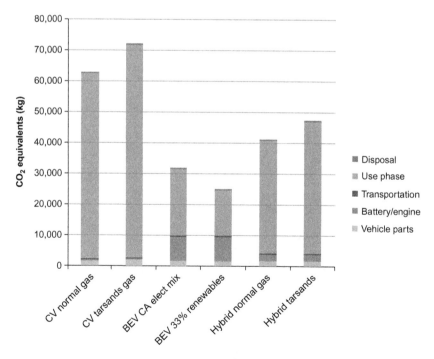

Fig. 1.12.3 Emission of CO_2 equivalents lifecycle comparison.
Aguirre, K., Eisenhardt, L., Lim, C., Nelson, B., Norring, A., Slowik, P. and Tu, N. (2012). Lifecycle analysis comparison of a battery electric vehicle and a conventional gasoline vehicle. pp. 7–24.

Table 1.12.2 Electric vehicle energy and emissions intensity for different electricity mixes

	Energy (MJ/km)	Emissions (kg CO_2 eq/km)
California—33% renewables (2020)	1.45	0.09
California mix (base case)	1.76	0.11
US mix	2.27	0.18
China mix	2.5	0.25

Aguirre, K., Eisenhardt, L., Lim, C., Nelson, B., Norring, A., Slowik, P., & Tu, N. (2012). Lifecycle analysis comparison of a battery electric vehicle and a conventional gasoline vehicle. pp. 7–24.

vehicle and hybrid vehicles would release 15% more CO_2 emissions. However, the emissions from electric vehicles would fall by 31%.

If, for example, electricity had a higher intensity, this would mean that electric vehicles would be worse than conventional gasoline vehicles from the point of view of emissions. If they were charged with coal-fired plants, they would contribute to higher emissions.

Table 1.12.2 shows how the energy and emissions intensity would vary depending on the different electricity mixes considered. California electricity mix, California projected mix with 33% electricity generated from the renewables, US electricity mix, and Chinese mix were considered in this study (Aguirre et al., 2012).

If California carries out these plans in the future, this would lead to a decrease of 18% in energy intensity and of 22% in emissions intensity compared with the base case. As the table shows, China has the highest energy and emissions intensity, 43% and 122%, respectively, if compared with the California results.

Therefore, electric vehicles charged with green energy would lead to significant improvements in the greenhouse gases balance; that is why the penetration of electric vehicles based on the use of renewable sources is so important.

1.13   The re-emergence of electric vehicles

Electric vehicles are not a new invention; the first manufactured EV was made in the 1880s. However, by that time, the popularity of electric vehicles declined for different reasons; for example, in 1912, an electric car cost $1750, while a gasoline car cost $650 (IEEE, 2013); there were also improvements in the internal combustion engine technologies, the growing petroleum infrastructure that allowed for reduced gasoline prices, quicker refuelling times, mass production of gasoline vehicles, and better road system to connect cities were some of the reasons why the deployment of EV was not successful.

Nevertheless, nowadays, electric cars are rising in popularity. Thanks to the ambitious targets set and the policy support, electric vehicle costs have been lowered, the vehicle range has been extended, and the consumer barriers have been reduced in several countries.

Table 1.13.1 shows the pros and cons of having an electric car in the present year 2016.

Table 1.13.1 **Electric cars pros and cons**

Advantages	Disadvantages
Quiet and quick	Limited range
Home recharging	Long refuelling time
Cheaper to operate	Higher cost
No tailpipe emissions	Lack of consumer choice
Higher efficiencies	Lower energy density

Based on Berman, B., & Shenoi, R. (2016). Electric cars pros and cons [online]. PluginCars.com. Available at: http://www.plugincars.com/electric-cars-pros-and-cons-128637.html Accessed 23.12.16.

Driving an electric vehicle offers a lot of benefits, making the driver feel more calm and *relaxed as they are very* quiet and the driving is very smooth. Also, due to the high torque, the power is delivered directly to the wheels, increasing the efficiency; electric vehicles are many times more efficient than internal combustion engines (ICE), around 85% versus 25% based on a well-to-wheels analysis (Ryan, 2016). In another research conducted by Muneer T. shows that the hydrocarbon fuelled vehicle has an overall tank-to-wheel efficiency of 13%, whereas electric vehicles charged with power plants would have an overall efficiency much higher, around 22% (Muneer, Celik, & Caliskan, 2011). Besides, installing a plug to charge your car in your garage would be very convenient and comfortable as you never will have to go to a petrol station again.

Electric vehicles would also benefit from the price of electricity; in most parts of the world, this is cheaper than petroleum; for example, the cost per mile to fuel an electric vehicle is around one-third to one-quarter the cost of gasoline (Berman & Shenoi, 2016).

Nevertheless, some of the drawbacks of EVs—which, however, are improving along the years—are range anxiety, having only about 80–100 mi of range at the moment and taking hours to fully refuel. But this is changing; every year, the range and cost of electric car batteries are improving; as an example, by 2017, Tesla will launch a new model of EV, Model 3, which will be able to drive nearly 350 km with only one charge (Tesla, 2016).

Besides, the initial costs of electric vehicles are usually higher than ICE vehicles, but thanks to the incentives provided by governments and their low maintenance costs, the overall cost would be lower. For example, in the United Kingdom, some incentives provided by the government offer up to a maximum of either £2500 or £4500 depending on the model for the purchase of a new electric vehicle (Gov.uk., 2016b). Furthermore, some grants to install charging points at home, on the streets, or in station car parks are also offered (Gov.uk., 2016c).

Since 2010, the global electric car fleet has been growing; in 2005, the number of electric vehicles was still measured in hundreds. According to the report Global EV Outlook (IEA, 2016d), by 2015, there were around 1.26 million electric vehicles on the road. Fig. 1.13.1 shows that in 2015 the United States, China, Japan, the Netherlands, and Norway accounted for 80% of the electric car fleet in the world.

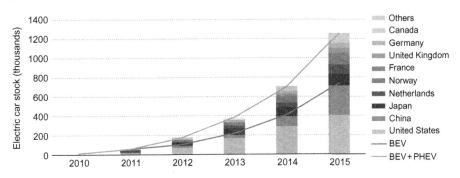

Fig. 1.13.1 Evolution of the global electric vehicle stock from 2010 to 2015.
Information resourced from IEA analysis based on EVI country submissions, complemented by
EAFO (2016), IHS Polk (2014), MarkLines (2016), Acea.be. (2016). Passenger cars | ACEA—
European Automobile Manufacturers' Association [online]. Available at: http://www.acea.be/
automobile-industry/passenger-cars Accessed 07.12.16; European Environment Agency
(EEA). (2015). Air quality in Europe—2015 report [online]. Luxembourg, pp. 49–53. Available
at: http://www.eea.europa.eu/publications/air-quality-in-europe-2015 Accessed 05.12.16;
IA-HEV (2015) as reported in The International Energy Agency (IEA). (2016d). Global EV
outlook 2016—Beyond one million electric cars (pp. 4–34). Paris: IEA. Available at: https://
www.iea.org/publications/freepublications/publication/Global_EV_Outlook_2016.pdf
Accessed 20.12.16.

In the United Kingdom, all the governments of all parties have been trying to promote the growth/sale of ultralow emission vehicles. With this, the United Kingdom wants to achieve the decarbonization of the cars in order to preserve the environment, reduce the climate change impacts, and improve air quality (Butcher, 2016).

The sales of electric car sales in the United Kingdom have risen dramatically during the past 3 years, especially in 2016. In 2014, around 16,500 electric cars were licensed, and in 2015, this figure increased up to more than 25,000 vehicles (see Table 1.13.2). That table shows the cars licensed by fuel type and as anticipated vehicles fuelled by petrol and diesel are the prevailing ones.

Now, in 2016, the electric vehicle fleet has already exceeded the amount of 75,000 electric vehicles in the United Kingdom, accounting for approximately 1.4% of the total market in the United Kingdom (see Fig. 1.13.2). According to National Grid, by 2022, there will be over one million of EVs on the road (Jackson, 2016). The United Kingdom's target is 350,000 electric vehicles on the road by 2020 (Azadfar, Sreeram, & Harries, 2015).

With regard to charging points, in the United Kingdom, there are currently 4037 charging locations, 6000 charging devices, and around 10,000 charging point connectors compared with only a few hundred charging points in 2011.

Fig. 1.13.3A shows the increase in charging points over the last 2 years and how many of them are used depending on the charger speed, the fast chargers being the most abundant.

Fig. 1.13.3B presents the charging profile of those chargers showing the remarkable developments related to DC rapid charging.

Table 1.13.2 **Licensed vehicles by body type, by local authority, the United Kingdom, annually, 2015**

Year/2015	Cars licensed by propulsion/fuel type						
	Petrol	Diesel	Hybrid electric	Gas	Electric	Other	Total
Total number of cars	18,929,438	11,927.78	247.659	40.276	25.13	0.421	31,170,701
Percentage of cars	60.7	38.3	0.8	0.1	0.1	–	100%

Note that gas includes gas bifuel, petrol/gas and gas–diesel, and other includes vehicles propelled by steam Cars licensed by propulsion/fuel type.Gov.uk. (2016a). Cars (VEH02) - GOV.UK [online]. Available at: https://www.gov.uk/government/statistical-data-sets/veh02-licensed-cars Accessed 24.11.16.

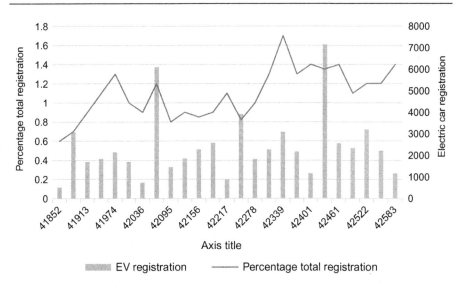

Fig. 1.13.2 Electric car registrations in United Kingdom, 2014–15.
Based on data from Society of Motor Manufacturers and Traders (2016).

Fig. 1.13.4 shows how they are spread over the United Kingdom depending on the area.

However, the market of EVs still needs policy support in order to be broadly adopted and deployed. The consumer acceptance will play a very important role for the growth of the EVs, but to this effect, improvements like increasing the charging infrastructure to avoid this range anxiety or reducing the costs to produce EVs are needed.

Battery cost represents a large portion of the total EV production cost. Fig. 1.13.5 shows how these costs have started to decrease since 2008, while the battery energy density has been increasing along the years, which makes electric cars more attractive for the consumers in the future.

The source where electricity to charge the electric vehicle is obtained from is very important, as the CO_2 emissions reduction potential of electric vehicles will depend on the way that electricity is generated.

Fig. 1.13.6 below shows the primary sources of electricity generation in several countries. China and India have the highest carbon intensity for electricity, 79% and 69%, respectively, of the electricity is generated with coal. If, in these countries, the electricity to charge EV's was taken from the grid, it would be a catastrophe for climate change, as there would be an increase in CO_2 emissions; therefore, cleaner sources should be applied. As a curious fact, in India, the carbon intensity in 2008 was very high, around 1000 g CO_2/kWh (see Fig. 1.13.7 (IEA, 2011)). However, in other countries like Brazil, the use of renewable energy to generate electricity represents a high percentage, 84% (see Fig. 1.13.6).

Therefore, the use of electric vehicles could help to tackle climate change only if electricity was generated through renewable energy. For example, according to the US

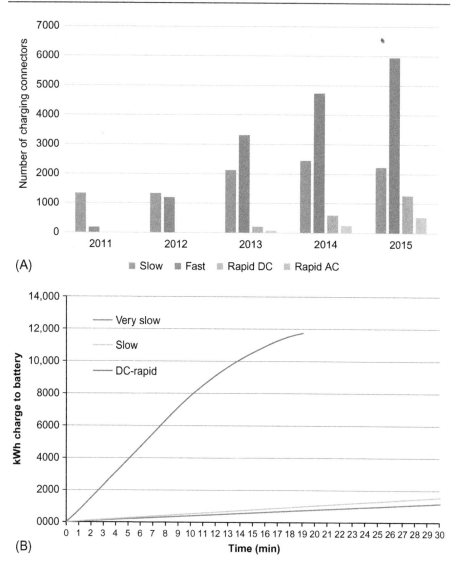

Fig. 1.13.3 (A) Number of type of charging connectors in United Kingdom from 2011 to 2015. (B) Mitsubishi iMiev (16 kWh) battery charging profile data for three different charges.
A: Lane, B. (2016). Electric vehicle market statistics 2016 - How many electric cars in UK. [online] Nextgreencar.com. Available at: http://www.nextgreencar.com/electriccars/statistics/; B: Muneer, T., Milligan, R., Smith, I., Doyle, A., Pozuelo, M., & Knez, M. (2015). Energetic, environmental and economic performance of electric vehicles: Experimental evaluation. *Transportation Research Part D: Transport and Environment, 35*, 40–61.

Department of Energy, on average in the United States, the hybrid vehicle Toyota Prius and the electric vehicle Nissan Leaf produce the same amount of GHG pollutants, around 200 g/mi. In Minnesota, the emission values are worse, as fossil fuels are the main source of electricity; an electric vehicle would emit 300 g of GHG emissions

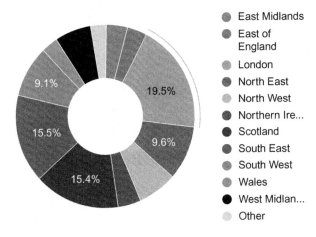

East Midlands
East of England
London
North East
North West
Northern Ire...
Scotland
South East
South West
Wales
West Midlan...
Other

Fig. 1.13.4 Profile of charging locations across the UK regions. From Lane, B. (2016). Electric vehicle market statistics 2016 - How many electric cars in UK &#63. [online] Nextgreencar.com. Available at: http://www. nextgreencar.com/ electriccars/statistics/.

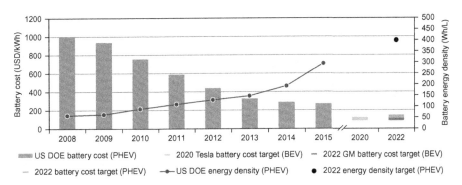

US DOE battery cost (PHEV) — 2020 Tesla battery cost target (BEV) — 2022 GM battery cost target (BEV)
— 2022 battery cost target (PHEV) —•— US DOE energy density (PHEV) ● 2022 energy density target (PHEV)

Fig. 1.13.5 Evolution of battery energy density and cost.
Information resourced from US DOE (2015, 2016) for PHEV battery cost and energy density estimates; EV Obsession (2015); HybridCARS (2015) as reported in The International Energy Agency (IEA). (2016d). Global EV outlook 2016—Beyond one million electric cars (pp. 4–34). Paris: IEA. Available at: https://www.iea.org/publications/freepublications/publication/Global_EV_Outlook_2016.pdf Accessed 20.12.16.

per mile, leading to higher emissions. However, in California, the city with the highest percentage of clean energy in the country, an electric vehicle would produce only 100 g/mi (Biello, 2016).

Companies are starting to be really interested in expanding charging networks around the world with stations powered by solar photovoltaic technologies (REN21, 2016).

In 2015, the use of renewable energy in the transport sector drew more and more international attention. Globally, the sale of electric vehicles has increased over the past, and major improvements associated to the integration of renewable energy into EV charging infrastructure have been made.

For example, in 2015, China opened its largest solar PV charging station able to charge 80 EVs per day, and in Shanghai, a pilot project to test the ability of EVs to

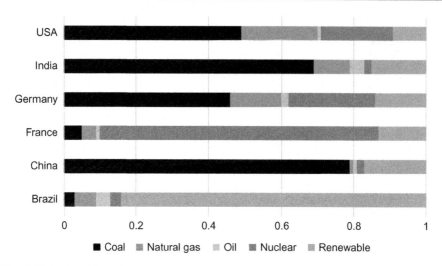

Fig. 1.13.6 Electricity mix in various countries in 2008.
Based on The International Energy Agency (IEA). (2016d). Global EV outlook 2016—Beyond one million electric cars (pp. 4–34). Paris: IEA. Available at: https://www.iea.org/publications/freepublications/publication/Global_EV_Outlook_2016.pdf Accessed 20.12.16.

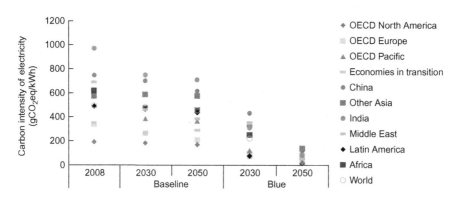

Fig. 1.13.7 CO_2 intensity of electricity generation by region, year, and scenario.
The International Energy Agency (IEA). (2011). Technology roadmap: Electric and plug-in hybrid electric vehicles [online]. Available at: http://www.iea.org/publications/freepublications/publication/EV_pHEV_roadmap.pdf Accessed 30.11.16.

help to integrate renewable energy into the electricity grid was launched (Ho, 2015 and The Huffington Post, 2016).

The integration of electric vehicles and the energy system is very important; if adequate charging strategies are not applied, the growth in electricity demand by EVs would have important effects on the grid with regard to load balancing (Kaschub, Mültin, Fichtner, Schmeck, & Kessler, 2010). A significant number of

Fig. 1.13.8 EV ARC able to generate between 3800 to 7000 kWh of solar electricity annually. Movellan, J. (2015). 100 Percent renewable energy charged EV stations allow driving on sunshine [online]. Renewable Energy World. Available at: http://www.renewableenergyworld. com/articles/2015/08/100-percent-renewable-energy-charged-ev-stations-allow-driving-on-sunshine.html Accessed 30.12.16.

electric vehicles that interact with smart grids could help to overcome the problem of load leveling and introduce considerable energy-storage capacity to the grids.

Smart grids could benefit from the storage capacity of electric vehicles, employing price signals to regulate the demand and avoiding the intermittent supply typical of renewable energy sources, keeping that way an optimal balance between the supply and the demand (Mitchell et al., 2010).

Also, in the United States, California has developed an electric vehicle autonomous renewable charger (see Fig. 1.13.8). This charging station is an off-grid system that can provide 100% of renewable electricity for EVs (Movellan, 2015).

The deployment of EVs will play a fundamental role in reducing the dependency on oil fuels. Again, the only route to sustainability will only be achieved through renewable electricity, helping that way to mitigate the greenhouse gases emissions and CO_2 emissions. It is estimated that, to achieve the world emission targets, around 20 million EVs will be needed by 2020 (Gomez et al., 2016).

1.14   The rapid rise of renewable energy

Solar radiation striking the earth on an annual basis is equivalent to 15,000 times that of current global energy consumption. Although photosynthetic energy capture is estimated to be 10 times that of global annual energy consumption, only a small part of this solar radiation is used for photosynthesis. Approximately two-thirds of the net global photosynthetic productivity worldwide is of terrestrial origin, while the remainder is produced mainly by phytoplankton (microalgae) in the oceans that cover approximately 70% of the total surface area of the earth. Since biomass originates from plant and algal photosynthesis, both terrestrial plants and microalgae are appropriate targets for scientific studies relevant to biomass energy production. Any analysis of biomass energy production must consider the potential efficiency of the processes involved. Although photosynthesis is fundamental to the conversion of solar radiation into stored biomass energy, its theoretically achievable efficiency is limited both by the limited wavelength range applicable to photosynthesis and the quantum requirements of the photosynthetic process. Only light within the wavelength range of 400–700 nm (photosynthetically active radiation, PAR) can be utilized by plants, effectively allowing only 45% of total solar energy to be utilized for photosynthesis. On the basis of these limitations, the theoretical maximum efficiency of solar energy conversion is approximately 11%. In practice, however, the magnitude of photosynthetic efficiency observed in the field is further decreased by factors such as poor absorption of sunlight due to its reflection, respiration requirements of photosynthesis, and need for optimal solar radiation levels, the net result being an overall photosynthetic efficiency of between 3% and 6%.

In the 1980s, leading consultants were skeptical about cellular phones. McKinsey & Company noted that the handsets were heavy, batteries did not last long, coverage was patchy, and the cost per minute was exorbitant. It predicted that in 20 years the total market size would be about 900,000 units. Likewise, in the early 1970s, when the IBM 360 mainframe computer was introduced, the forecast was that only five units would be sold worldwide.

The experts are saying the same about solar energy now. They note that after decades of development, solar power hardly supplies 1% of the world's energy needs. They say that solar is inefficient, too expensive to install, and unreliable and will fail without government subsidies. Ray Kurzweil notes that solar power has been doubling every 2 years for the past 30 years. He says solar energy needs only six more doublings, or less than 14 years, to meet most of today's energy needs. Energy usage will keep increasing, so this is a moving target. But, by Kurzweil's estimates, inexpensive renewable sources will provide more energy than the world needs in less than 20 years.

The prices of solar technology are decreasing, and this technology is becoming more efficient every year, which makes it very competitive. According to the International Energy Agency, by 2050, the costs of electricity from photovoltaic energy will be reduced by 65% (IEA, 2014).

In places such as Germany, Spain, Portugal, Australia, and the southwest United States, residential-scale solar production has already reached grid parity with average residential electricity prices. In other words, it costs no more in the long term to install solar panels than to buy electricity from utility companies. The prices of solar panels have fallen 75% in the past 5 years alone and will fall much further as the technologies to create them improve and scale of production increases. By 2020, solar energy will be price-competitive with energy generated from fossil fuels on an unsubsidized basis in most parts of the world. Within the next decade, it will cost a fraction of what fossil-fuel-based alternatives do.

According to Renewable Energy Policy Network for the twenty-first century, 2015 was one of the best years for renewable energy (REN21, 2016), and the International Energy Agency (IEA) stated that 'the sun could be the world's largest source of electricity by 2050, ahead of fossil fuels, wind, hydro, and nuclear', saying as well that solar photovoltaic systems and solar thermal electricity could generate up to 27% of the world's electricity by 2050 (IEA, 2016e).

This is true to a considerable degree, as the solar photovoltaic capacity in the world is increasing every year. For example, last year, the solar PV capacity experienced another year of record growth—25% more than in 2014, around 50 GW were added, bringing a total solar PV global capacity to about 227 GW in the world (see Figs. 1.14.1 and 1.14.2).

Due to its high potential, solar energy is one of the most investigated energy sources. These are some of its advantages compared with fossil fuels (IEA, 2016e):

- Improved security of supply
- Stabilized costs of electricity generation
- Avoidance of fuel price risks or constraints
- No greenhouse gas (GHG) or other pollutants emissions

Despite the widespread belief that investing in photovoltaic energies in countries with many cloudy days such as the United Kingdom is unprofitable, the fact is that the PV modules installed in the United Kingdom work really well due to the temperate climate; even in cloudy days, the PV modules can generate an appropriate amount of energy, and actually, PV modules work more efficiently in this kind of climate.

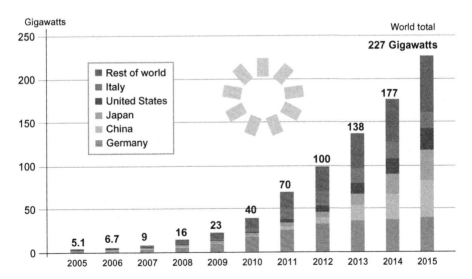

Fig. 1.14.1 Solar PV global capacity by country/region, 2005–15.
Renewable Energy Policy Network for the 21st Century (REN21). (2016). Renewables 2016 global status report [online]. Paris, pp. 27–29. Available at: http://www.ren21.net/wp-content/uploads/2016/06/GSR_2016_Full_Report1.pdf Accessed 03.07.16.

Gigawatts

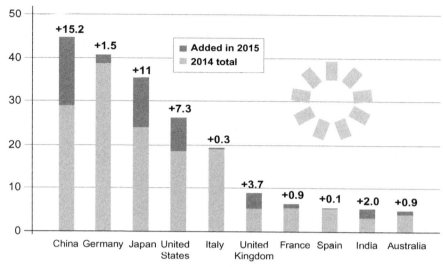

Fig. 1.14.2 Solar PV global capacity and additions, top 10 countries, 2015.
Renewable Energy Policy Network for the 21st Century (REN21). (2016). Renewables 2016
global status report [online]. Paris, pp. 27–29. Available at: http://www.ren21.net/
wp-content/uploads/2016/06/GSR_2016_Full_Report1.pdf Accessed 03.07.16.

The chart below shows the cell temperature in the location of Scotland, in Edinburgh. PV modules start losing efficiency when the cell temperature is in excess of 25°C. As Edinburgh has low temperatures, cell temperatures above 25°C are only reached at certain times of the day in some months. Fig. 1.14.3 shows the cell efficiency in the warmest month in Edinburgh, the green line represents the cell efficiency according to the manufacturer's data, 15.2%. Most of the time, the system will work above this efficiency, generating overall an increase in its power output.

It isn't just solar production that is advancing at a rapid rate. Research projects all over the world are actively seeking to exploit the power of wind, biomass, thermal, and tidal energy. Wind power, for example, has also come down sharply in price and is now competitive with the cost of new coal-burning power plants in the United States.

Table 1.14.1 below shows the GW install worldwide in 2014 and in 2015 according to different types of renewable energies. In general, the capacity install has increased with regard to the past year 2014 (REN21, 2016).

Coal is one of the world's largest sources of greenhouse gas emissions and a major climate change contributor. So why are we still using it? For the same reasons, we always have: it's cheap, plentiful, easy to transport, and easy to get.
(Jorge Ribas and Julio Negron/The Washington Post)

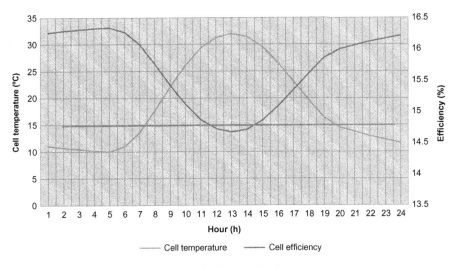

Fig. 1.14.3 Influence of the temperature in the month of August.
Author's illustration.

Table 1.14.1 **Renewable energy indicators 2015**

Power	Unit	2014	2015
Renewable power capacity (not including hydro)	GW	665	785
Renewable power capacity (including hydro)	GW	1.701	1.849
Hydropower capacity	GW	1.036	1.064
Biopower capacity	GW	101	106
Biopower generation (annual)	TWh	429	464
Geothermal power capacity	GW	12.9	13.2
Solar PV capacity	GW	177	227
Concentrating solar thermal power capacity	GW	4.3	4.8
Wind power capacity	GW	370	433

Renewable Energy Policy Network for the 21st Century (REN21). (2016). Renewables 2016 global status report [online]. Paris, pp. 27–29. Available at: http://www.ren21.net/wp-content/uploads/2016/06/GSR_2016_Full_Report1. pdf Accessed 03.07.16.

The environment will surely benefit from the elimination of fossil fuels, which will also boost most sectors of the economy. Electric cars will become cheaper to operate than fossil-fuel-burning ones, for example. We are heading into the era of abundance that Peter Diamandis has written about—the era when the basic needs of humanity are met through advancing technologies.

In 31 Aug. 2016 edition of the London-based 'i' newspaper, it was reported that in the year 2015 Denmark with its 700 W/capita installation of wind turbine capacity, the highest in the world, generated 140% of its electricity demand. Between 2010 and 2014, the renewable energy consumption of the top 15 countries in the world has

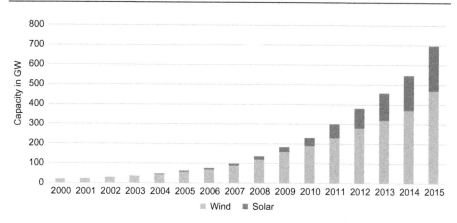

Fig. 1.14.4 Globally installed wind and solar capacity from 2000 to 2015.
Akuoko, A. (2016). Increasing solar energy system & challenges to the UK grid—The power
academy essay challenge 2016 [online]. Loughborough University, p. 3. Available at: http://
file:///C:/Users/irene/Downloads/Akuoko.%20A.pdf Accessed 10.01.17.

doubled. Fig. 1.14.4 provides information on the rapid rise of wind and solar power
plants across the world ('i' newspaper, London, England, 31 Aug. 2016, p. 2).

On 9 Jun. 2016, the Fraunhofer ISE research institute announced that Germany set
a record high for solar use. On that day, the country's solar power output rose to
23.1 GW, nearly 51% of all electricity demand. Despite not having a generally sunny
climate, Germany has been pushing solar energy through rooftop solar modules. Over
90% of mounted solar panels within Germany are on rooftops. The country broke two
other records around the same time, producing 24.24 GW of solar-generated power
between 1 p.m. and 2 p.m. on 6 Jun. of the same year, and over that entire week,
the country produced 1.26 TWh of electricity from solar power. The total weekly electricity consumption in Germany is around 12 TWh (AGEB, 2016).

In the year 2015, France approved a decree that new rooftops must be covered in
plants or solar panels. Any new buildings in commercial zones across the country must
comply with new environmental legislation. Green roofs have an isolating effect,
helping reduce the amount of energy needed to heat a building in winter and cool
it in summer. The law was also made less onerous for businesses by requiring only
part of the roof to be covered with plants, and giving them the choice of installing solar
panels to generate electricity instead.

Within the United States, renewable energy sources, such as wind and solar energy,
are continuing to grow significantly each year. Renewable energy in the United States
through the first half of 2016, including hydroelectric power, biomass, geothermal,
wind, and solar (including distributed solar), provided nearly 17% of electricity
generation. In all of 2015, that number was 14%. Nonhydro renewable energy was
9.2% of US electric generation through the first half of 2016. For the entire year
2015, it was 7.6%.

Within the Unites States, coal-fired power is down 20%. Wind and solar powers (increasing 23% and 31%, respectively) are making up for most of that lost generation from coal. Nuclear power is up by about 1%. However, the longer term trend for nuclear power looks uncertain with facilities in Illinois, Massachusetts, Nebraska, and New Jersey to close this decade. Only two new nuclear plants are under construction in the United States with commercial operation expected in 2019 and 2020 if no further delays occur.

For renewable energy, growth is occurring in virtually all 50 states within the United States, as new wind and solar installations are placed into service. Every region is enjoying this growth. Requests for interconnection to utilities throughout the country are now primarily for wind and solar projects.

For wind power, the states with the most growth to date (in terms of actual generation) are Texas, Oklahoma, Kansas, Iowa, and Colorado. Just those five states alone have added enough generation in the first 6 months of the year to power an additional 3 million homes. For solar power, the states with the most growth (in terms of actual generation) are California, North Carolina, Nevada, Arizona, and Georgia. Just those five states alone have added enough generation in the first 6 months of the year to power nearly an additional one million homes. Utah has increased solar power generation over 700%. In Iowa, wind power is rivalling coal as the top source of electric generation. By 2017, Iowa may become the first US state to generate a majority of its power from wind.

California is generating nearly 30% of its generation from nonhydro renewable energy. In general terms, wind energy generation doubled between 2010 and 2015. Solar energy generation increased by more than 20 times between those years. By the end of 2017, solar energy will likely double 2015 generation. The intermittency of renewable energy, however, is an issue, and in this respect, technological developments related to battery storage offer good solutions.

The following Table 1.14.2 shows the total capacity of the top five countries who have invested in different renewable energies at the end of 2015.

China installed a world record of 32.5 GW of wind power last year and a world record 18.3 GW of solar power, according to official figures from the National Bureau of Statistics of China on 29 Feb. 2015. Coal consumption fell 3.7%, nuclear power grew 30%, and natural gas 3.3%. These trends mark a rapid diversification of China's electricity generation capacity with reduced dominance of coal. Some even believe China's CO_2—emissions have peaked—15 years before the 2030 target.

'The latest figures confirm China's record-breaking shift toward renewable power and away from coal', said Tim Buckley, Director of Energy Finance Studies at the Institute for Energy Economics and Financial Analysis (IEEFA). Note though that the Chinese coal consumption still accounted for 64.0% of total energy consumption.

Tesla's Powerwall 2 is the most public step towards solar storage inverters, but it's not the only one. Leading inverter company SMA has combined forces with battery leader LG Chem to create a complete inverter battery system that will launch soon in Europe and Australia. Sonnen also has an inverter battery solution, leveraging its market-leading position in Germany. Enphase Energy has what it calls an AC battery that aims to make battery installations seamless.

Table 1.14.2 Leading countries for a given renewable energy source in 2015

Power	1	2	3	4	5
Renewable power (including hydro)	China	The United States	Brazil	Germany	Canada
Renewable power (not including hydro)	China	The United States	Germany	Japan	India
Biopower generation	The United States	China	Germany	Brazil	Japan
Geothermal power capacity	The United States	Philippines	Indonesia	Mexico	New Zealand
Hydropower capacity	China	Brazil	The United States	Canada	Russia
Hydropower generation	China	Brazil	Canada	The United States	Russia
CSP	Spain	The United States	India	Morocco	South Africa
Solar PV capacity	China	Germany	Japan	The United States	Italy
Solar PV capacity per capita	Germany	Italy	Belgium	Japan	Greece
Wind power capacity	China	The United States	Germany	India	Spain
Wind power capacity per capacity	Denmark	Sweden	Germany	Ireland	Spain

Renewable Energy Policy Network for the 21st Century (REN21). (2016). Renewables 2016 global status report [online]. Paris, pp. 27–29. Available at: http://www.ren21.net/wp-content/uploads/2016/06/GSR_2016_Full_Report1.pdf Accessed 03.07.16.

We are still in the very early phases of energy storage in the home, but it's clear that batteries will start to creep their way into more and more installations. That will open up a new world of innovations.

It is easy to see how a battery/inverter combination would make sense, and how it will become a standard component. Electricity generated during the day can be used to charge electric vehicles when people are home. Alternatively, the e-vehicle batteries can even be used to help balance demand on the grid.

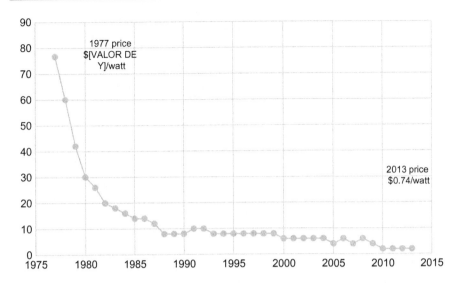

Fig. 1.14.5 The swanson effect—Price of crystalline silicon photovoltaic cells, $/watt. Information resourced from Bloomberg, New Energy Finance (2014) as reported in Solar Action Alliance. (2016). Solar panels for New Jersey homes: Tax incentives, prices, info [online]. Available at: https://solaractionalliance.org/new-jersey/ Accessed 15.01.17.

The US states of Hawaii and Nevada have put rules in place that make energy arbitrage more profitable and California utilities are moving towards time of use rates that makes more financial sense for smart energy devices like storage integrated into solar systems.

Fig. 1.14.5 below shows the Swanson effect representing the decrease of the solar power technologies. Costs are dropping by 20% every 10 years, the figure shows how solar panel costs have decreased from $76.67/W in 1977 to $0.74 in 2013.

1.15 Conventional energy matrix

By way of demonstration in this section, we present the energy resourcing attempts being made a fast developing economy. The analysis will demonstrate that using modern technological developments conventional technologies will only offer only incremental fuel and carbon savings. The solution therefore is to switch to renewables in an organized and wholesome way.

On 26 Apr. 1986, a massive explosion ripped the roof off the fourth reactor at the Soviet nuclear power station, located 95 km from Kiev, Ukraine, spreading a radioactive dust cloud over much of Europe. Since then the exclusion zone, comprising all land within a 30 km radius of the plant has been deemed unevenly contaminated.

Thirty years after Chernobyl became the deadliest nuclear disaster in history, a Chinese clean energy giant has been selected to build a 1 GW solar power plant in its exclusion zone to help revive the region. Chernobyl is considered the deadliest

Table 1.15.1 **China's electricity generation capacity, GW**

	2014	2015	% Increase
Thermal	918.6	990.2	7.80
Nuclear	20.1	26.1	29.90
Hydro	304.5	319.4	4.90
Wind	96.9	129.3	33.50
Solar	24.9	43.2	73.70
Total	1365	1508.2	10.50

nuclear disaster in history, more so, even than the 2011 Fukushima Daiichi event. The present energy generation mix for China is shown in Table 1.15.1. Countries such as China, whose Carbon emission is of concern to the global community, are indeed increasing their coal-fired electricity generation efficiency by employing supercritical and ultrasupercritical boilers.

China has invested heavily in its electrical generation infrastructure, including coal. All four 1000 MW coal-fired ultrasupercritical pressure boilers at Yuhuan have come online. Located on the coast of east China's Zhejiang Province, the last unit began commercial operation in Nov. 2007. The plant cost ¥9.6bn (€900 m), and the units run at about 45% efficiency. Yuhuan has China's first 1000 MW ultrasupercritical pressure boilers. The site is annually generating 22 billion kWh of electricity. Siemens reports that just a 1% gain in efficiency for a typical 700 MW plant reduces 30-year lifetime emissions by 2000 t NO_x, 2000 t SO_2, 500 t particulates, and 2.5 million t of CO_2.

About 70% of China's total energy consumption comes from coal, and the country still has huge reserves. Burning it, however, has severely damaged the environment. In northern China, cities like Beijing and Shenyang have some of the highest readings for total suspended particulates and SO_2 in the world, with coal burning being a major source of this pollution. In southern China, large areas have growing acid rain problems (http://www.power-technology.com/projects/yuhuancoal/).

A typical subcritical coal power plant with an efficiency of 36% emits around 910 g CO_2/kWh_e. Thus, for each 10% increase in efficiency of a coal-fired power plant the per km tailpipe reduction of carbon emission from an electricity-charged vehicle is reduced by only 22% to 712 gCO_2.

At a considerably high pressure of 220.9 bar, water heated to a temperature of 374°C instantly flashes into vapour. This is known as the critical point. The older power plants all operated at subcritical pressures and had much lower thermal efficiencies of around 40%. Supercritical operation of large thermal baseload power plants during the 1980s used steam temperatures of typically 550°C, leading to around higher thermal efficiencies. Ultrasupercritical steam conditions now use supercritical pressures up to 300 bar, with 600°C steam and reheat steam temperatures. This gives a net efficiency of 46%. Table 1.15.2 presents the relevant information. What is clear that despite using the latest technological solutions, there in only a marginal increase in thermal efficiency.

Table 1.15.2 **Lower heating value efficiencies for coal-fired power plants**

Design	Maximum pressure (bar)	Maximum temperature (celsius)	Efficiency (%)
Subcritical	167	538	40.5
Supercritical	250	560	42
Supercritical	250	566	42.5
Supercritical	270	600	44
Ultrasupercritical	285	620	44.7

Acknowledgements

In the preparation of this chapter, the authors have sourced data from a number of references a list of which is provided below. The author would like to express their thanks to the authors of those references.

Fig. 1.2.1	http://www.statista.com/graphic/5/200002/international-car-sales-since-1990.jpg
Fig. 1.2.2	www.bbc.co.uk/news/world-asia-china-20069627
Fig. 1.2.3	https://www.gov.uk/government/statistical-data-sets/veh02-licensed-cars
Fig. 1.2.4	http://www.economist.com/node/21563280
Fig. 1.3.1	https://ourworldindata.org/world-population-growth/
Fig. 1.3.2	https://ourfiniteworld.com/2012/03/12/world-energy-consumption-since-1820-in-charts/
Fig. 1.3.3	https://www.iea.org/publications/freepublications/publication/ETP2012_free.pdf
Fig. 1.3.4	http://www.worldenergyoutlook.org/media/weo2010.pdf
Fig. 1.3.5	http://www.eea.europa.eu/publications/air-quality-in-europe-2015
Fig. 1.3.6	http://data.worldbank.org/indicator/EN.ATM.CO2E.PC
Fig. 1.3.7	http://climate.nasa.gov/scientific-consensus/
Fig. 1.4.1	https://www.theguardian.com/environment/2016/may/12/air-pollution-rising-at-an-alarming-rate-in-worlds-cities
Fig. 1.4.2	https://www.theguardian.com/environment/2016/may/12/air-pollution-rising-at-an-alarming-rate-in-worlds-cities
Fig. 1.4.3	http://www.who.int/mediacentre/news/releases/2014/air-pollution/en/ and https://www.iea.org/publications/freepublications/publication/weo-2016-special-report-energy-and-air-pollution.html
Fig. 1.4.4	Ryan, D., Edinburgh Napier University (personal communication)
Fig. 1.4.5	http://www.environment.ucla.edu/media/files/BatteryElectricVehicleLCA2012-rh-ptd.pdf
Fig. 1.4.6	http://www.unep.org/transport/gfei/autotool/case_studies/northamerica/canada/cs_cn_0.asp

Continued

Fig. 1.5.1	http://www.gov.scot/Publications/2015/11/5671/6
Fig. 1.5.2	http://www.eea.europa.eu/publications/air-quality-in-europe-2015
Fig. 1.6.1	https://www.researchgate.net/figure/233952036_fig2_Figure-2-Global-production-of-fossil-energy-from-1800-to-2010-Adapted-from-Hook-et-al
Fig. 1.6.2	http://www.eia.gov/outlooks/ieo/transportation.cfm
Fig. 1.6.3	https://www.iea.org/publications/freepublications/publication/WEO2013.pdf
Fig. 1.6.4	https://www.iea.org/publications/freepublications/publication/KeyWorld2016.pdf
Fig. 1.6.5	BP Statistical Review, 2012
Fig. 1.6.6	https://www.iea.org/publications/freepublications/publication/WEO2013.pdf
Fig. 1.6.7	Bureau of Labor Statistics, the United States, 2016
Fig. 1.6.8	https://www.iea.org/publications/freepublications/publication/KeyWorld2016.pdf
Fig. 1.7.1	https://www.statista.com/statistics/262747/worldwide-automobile-production-since-2000/
Fig. 1.7.2	http://www.oica.net/category/economic-contributions/auto-jobs/
Fig. 1.7.3	http://www.transport-research.info
Fig. 1.7.4	http://publications.jrc.ec.europa.eu/repository/bitstream/JRC93302/move%20jobs%20%20jrc%20final%20report%20final%2020150113.pdf
Fig. 1.7.5	http://publications.jrc.ec.europa.eu/repository/bitstream/JRC93302/move%20jobs%20%20jrc%20final%20report%20final%2020150113.pdf
Fig. 1.8.1	https://www.statista.com/statistics/264753/ranking-of-countries-according-to-the-general-quality-of-infrastructure/
Fig. 1.8.2	https://www.iea.org/publications/freepublications/publication/TransportInfrastructureInsights_FINAL_WEB.pdf
Fig. 1.8.3	https://www.iea.org/publications/freepublications/publication/TransportInfrastructureInsights_FINAL_WEB.pdf
Fig. 1.8.4	https://www.iea.org/publications/freepublications/publication/TransportInfrastructureInsights_FINAL_WEB.pdf
Fig. 1.8.5	https://www.iea.org/publications/freepublications/publication/TransportInfrastructureInsights_FINAL_WEB.pdf
Fig. 1.9.1	http://www.who.int/violence_injury_prevention/road_safety_status/2015/en/
Fig. 1.9.2	http://www.who.int/healthinfo/global_burden_disease/estimates/en/index1.html
Fig. 1.9.3	http://www.who.int/violence_injury_prevention/road_safety_status/2015/en/
Fig. 1.9.4	http://www.who.int/violence_injury_prevention/road_safety_status/2015/en/
Fig. 1.10.1	http://www.efe.com/efe/espana/sociedad/barcelona-ofrecera-transporte-publico-gratis-a-cambio-del-coche-contaminante/10004-3121846

Fig. 1.10.2	http://www.citylab.com/housing/2015/03/how-much-sprawl-costs-america/388481/
Fig. 1.10.3	http://visual.ly/cost-efficiency-transportation
Fig. 1.11.1	https://www.amazon.es/Reinventing-Automobile-Personal-Mobility-Century/dp/0262013827
Fig. 1.12.1	http://www.environment.ucla.edu/media/files/BatteryElectricVehicleLCA2012-rh-ptd.pdf
Fig. 1.12.2	http://www.environment.ucla.edu/media/files/BatteryElectricVehicleLCA2012-rh-ptd.pdf
Fig. 1.12.3	http://www.environment.ucla.edu/media/files/BatteryElectricVehicleLCA2012-rh-ptd.pdf
Fig. 1.13.1	https://www.iea.org/publications/freepublications/publication/Global_EV_Outlook_2016.pdf
Fig. 1.13.2	http://www.nextgreencar.com/electric-cars/statistics/
Fig. 1.13.3A	http://www.nextgreencar.com/electriccars/statistics/
Fig. 1.13.3B	Muneer, T., Milligan, R., Smith, I., Doyle, A., Pozuelo, M., and Knez, M. (2015). Energetic, environmental and economic performance of electric vehicles: Experimental evaluation. *Transportation Research Part D: Transport and Environment, 35*, 40–61
Fig. 1.13.4	http://www.nextgreencar.com/electriccars/statistics/
Fig. 1.13.5	https://www.iea.org/publications/freepublications/publication/Global_EV_Outlook_2016.pdf
Fig. 1.13.6	https://www.iea.org/publications/freepublications/publication/Global_EV_Outlook_2016.pdf.
Fig. 1.13.7	http://www.iea.org/publications/freepublications/publication/EV_pHEV_roadmap.pdf
Fig. 1.13.8	http://www.renewableenergyworld.com/articles/2015/08/100-percent-renewable-energy-charged-ev-stations-allow-driving-on-sunshine.html
Fig. 1.14.1	http://www.ren21.net/wp-content/uploads/2016/06/GSR_2016_Full_Report1.pdf
Fig. 1.14.2	http://www.ren21.net/wp-content/uploads/2016/06/GSR_2016_Full_Report1.pdf
Fig. 1.14.4	http://file://C:/Users/irene/Downloads/Akuoko.%20A.pdf
Fig. 1.14.5	https://solaractionalliance.org/new-jersey/
Tables 1.2.1, 1.2.2, 1.2.5, and 1.2.6	Jeekel (2013)
Table 1.2.3	http://ec.europa.eu/eurostat/statisticsexplained/index.php/Passenger_cars_in_the_EU https://uk.pinterest.com/pin/437623288774015471/
Table 1.2.4	http://ec.europa.eu/eurostat/statisticsexplained/index.php/Passenger_cars_in_the_EU https://uk.pinterest.com/pin/437623288774015471/
Table 1.4.1	http://www.who.int/mediacentre/news/releases/2014/air-pollution/en/
Table 1.7.1	http://publications.jrc.ec.europa.eu/repository/bitstream/JRC93302/move%20jobs%20%20jrc%20final%20report%20final%2020150113.pdf
Table 1.10.1	http://oro.open.ac.uk/4378/1/PT_Energy_and_Emissions.pdf

Continued

Table 1.10.2	http://oro.open.ac.uk/4378/1/PT_Energy_and_Emissions.pdf
Table 1.11.1	https://www.amazon.es/Reinventing-Automobile-Personal-Mobility-Century/dp/0262013827
Table 1.11.2	https://www.amazon.es/Reinventing-Automobile-Personal-Mobility-Century/dp/0262013827
Table 1.12.1	http://www.environment.ucla.edu/media/files/BatteryElectricVehicleLCA2012-rh-ptd.pdf
Table 1.13.1	http://www.plugincars.com/electric-cars-pros-and-cons-128637.htm
Table 1.13.2	https://www.gov.uk/government/statistical-data-sets/veh02-licensed-cars
Table 1.14.1	http://www.ren21.net/wpcontent/uploads/2016/06/GSR_2016_Full_Report1.pdf
Table 1.14.2	http://www.ren21.net/wpcontent/uploads/2016/06/GSR_2016_Full_Report1.pdf

References

Acea.be. (2016). Passenger Cars | ACEA - European Automobile Manufacturers' Association. [online] Available at: http://www.acea.be/automobile-industry/passenger-cars Accessed 07.12.16.

Ag-energiebilanzen e.V (2016). AG Energiebilanzen e.V. | Arbeitsgemeinschaft. [online] Available at: http://ag-energiebilanzen.de/index.php Accessed 15.01.17.

Aguirre, K., Eisenhardt, L., Lim, C., Nelson, B., Norring, A., Slowik, P., & Tu, N. (2012). *Lifecycle analysis comparison of a battery electric vehicle and a conventional gasoline vehicle—semantic scholar.* [online] Semanticscholar.org. Available at: https://www.semanticscholar.org/paper/Lifecycle-Analysis-Comparison-of-a-Battery-Aguirre-Eisenhardt/2913dc7927c411ba6a922f426a67f01108699fd3 Accessed 12.11.16.

Austrian Institute for SME Research. (2012). *WIR: Women in Rail.* Brussels: Austrian Institute for SME Research.

Azadfar, E., Sreeram, V., & Harries, D. (2015). The investigation of the major factors influencing plug-in electric vehicle driving patterns and charging behaviour. *Renewable and Sustainable Energy Reviews, 42,* 1065–1076.

Bekker, H. (2016). 2015 (Full Year) International: Worldwide car sales [online]. Car sales statistics. Available at: http://www.best-selling-cars.com/international/2015-full-year-international-worldwide-car-sales/ Accessed 19.11.16.

Berman, B. and Shenoi, R. (2016). Electric cars pros and cons [online] PluginCars.com. Available at: http://www.plugincars.com/electric-cars-pros-and-cons-128637.html Accessed 23.12.16.

Bertaud, A., & Malpezzi, S. (2003). *The spatial distribution of population in 48 world cities: Implications for economies in transition.* University of Wisconsin.

Biello, D. (2016). Electric cars are not necessarily clean [online]. *Scientific American.* Available at: https://www.scientificamerican.com/article/electric-cars-are-not-necessarily-clean/ Accessed 30.12.16.

Butcher, L. (2016). *Electric vehicles and infrastructure. Number CBP07480 [online].* House of Commons Library. pp. 1–3. Available at: http://researchbriefings.parliament.uk/ResearchBriefing/Summary/CBP-7480 Accessed 30.12.16.

Casey, M. (2014). The 10 biggest R&D spenders worldwide [online]. *Fortune*. Available at: http://fortune.com/2014/11/17/top-10-research-development/ Accessed 08.12.16.

Climate.nasa.gov. (2016). Scientific consensus: Earth's climate is warming. [online] Available at: http://climate.nasa.gov/scientific-consensus/ Accessed 10.11.16.

Deuse, K. (2014). German companies invest billions in research [online] DW.COM. Available at: http://www.dw.com/en/german-companies-invest-billions-in-research/a-17322882 Accessed 08.11.16.

EAFO (European Alternative Fuels Observatory) (2016). Available at: www.eafo.eu.

EC. (2013a). *Commission staff working document. Proposal for a legislative instrument on control of emissions from medium combustion plants—Impact assessment, SWD (2013) 531.* Brussels: European Commission. Available at: http://ec.europa.eu/environment/archives/air/pdf/Impact_assessment_en.pdf Accessed 04.12.16.

EC. (2013b). *The clean air package.* Brussels: European Commission. Available at: http://ec.europa.eu/environment/air/clean_air_policy.htm Accessed 04.12.16.

Ec.europa.eu. (2016a). Passenger cars in the EU—Statistics explained [online]. Available at: http://ec.europa.eu/eurostat/statistics-explained/index.php/Passenger_cars_in_the_EU Accessed 15.11.16.

Ec.europa.eu. (2016b). Reducing emissions from transport—European Commission [online]. Available at: https://ec.europa.eu/clima/policies/transport/index_en.htm Accessed 24.11.16.

ETC/ACM. (2015). European air quality maps of PM and ozone for 2012 and their uncertainty, Horálek, J., de Smet, P., Kurfürst, P., de Leeuw, F., & Benešová, N., ETC/ACM Technical Paper 2014/4.

European Environment Agency (EEA). (2015). Air quality in Europe—2015 report [online]. Luxembourg, pp. 49–53. Available at: http://www.eea.europa.eu/publications/air-quality-in-europe-2015 Accessed 05.12.16.

Eurostat. (2015). *Energy, transport and environment indicators—2015 edition [online].* Luxembourg: Marcel Jortay. p. 16. Available at: http://ec.europa.eu/eurostat/documents/3217494/7052812/KS-DK-15-001-EN-N.pdf/eb9dc93d-8abe-4049-a901-1c7958005f5b Accessed 24.11.16.

Gomez Vilchez, J., & Jochem, P. (2016). *The impact of electric vehicles on the global oil demand and CO2 emissions.* Karlsruhe: Institute for Industrial Production (IIP) and Graduate School of Scenarios Karlsruhe Stuttgart, Karlsruhe Institute of Technology (KIT). pp. 1–16.

Gov.uk. (2016b). Plug-in car and van grants—GOV.UK [online]. Available at: http://www.gov.uk/plug-in-car-van-grants/what-youll-get Accessed 23.12.16.

Gov.uk. (2016c). Plug-in vehicle charge point grants—GOV.UK [online]. Available at: http://www.gov.uk/government/collections/plug-in-vehicle-chargepoint-grants Accessed 23.12.16.

Ho, V. (2015). China starts building its largest electric car solar charging complex [online] Mashable. Available at: http://mashable.com/2015/10/21/china-electric-car/#_dGbxpx8tPq1 Accessed 30.12.16.

Höök, M., & Tang, X. (2013). Depletion of fossil fuels and anthropogenic climate change—A review. *Energy Policy, 52*, 797–809. Elsevier.

IEEE. (2013). The rise & fall of electric vehicles in 1828–1930: Lessons learned [online]. *Proceedings of the IEEE, 101*(1), 206–212. IEEE. Available at: http://www.eee.hku.hk/doc/ccchan/CC_Chan_IEEE%20Proceedings%20The%20rise%20&%20fall%20of%20EVs.pdf Accessed 27.12.16.

IHS Polk (2014), extraction from IHS Polk databases on vehicle registrations and other characteristics, 2005, 2008 and 2010–2014 figures.

Jackson, C. (2016). Solar car parks—A guide for owners and developers. BRE National Solar Centre [online] (No. 1092193). Available at: http://www.brc.co.uk/filelibrary/nsc/Documents%20Library/NSC%20Publications/BRE_solarcarpark-guide.pdf Accessed 30.12.16.

Jeekel, H. (2013). *The car dependent society* (1st ed.). Burlington, VT: Ashgate Pub. pp. 40–63, 236–244.

Joint Research Centre of the European Commission. (2014). *Analysis of labour supply and demand. Future employment in transport.* Luxembourg: Joint Research Centre of the European Commission. pp. 9–38, 94–101.

Kaschub, T., Mültin, M., Fichtner, W., Schmeck, H., & Kessler, A. (2010). *Intelligentes Laden von batterieelektrischen Fahrzeugen im Kontext eines Stadviertels. E-Mobility: Technologien—Infrastruktur—Märkte.* Leipzig, Deutschland: Kongressbeiträge des VDE-Kongress, 8–9 November 2010, 6 S., VDE-Verl., Leipzig.

Mitchell, W., Borroni-Bird, C., & Burns, L. (2010). *Reinventing the automobile* (1st ed.). Cambridge, MA: Massachusetts Institute of Technology.

Movellan, J. (2015). 100 Percent renewable energy charged EV stations allow driving on sunshine [online]. Renewable Energy World. Available at: http://www.renewableenergyworld.com/articles/2015/08/100-percent-renewable-energy-charged-ev-stations-allow-driving-on-sunshine.html Accessed 30.12.16.

Muneer, T., Celik, A., & Caliskan, N. (2011). Sustainable transport solution for a medium-sized town in Turkey—A case study. *Sustainable Cities and Society, 1*(1), 29–37.

Newman, P., & Kenworthy, J. R. (1999). *Sustainability and cities: Overcoming automobile dependence.* Washington, DC: Island Press.

Oecd.org. (2014). The cost of air pollution—Health impacts of road transport - en - OECD [online]. Available at: http://www.oecd.org/env/the-cost-of-air-pollution-9789264210448-en.htm Accessed 26.11.16.

Oica.net. (2016). Auto Jobs | OICA [online]. Available at: http://www.oica.net/category/economic-contributions/auto-jobs/ Accessed 07.12.16.

Ortego, L. (2013). El tamaño contra el consumo ¿De verdad necesitamos coches tan grandes? [online] Tecmovia. Available at: http://www.diariomotor.com/tecmovia/2013/05/05/el-tamano-contra-el-consumo/ Accessed 29.12.16.

Potter, S. (2000). *Travelling light, theme 2 of T172 working with our environment: Technology for a sustainable future.* Milton Keynes: The Open University.

Potter, S. (2003). Transport energy and emissions: Urban public transport. In D. Hensher & K. Button (Eds.), *Handbooks in transport: Vol. 4. Handbook of transport and the environment* (pp. 247–262). Amsterdam: Elsevier.

PWC. (2016). 2016 Auto industry trends. Automakers and suppliers can no longer sit out the industry's transformation [online]. PWC, p. 11. Available at: http://www.strategyand.pwc.com/media/file/2016-Auto-Trends.pdf Accessed 03.01.17.

Renewable Energy Policy Network for the 21st Century (REN21) (2016). Renewables 2016 global status report [online]. Paris, pp. 27–29. Available at: http://www.ren21.net/wp-content/uploads/2016/06/GSR_2016_Full_Report1.pdf Accessed 03.07.16.

Ryan, D. (2016). *Sustainable technologies notes and lecture materials: Discussion questions about transport.* Edinburgh: Napier University.

Statista (2016a). Number of cars sold worldwide from 1990 to 2016 (in million units) [online]. Available at: http://www.statista.com/graphic/5/200002/international-car-sales-since-1990.jpg Accessed 18.11.16.

Statista. (2016c). Quality of infrastructure: countries with best infrastructure 2016. Statista [online]. Available at: http://www.statista.com/statistics/264753/ranking-of-countries-according-to-the-general-quality-of-infrastructure/ Accessed 11.12.16.

Sustainable Energy Ireland (SEI). (n.d.). A guide to vehicle aerodynamics [online]. Available at: http://www.seai.ie/Your_Business/Technologies/Transport/Aerodynamics_Transport_Guide.pdf Accessed 03.01.17.

Tesla (2016). Model 3. Tesla [online]. Available at: https://www.tesla.com/model3 Accessed 30.12.16.

The Huffington Post. (2016). Big plans for integrating renewable energy into China's electricity Grid [online]. Available at: http://www.huffingtonpost.com/barbara-afinamore/big-plans-for-integrating_b_9421864.html Accessed 30.12.16.

The International Energy Agency (IEA). (2010). World energy outlook, 2010 [online]. France, pp. 95–96. Available at: http://www.worldenergyoutlook.org/media/weo2010.pdf Accessed 30.06.16.

The International Energy Agency (IEA). (2011). Technology Roadmap: Electric and plug-in hybrid electric vehicles [online] Available at: http://www.iea.org/publications/freepublications/publication/EV_pHEV_roadmap.pdf Accessed 30.11.16.

The International Energy Agency (IEA). (2013a). World energy outlook 2013 [online]. France, p. 71 and p. 501. Available at: http://www.iea.org/publications/freepublications/publication/WEO2013.pdf Accessed 01.12.16.

The International Energy Agency (IEA). (2013b). Global land transport infrastructure requirements. Estimating road and railway infrastructure capacity and costs to 2050 [online]. France, pp. 8–11, 16, 34. Available at: http://www.iea.org/publications/freepublications/publication/TransportInfrastructureInsights_FINAL_WEB.pdf Accessed 10.12.16.

The International Energy Agency (IEA). (2014). Technology roadmap solar photovoltaic energy. Energy technology prospective [online]. Paris, p. 5. Available at: http://www.iea.org/publications/freepublications/publication/TechnologyRoadmapSolarPhotovoltaicEnergy_2014edition.pdf Accessed 20.12.16

The International Energy Agency (IEA). (2016a). *Key world energy statistics [online]*. France: International Energy Agency. p. 6. Available at: https://www.iea.org/publications/freepublications/publication/KeyWorld2016.pdf Accessed 02.12.16.

The International Energy Agency (IEA). (2016b). *Global EV outlook 2016—Beyond one million electric cars*. Paris: IEA. pp. 4–34. Available at: https://www.iea.org/publications/freepublications/publication/Global_EV_Outlook_2016.pdf Accessed 20.12.16.

The International Energy Agency (IEA). (2016a). Publication: World energy outlook special report 2016c: Energy and air pollution [online]. Available at: https://www.iea.org/publications/freepublications/publication/weo-2016-special-report-energy-and-air-pollution.html Accessed 01.12.16.

The International Energy Agency (IEA). (2016b). International energy outlook 2016d—Transportation sector energy consumption—Energy Information Administration [online]. Available at: http://www.eia.gov/outlooks/ieo/transportation.cfm Accessed 01.12.16.

The International Energy Agency (IEA). (2016e). How solar energy could be the largest source of electricity by mid century [online]. Available at: http://www.iea.org/newsroomandevents/pressreleases/2014/september/how-solar-energycould-be-the-largest-source-of-electricity-by-mid-century.html Accessed 20.12.16.

The Natural Hazards Partnership. (2015). Air pollution [online]. pp. 1–7. Available at: http://www.metoffice.gov.uk/nhp/media.jsp?mediaid=17374&filetype=pdf Accessed 05.12.16.

Transport Research and Innovation Portal (TRIP). (2013). Employment in the EU transport sector. Communicating Transport Research and Innovation [online]. pp. 2–24. Available at: http://www.transport-research.info Accessed 08.12.16.

Tverberg, G. V. (2012). World energy consumption since 1820 in charts [online]. Our Finite World. Available at: https://ourfiniteworld.com/2012/03/12/world-energy-consumption-since-1820-in-charts/ Accessed 15.11.16.

US DOE (2015). *Vehicle technologies office: batteries*. Available at: US DOE, http://energy.gov/eere/vehicles/vehicle-technologies-office-batteries.

US DOE (United States Department of Energy) (2016), personal communication with David Howell developed in the framework of the US data submission for the EVI "Global EV outlook 2016.

Vda. (2016). VDA [online]. Available at: https://www.vda.de/en/press/press-releases/20160102-german-automotive-industry-invests-34-billion-euro-in-research-and-development.html Accessed 08.12.16.

Vidal, J. (2016). Air pollution rising at an 'alarming rate' in world's cities [online]. The Guardian. Available at: https://www.theguardian.com/environment/2016/may/12/air-pollution-rising-at-an-alarming-rate-in-worlds-cities Accessed 25.11.16.

World Health Organization. (2015). *Global status report on road safety 2015*. Geneva: World Health Organization Available at: http://www.who.int/violence_injury_prevention/road_safety_status/2015/en/ Accessed 01.12.16.

www.efe.com. (2016). Barcelona ofrecerá transporte público gratis a cambio del coche contaminante [online] Available at: http://www.efe.com/efe/espana/sociedad/barcelona-ofrecera-transporte-publico-gratis-a-cambio-del-coche-contaminante/10004-3121846 Accessed 14.12.16.

Zaragoza, S. (2015). *Meeting global air quality guidelines could prevent 2.1 million deaths per year [online]*. UT News, The University of Texas at Austin. Available at: http://news.utexas.edu/2015/06/16/clean-air-globally-could-yield-large-health-benefits Accessed 26.11.16.

Further Reading

Akuoko, A. (2016). *Increasing solar energy system & challenges to the UK grid—The Power Academy Essay Challenge 2016 [online]*. Loughborough University. p. 3. Available at: http://file://C:/Users/irene/Downloads/Akuoko.%20A.pdf Accessed 10.01.17.

BBC News. (2012). China's economic miracle [online] Available at: http://www.bbc.co.uk/news/world-asia-china-20069627 Accessed 18.11.16.

Data.worldbank.org. (2016). CO2 emissions (metric tons per capita) | Data [online]. Available at: http://data.worldbank.org/indicator/EN.ATM.CO2E.PC Accessed 18.11.16.

Davis, S., Diegel, S., & Boundy, R. (2010). *Transportation energy data book* (28th ed.). Oak Ridge, TN: Oak Ridge National Laboratory.

Ec.europa.eu. (2016c). Reducing CO2 emissions from passenger cars—European Commission [online]. Available at: https://ec.europa.eu/clima/policies/transport/vehicles/cars/index_en.htm Accessed 26.11.16.

European Commission. (2012). *EU Transport in figures—Statistical pocket book 2012*. Brussels: European Commission.

Gov.scot. (2015). Cleaner air for Scotland—The road to a healthier future [online]. Available at: http://www.gov.scot/Publications/2015/11/5671/6 Accessed 30.11.17.

Gov.uk. (2016a). Cars (VEH02)—GOV.UK [online]. Available at: https://www.gov.uk/government/statistical-data-sets/veh02-licensed-cars Accessed 24.11.16.

Grove, J. (2016). Vehicle licensing statistics: Quarter 4 (Oct.–Dec.) 2015 [online]. Department of Transport. Available at: https://www.gov.uk/government/uploads/system/uploads/attachment_data/file/516429/vehicle-licensing-statistics-2015.pdf Accessed 23.11.16.

IRF (International Road Federation). (2012). *World road statistics.* Geneva: IRF.

Misra, T. (2015). You won't believe how much sprawl costs America [online]. CityLab. Available at: http://www.citylab.com/housing/2015/03/how-much-sprawl-costs-america/ 388481/ Accessed 11.12.16.

Muneer, T., Milligan, R., Smith, I., Doyle, A., Pozuelo, M., & Knez, M. (2015). Energetic, environmental and economic performance of electric vehicles: Experimental evaluation. *Transportation Research Part D: Transport and Environment, 35,* 40–61.

Reed, S. (2017). *One hundred years of price change: the Consumer Price Index and the American inflation experience: Monthly Labor Review: U.S. Bureau of Labor Statistics.* [online] Bureau of Labor Statistics. Available at: https://www.bls.gov/opub/mlr/2014/article/one-hundred-years-of-price-change-the-consumer-price-index-and-the-american-inflation-experience.htm Accessed 1.11.16.

Solar Action Alliance. (2016). Solar panels for New Jersey homes: Tax incentives, prices, info [online]. Available at: https://solaractionalliance.org/new-jersey/ Accessed 15.01.17.

Statista. (2016b). Car production: Number of cars produced worldwide 2015. Statista [online]. Available at: https://www.statista.com/statistics/262747/worldwide-automobile-production-since-2000/ Accessed 08.12.16.

The Economist. (2012). Seeing the back of the car [online]. Available at: http://www.economist.com/node/21563280 Accessed 25.11.16.

The International Energy Agency (IEA). (2012). Energy technology perspectives 2012—Pathways to a clean energy system [online]. Available at: https://www.iea.org/publications/freepublications/publication/ETP2012_free.pdf Accessed 26.12.16.

Unep.org. (n.d.). Canadian automotive fuel economy policy [online]. Available at: http://www.unep.org/transport/gfei/autotool/case_studies/northamerica/canada/cs_cn_0.asp Accessed 30.11.16.

World Health Organization. (2014). *Global health estimates.* Geneva: World Health Organization. Available at: http://www.who.int/healthinfo/global_burden_disease/estimates/en/index1.html Accessed 01.12.16.

World Health Organization, (2016). WHO | 7 million premature deaths annually linked to air pollution [online]. Available at: http://www.who.int/mediacentre/news/releases/2014/air-pollution/en/ Accessed 26.11.16.

Our World In Data. (2016). World population growth [online]. Available at: https://ourworldindata.org/world-population-growth/ Accessed 15.11.16.

Worldbank.org. (2016). Transport overview [online]. Available at: http://www.worldbank.org/en/topic/transport/overview Accessed 05.12.16.

Traction energy and battery performance modelling

Aisling Doyle, Tariq Muneer
Edinburgh Napier University, Edinburgh, Scotland, United Kingdom

2.1 Introduction

This chapter presents a technical review of the electric vehicle's (EV) traction energy. Traction energy is the energy required to propel a vehicle. Factors such as friction, wind drag, acceleration, and hill climb are forces that the vehicle must overcome. Traction energy is the name referring to the energy required to overcome these collective forces. Traction energy may be supplied from various energy sources such as a fossil-fuel engine, electricity, fuel cell, ultra capacitor, or hydrogen engine to name a few. This chapter will focus on the battery-powered electric vehicle (BEV) and the vehicle's traction energy produced with battery technology as the main energy source.

This chapter aims to present an overview of the present battery technology in the EV automobile market in a clear and concise manner for readers of all experience levels and backgrounds. The information discussed will be useful to various stakeholders from experienced industrial companies to local governments and the general public who have interest in supporting the growth of this niche market.

It is important that the reader understands the various categories that exist in the sustainable transport sector. This section will guide the reader through the various vehicle energy storage compositions related to the latest low-emission vehicles, focusing on the BEV, to fully understand how the technology operates and to demonstrate how they can be a potential alternative to the conventional internal combustion engine vehicle (ICEV). The success of the EV dominating the automobile industry is when technology can compete with the ICEV without jeopardizing the drivers' experience.

It is widely accepted globally that renewable sources of energy is a potential alternative to economies depending on fossil-fuel-based energy sources. Wind, solar, hydro, and biomass have been developed and implemented over recent decades; however, sustainable transport is a relatively foreign concept in comparison as confidence in electric propulsion vehicles has still a long way to go. In some societies around the world, the automobile is used as a statement not only as a mode of transport. The vehicle expresses one's identity or social status that poses a challenge when encouraging the public to go electric. For this reason, it is important that appropriate technology develops alongside the aesthetic look of the vehicle.

There has been a change in the automobile industry in recent years with several factors contributing to this change. These include the 2008 financial crisis, the instability of oil prices, and a shift in cultural awareness of environmental impacts, which

Electric Vehicles: Prospects and Challenges. http://dx.doi.org/10.1016/B978-0-12-803021-9.00002-1

has resulted in an increased use of BEVs. Each of these factors has influenced the automobile industry in many different ways, but the EV may be seen as a solution to replace the conventional ICEV that has taken the centrepiece of our everyday lives.

Without a combustion fuel that is suitable for the automobile industry, it is inevitable that the industry will shift from ICEV to EV. It is important that the reader understands the various components that gives the EV its exciting potential to play a vital role in the future to attain private mobility that is efficient and reliable to take the place of the ICEV.

This chapter will discuss the components that are considered in the understanding of traction energy of the EV. How battery technology operates and charging and discharging will be discussed. Regenerative braking system is a technique that utilizes potential and kinetic energy loss that would have been conventionally lost. This braking system will be reviewed in this section to illustrate how 'waste' kinetic energy can be harnessed and converted to traction energy to extend the driving range of the vehicle. A software program is additionally presented in this chapter to compute traction and regenerative energy for various drive cycles. This software is complimentary on purchase of this book. The software may be downloaded from:

https://www.dropbox.com/sh/xfnhg5nngzcu4pd/AADukvT6HqXq3jXOW8re6k2 0a?dl=0.

2.2 What is meant by an EV?

The EV was one of the first successful modes of transport known to man with the exception to steam- and horse-powered modes until the ICEV dominated the automobile production sector. However, the EV concept has returned to the market in the hope of working towards a more sustainable future. It is important that the reader understands the term EV so that they can get a clear picture of what technology is being discussed.

There are a wide range of low-emission hybrid vehicles in the automobile market. Many of these technologies come under the term EV but operate quite differently. A battery electric vehicle (BEV) is a vehicle that runs on 100% electric propulsion. The most common energy storage system used in a BEV is lithium-ion batteries albeit other materials are used (when referring to a BEV, this book will assume that lithium-ion batteries are used if not stated otherwise).

Other vehicles may use a combination of fossil-fuelled vehicles and electricity as a source of traction energy. These vehicles that use a combination of both of the above sources are known as hybrid electric vehicles (HEV). The hybrid vehicle will vary in how it operates depending on the relationship between the electric motor and the engine and how they are connected. The system can be a series, parallel, or series–parallel hybrid system.

Parallel hybrid vehicles operate where the engine and electric motor can directly power the vehicle. The vehicle may be powered by combustion fuel, electric propulsion, or both. According to Serra (2013), this simultaneous or parallel power transmission relationship between the ICE and electric motor has many benefits. These benefits include the simplistic design of the improved power source and its economic savings due to the downsizing of the ICE, without jeopardizing the vehicle's performance, leading to reduction in emissions. This system does not rely solely on battery technology, and so the longevity of the vehicle performance is not a concern. However, there are also some drawbacks to the suggested technology (Serra, 2013). The parallel hybrid vehicle continues to rely heavily on combustion-fuelled engines. This dependency results in high maintenance and service costs due to the many moving parts of the ICE. Serra (2013) claims that in a parallel hybrid vehicle 80% of the vehicle propulsion remains dependent on the ICE, therefore not optimizing the energy efficiency that the electric motor can add to a vehicle.

The series hybrid is also known as a serial system. In this scenario, the ICE is used to support the electric motor. An electric motor will use electrical energy and convert it into kinetic energy to supply power to the vehicle wheels. The motor operates the opposite to a generator, and this characteristic is taken advantage of in vehicles that has an electric generator to capture kinetic energy when braking. This concept is referred to as regenerative braking and will be discussed later in this chapter. The hybrid vehicle in series acknowledges this principle. The ICE is used to turn the chemical reaction from combusting a fossil fuel and converts it into kinetic energy that is used to recharge the battery. In a serial hybrid, the battery is the primary energy source. A drawback of the hybrid vehicle in series is that the ICE's high power capacity is not utilized to its full potential as it supports the electric motor and does not act as the primary energy source at any stage.

In terms of thermal efficiency, petrol-based four-stroke piston engines are around 33% efficient. Diesel engines are around 42% efficient. Gas turbines have higher efficiencies when deployed for large-scale power generation when they are operated at high loads. However, at varying loads, they will have much poorer efficiency than a piston engine. The limiting factor for thermal efficiency of the above internal combustion engines may be explained on the basis of the second law of thermodynamics. According to Carnot's principle, the maximum thermal efficiency of any given heat engine operating within the temperature limits of T_L and T_H is

$$\eta = 1 - T_L/T_H \tag{2.2.1}$$

In view of the material fatigue subjected to repeated thermal cycling, the average T_H can hardly be expected to be over 1500 K for any of the three power plants discussed above. The advantage offered by the EV is that it has the potential to be charged with sustainable electricity that offers much higher efficiency if it is generated by means of wind turbines or even with solar photovoltaic plants. Note that using Eq. (2.2.1) and noting that the source of solar radiation, the sun, has a surface temperature of 5800 K, there is potential for much higher efficiencies than could ever be expected from heat engines.

Both systems have their own positive and negative attributes. The parallel–series hybrid vehicle incorporates both connections and mitigates as many hindrances as possible. This setup takes advantage of the benefits of both systems in appropriate driving situations. This vehicle is known as the parallel–series hybrid vehicle. However, this system can be more cost-intensive due the complexity of the system.

The BEV's recharging infrastructure is supplied through the main electricity distribution network, and additionally, HEVs can be plug-in vehicles. Some HEVs are independent of recharging infrastructure and rely on the ICE to recharge the battery or, in other words, are not plugged into the mains.

2.3 Lithium and its use in BEV

Lithium is used in industry for many different purposes with the most potential for growth in the battery market. Lithium is a soft, silvery-white alkali metal that is being used as a propulsion energy source in the electrified automobile market as a replacement to petrol and diesel. The use of lithium has increased in recent years with the growth of electrified vehicles. By 2009, global lithium consumption grew by 31% due to the growth of the battery production industry (Eason, 2010). With an increased interest in the automobile industry heading towards battery-powered vehicles, the question stands, is there sustainably enough lithium resource for the future or will this be another mineral we are soon to exhaust like we have seen in the oil industry (Ultra Lithium, 2013)? To evaluate the world's lithium resources (see Table 2.3.1), it is important not only to consider the supply of the raw material compared with demand but also to consider the potential lithium that has to be used as a high-quality recyclable resource.

Table 2.3.1 shows the global resources of lithium. Lithium by nature is not available in its elemental form. Industrial processes are required to extract lithium found in brine pools (large area of dense water on ocean basins containing salt deposits) or from spodumene (a translucent, typically greyish-white aluminosilicate mineral). Lithium extracted from brine requires less energy-intensive activities and therefore is cheaper to produce than spodumene-sourced lithium. The largest producer of lithium is Chile

Table 2.3.1 **Lithium world reserve base (tonnes)** **(Gaines & Nelson, 2014)**

Country	Lithium reserve (tonnes)
Bolivia	5,400,000
Chile	3,000,000
China	1,100,000
Brazil	910,000
The United States	410,000
Canada	360,000
Australia	220,000
Zimbabwe	27,000
Argentina	Not available
Portugal	Not available
World total	11,000,000

along with Argentina, Bolivia, Australia, China, and the United States. Chile leads in brine-sourced lithium, whereas Australia is the largest producer from spodumene-sourced lithium.

A study carried out in the United States evaluated the possible material demand issues for the lithium-ion battery, should lithium be extensively extracted from the earth to supply energy to the automobile industry (Dunn et al., 2012). The aforementioned study estimated how much material would be needed if batteries were made from various chemistries using different material. The most lithium-intensive analysis showed that 13 kg of material was required. This figure represents <3% of the battery mass. The Argonne National Laboratory presents findings that show that the recyclable nature of the lithium-ion battery will reduce the US material demand in 2050 from over 50,000 to about 12,000 tonnes. Recycling the battery is a valuable property when considering the sustainability of the lithium resource. It is predicted that lithium-ion battery production will rise over 20 times of its current level should electric propulsion vehicles dominate the automobile market (Gaines & Nelson, 2014). The latter study also predicts that by 2050 the material demand would then reduce due to the availability of recycled materials and would be four times in today's current demand levels. Gaines and Nelson (2014) state that in the foreseeable future lithium reserves will not be dangerously depleted and will meet the world's demand for 2050. Concerns that lithium reserves will be exhausted are immature. The EV industry is moving at a rapid pace, and these predictions may soon be outdated as battery technology and addition exploration for lithium will extend supply. It is important that consumers are aware by what is meant by 'zero-emission' vehicles. EVs have no tailpipe emissions, but the production of the vehicle's battery and various components requires energy in the production stage and thus creates greenhouse gases. Furthermore, the sourcing of electricity required to recharge the battery will add carbon to the environmental audit. Interesting findings are reported by Dunn et al. (2012) who analysed the cradle-to-gate impact analysis of excavating lithium from Chile or Nevada for battery assemble. These findings included a transportation factor to material extracted in Chile and Nevada and transported to or within the United States for further production, to be 13% and 6% of the total energy consumption, respectively. However, with higher transportation impact, producing lithium-ion batteries from material harvested in Chile is still 40% less energy-intensive than if production was to transport material from Nevada. Factors increasing the energy consumption in Nevada compared with Chile are due to the lithium being seven times more concentrated natural resource. The study went further and compared the two resources should they have equal lime requirements, and again, Nevada was 25% more energy-intensive. It is an important concept to consider how available the material is considering the processes that are required for various resources getting from the 'cradle-to-gate'. This concept will be touched again later in the chapter when the well-to-wheel analysis of the EV will be discussed. The well-to-wheel analysis is in terms of considering how the vehicle recharges, and energy intensity of various sources of electricity (e.g. coal, nuclear, renewable, and mixed-based electricity) will be explained.

Eason (2010) calculates the quantity of lithium required for battery production. The latter calculated that 3861 Ah/kg is the charge capacity of lithium. Considering a voltage of 3.6 V and an efficiency factor of 73%, the maximum realistic energy capacity for the lithium-ion battery was computed as 10.1 kWh/kg. In other words, for every 10 kWh of battery-stored energy, 1 kg of lithium is required. A 24 kWh battery-powered electric vehicle requires 2.4 kg of lithium. The latter is the capacity of a Nissan Leaf BEV. Additionally, the study analysed a scenario where there is one vehicle for every two people. Assuming a North American standard of living across the globe would result in the production of 3.4 billion vehicles, this would require 32% of identified lithium resources. Using all the known lithium resources globally, 10.6 billion 24 kWh vehicles can be produced (Eason, 2010).

2.4   Battery technology

This section will show the potential of lithium ion over lead-acid batteries. We are also proposing the use of lead-acid battery technology to operate auxiliary ventilation and heating.

2.4.1   Measuring state of charge of a lead acid battery

There seems to be a dearth of information with respect to the behaviour of a battery-charging profile, and so, measuring the state of charge (SoC) is still unclear. The following section discusses five different methods that propose to compute an approximate SoC of a battery.

(a) Voltage method

The most common and accepted way to estimate a battery's SoC is to take a voltage reading across the two terminals. Table 2.4.1 shows the various voltages that correspond with the various approximate SoC levels.

Table 2.4.1 **BCI standard for SoC estimate of a lead acid battery (Event Horizon Solar and Wind Inc, 2015)**

% SoC	12 V DC system	24 V DC system	48 V DC system
100	12.7	25.4	50.8
90	12.6	25.2	50.4
80	12.5	25.0	50.0
70	12.3	24.6	49.2
60	12.2	24.4	48.8
50	12.1	24.2	48.4
40	12.0	24.0	48.0
30	11.8	23.6	47.2
20	11.7	23.4	46.8
10	11.6	23.2	46.4
0	<11.6	<23.2	<46.4

Temperature has a role to play in measuring voltages across the two terminals. If the battery is evaluated at lower than the optimum value (25°C), larger voltage reading would be expected and vice versa for higher temperatures showing a misrepresentation of the battery's SoC. According to Event Horizon Solar and Wind Inc. (2015), readings should be obtained early morning before sunrise or late in the evening. The reading should be taken when the battery has been in a settled state and not while charging or discharging. This method can be inaccurate as cell types have different chemical composition and therefore will have varied voltage profiles.

While a battery is charging or discharging, the battery is being agitated and will not show a true representation of the SoC. The voltage reading should be recorded 3–4 h after being in a charge or discharge state. Manufactures recommend allowing the battery to settle 24 h to get an optimum reading (Buchmann, 2011). The aforementioned SoC reading method works well for lead-acid batteries. However, Buchmann (2011) states that this method is impractical for nickel- and lithium-based batteries. It must be noted that chemicals, such as calcium, introduced into the lead-acid battery to give it its maintenance free properties, causes a rise in the battery's voltage by ~5%–8%. After charging a battery, the terminals will show a raised voltage reading. To get a more realistic SoC reading, a brief discharge before recording the voltage value will counteract this error. However, it is recommended that the battery should be left unused for at least 3 h or for 24 h for a more accurate value. The voltage-based SoC is most popular because of its simplicity in comparison with other expensive alternatives that also require calibration.

(b) Hydrometer

The hydrometer is an alternative method used to measure SoC of a battery. This method can only be used for lead-acid and flooded nickel-cadmium-based batteries (Buchmann, 2011). As the battery is being charged, the sulphuric acid in the battery gets heavier causing the specific gravity (SG) of the battery to increase (see Table 2.4.2). As the battery is being discharged, the sulphuric acid is removed from the electrolyte and binds to the plate to form lead sulphate. This subsequently reduces the weight of the electrolyte resulting in a reduction in the SG of the battery.

Table 2.4.2 **BCI standard for SoC estimate using the hydrometer method at 26°C and rested for 24 h postcharge or discharge state of a lead-acid battery (Buchmann, 2011)**

% SoC	Specific gravity
100	1.265
75	1.239
50	1.200
25	1.170
0	1.110

The SG of a battery may change due to the fluid levels in the battery. Reduced fluid levels will result in higher concentration of the electrolyte and will not give a true representative of the battery's SoC. Alternatively, should the battery be overfilled, the concentration will be reduced and give a lower SG reading. It is important that, if water is added, adequate time should be given to allow for mixing before SG readings are taken.

The SG readings will also vary according to the battery type. Deep-cycle batteries have an electrolyte with 100% SoC and SG of up to 1.33, aviation batteries can reach 1.285, traction batteries (e.g. in forklifts) are around 1.28, starter batteries have an SG of 1.265, and stationary batteries as low as 1.225. The lower the SG, the less corrosion occurs. For stationary batteries, longevity of the battery is a priority over a higher SG value.

The SG given in Table 2.4.3 is not absolute and SG may vary from battery to battery. Temperature affects the SoC of a battery. In cooler temperatures, batteries will experience a higher or a denser SG value as seen in Table 3.4. The stratification of the acid will also give misleading values when lighter particles are at the top and heavier on the bottom. Similar to the voltage method, the battery should be allowed to settle before measurements are taken.

(c) Coulomb counting

Many appliances such as laptops, medical equipment, and professional portable devices use coulomb counting to measure SoC. Measuring current flowing in and out of the battery will give a reading for the SoC of the battery. However, it is important to note that it is not assumed that all charge entering the battery is accepted especially in the later stages of charging where charging is much slower and will require more current (Buchmann, 2011). Inefficiencies from discharge losses may also be experienced that will not give a true value of the battery's SoC. The energy available will always be less than the energy provided to the battery. However, this method gives reasonable results and works especially well for the lithium-ion batteries. Manufactures claim that this method will give readings with accuracy varying by 1%, although this may only be the case for new batteries. The age of a battery will skew the reading further and is more likely to obtain a reading with accuracy wavering

Table 2.4.3 The relationship between temperature and SG of a deep-cycle lead-acid battery (Buchmann, 2011)

Temperature (°C)	Specific gravity at full charge
40	1.266
30	1.273
20	1.280
10	1.287
0	1.294

around 10% error. In this method temperature, surface charge and acid stratification will not give notable discrepancies in the SoC readings.

(d) Impedance spectroscopy

Although resistance and impedance are measured in the same units, ohms (Ω), they are differentiated by reactance (Buchmann, 2011). Resistance is the ratio of voltage to the amount of current able to flow in the circuit and can be defined by Ohms law as $R = \dfrac{V}{I}$ where R is resistance, V is voltage, and I is the current. Impedance is the total effective resistance when an AC voltage source is applied to the circuit. A resistor will remain unaffected by the circuit's frequency due to its AC nature. However, capacitors and inductors will not behave the same, and their reactance will depend on the AC frequency. Capacitive resistance will decrease with frequency, whereas inductive resistance will increase with frequency. Impedance is the term given to the sum of all these resisting forces in an AC circuit. The impedance spectroscopy method to measure SoC is based on measuring the impedance values of batteries. This method works on flooded and sealed lead-acid batteries. Unlike methods previously mentioned, the battery does not need to rest to take measurements as parasitic loads will not affect the measurement of an accurate impedance value.

(e) Quantum magnetism

Quantum magnetism is the newest method to measure SoC (Buchmann, 2011). While discharging, a battery's plate undergoes a chemical change, it changes lead to lead sulphate that have different magnetic properties. This change in magnetism can be measured using a sensor and can give an SoC reading. This method can be used with lead-acid batteries but works particularly well with lithium-ion batteries.

This technology uses two metal alloys separated by a very thin insulator. This dimension is the thickness of a few atoms. Electrons will move more easily in a charged state through the insulator than when the battery is in a neutral state. The magnetic flow or change in resistivity through the insulator is observed. These measurements are coupled with a mathematical model to measure the battery's SoC. This method is not affected by voltage distortion as a result of battery loading. The measurement is reliable because it is independent from the characteristics of the battery voltage. The applications of this method vary from improved charging methods of the battery, to the diagnosis of battery deficiencies, and to predicting the battery's life.

2.5 Recharging of the EV

This section will discuss the current infrastructure available to support BEV. Route planning plays an important role in the success of the BEV; therefore, it is important that the driver is aware of the time required to recharge the vehicle's battery. Charging profiles for the Nissan Leaf and Renault Zoe are presented in this section.

2.5.1 Supporting infrastructure

Recharging of the EV is the bone of contention in the electric automobile world. The charging infrastructure has developed dramatically, and improvements have seen to reduce charging time. The refuelling experience of an EV requires the driver to be aware of their vehicle's range until widespread chargers are deployed nationwide. Planning journeys is the compromise that the driver has to make when going electric. There are various options available for the consumer to charge their battery ranging from slow to rapid charging. Many of these have been installed by governments with more promised in the pipeline. The supporting infrastructure is available domestically with a single-phase slow or fast charger. Single-phase fast and three-phase rapid chargers are currently free to avail of at public charging stations. To calculate the charging time of various batteries, Eq. (2.5.1) is used to compute Table 2.5.1:

$$\text{Charging time (h)} = \frac{\text{Battery capacity (kWh)}}{\text{Power (kW)}}. \tag{2.5.1}$$

Note the 22 and 24 kWh batteries represent the Renault Zoe BEV and Nissan Leaf BEV, respectively. Another option has been explored by Tesla in California where owners can avail a battery-swapping service that takes less than 90 s. Innovation such as this will give confidence to the public that the industry is confident that this technology will play an important role in the private transport sector for years to come and overcome major barriers in regards to vehicle range.

2.5.2 Charging profile

Muneer et al. (2015) illustrate the charging profiles of EVs to complement (Table 2.5.1). These profiles are displayed in Fig. 2.5.1. It is important that the reader understands the basics of how a battery works.

There are many different batteries albeit all operate under the same concepts where chemically stored energy is converted into electrical energy and vice versa. A battery

Table 2.5.1 Charging times for 22 and 24 kWh EV battery

Authors' classification	Phases	Current (A)	Voltage (V)	Power (kW)	Charge time for 22 kWh battery (h)	Charge time for 24 kWh battery (h)
Very slow	1	10	230	2.3	9.6	10.4
Slow	1	16	230	3.7	6	6.5
Fast	1	32	230	7.4	3	3.3
AC rapid[a]	3	32	230	22.1	1	1.1
DC rapid[a]	3	63	230	43.5	0.5	0.6

[a]Rapid charging is generally realized until 80%.

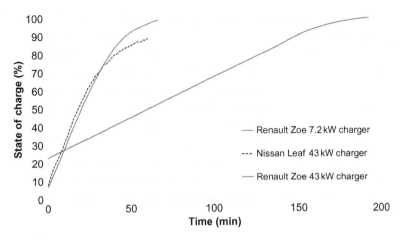

Fig. 2.5.1 Charging profiles of Nissan Leaf and Renault Zoe (Muneer et al., 2015).

cell is composed of two terminals or electrode plates, one positive (cathode) and one negative (anode). Both of these plates are immersed in a chemical medium or electrolyte separating both electrodes. Take the lithium-ion battery, for example, as it is a popular choice in the electric automobile industry, when discharging the anode releases electrons and ions. The electrolyte restricts the flow of electrons allowing only the lithium ions to travel to the cathode. The electrons have to take an external route to reach the cathode, and any load (motor to move the wheels in the case of an EV) applied to this external circuit is powered by these moving electrons. When a recharger is externally added to the circuit, the reverse happens. The ions and electrons flow in the opposite direction moving from the cathode to the anode.

A basic understanding of how a battery operates can now be applied to understanding the charging profile of the EV's lithium-ion battery. Fig. 2.5.1 demonstrates that the reaction is linear for the majority of the time, while the battery is being recharged. The chemical reaction progresses uniformly. However, the final period of recharging during fast or rapid charging shows a nonlinear relationship between time of charging and SoC. Fig. 2.5.1 illustrates that in the final 10%–20% of charge, the chemical reaction in the battery slows down. This can be explained through a 'mating analogy' shown in Fig. 2.5.2 (Serra, 2013). This analogy states that the reaction progresses relatively quickly at the beginning of charging as there are several ions to react with at the opposite terminal or a significant amount of suitors to pair up with initially. As charging continues, suitable matches become fewer and fewer, and it takes ions at the terminal longer to 'find' or react with one another. This analogy gives a simple explanation of why a battery recharging slows down; that is, it becomes nonlinear.

Reports of battery-charging efficiency vary between 90% and 95% for charging infrastructure of 3 kW single-phase supply (AC Propulsion, 2011; BRUSA Elektronik AG, 2012). The value for battery-charging efficiency ($_{bc}$) will be used later in this chapter in calculating the total trip energy.

Fig. 2.5.2 Battery technology charging and discharging, the movement of ions (Serra, 2013).

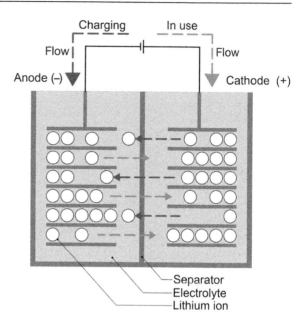

2.6 Regenerative braking

Regenerative braking is a unique technique that is used in EVs to capture energy that the vehicle has due to its motion or, in other words, its kinetic energy that would have been wasted when the vehicle decelerates or comes to a standstill while braking. By taking a measure of the initial and final vehicle velocity, the amount of kinetic energy that is lost to braking can be calculated.

Urban drive cycles have a considerable amount of acceleration and decelerating periods due to traffic control systems in place around towns and cities, and therefore, when decelerating, significant energy is lost. However, with regenerative braking, this energy can be captured, and 'waste' energy can be harnessed and utilized for vehicle propulsion. Taking the Renault Zoe EV as an example with one occupant with the vehicle mass as ~1600 kg with a speed of 120 km/h (33.33 m/s), the kinetic energy has a value of 0.25 kWh. In a conventional ICEV, on approaching a halt, this energy is totally wasted. Data obtained over a number of drive cycles in the Renault Zoe show that this vehicle has an energy efficiency of ~0.2 kWh/km. An analysis of the energy efficiency of the vehicle and kinetic energy the vehicle has lost due to braking shows that the energy equivalent for the vehicle to travel 1.25 km is wasted when brought to a complete stop. This range loss relates to a full cycle of the battery from 100% capacity to a completely discharged battery.

In an EV, the power source is the battery that supplies electric energy to operate the motor. The motor supplies energy to rotate the vehicle wheels producing kinetic energy. The motor can operate in reverse. When a motor operates in reverse it acts as a generator. When the vehicle slows down, the generator converts the kinetic energy into electrical energy to charge the vehicle's battery. When the conventional vehicle brakes, the energy is lost to heat energy resulting from the friction between the brake pads and wheels. Regenerative braking allows the range of the EV to be extended; however, the efficiency of capturing this energy is reported to vary from 16% to 70% (Boretti, 2013). The reason for this significant difference in efficiency will depend on the driver's style of driving whether they brake gradually or severely. Furthermore, temperature of the system and outside ambient temperature affect the efficiency greatly. The driving technique of the driver is a considerable factor when calculating the efficiency of the regenerative braking system. Friction brake pads are installed in the EV to allow for more rapid braking as opposed to slow braking with the regenerative braking system. The utilization of the friction brakes results in the efficiency of the regenerative system to vary. This property confirms that consumer education is required to inform the EV owner of different driving techniques of how to maximize the efficiency of the vehicle energy recapturing system to extend the longevity of the battery.

Not only does regenerative braking improve fuel efficiency in EVs, but also it can be adapted for the ICEV to help lower vehicle emissions (Clarke, Muneer, & Cullinane, 2010). There are various energy capturing devices that are suitable to be used in regenerative braking systems. The flywheel is a device that when rotated, can store kinetic energy during braking. The ultracapacitor is the most commonly adopted device in regenerative braking systems. The ultracapacitor temporarily stores electrical charge. This short charge and discharge period is cheaper than the flywheel system and additionally has a higher energy density. This device also has fewer hazardous materials that will have negative effects on the environment compared with other storage appliances. The ultracapacitor is a better alternative to an electrical battery for short journey times owing to very high associated efficiencies (McCluer & Christin, 2008). Reports have stated that ultracapacitor's coulomb efficiency can reach efficiencies over 99% (Maxwell, 2010).

It cannot be assumed that because the motor acts as a generator during regenerative braking the motor and generator efficiencies are equal. An experimental set of data published by a US laboratory in 2001 on the efficiency of motors or likewise generators is displayed in Fig. 2.6.1 relevant to engines or generators that are rated 75–100 hp. or 56–75 kW. Fig. 2.6.1 illustrates that when the motor or generator is subjected to a load factor above 0.2, efficiency is in excess of 97%. Loads below this show a significantly fall in the motor/generator efficiency. Applying these findings to the vehicle shows that when the vehicle is driven at low speed or in other terms is operated with low loading the motor efficiency is low. Hence, in EVs such as the Renault Zoe, the car will not capture recovered kinetic energy at speeds below 9 mph or 15 kmph in the control algorithm.

2.7   Energy usage in the electric vehicle

A study carried out in the United States explored the energy consumption of the most recent conventional internal combustion passenger vehicle on the market (Holmberg, Andersson, & Erdemir, 2012). The latter study tested four different types of passenger

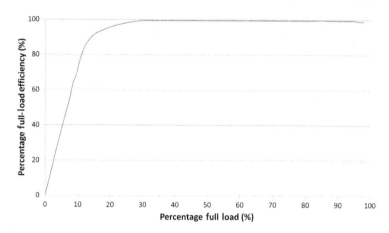

Fig. 2.6.1 Three-phase induction motor/generator efficiency profile (US Department of Energy, 2001).

vehicles to get a breakdown of the various components of energy consumption. Holmberg et al. (2012) stated that overall the ICEV uses 21.5% of total fuel used to propel the vehicle illustrated in Fig. 2.7.1. Energy lost due to exhaust, engine cooling, and transmission losses equate to 78.5% of the total fuel energy.

Fig. 2.7.2 illustrates the breakdown of a comparable passenger EV's energy consumption during winter conditions when heating of cabin space is required. The EV eliminates exhaust losses and has fewer mechanical losses due to fewer moving parts in the vehicle. Around 86% of the battery's energy is contributed to traction energy, significantly higher when compared with 38% from the ICEV. An additional factor

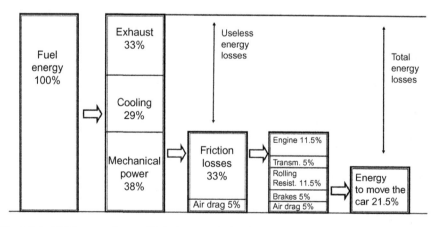

Fig. 2.7.1 ICEV energy losses (Holmberg et al., 2012).

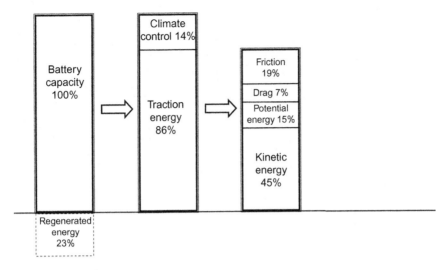

Fig. 2.7.2 EV energy losses for cooling periods.

unique to the EV illustrated is the regenerative braking system to extend the battery capacity by 21%.

2.8 The environmental impact of the EV

There are various analyses to assess the efficiency or environmental impacts of a technology when it is projected into the market. A well-to-wheel analysis was considered the most holistic to give the consumer a true and fair picture of the environmental impacts. This analysis is based on the work of Acha, Green, and Shah (2011).

A well-to-wheel analysis not only looks at the vehicle environmental impacts but also considers the impacts of the process of extracting the fuel (oil, coal, etc.) used to power the vehicle's engine. In the case of the EV, the well-to-wheel analysis will incorporate the fuel used in power stations that is used to recharge the battery of the electric automobile should it be a coal, nuclear, or renewable based station. This analysis illuminates the advantage the EVs have, over conventional vehicles, to be advertised as a 'zero'-emission vehicle as no vehicle has true green value while being recharged by electricity supply. The analysis is broken up into two parts. Firstly, the energy efficiency of the vehicle considers the vehicles performance in terms of well-to-vehicle efficiency, and then, vehicle-to-wheel performance is added. The second step takes into account the CO_2 efficiency of the vehicle that is calculated from the vehicle energy efficiency and carbon content of the fuel being used in each technology. The performance of the vehicle is given as

$$P_{W2W} = \eta_{W2V} \cdot P_{V2W} \tag{2.8.1}$$

where

η_{W2V} is the well-to-vehicle efficiency (dimensionless)

P_{V2W} is the vehicle-to-wheel performance (km/kWh)

P_{W2W} is the well-to-wheel performance (km/kWh)

Table 2.8.1 illustrates that the HEV, plug-in hybrid electric vehicles (PHEV), and EV all outperform the conventional ICEV. From the above table, it is evident that the Nissan Leaf is 2.3 times more efficient than the Toyota Camry getting 1.32 more kilometres for every kWh of energy used. From this analysis, the future of the EV has potential to grow as an energy-efficient mode of transport. Table 2.8.2 shows the well-to-wheel CO_2 efficiency analysis of the above vehicle models. The CO_2 efficiency is calculated by

$$W2W_{CO2} = \frac{CO_2}{P_{W2W}}$$

(2.8.2)

Table 2.8.1 **Well-to-wheel energy efficiency of various automobile models (Acha et al., 2011)**

Technology	Model	Fuel	η_{W2V}	P_{V2W}	P_{W2W}
ICE	Toyota Camry	Crude oil	0.82	1.23	1.01
ICE	Honda Civic	Crude oil	0.82	2.27	1.86
HEV	Toyota Prius	Crude oil	0.82	2.47	2.03
PHEV	Chevrolet Volt	Coal	0.35	4	1.4
EV	Tesla Roadster	Coal	0.35	6.1	2.14
EV	Nissan Leaf	Coal	0.35	6.66	2.33
EV	Renault Zoe	Coal	0.35	6.4	2.24

Table 2.8.2 **Well-to-wheel CO_2 efficiency of various automobile models (Acha et al., 2011)**

Technology	Model	Fuel	CO_2	P_{W2W}	$W2W_{CO2}$
ICE	Toyota Camry	Crude oil	0.292	1.01	0.289
ICE	Honda Civic	Crude oil	0.292	1.86	0.157
HEV	Toyota Prius	Crude oil	0.292	2.03	0.144
PHEV	Chevrolet Volt	Coal	0.87	1.4	0.621
EV	Tesla Roadster	Coal	0.87	2.14	0.407
EV	Nissan Leaf	Coal	0.87	2.33	0.373
EV	Renault Zoe	Coal	0.87	2.24	0.388
EV	Renault Zoe	Electricity UK[a]	0.410	2.24	0.183

[a]Department of Energy and Climate Change figures (DEFRA, 2016).

where

CO$_2$ is the carbon content of the fuel (kg/kWh)

W2W$_{CO2}$ is the carbon emitted per vehicle (kg/km)

By looking at the results from this holistic analysis, Table 2.8.2 shows that a zero-carbon emission vehicle is still a long way away. The Toyota Prius performs the best in terms of CO$_2$ efficiency; that is. 144 g of CO$_2$ emissions is released for every kilometre travelled compared with the Toyota Camry excreting 289 g of CO$_2$. When examining the EV with a coal-based power station as its primary source of electricity for recharge of the vehicles' battery, the emissions supersede that of the ICEV by a considerable amount; the most extreme case is found when comparing the Volt and Civic models, with a difference of 464 g/km. The Volt is almost four times more CO$_2$-inefficient than Civic. The results also show data presented by the DEFRA for the United Kingdom's existing electricity composition using 2016 figures (DEFRA, 2016). There is an improvement in CO$_2$ efficiency of the EV when the existing electricity grid's nature was considered (over a 200 g reduction of CO$_2$ for every kilometre travelled) in comparison with a purely coal-based electricity source. However, it should be noted that although improvements are evident in moving away from solely coal-based electricity it is vital that installation of an adequate infrastructure is undertaken to provide EV's access to cleaner electricity.

It is important not to assume that there is a relationship between energy efficiency and CO$_2$ efficiency. The latter study shows that high energy efficiency will not translate directly to high CO$_2$ efficiency; other factors will also play their role in CO$_2$ performance.

2.9 Understanding and optimizing the EV performance

The EV is a new technology that has been developed in the hope that electric-sourced propulsion vehicles will substitute the conventional fossil-fuelled vehicles. Unfortunately, with existing battery technology, the drivers experience is somewhat jeopardized in terms of 'range anxiety', and people are being asked to adopt new driving behaviours rather than simply replacing one vehicle for another. EV on the market requires a more conscientious driver who is aware of his/her vehicle limitations. Imposing a switch from fossil-fuelled to EVs on people, to reach government targets, will not lead to the success of EV penetration in market. The industry, governments, and researchers are trying to work together to overcome barriers that exist in the EV technology so that the EV will not be seen as 'different' or a compromise. Existing technology requires the driver to be considerate and prepared to plan trips considering factors like range or charging points along the way and to reach their destination within a reasonable time considering all of the above factors.

Tackling misconceptions that the wider public have towards the EV is also a challenge that needs to be addressed. The vehicle needs to be presented as a well-established technology that is fit for purpose and instil confidence in the driver that the vehicle is just as reliable as an ICE but requires education on how to get the optimum from your vehicle.

2.9.1 Factors that affect energy consumption

Before looking at the various driving styles, it is important to understand the various factors that will affect energy consumption. A simulation program will be presented later in this chapter to calculate the energy consumption of the vehicle. The components of the program are as follows:

- Energy used to propel the vehicle. This can be in regard to energy to maintain the speed of the vehicle and also the energy the vehicle consumes to accelerate. To propel the vehicle from standstill is a lot more energy-intensive than energy required to maintain the vehicle at constant speed. However, as will be discussed later in the chapter, higher constant speeds will have higher energy consumption.
- The gradient the vehicle has to overcome. This will incorporate a change in potential energy.
- Climate control system (cooling of the vehicle is only in relation to ICEV as will be discussed further in Chapter 4; the ICEV uses 'waste' energy from the engine to heat the cabin space).
- Weight of the vehicle including the number of passengers on board.
- Friction between tyres and road surface.
- The drag effect. This factor will vary from vehicle to vehicle depending on the vehicles design that determines the coefficient of drag.
- The impact of the regenerative braking system.

2.9.2 Categorizing various types of driving

When a new EV is released in the market, the first question that the public and competitors are interested in is 'what is the vehicle's range?' The range of the vehicle set out by manufactures is a guideline to the users of how far they may travel in their EV. However, a new driver should note that this is not guaranteed, and to get this alleged range, the driver must be aware of how battery energy depletes. The EV driver is forced to be conscious of how their vehicle consumes energy. This section presents an analysis of how the energy is used up by an EV and how the manufactures set their vehicle's range. Three different drive cycles (a) urban, (b) rural, and (c) motorway driving shall be presented.

(a) Urban driving

Due to range constraints, the EV has been suggested as an appropriate second vehicle, as the culture that modern households having access to more than one vehicle is growing. It is suggested that the EV would be a good city car, and so, this section will demonstrate how the vehicle performs real-life city driving cycles. The test vehicle was a 24 kWh battery-powered vehicle (Nissan Leaf). No heating or cooling was utilized during the test run so all energy was used for traction (energy used to propel the vehicle). The route varied in altitude and road surfaces to give real-life data as seen in Fig. 2.9.1. Real-life data were preferred to be obtained as usually manufactures will test in laboratories with ideal conditions that can at times be difficult to simulate in real-life conditions and make fair comparisons. The driver was asked to drive in a normal fashion and not be cautious of the energy usage. Table 2.9.1 represent the data

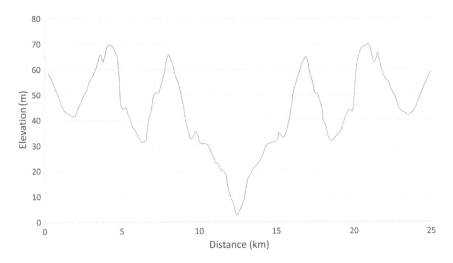

Fig. 2.9.1 Return trip topography.

Table 2.9.1 **Urban driving drive cycles, outgoing trips**

Trial	Duration (min)	Number of times vehicle stopped	Distance travelled (miles)	Energy used by battery (kWh)	Energy consumed by battery (miles; deducted from the display range)
1	65	32	15.5	3.98	15
2	62	31	15.5	3.29	13
3	59	31	15.5	3.85	16

collected from the three test runs. The energy consumed by the battery is displayed in miles. This is the amount of energy, the vehicle algorithm computed, that the driver's driving style consumed from the battery range for that trip.

The routes were repeated again three times, and an experienced EV driver was asked to be aware to use regenerative braking and optimize the vehicle range. Table 2.9.2 illustrates the results from those test runs.

Driving styles in urban areas are highly influenced by traffic management infrastructure in place for a particular town or city. As can be seen in the tables above, the driver experienced a very disruptive drive as the vehicle was at a complete stop a significant amount of times. On average, every 2 min, the vehicle came to a complete stop at traffic lights. To move the vehicle from standstill consumes a considerable amount of energy. The experiment to demonstrate urban driving showed little difference between experienced and inexperienced EV drivers; although the experienced driver experienced more stops, the results remained very similar to those

Table 2.9.2 **Urban driving drive cycles, return trips**

Trial	Duration (min)	Number of times vehicle stopped	Distance travelled (miles)	Energy used by battery (kWh)	Energy consumed by battery, miles (deducted from the display range)
4	65	35	15.7	4.51	17
5	70	38	15.5	4.29	15
6	64	28	15.5	3.92	15

obtained from the driver not conscious of energy consumption. Additionally, it was noted that both drivers' results were similar to that of the actual distance travelled, taking an overall six routes the ability the driver had to extend their range was limited to a value of 4% in range extension. These figures show that urban areas control the driver's behaviour with the EV driver having less control to optimize and potentially extend their battery range. When the vehicle is 100% charged, the vehicle display shows that the vehicle had a range of 96 miles however, the manufactures claim 110 mile range. The manufactures rely on low-speed driving and the use of regenerative braking when computing approximate vehicle range.

(b) Rural driving

Rural driving cycles demonstrate how the BEV performs over a long stretch of road undisrupted. Some sections of the route pose challenges for energy efficiency of the BEV in terms of unfavourable uphill climb. Figs 2.9.2 and 2.9.3 illustrate the

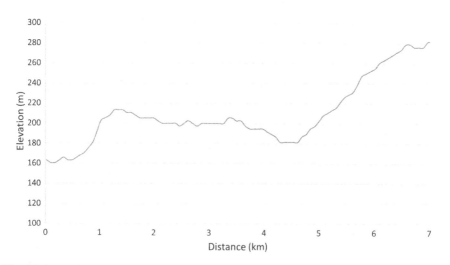

Fig. 2.9.2 Rural drive cycle route, outgoing trip.

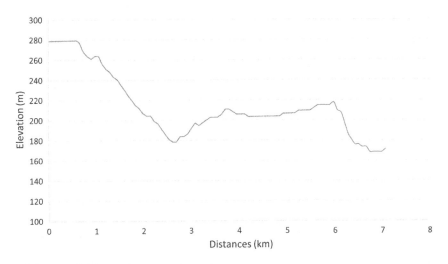

Fig. 2.9.3 Rural drive cycle route, return trip.

topography of the chosen route (1 Out) and its return (1 Ret), respectively. By looking at these graphs, it can be assumed that more traction energy will be required for the first part of the journey. The return trip (1 Ret) gradient is downhill and will require less energy to drive the vehicle. These trips were repeated six times to get a clear picture of energy consumption in rural conditions. Results are illustrated in Table 2.9.3.

The route chosen to demonstrate rural driving illustrated two very different outcomes. The return trips illustrate minus energy consumption. Fig. 2.9.3 illustrates that on the return trip the vehicle travelled 4.5 miles; albeit, the regenerative braking outperformed traction, thus extending vehicle range. The vehicle experienced an extension of range between 1 and 4 miles over the six trips. This is linked to the vehicle taking advantage of the return trips that have a favourable downhill gradient. However, outgoing trips consumed a significant amount of energy to overcome the uphill gradient.

(c) Motorway driving

Fig. 2.9.4 illustrates how the speed at which the vehicle travels can affect the BEV range. This experiment used a passenger EV with a battery of 24 kWh (Nissan Leaf). The vehicle travels at constant speed eliminating the use of regenerative braking simulating driving techniques similar to motorway driving where braking is rarely used, and majority of the journey is based on constant propulsion. The test was carried out on public roads to get realistic readings in terms of ground friction, aerodynamic drag, and temperature. These real-life data give a better understanding of how to optimize the BEV battery discharge rate. The range factor shown in Fig. 2.9.4 was developed to predict the vehicle range depletion at various speeds. With a range factor of 1 for a certain speed (~53 mph), the vehicle will display range like-for-like with distance travelled. In other words, at this speed, the distance travelled is equal to the distance

Table 2.9.3 Rural drive cycle outgoing (Out) and return (Ret) trips

Trial	Distance travelled (miles)	Energy used by battery (kWh)	Energy consumed by battery (miles) (deducted from the display range)
1 Out	4.4	0.36	11
2 Out	4.5	0.37	15
3 Out	4.4	0.31	11
4 Out	4.5	0.37	15
5 Out	4.5	0.37	13
6 Out	4.4	0.31	11
1 Ret	4.5	−0.09	−3
2 Ret	4.5	−0.11	−4
3 Ret	4.4	−0.06	−2
4 Ret	4.4	−0.06	−2
5 Ret	4.4	−0.02	−1
6 Ret	4.5	−0.06	−2

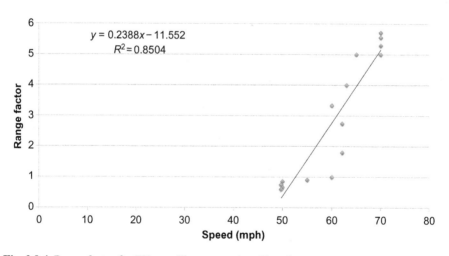

Fig. 2.9.4 Range factor for EV travelling at speeds >50 mph.

consumed as shown by the battery. For a range factor <1, the driver will travel further and notice that the distance travelled is more than the vehicle's built-in algorithm's estimation. The range factor predicts that, when the EV travels at speeds of around 70 mph, speeds typically experienced when driving on a motorway, the range factor is 5 or above. The latter result means that, for every mile travelled at this speed, the battery will consume the energy at five times more than the algorithm's prediction, thus reducing the range dramatically. Fig. 2.9.4 is applicable to motorway driving only when speeds are above 50 mph.

Example 2.9.1

If a vehicle is to travel from Edinburgh to Newcastle ($\sim$120 miles, 193 km) assuming the vehicle is a 24 kWh vehicle (Nissan Leaf) and the topography of the route is level (ideal conditions), find the following:

(a) What capacity of the battery will be needed to make the journey if the vehicle travels at a constant speed of 55 mph (in miles)?
(b) At what speed will the vehicle use 120 miles of the battery capacity?
(c) At what speed will the vehicle be optimized and extend the range of the vehicle battery?

Solution

(a) Using Fig. 2.9.4, the range factor that the vehicle would have to travel at 55 mph is 1.582, and therefore, the driver would have to ensure that they had at least 190 miles battery capacity. However, in a 24 kWh, this exceeds the driver's range capacity, and therefore, the driver must be aware that they have to plan a stop to recharge the vehicle to reach their destination.
(b) This is where the vehicle will use the same battery range capacity as to distance travelled or in other words when the range factor is equal to 1. This happens when the vehicle travels at 52.6 mph.
(c) Speeds under 52.6 mph the vehicles range will be extended.

2.9.2.1   Author's comments on motorway driving

The EV does not perform well during motorway driving, and as illustrated above, at speeds as high as 70 mph, the vehicle consumes over five times more energy than expected energy consumption in comparison with when the vehicle is travelling at speeds at or below 50 mph.

2.9.3   Concluding comments of where EV's are best suited

The BEV has potential to be to a very successful alternate to the ICEV in the automobile industry. Informed route choice plays a significant role in vehicle efficiency. These vehicles perform best on level or descending gradient and performing best at speeds around 50 mph or below to extend vehicle range. Rough road surfaces consume a lot of energy, due to increased friction between tyres and road surface. Urban driving where there are fewer stops and lower speeds are ideal for this vehicle depending on topography. Change in direction, in rural driving, gives the BEV the opportunity to decelerate and avail of regenerative braking. A 4 mile drive cycle carried out in Edinburgh (see Table 2.9.4) demonstrates that battery energy consumption is proportionally linked to vehicle speed.

Table 2.9.4 Traction and regenerated energy from drive cycle (city of Edinburgh, Newington to Chesser ASDA, 4 miles) at various speeds

Speed (mph)	Energy$_{traction}$ (kWh)	Energy$_{recovered}$ (kWh)	E_{nett} (kWh)
30	1.4	0.4	1
25	1.1	0.3	0.8
20	0.8	0.2	0.6

2.10   Traction modelling software

This section will discuss the traction model that incorporates the factors aforementioned. This model has the potential for drivers to use as an aid and optimize route planning.

2.10.1   Previous work's contribution to traction energy modelling

Extensive research of previous work ensures that appropriate assumptions were applied to create a traction and regenerative energy model. Holmberg et al. (2012) claim that one-third of the fuel energy in a conventional ICEV is used to overcome mechanical friction losses. These friction losses include friction between moving parts in the engine, gears, bearings, seals, forks, transmission, tyres, and brakes. Viscous losses in the oil tank are additionally considered under friction losses. When braking is excluded, 28% of fuel energy is lost to direct friction losses aforementioned (Holmberg et al., 2012). These losses, excluding braking, are exempt and are halved in an electric propulsion vehicle due to the nature of the vehicle with fewer components or moving parts and the instantaneous power transmission from chemical energy to electrical energy. The latter study claims that 21.5% of fuel energy in an ICEV is used in the traction of the vehicle. Fuel energy is lost through exhaust emissions (33%), cooling of the vehicle (29%), and mechanical losses (38%) as illustrated in Fig. 2.7.1. These mechanical losses can be further broken down into 5% for air drag and 33% to overcome friction. The figures mentioned apply to an average-sized ICE passenger vehicle (year 2000 model). The friction losses of an average-sized passenger vehicle can be further subdivided into 35% to overcome tyre's rolling friction, 35% to overcome friction of the moving parts in the engine, 15% to overcome friction in transmission, and 15% to overcome friction created during brake contact. Holmberg et al. (2012) presented data for tyre rolling friction coefficients of 0.013, 0.007, and 0.001 for vehicles manufactured in 2000, 2010, and to be produced in 2020, respectively. In the traction energy model, presented in this chapter, a conservative friction coefficient value of 0.013 was used.

Howey et al. (2011) presented the following equation to calculate the total energy consumption for EVs:

$$\text{Total energy} = \frac{1}{\eta_{\text{bc}}}\left[\frac{1}{\eta_{\text{d}}}E_{\text{Traction}} - \eta_{\text{c}}E_{\text{Regeneration}}\right] \tag{2.10.1}$$

where η_{d} and η_{c} are the discharge and charging efficiency for the battery, respectively, E_{Traction} is energy consumed during the discharge of the battery, and $E_{\text{Regeneration}}$ is energy recovered from regenerative braking. As mentioned previously, the battery-charging efficiency lies between 90% and 95%. When developing the present traction energy model, this study used 0.925 as a value for the latter efficiency, η_{bc}.

2.10.2 The vehicle dynamic and energy consumption simulation equations

Muneer et al. (2015) present a vehicle dynamic and energy consumption (VEDEC) simulation that has the ability to compute power and energy of any vehicle during a drive cycle. This software can be adapted to the ICEV and the BEV. The model could be used as a valuable tool to demonstrate how much energy is saved when moving towards a BEV-dominant automobile market. The one advantage that the BEV has over the ICEV is its ability to capture energy that would be conventionally lost due to braking as aforementioned in Section 2.6. The software computes energy savings that is recaptured from regenerative braking system when compared directly with the energy requirements of the same vehicle without such system (Muneer et al., 2015). Various modes of driving styles such as cruise, acceleration, and change of gradient are all considered in the software.

(a) Tyre friction component

Friction is a force that resists motion. Friction has been a topic that has attracted a lot of work in the research community. Research is continuously looking at how to reduce tyre friction to an optimum yet safe value so that with adequate contact to the ground friction provides motion to propel the vehicle and also gradual deceleration when no traction energy is applied:

$$F_{\text{friction}} = \mu W = \mu mg \tag{2.10.2}$$

where friction is dependent on the weight of the vehicle (W) and μ is the friction coefficient of specific tyres used. Where m is the of the vehicle and g is the acceleration due to gravity.

(b) Potential force component

The energy that an object has due to its position with respect to a datum is incorporated into the simulation. The vehicle's position with respect to a given datum will have a positive or negative gradient angle depending on whether the vehicle is ascending or descending. This force is also dependent on the vehicle weight and is computed as follows:

$$F_{\text{potential}} = W \sin \theta = mg \sin \theta \qquad (2.10.3)$$

where θ is the angle relating to the gradient of the road on which the vehicle is travelling.

(c) Aerodynamic drag component

Air drag is another force that researchers and automobile designers are trying to combat as effectively as possible. The shape of the vehicle has changed dramatically from the very first steam-powered locomotive to the various automobiles on the road today. Altering the shape of the vehicle's shell can increase or reduce the air-drag coefficient dramatically. The air-drag component considered in the present VEDEC simulation is calculated as follows:

$$F_{\text{drag}} = \frac{1}{2}C_d A \rho v^2 = \frac{1}{2}C_d A \rho \frac{v_f^2 + v_i^2}{2} = \frac{1}{4}C_d A \rho \left(v_f^2 + v_i^2\right) \qquad (2.10.4)$$

where C_d is the drag coefficient of a specific vehicle, A is the frontal cross-sectional area of the vehicle, ρ is the density of air (standard value $=1.225$ kg/m^3), v_f is final velocity, and v_i is the initial velocity.

(d) Kinetic energy component

While the vehicle is moving, it builds up kinetic energy. This is the energy a vehicle possesses by virtue of its motion. This energy can be positive or negative depending if the vehicle is accelerating or decelerating. The two variables that are considered in kinetic energy are the mass and the velocity of the vehicle (v).

$$F_{\text{kinetic}} = \frac{1}{2}mv^2 = \frac{1}{2}m\left(v_f^2 - v_i^2\right) \qquad (2.10.5)$$

(e) The VEDEC equation

Eq. (2.10.6) is used in this software to calculate power and energy requirements of a vehicle. All of the relevant forces, friction, potential energy, aerodynamic drag, and change in kinetic energy are considered respectively on the right hand side of the equation where Δd represents the distance travelled by the vehicle in a given time interval:

$$E = \left[\mu mg \cos \theta + mg \sin \theta + \frac{1}{4}C_d A \rho \left(v_f^2 + v_i^2\right)\right]\Delta d + \frac{1}{2}m\left(v_f^2 - v_i^2\right) \qquad (2.10.6)$$

Many of the variables (such as μ, m, C_d, and A) are obtained from manufacturer specifications for specific vehicle, and others are known constants (such as g and ρ) or can be obtained from measurements and or surveying maps (such as θ, v, and d). Note that the rapid developments in GPS technology also enable acquisition of these data.

2.10.3 Supporting software

(a) To obtain angle of gradient and distance, θ and d

Presently, we describe details of experiments undertaken by the present authors. For the purpose of energy auditing, topography of routes being analysed is recorded using a GPS tracking system provided by 'Masternaut', a Leeds-based company in the United Kingdom. 'Mapometer', an online altimeter available to the public, provides an accurate measure of topography. The validation of this software is illustrated in Table 2.10.1, showing confidence that this supporting software will compute an accurate value in the simulation to compute energy used by the vehicle in uphill, downhill, or a level road scenario. The angle θ and distance travelled, d, are obtained by the 'Mapometer' software (see Fig. 2.10.1). Alternative methods to obtain the angle of altitude and distance travelled are by using an onboard altimeter or by using topography maps if available.

Table 2.10.1 **Validation of the supporting software program Mapometer (www.mapometer.com)**

	Measured values		Mapometer values		Error	
Experiment	Distance (m)	Altitude (m)	Distance (m)	Altitude (m)	Distance (%)	Altitude (%)
1	1276.5	73.8	1280	78	0.3	5.7
2	421.6	33.4	430	33	2.0	1.2

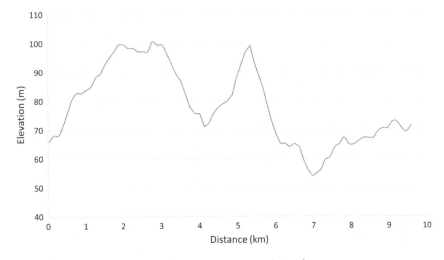

Fig. 2.10.1 Sample route display on Mapometer supporting software.

2.10.4   Validation of simulations

To validate the model, energy simulated is compared with energy measured by the test vehicle's onboard display of energy consumption data for traction energy, recovered energy due to braking, and climate control at the end of each trip. During periods of hard braking when the driver applies force to the brakes, friction between the brake pads and tyre will cause the vehicle to decelerate. Not all energy will be regenerated by the braking system as frictional braking will account for the majority of braking. Therefore, there will be an increased error in the simulation when calculating recovered energy as efficiency will vary depending on drivers braking style.

Sixteen routes were analysed on the experimental vehicle (Renault Zoe) to validate the simulation as shown in Table 2.10.2. The test vehicle's built-in algorithm that computes energy consumption displays to only one decimal place.

Fig. 2.10.2 illustrates traction error as circles and regenerative error as squares. Results of a study undertaken in the United States regarding the efficiency of an electric motor were discussed in Section 2.6. Based on the latter study, a full-load motor and generator efficiency of 85% and 55% was, respectively, assumed for the simulation. Experiments have determined that the maximum speed on a level motorway was recorded as 88 mph, and this was taken as full load for the motor. Table 2.10.2 displays the 16 routes that were analysed. The average speeds for these routes varied from 16 to 35.6 mph, equating to 18%–40% of the battery's full load, which in turn corresponds to an efficiency from 93% to 98%. Losses exist at battery connectors, transmission, and mechanical losses.

Fig. 2.10.2 illustrates a decreasing trend in accuracy of the simulation when calculating regenerated energy compared with increasing speeds or increased loading. As discussed in Section 2.6, the EV will stop regenerating energy during low-speed episodes (below 9 mph). Thus, in reality, assuming a constant efficiency for the generator is not foolproof and will vary with various loading levels on the battery. Note that the computation of regenerative efficiency is further complicated by the fact that the manufacturers limit the capture of braking energy to avoid severe braking; that is, many drivers prefer to 'coast' rather than decelerate even if the latter results in increased efficiency. Therefore, when modelling the regenerative braking system, there is a balance between efficiency and driver safety and comfort to be obtained. The BEV industry has improved rapidly with earlier models of the Nissan Leaf having a 30% efficient regenerative energy system set by their manufacturers, and today, the present VEDEC model sets the optimum regenerative system efficiency as 55% in consultation with research carried on regenerative braking systems (Boretti, 2013; US Department of Energy, 2001). It is suspected that the manufactures algorithm reduces the efficiency at higher speeds for the sake of drive comfort. Due to company confidentiality, attempts to obtain further information regarding the manufacturer's algorithm were futile. Using Fig. 2.10.2, the decreasing accuracy of regeneration can be explained as follows: during times of braking, where energy is being recovered, speed or loading on the battery will reduce significantly. Assuming regenerative efficiency of 55% would thus be in excess of the actual generated quantity. The computational errors, illustrated in Table 2.10.2, for the simulation would be positive at these lower speeds. The same theory applies when the vehicle is driven at higher speeds and vice versa.

Table 2.10.2 Test runs undertaken by Renault Zoe (Edinburgh based)

Experiment number	Route	Average Speed (mph)	Simulation		Experiment		Computational accuracy	
			Energy used[a] (kWh)	Energy regenerated[b] (kWh)	Energy used (kWh)	Energy regenerated (kWh)	Traction program error (%)	Regeneration program error (%)
1	Morningside to Leith	17	1.14	0.36	1.1	0.3	4	21
2	Leith to Morningside	16	1.41	0.23	1.5	0.2	−6	15
3	Home to Sighthill	25	2.12	0.47	2.2	0.5	−3	−6
4	Sighthill to Home	36	3.32	0.48	3	0.7	11	−32
5	Home to Greens	23	1.33	0.33	1.4	0.4	−5	−18
6	Greens to Home	20	1.49	0.33	1.4	0.4	6	−18
7	Home to Costco	24	1.67	0.22	1.7	0.2	−2	12
8	Costco to ESR	26	1	0.37	0.9	0.4	12	−7
9	Napier to Sighthill	25	1.21	0.31	1.1	0.3	10	3
10	Sighthill to Napier	25	1.33	0.23	1.3	0.2	2	14
11	Napier to Dalkeith	26	2.96	0.8	2.9	0.9	2	−11
12	Home to Arthur	23	0.65	0.18	0.6	0.2	9	−8
13	Arthur to Arthur (slow)	23	1.08	0.26	0.9	0.2	20	32
14	Arthur to Arthur (fast)	29	1.31	0.35	1.2	0.4	10	−13
15	Arthur to Shop	20	0.91	0.12	0.9	0.1	1	18
16	Bruntsfield to Juniper Green to Bruntsfield	29	2.42	0.51	2.5	0.6	−3	−15

[a]Energy used for traction.
[b]Energy recovered by regenerative braking.

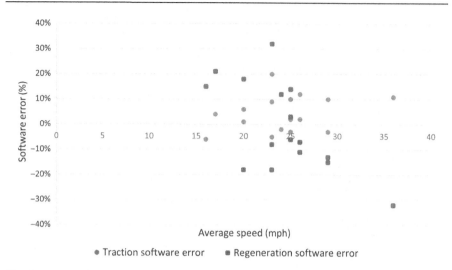

Fig. 2.10.2 Traction and regenerated program error.

Further research to this VEDEC model will develop a robust model to improve the accuracy of traction and recovered energy for the use by the general public. The macros used in this simulation are provided on this book's companion website.

2.11   Conclusion

This chapter presents real-life data and information on BEV traction energy and battery technology that supports the vehicle. The EV can be categorized depending on how traction energy is supplied to the vehicle through 100% electric, series hybrid, parallel hybrid, or parallel–series hybrid connection.

Lithium extraction has increased due to the growth of battery production. Concerns exist around lithium reserves are being exhausted. Immature fears exist that soon lithium will follow similar trends seen in the oil and gas industry when fossil-fuelled vehicles are replaced by lithium-based batteries to propel future vehicles. However, studies illustrate that there is enough lithium resources to support the growth in the battery production industry as the quantity of lithium used in the battery is relatively small. With the prospects of recycling batteries and further lithium exploration, this will further extend global lithium resources. Although lithium extracted in Chile has a higher environmental footprint in terms of transporting it to the United States in comparison with lithium harvested in Nevada, lithium in Chile is 40% less energy-intensive than in Nevada and, thus, is more favourable in battery production.

A VEDEC simulation program was developed to predict vehicle energy consumption for a given trip. The software program showed an average error of 4.4% for calculating traction and 0.6% for regenerative energy.

A passenger vehicle travelling at 30 and 20 mph will consume 1.0 and 0.6 kWh of energy, respectively, for a 4 mile drive cycle. To improve the efficiency and range of the EV, driver education is required.

A range factor was developed to predict the vehicle range depletion at various speeds. With a range factor of 1 for a certain speed (~53 mph), the vehicle will display range like-for-like with distance travelled. In other words, at this speed, the distance travelled is equal to the distance consumed as shown by the battery. For a range factor < 1, the driver will travel further and notice that the distance travelled is more than the vehicle's built-in algorithm's estimation. The range factor predicts that, when the EV travels at speeds of around 70 mph, speeds typically experienced when driving on a motorway, the range factor is 5 or above.

Tests on a popular 24 kWh passenger vehicle on the EV market (Nissan Leaf) illustrated that the city driving can be difficult for the EV to extend its range or reach optimum range due to the frequent stop-and-starting nature of city driving. The energy used to accelerate from nil is significantly high, and drivers found it difficult to avail of the advantages of regenerative braking efficiently. The vehicle was tested in rural conditions where the vehicle had to overcome challenging topographies, but unlike city driving, the driver was not interrupted by traffic control management. The vehicle showed poor performance on ascending gradients. However, under downhill conditions, the vehicle recharged its battery extending its range. Additionally, motorway driving illustrated by high speeds and mostly consistent gradient conditions illustrated that the vehicle at speeds reaching 70 mph will use up to five times more energy than at speeds of around 50 mph. BEVs perform best when gradient is kept constant and at speeds of ~50 mph. Rural conditions are ideal for optimizing driving, as periods when the vehicle is at a complete stop are minimal and opportunity to decelerate and optimize regenerative braking is more frequent than motorway driving and city driving when the vehicle is restricted to travel at higher and lower speeds, respectively.

References

AC Propulsion (2011). .*Creating electric vehicles that people want to drive. Available at http://www.acpropulsion.com/faqs.html (Accessed July 24, 2015).*

Acha, S., Green, T. C., & Shah, N. (2011). Optimal charging strategies of electric vehicles in the UK power market. In *Proceedings of the 2011 IEEE PES innovative smart grid technologies (ISGT)* (pp. 1–8).

Boretti, A. (2013). *Analysis of the regenerative braking efficiency of a latest electric vehicle.* (No. 2013-01-2872), SAE Technical Paper.

BRUSA Elektronik AG (2012). *Battery charger NLG5*. Available at http://www.brusa.biz/_files/drive/02_Energy/Chargers/NLG5/NLG5_SW-Manual_EN.pdf [Accessed April 21, 2017].

Buchmann, I. (2011). *Batteries in a portable world—Chemistry comparison.* Richmond, VA: Cadex Ekectronis Inc.

Clarke, P., Muneer, T., & Cullinane, K. (2010). Cutting vehicle emissions with regenerative braking. *Transportation Research Part D, 15*(3), 160–167. Available at: http://linkinghub.elsevier.com/retrieve/pii/S1361920910000039 [Accessed February 7, 2014].

DEFRA. (2016). *Greenhouse gas reporting—Conversion factors 2016. Available at https:// www.gov.uk/government/publications/greenhouse-gas-reporting-conversion-fac tors-2016 [Accessed August 16, 2016].*

Dunn, J. B., et al. (2012). Impact of recycling on cradle-to-gate energy consumption and greenhouse gas emissions of automotive lithium-ion batteries. *Environmental Science and Technology, 46*, 12704–12710.

Eason, E. (2010). *World lithium supply*. Stanford University. Available at *http://large.stanford. edu/courses/2010/ph240/eason2/* [Accessed October 26, 2015].

Event Horizon Solar and Wind Inc (2015). *Solar and wind power systems, components and design assistance. Available at http://eventhorizonsolar.com/210.12-solarsizing.html [Accessed May 13, 2015].*

Gaines, L. & Nelson, P., 2014. Lithium-ion batteries: Possible material demand issues. Argonne: Argonne National Labratory. Available at http://www.sciencedirect.com/sci ence/article/pii/B9780444595133000182.

Holmberg, K., Andersson, P., & Erdemir, A. (2012). Global energy consumption due to friction in passenger cars. *Tribology International, 47*, 221–234. Available at http://linkinghub. elsevier.com/retrieve/pii/S0301679X11003501 [Accessed March 19, 2014].

Howey, D. A., et al. (2011). Comparative measurements of the energy consumption of 51 electric, hybrid and internal combustion engine vehicles. *Transportation Research Part D, 16* (6),   459–464.   Available   at: http://linkinghub.elsevier.com/retrieve/pii/ S1361920911000459 [Accessed December 15, 2015].

Maxwell (2010). *Top 10 reasons for using ultracapacitors in your system designs.* Available at http://www.maxwell.com/images/documents/whitepaper_top_10_reasons_for_ultracaps. pdf [Accessed April 21, 2017].

McCluer, S., & Christin, J. -F. (2008). *Comparing data center batteries, flywheels, and ultracapacitors*: (pp. 1–17). Available at http://it-resource.schneider-electric.com/i/ 482926-wp-65-comparing-data-center-batteries-flywheels-and-ultracapacitors/4 [Accessed April 21, 2017].

Muneer, T., et al. (2015). Energetic, environmental and economic performance of electric vehicles: Experimental evaluation. *Transportation Research Part D: Transport and Environment, 35*, 40–61. Available at http://linkinghub.elsevier.com/retrieve/pii/ S1361920914001783 [Accessed January 12, 2015].

Serra, J. V. F. (2013). *Electric vehicles: Technology, policy and commercial development.* New York: Routledge.

Ultra Lithium (2013). *Lithium.* Ultra Lithium Inc, Vancouver, British Columbia Canada, Available at http://www.ultralithium.com/lithium_uses/lithium/ [Accessed October 26, 2015].

US Department of Energy (2001). *Determining electric motor load and efficiency.* Available at https://energy.gov/sites/prod/files/2014/04/f15/10097517.pdf [Accessed April 21, 2015].

Parasitic energy consumption for heating and cooling

Aisling Doyle
Edinburgh Napier University, Edinburgh, Scotland, United Kingdom

3.1 Introduction

Human thermal comfort is a complex value to determine. Depending on human physiological factors, thermal comfort state will vary in individuals. It is a state of mind, satisfied or dissatisfied with one's thermal environment. Human productivity level will be affected when the individual's environment varies too far from their thermal comfort range. For this reason, thermal comfort modelling of electric vehicles (EVs) is vital for drivers when operating a vehicle. The conventional internal combustion engine vehicle (ICEV) uses waste heat from the vehicle's engine to heat the vehicle's cabin space. The EV does not produce significant heat from the battery to avail of this recyclable heat energy. A heating system has to be powered by the battery to achieve thermal comfort in the vehicle. For the EV to successfully penetrate the automobile market as a worthy competitor to the ICEV, it is vital the climate control system does not limit the vehicle's range to an extent that will require the driver to switch off the climate control system. Successful uptake will occur when the driver's experience is not jeopardized and the driver's experience is similar or even surpasses the ICEV. A driver with 'range anxiety' and driving at cabin temperatures above or below their thermal comfort level will result in reduced concentration, ultimately dangerous for other road users, and loses their confidence in the technology.

An individual's thermal comfort level will depend on the environment they are exposed to and the person's physiological characteristics. Comfort level will vary depending on the individuals clothing factor, activity level, and their metabolic rate. Thermal comfort in a vehicle will differ from building comfort level. The difference between a building and a vehicle is the glazing-to-solid fabric ratio (e.g., building's brick walls or vehicle's aluminium-sheeted shell). The vehicle has a higher glazing ratio and is exposed to more solar irradiance influences. Additionally, thermal comfort will vary depending on the climate the individual inhabits. Humans can become 'used to' or acclimatized to the temperatures they inhabit. For example, people from warmer climates will have more tolerance to warmer climates compared with those used to cooler more temperate climates. The American Society of Heating, Refrigerating, and Air-Conditioning Engineers (ASHRAE) standard 55 has tried to capture the latter concept through the following equation, with occupant thermal comfort range dependant on the locations outside temperature (de Dear & Brager, 2002):

Electric Vehicles: Prospects and Challenges. http://dx.doi.org/10.1016/B978-0-12-803021-9.00003-3

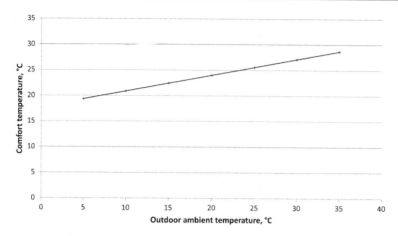

Fig. 3.1.1 Thermal comfort range in Edinburgh stated by ASHRAE standard 55.

$$T_{com} = 0.31T_{ommt} + 17.8 \tag{3.1.1}$$

where T_{com} is comfort temperature and T_{ommt} is outside monthly mean temperature. From the aforementioned equation, Edinburgh's occupant thermal comfort levels range from 18°C to 26°C (see Fig. 3.1.1).

A 5-year thermal analysis of the city of Edinburgh, as seen in Fig. 3.1.2, shows that 96% of the time temperatures in Edinburgh are below 18°C. The majority of the time heating will be required to heat the vehicles cabin space, a factor that previously was created from free energy in the ICEV. The analysis shows that 0.05% of the time

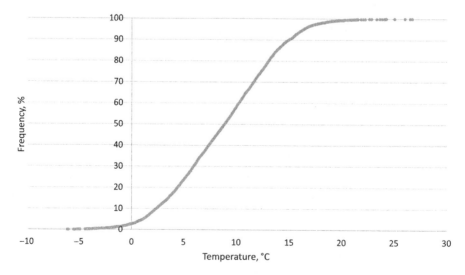

Fig. 3.1.2 A 5-year thermal analysis of Edinburgh.

outdoor ambient temperatures for a 5-year period exceed 26°C. Due to the fabric-to-glazing ratio, the vehicle losses a lot of heat radiation through the vehicle glazing. The vehicle's indoor cabin temperature follows similar trends to outside ambient temperatures during the cooler months when solar radiation levels are low. In warmer conditions, the glazing allows for higher transmittance and experiences solar space heating. The periods when outdoor temperature exceeds 26°C may not mean the vehicle's cabin space has a similar temperature. In these cases, the vehicle's cabin space may experience greenhouse-like effects, and a build-up of temperature may result in temperatures reaching unbearable levels of up to 60°C or even 70°C. A study in Kuwait will examine this later in Section 3.2.4.

3.2   Previous work on thermal comfort

Thermal comfort indices such as the predictive mean vote (PMV) are recognized by ASHRAE. The PMV takes into account four physical variables (air temperature, air velocity, mean radiant temperature, and relative humidity) and also two personal variables (clothing insulation and activity level of the occupant). The discrepancies between the predicted and actual occupant thermal comfort level are suggested to arise from inaccurate measurements of the person's characteristics aforementioned (Charles, 2003). The PMV model developed in the 1970s by Fanger gives satisfactory results for HVAC spaces due to the ability to accurately monitor and control the air speeds and an accurate measurement of clothing insulation. The average journey trip in a passenger vehicle in Scotland is 12.1 km, and with a time duration between 15 and 30 min, it is important that thermal comfort is achieved in the early stages of the trip to satisfy the driver's thermal acceptability (Transport Scotland, 2013).

3.2.1   Predictive mean value (PMV) index

The automobile is an essential commodity in many people's daily activities that varies from commuting, business, and recreational purposes. The amount of time that people spend in an automobile is significant, and passenger thermal comfort must be a consideration in the vehicle's design. Research and analysis of occupant thermal comfort has led to the development of indices like PMV. With this index, an analysis of the thermal performance of the vehicle can be carried out to achieve optimum energy efficiency and performance. PMV can help the development of an efficient climate control systems at an effective cost by developing a model that will predict if occupants will be comfortable in a controlled environment of a laboratory or climate chamber. For the development of the PMV model, the occupants with standard dress were exposed to various conditions and asked to note their thermal sensation on a seven-point psychophysical ASHRAE standard scale that varied from cold to hot and Fanger giving them a value between −3 and +3 where −3 was cold, 0 neutral (neither hot nor cold), and +3 hot (Fanger, 1967). Another element of Fanger's PMV experiments allowed the subject to individually control the climate conditions as desired to achieve neutral or voting with a 0 value or neutral on the ASHRAE scale (Charles, 2003).

This internationally accepted standard evaluates human observation of their personal thermal comfort for a large group of people who are exposed to the same homogeneous environmental conditions (Alahmer et al., 2011). The model considers four physical variables: air temperature, air velocity, radiant temperature, and relative humidity (Charles, 2003). This heat transfer and human sensation index are represented by the equation:

$$PMV = [0.303 \exp\{-0.036M + 0.028\} \times E_{st} \qquad (3.2.1)$$

where M is metabolic energy of the human body (W m^{-2}) and E_{st} is the rate of change of energy storage in the human body (W m^{-2}). The value for M may vary with different studies from 52.8 W m^{-2} (Martinho, da Silva, & Ramos, 2004) and 58 W m^{-2} (Chakroun & Al-fahed, 1997) to 60 W m^{-2} (Mei, 2008) for an occupant sitting in a car. Alahmer et al. (2011) stated that when used in vehicular cabin, analysis from the PMV index showed poor results.

3.2.2 Predicted percentage dissatisfaction index

The predicted percentage dissatisfaction (PPD) is an accepted index that calculates the number of people in a particular space thermally dissatisfied. This index was created to bridge the gap in the PMV model that analyses the mean thermal comfort level of a group of people. An individual observed thermal comfort level is scattered around the mean that Fanger (1973) acknowledges as thermal satisfaction that is very individual and will vary in a large group of people. The PPD is implemented to predict the number of individuals that will be dissatisfied in a particular environment. The percentage of dissatisfaction is determined by the following equation and derived using a value for PMV of a particular environment:

$$PPD = 100 - 95 \exp\left[-\left(0.03353 \times PMV^4 + 0.2179 \times PMV^2\right)\right] \qquad (3.2.2)$$

As stated by Charles (2003), the PMV and PPD have a U-shaped relationship, where the number of people thermally dissatisfied increased when the PMV values vary above or below 0 or neutral observed sensation. A rule of thumb indoor thermal comfort is achieved when PMV is between -0.5 and $+0.5$ corresponding to a value of 10% for PPD (Chakroun & Al-fahed, 1997; Fig. 3.2.1).

3.2.3 Han et al.'s (2001) thermal comfort model

A study by Han et al. (2001) developed a model to predict passenger thermal comfort in an automobile. This study looks at 16 segments of the human body. Each segment has individual properties at four body layers: the core, muscle, fat, and skin tissue as well as a clothing layer that is considered (Han et al., 2001). An automobile is thermally nonhomogeneous in nature, and using indices such as PMV will not give an adequate analysis of thermal comfort. Han et al. (2001) considered their analysis as

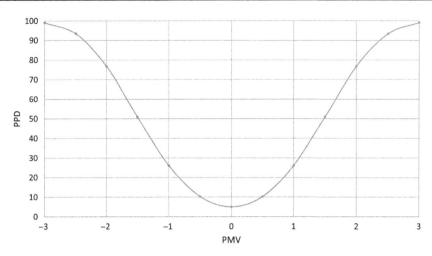

Fig. 3.2.1 The relationship between PMV and PPD.

a function of ambient air temperature, cabin air temperature, air velocity, humidity, direct solar flux, level of individual activity, and clothing type of the occupant. This model uses a technique the authors have called virtual thermal comfort engineering (VTCE) that allows the user to explore different climate control conditions in a vehicle and their effects on the human body in a quick and inexpensive manner. The latter study focuses on reducing the thermal load on the vehicle and to achieve thermal occupant comfort in warmer climatic conditions. The VTCE recognizes that solar radiation at various angles of incidence, glass properties, ambient air temperatures, and air velocity are properties that contribute to human thermal comfort, which are dependent on each other. The complex relationship between all of these variables has yet to be adequately investigated. However, Han et al. (2001) claimed their VTCE tool can adjust one variable and be analysed without influencing other parameters. This study is known as the Berkeley comfort model. Many studies have followed on from the Berkeley study (Kaynakli & Kilic, 2005; Martinho et al., 2004) to further develop an improved mathematical models to predict temperatures towards achieving an adequate understanding of thermal comfort.

3.2.4 The study of thermal comfort in warmer climates

Chakroun and Al-Fahed (1997) analysed the behaviour of in-cabin air temperature when the vehicle is exposed to the sun through an experimental-based study. The aforementioned study was carried out on a hot desert climate in Kuwait where summer temperatures can reach up to and exceed 50°C and subsequently in-cabin temperatures may reach 75°C (Chakroun & Al-fahed, 1997). Fig. 3.2.2 illustrates the temperature behaviour inside the cabin of a parked vehicle exposed to the sun with increased outside ambient temperatures.

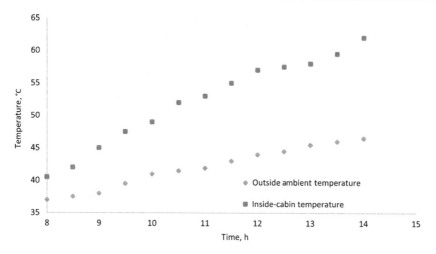

Fig. 3.2.2 Temperature variation inside and outside a parked vehicle exposed to the sun in a hot desert climate (Chakroun & Al-fahed, 1997).

These extreme temperatures concern vehicle drivers in such climates should their automobile be sitting exposed to the sun. This study proposes a number of solutions to these issues from covering front windshield, covering front and side windows, and implementing solar-powered fans to extract heat from the cabin as a result of solar radiation exposure.

3.3 Thermal comfort research by the present research team

A number of trips were recorded by the present study, and energy consumption results are shown in Fig. 3.3.1. This figure plots the energy consumption of the EV's built-in heating system for various temperature differences. The temperature difference plotted is the difference in temperature between outdoor ambient temperature and the desired indoor cabin temperature. Higher values in ΔT indicate lower outdoor ambient temperature. Fig. 3.3.1 illustrates that the lower outside temperature (high ΔT values) indicates higher energy consumption. This figure includes daytime and nocturnal data. When the data are separated, it can be seen that the scatter of the plot is reduced in the nocturnal data. A plot for daytime driving trips shows that the energy consumption by the heat pump is still quite scattered. The reason for this is because solar radiation is an additional factor that is not considered when driving at night. It can be seen that even with high values of ΔT, in the daytime trips, energy consumption can be relatively low. The reason for this is due to the contribution of solar space heating requiring less energy from the climate control system to achieve thermal comfort. This concludes that taking the opportunity to maximize solar space

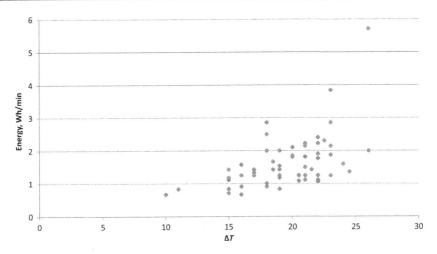

Fig. 3.3.1 Energy consumption of the built-in heat pump for various trips in Edinburgh.

heating will reduce the electrical loading on the vehicle's primary battery to operate the heating system.

3.4 Energy consumption of the EV in comparison to the ICEV

As discussed in Section 2.7, the EV's energy consumption varies greatly when compared with the ICEV. Fig. 2.7.1 illustrated the ICEV energy consumption and how heating is not included in the energy consumption sync diagram. As previously aforementioned, 38% of the ICEV fuel energy is consumed through mechanical requirements to propel the vehicle and 29% to cool the vehicle. Fig. 2.7.2 represents the energy consumption of an EV during warmer periods when cooling is required to compare similar energy consumptions with the ICEV. The EV uses 14% of its lithium-ion battery energy capacity to cool the vehicle. Fig. 3.4.1 illustrates the energy consumption of the EV's lithium-ion battery during periods when heating is required for cabin space heating. The latter illustrates that 18% of the battery's full capacity is used to heat the vehicle's cabin space. These heating requirements do not appear in the ICEV's energy consumption. With a 22 kWh battery having a range of around 145 km, the operation of the heating system would limit the vehicles range by around 26 km. With the average Scottish trip being 12.1 km, this would reduce the vehicle's range by two trips every charge when heating the cabin space. Energy consumption from the climate control system can therefore not be considered as negligible as previously considered with the ICEV. A study by Alahmer, Abdelhamid, and Omar (2012) stated that currently 30% of the mile per gallon of a vehicle's fuel is consumed by the climate control system. The optimization of the climate control system in the EV will reduce this figure and subsequently improve vehicle range.

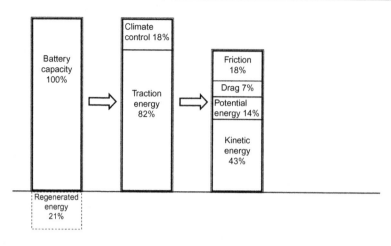

Fig. 3.4.1 Energy consumption of the EV's heating system.

3.5 Optimizing the climate control systems

It is clear the climate control system may constrain the vehicle's range and contribute to 'range anxiety'. To optimize the vehicle's energy efficiency, some manufactures have implemented auxiliary systems into the EV to mitigate or reduce these concerns. The Toyota Prius introduced solar panels onto the vehicles roof (Electreck, 2016). The 2017 Prius Prime model has an 8.8 kWh battery. This battery has a range of 35 electric kilometres. Previous Toyota Prius models connected the integrated solar-roofed panels to an auxiliary battery and powered a ventilation system to acclimatize the vehicle's cabin space prior to the driver entering the cabin. This would reduce the thermal requirements of the main lithium-ion battery to provide heat to the cabin space. The 2017 model solar panel is connected to the primary battery and is said to extend the vehicle's range by 10% adding nearly 4 km onto the vehicle's electric range. The vehicle is available to customers in Europe and Asia; however, the vehicle does not meet the American crash-test standards and so will not be available in America until these issues are resolved.

A study carried out in Oman is an example of how temperature of a sedan car's cabin space is influenced by high levels of solar irradiance and outside ambient temperature of a hot desert climate (Vishweshwara, Marhoon, & Dhali, 2013). The latter installed a simple ventilation system to reduce the greenhouse effect on the vehicle's cabin space. The system consisted of two fans: the first fan to mitigate heat build-up of the space and a second fan that would cool the space with fresh air from outside for a limited period of time. Vishweshwara et al. (2013) used photovoltaic panels to power the ventilation system independently of the vehicle's power system. The study's results showed that for days with high solar irradiance, the cabin space reached temperatures approximately 22°C higher than outside ambient temperatures. With the installation of a simple ventilation system, the temperature difference between inside

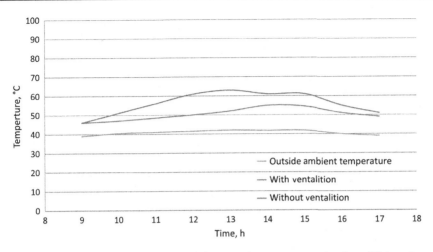

Fig. 3.5.1 Temperature of the front of a cabin space for two consecutive days (Vishweshwara et al., 2013).

and outside temperature can be reduced by approximately 50%. Fig. 3.5.1 illustrates the recorded temperatures for two consecutive days showing the effects of an installed ventilation system reducing indoor cabin temperature. The aforementioned study also acknowledges that a vehicle cabin space is nonhomogeneous. Temperatures recorded in the front of the vehicle's cabin were higher than in the rear. The vehicle front windshield allows for a larger quantity of solar radiation to enter the vehicle, thus a higher temperature in the front of the car. Although the developed system aided in the reduction of cabin temperature, temperatures recorded in the cabin space continue to lie outside the human thermal comfort range. The system alone will not reach acceptable levels of thermal comfort in Oman and will require an additional source of energy climatize the cabin space.

Adhikari (2014) carried out a similar experiment evaluating the effectiveness of solar-powered extractor fans in reducing cabin temperature in a cooler climate to that of Oman. This experiment used four 5.3 W extractor fans installed in the vehicles glazing (see Fig. 3.5.4) and operated by a 150 W photovoltaic modules for a stationary vehicle parked in Edinburgh. Additionally, Adhikari (2014) investigated the impact of the solar panel powering the ventilation system at various angles. The study analysed the optimum angle of 23° and in a horizontal plane (180°). Figs 3.5.2 and 3.5.3 can be read together. Fig. 3.5.3 is the resulting cabin temperatures as a result of the solar irradiance level experiences on that day, as seen in Fig. 3.5.2. In Fig. 3.5.3, temperatures cool at around 13:12; this is a result of the low solar irradiance experienced at this time in Fig. 3.5.2. Albeit solar irradiance increased post 13:12, cabin space temperatures are expected to rise as a result. The operation of the extractor fans hinders the build-up of heat in the cabin space; thus, a decrease in temperature is observed. With the installation of extractor fans, the temperatures in the cabin space can be reduced to levels near ambient air temperature. The utilization of systems such as these reduces temperatures in the cabin space, so that when the driver enters the vehicle, the

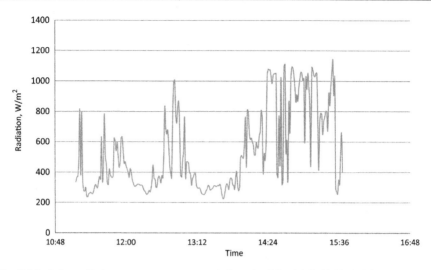

Fig. 3.5.2 Solar radiation measurements recorded on the 9th of Jul. 2014 (Adhikari, 2014).

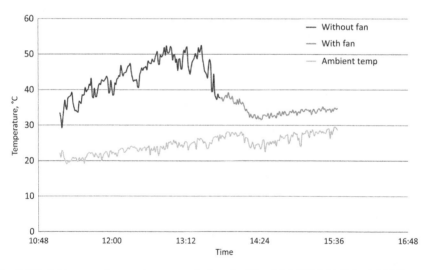

Fig. 3.5.3 Cabin space temperatures recorded on the 9th of Jul. 2014 (Adhikari, 2014).

vehicle's cooling system will not require as much energy to acclimatize the vehicle's cabin space.

Adhikari (2014) developed a thermal model based on a linear relationship between solar irradiance and the temperature difference between the cabin space and outside temperature. The model was developed under three assumptions: (a) cabin temperature is dependent on solar radiation falling on the body's surface; (b) the ventilation system will not be as effective for conditions where ambient temperatures are high as temperature will not fall below outside ambient temperature; thus, temperature drop is

Fig. 3.5.4 Experimental set-up of installed extractor fans in glazing.

reduced with increased outside temperatures; and (c) the analysis of the vehicle considers a north-facing vehicle. The temperature build-up was modelled by the following equation:

$$M_t \frac{Tc(t)}{dt} = S(t) - C(T_c(t) - T_o(t))$$

where M_t is the thermal mass of the vehicle's contents, $\frac{Tc(t)}{dt}$ is the rate of change of temperature, $S(t)$ is the solar radiation falling on the vehicle, C is the specific heat absorption capacity ($C = 1/mxCp$), $T_c(t)$ is the cabin temperature, and $T_o(t)$ is the outside ambient temperature. The model is taken as a steady-state condition; thus, $M_t \frac{Tc(t)}{dt} = 0$. This equation illustrates that cabin temperature is dependent on solar radiation, the product of thermal conductivity of different parts of the car, and temperature difference between inside and outside temperatures.

The present research team has developed thermal models to predict indoor cabin temperature for various weather conditions. Fig. 3.5.5 shows the auxiliary heating system that is used during colder months in Edinburgh and is used to increase cabin

Fig. 3.5.5 Experimental set-up, heating system.

temperatures prior to the driver entering the vehicles cabin space to reduce the electrical loading on the EV's battery as a result of a reduced energy demand required by the built-in heat pump. The developed thermal modal is available for use from the link given at the end of this chapter on the publication of this book.

3.6 Conclusion

This chapter discusses the parasitic energy losses from the heating and cooling system in the EV. Thermal comfort is a state of mind where the person is satisfied with the thermal properties of their environment and thus is very individual, from one person to another. Standards and indices are presented in this chapter to help understand how thermal comfort is defined and how to determine acceptable temperature levels. Due to its high glazing-to-fabric ratio, the vehicle will have different thermal comfort sources to a building as solar irradiance is highly influential in achieving thermal comfort. The ICEV uses waste heat from its engine to heat the vehicle's cabin space. However, the EV cannot avail of this recyclable energy, and energy must be provided to operate the built-in climate control system. The primary battery uses 18% of it's capacity to heat the vehicle's cabin space. The operation of the climate control system will limit the range of the vehicle and thus contribute further to 'range anxiety', the primary barrier hindering the penetration of EVs into the automobile market. Toyota is an example of a vehicle manufacturer that considered integrating solar energy to power a ventilation system to optimize the vehicle's built-in climate control system. The information presented in this chapter states that climate control requirements of an EV should not be considered as negligible.

The present research team has developed a Visual Basic Application for thermal modelling that will be available through this link on publication: https://www.dropbox.com/sh/hbtd0pr1narypzz/AADq4NJ4szU2Skoz2odHXxP3a?dl=0.

References

Adhikari, S. K. (2014). *Reducing the temperature inside cabin of the car using fans.* PV Technology (MS), Edinburgh Napier University (MS: manuscript).

Alahmer, A., Abdelhamid, M., & Omar, M. (2012). Design for thermal sensation and comfort states in vehicles cabins. *Applied Thermal Engineering, 36*, 126–140. Available at: http://linkinghub.elsevier.com/retrieve/pii/S1359431111006843. Accessed 14.11.14.

Alahmer, A., et al. (2011). Vehicular thermal comfort models; a comprehensive review. *Applied Thermal Engineering, 31*(6–7), 995–1002. Available at: http://linkinghub.elsevier.com/retrieve/pii/S135943111000520X. Accessed 14.11.14.

Chakroun, W., & Al-fahed, S. (1997). Thermal comfort analysis inside a car. *International Journal of Energy Research, 21*, 327–340.

Charles, K. E. (2003). *Fanger's thermal comfort and draught models. Institute for Research in Construction 29.* Ottawa, Canada: National Research Council of Canada, Available at: http://www.nrccnrc.gc.ca/obj/irc/doc/pubs/rr/rr162/rr162.pdf.

de Dear, R. J., & Brager, G. S. (2002). Thermal comfort in naturally ventilated buildings: Revisions to ASHRAE standard 55. *Energy and Buildings, 34*(6), 549–561. Available at: http://linkinghub.elsevier.com/retrieve/pii/S0378778802000051.

Electreck (2016). *Toyota brings back the solar panel on the Plug-In Prius Prime—but now it powers the car. Available at: https://electrek.co/2016/06/20/toyota-prius-plug-prime-solar-panel/. Accessed 14.11.16.*

Fanger, P. O. (1967). Calculation of thermal comfort: Introduction of a basic comfort equation. *ASHRAE Transactions, 73*(2).

Fanger, P. O. (1973). Thermal comfort. New York: McGraw-Hill Book Company.

Han, T., et al. (2001). *Virtual thermal comfort engineering.* SAE Technical Paper Series Detroit, Michigan: SAE International, SAE 2001 World Congress.

Kaynakli, O. & Kilic, M., 2005. An investigation of thermal comfort inside an automobile during the heating period Applied Ergonomics, 36(3), pp. 301–312. Available at http://www.ncbi.nlm.nih.gov/pubmed/15854573. Accessed November 14, 2014.

Martinho, N., da Silva, M. C. G., & Ramos, J. E. (2004). Evaluation of thermal comfort in a vehicle cabin. *Proceedings of the Institution of Mechanical Engineers, Part D: Journal of Automobile Engineering, 218*(2), 159–166. Available at: http://pid.sagepub.com/lookup/doi/10.1243/095440704772913936. Accessed 14.11.14.

Mei, Y. (2008). 3-D numerical and experimental analysis for airflow within a passenger compartment. *International Journal of Automotive Technology, 9*(4), 437–446.

Transport Scotland, 2013, Switched on Scotland: A roadmap to widespread adoption of plug-in vehicles, Glasgow, Scotland, Available at: https://www.transport.gov.scot/media/30506/j272736.pdf. Accessed 26.04.2017.

Vishweshwara, S. C., Marhoon, J., & Dhali, A. L. (2013). Study of excessive cabin temperatures of the car parked in Oman and its mitigation. *International Journal of Multidisciplinary Sciences and Engineering, 4*(9), 18–22. Available at: http://www.ijmse.org/Volume4/Issue9/paper4.pdf.

Battery technologies for electric vehicles

Koki Ogura*, Mohan Lal Kolhe†
*Kawasaki Heavy Industries, Ltd., Kobe, Japan, †University of Agder, Kristiansand, Norway

4.1   Introduction

Climate change, diminishing reserves of fossil fuels, and energy security demand alternatives to our current course of energy usage and consumption. A broad consensus concurs that implementing energy efficiency and sustainable energy technologies are necessities now rather than luxuries to be deferred to some distant future. Of particular interest for the sustainable energy technology are the following:

(1) Powering electric vehicles that can compete with car powered by internal combustion (IC) engines and resolved the issue with CO_2 emission.
(2) Stationary storage of electrical energy from renewable energy sources.

Investments for the exploitation of renewable energy resources are increasing worldwide, with particular attention to wind and solar power energy plants as the most mature technologies. However, a major barrier to wide-ranging application of renewable energy sources is the continuity of supply; the generation of power by solar and wind sources, for example, can be hampered by the time of day, dust, cloud, and weather conditions. Thus, utilization of these intermittence resources requires high-efficiency energy storage systems. Electrochemical systems, such as batteries and supercapacitors that can efficiently store and deliver energy on demand in stand-alone power plants and provide power quality and load levelling of the electrical grid in integrated systems, are playing a crucial role in this area. Selection of a storage system for a specific application is often made based on the amount of energy and power density that can be delivered from the system. Although specific energy and power are important, other factors must also be taken into account when selecting a system for a spin-off application, including reliability, safety, self-discharge, temperature, and even humidity. A useful means of representing the operational performance of energy storage and energy conversion devices is a graph of specific power density (W/kg) versus specific energy density (Wh/kg). This graph is known as a Ragone plot in Fig. 4.1.1; it is shown for supercapacitors and three common rechargeable batteries, namely, lead-acid battery, nickel-metal hydride (Ni-MH) battery, and lithium-ion (Li-ion) battery. Note that this plot shows specific energy and power on a cell level for batteries made for many different applications, from consumer electronic to vehicles. The following sections present the analysis of the above-mentioned electrochemical technologies for both the stationary and mobile applications.

Electric Vehicles: Prospects and Challenges. http://dx.doi.org/10.1016/B978-0-12-803021-9.00004-5

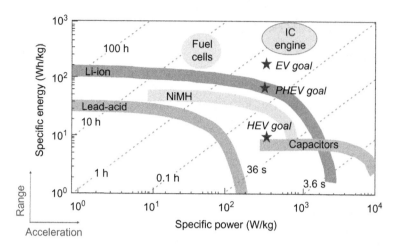

Fig. 4.1.1 Ragone plots for array of energy storage and energy conversion devices.

There are now four types of electric cars: battery electric vehicles (BEVs), plug-in hybrid electric vehicles (PHEVs), conventional hybrid electric vehicles (HEVs), and fuel-cell electric vehicles (FCEVs).

Battery electric vehicle (BEV): A BEV runs entirely on a battery and electric drive train, without a conventional internal combustion engine. These vehicles must be plugged into an external source of electricity to recharge their batteries. Like all electric vehicles, BEVs can also recharge their batteries through regenerative braking. In this process, the vehicle's electric motor assists in slowing the vehicle and recovers some of the energy normally converted to heat by the brakes.

Plug-in hybrid electric vehicles (PHEVs): PHEVs run mostly on batteries that are recharged by plugging into the power grid. They are also equipped with an internal combustion engine that can recharge the battery and/or to replace the electric drive train when the battery is low and more power is required. Because PHEVs can be recharges on the public network, they are often cheaper to run than tradition hybrids though the amount of savings depends on the distance driven on the electric motor alone.

Hybrid electric vehicles (HEVs): HEVs on the road today have two complementary drive systems: a gasoline engine and fuel tank and an electric motor, battery, and controls. The engine and the motor can simultaneously turn the transmission, which powers the wheels. HEVs cannot be recharged from the power grid. Their energy comes entirely from gasoline and regenerative braking.

Fuel-cell electric vehicles (FCEVs): The fuel cell is another type of electric vehicle expected to be widespread on the market in the next few years. The fuel-cell electric vehicles create electricity from hydrogen and oxygen. Because of these vehicles' efficiency and water-only emissions, some experts consider these cars to be the best electric vehicles, even though they are still in development phases. Toyota is slated to follow suit in Sep. 2015 named the Mirai.

Electrification is the most viable way to achieve clean and efficient transportation that is crucial to the sustainable development of the whole world. In the near future, electric vehicles will dominate the clean vehicle market. As shown in Table 4.1.1, the current major battery technology used in EVs is Li-ion batteries because of its mature technology. Due to the potential of obtaining higher specific energy and energy density, the adoption of Li-ion batteries is growing fast in EVs, particularly in PHEVs and BEVs. It should be noted that there are several types of Li-ion batteries based on similar but certainly different chemistry.

Table 4.1.1 Batteries used in electric vehicles of selected car manufacturers.

Company	Country	Vehicle model	Types of EV	Battery and capacity
Toyota	Japan	Prius PHV (Prime)	PHEV	Li-ion, 8.8 kWh
		Prius (fourth generation)	HEV	Ni-MH, 1.31 kWh
		Prius (fourth generation)	HEV	Li-ion, 0.75 kWh
		Aqua(Prius C)	HEV	Ni-MH, 0.94 kWh
Nissan	Japan	Leaf	BEV	Li-ion, 30 kWh
Honda	Japan	Accord Hybrid	HEV	Li-ion, 1.3 kWh
		Fit (Jazz) Hybrid	HEV	Li-ion, 0.86 kWh
Mitsubishi	Japan	i-MiEV	BEV	Li-ion, 16 kWh
		Outlander	PHEV	Li-ion, 12 kWh
BMW	Germany	i3	BEV	Li-ion, 33 kWh
		×5 xDrive40e	PHEV	Li-ion, 9.0 kWh
Mercedes-Benz	Germany	B250e	HEV	Li-ion, 28 kWh
Audi	Germany	A3 Sportback e-tron	PHEV	Li-ion, 8.3 kWh
Volkswagen	Germany	e-Golf	BEV	Li-ion, 35.8 kWh
Volvo	Sweden	XC90 T8	PHEV	Li-ion, 9.0 kWh
Fiat	Italy	500e	BEV	Li-ion, 24 kWh
Tesla	The United States	Model S	BEV	Li-ion, 60–100 kWh
General Motors	The United States	Chevrolet Volt	PHEV	Li-ion, 18.4 kWh
Ford	The United States	C-MAX Energi	PHEV	Li-ion, 7.6 kWh
Hyundai	Korea	Sonata Hybrid	HEV	Li-polymer, 1.6 kWh

Fig. 4.1.2 Toyota plug-in hybrid electric vehicle 'Prius PHV'.

The new Toyota Prius plug-in hybrid shown in Fig. 4.1.2 will gain additional electric driving range with a rooftop solar panel that helps recharge the batteries. Stretching nearly the entire length of the roof, the solar cells can charge the lithium-ion batteries when the car is parked and can boost efficiency by as much as 10%. When the car is being driven, the solar panels also supply power to accessories such as air conditioning, power window, and interior lights. Initially, the new technology will be offered only in Japan and Europe versions, where it is named the Prius PHV.

4.2   Electrochemical energy storage

4.2.1   Rechargeable battery

Batteries are devices that convert the chemical energy contained in an electrochemically active material directly into electrical energy by means of a redox reaction. For a rechargeable system, the battery allows the storage of a defined amount of chemical energy and can be recharged when the electrochemically active material has been transformed. Several types of rechargeable systems exist, from the mature lead acid to different newer technologies at various developmental stages.

4.2.1.1   Lead acid battery

The lead-acid battery was the first known type of rechargeable battery. It was suggested by French physicist Dr. Planté in 1860 for means of energy storage. Lead-acid batteries continue to hold a leading position, especially in wheeled mobility and stationary applications. The lead-acid battery is a combination of a lead, a lead dioxide, and an electrolyte composed of sulfuric acid and water.

Lead-acid battery is offered in two different types:

(1) The flooded type that is the cheapest and tends to be used in automotive and industrial applications.

(2) The sealed type, also called valve-regulated lead-acid (VRLA), that has been rapidly used in a wide range of applications including power supplies and stand-alone power supplies for remote areas.

Both the power and energy capacities of lead-acid batteries depend on the size and geometry of the electrodes, which makes it unfavourable for automotive industry. The power capacity can be improved by increasing the surface area of each electrode, which means greater quantities of thinner electrode plates in the battery.

Low cost, high power, and easy recyclability are among the advantages of the lead-acid batteries. One main drawback of lead-acid batteries is usable capacity decreases when high power is discharged. In addition, as shown in Fig. 4.1.1, lead-acid batteries have four times less specific energy than that offered by Li-ion batteries, and it is expected to be gradually displaced by Li-ion and Ni-MH, due to environmental impact concerns.

4.2.1.2   Ni-MH battery

Since 1990, when the first Ni-MH battery was commercialized in Japan, it has been adapted to a wide range of applications from hand-held electronics to HEVs. The Ni-MH battery consists of a positive nickel hydroxide, negative metal hydride, and nylon separator sheets. The electrodes that accommodate active materials required in the battery chemistry and the separator sheets have porous structures, and their pores are filled with a concentrated KOH solution as the electrolyte. Recently, Li-ion batteries are realized as the new competitor for Ni-MH cells, thanks to their outstanding energy density, minimal memory effect, and low self-discharge rate. Nonetheless, Ni-MH cells are expected to sustain their position in the market due to their environmental friendliness, relatively low price, and better thermal stability.

Ni-MH is not without drawbacks. Fig. 4.1.1 shows that the Ni-MH battery has twice less specific energy compared with Li-ion batteries that limits their application in electric power train. Ni-MH also has high self-discharge and loses about 20% of its capacity within the first 24 h and 10% per month thereafter. Modifying the hydride materials lowers the self-discharge and reduces corrosion of the alloy, but this decreases the specific energy. Moreover, Ni-MH batteries generate heat during fast charge and high-load discharge that can play a crucial role in their safety, performance, and longevity.

4.2.1.3   Li-ion battery

The Li-ion battery is used by millions of people around the world in mobile phones, laptops, tablets, hearing aids, cameras, power tools, and many other compact, lightweight mobile devices. Dr. Goodenough, Dr. Nishi, Dr. Yazami, and Dr. Yoshino each made substantial contributions to its development. In 1979, Dr. Goodenough showed that by using lithium cobalt oxide (LCO) as the cathode of a Li-ion rechargeable battery, it would be possible to achieve a high density of stored energy with an anode other than metallic lithium. This discovery led to the development of carbon-rich materials that allow for the use of stable and manageable negative electrodes in Li-ion batteries. Shortly after Dr. Goodenough's breakthrough, Dr. Yazami began exploring graphite compounds in which lithium could be reversibly inserted between graphite layers. This provided an alternative to the lithium metal negative electrode. Dr. Yazami's lithium-graphite is the most commonly used anode in commercial Li-ion batteries today. In 1985, Dr. Yoshino produced a rechargeable Li-ion battery prototype using a LCO cathode and a carbon anode, eliminating metallic lithium. This design significantly improved the safety of the battery while also providing practical energy output at a reasonable price. Dr. Yoshino's work resulted in the first safety-tested, commercially acceptable Li-ion battery.

Lithium is the lightest of all metals, has the greatest electrochemical potential, and provides the largest specific energy per weight. Thus, Li-ion batteries have characteristics of high energy and power density and have been shown to be less affected by memory effect, compared with other types of batteries; see Fig. 4.1.1. These traits make the Li-ion battery the most likely candidate to assist sustainable mobility such as EVs, HEVs, scooters, motorcycles, and electrical bicycles. Longevity, low self-discharge (less than half that of Ni-MH), and rapid charging and high-load capabilities are among the advantages offered by Li-ion batteries.

4.2.1.4 Supercapacitor

Supercapacitors are electrochemical devices that store energy by virtue of the separation of charge, unlike batteries, which store energy through chemical transformation of electrode materials. Supercapacitors continue to develop and mature as an energy storage technology, though somewhat still in the shadow of rechargeable batteries.

During the storage of electrochemical energy in a battery, chemical interconversions of the electrode materials occur usually with concomitant phase changes. Although the overall energy changes can be conducted in a relatively reversible thermodynamic route, the charge and discharge processes in a storage battery often involve irreversibility in interconversions of the chemical electrode-reagents. Accordingly, the cycle life of storage batteries is usually limited and varies with the battery type. By contrast, with energy storage by a capacitor, only an excess and a deficiency of electron charges on the capacitor plates have to be established on charge and the reverse on discharge, and no chemical changes are involved. Accordingly, a capacitor has an almost unlimited recyclability, typically between 10^5 and 10^6 times.

As shown in Fig. 4.1.1, electrochemical capacitors have superb specific power compared with batteries but modest specific energies. This translates, in transportation terms, as good acceleration but poor range, which is precisely opposite to batteries. From practical and technological viewpoints, electrochemical capacitors are robust devices with excellent cycle life that can improve the effectiveness of battery-based systems by shrinking the volume of batteries required and reducing the frequency of their replacement.

4.2.2 Advanced rechargeable battery technology

In addition to the above-mentioned energy storage systems that have been commercialized, a large number of battery technologies are still under development that can be categorized as follows:

(1) Rechargeable metal-air batteries
(2) Flow batteries

By virtue of removing much of the mass of the positive electrodes, metal-air batteries offer the best prospects for achieving specific energy that can be comparable with petroleum fuels. Lithium-air (oxygen) and sodium-air (oxygen) are two main categories of such advanced technology. Since this subject is not in the scope of this study, the readers are referred to references for more information in this regard.

A flow battery is a rechargeable battery where the energy is stored in one or more electroactive species dissolved into liquid electrolytes. Flow batteries can be fitted to a wide range of stationary energy storage systems, and they are not recommended for vehicular applications. Flow batteries are classified into redox flow and hybrid flow batteries where more detailed discussion can be found in some papers.

Overall, the high energy density, lightweight, no memory effect, and low self-discharge rate make Li-ion-based batteries superior to other energy storage devices for vehicular applications including HEVs, PHEVs, and BEVs, which is also the main focus of study.

4.3 Challenges in electric and hybrid electric vehicles

The strategy for electrifying vehicles with high-capacity batteries in order to reduce or remove the contribution of internal combustion engine into powertrain has attracted an intense attention. Li-ion batteries have become the dominant battery technology for automotive industry due to several compelling features such as high power and energy density, high voltage, long cycle life, excellent storage capabilities, and memory-free recharge characteristics. While Li-ion batteries are growing fast in popularity, safety issues and cost (related to cycle and calendar life) that all are coupled to thermal effects in batteries are the main barriers to the development of large fleets of vehicles on public roads equipped with Li-ion cells. It is evident that under high discharge conditions, which involve high rates of Joule heat generation and exothermic electrochemical reactions, batteries are prone to excessive temperature rise. The temperature profile presented in Fig. 4.3.1 is an evidence of temperature rise in the battery at different discharge rates of 1C, 2C, and 3C, which correspond to 20, 40, and 60 A discharge currents.

Fig. 4.3.1 illustrates that over a short 30 min time period (short from a vehicle operation viewpoint) for 2C and a 20 min time period for 3C discharge, enough heat is generated to increase the cell temperature to 50°C (for 2C) and 55°C (for 3C) from a 24°C start condition. This value is only for a single prismatic cell with free convection boundary condition, so even a greater temperature can result when extrapolated to the approximately 100 prismatic cells in a battery pack of PHEVs, HEVs, and BEVs, where there is no free boundary convection, but only conduction between pouch cells. This temperature evolution can initiate swelling; thermal runaway; electrolyte fire; and, in extreme cases, explosion. Moreover, exposure of Li-ion batteries to subfreezing temperatures drastically reduces their energy and power. Therefore, for optimum performance and longevity, the Li-ion battery should operate within 25–40°C; also it is desirable to maintain the temperature variation between battery modules in the battery pack <5°C.

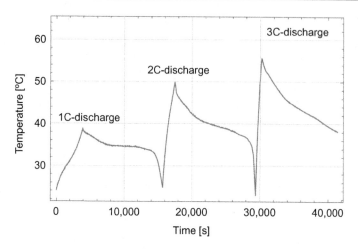

Fig. 4.3.1 Temperature rise at the surface of a prismatic Li-ion cell during 1C, 2C, and 3C discharge rates.

Hence, the development and implementation of Li-ion batteries, particularly in automotive applications, requires substantial diagnostic and practical modelling efforts to fully understand the thermal characteristics in the batteries across various operating conditions. Indeed, thermal modelling prompts the understanding of the battery thermal behaviour beyond what is possible from experiment, and it provides a basis for exploring thermal management strategies for batteries in HEVs and EVs. Once a battery is characterized to the extent that calculations can be used to provide a good estimation of electrochemical-thermal behaviour, it is possible to complete the design of the battery system on the vehicle and to size the heat exchangers and related thermal system components necessary to heat or cool the battery.

4.4   eVaro electric sports car

A Canadian company, future vehicle technologies (FVT), aims to introduce a high-performance electric sports car called *eVaro* as shown in Fig. 4.4.1. It requires an intelligent system to control and keep the temperature of their 21 kWh battery pack within the desirable range.

As shown in Fig. 4.3.1, during operation with high discharge rates, Li-ion batteries can heat up to 55°C or greater, particularly when cells are stacked into modules and then packs and when they are in a warm ambient temperature. Hence, an efficient battery thermal management system (BTMS) is required to maintain the battery temperature within the desirable range and ensure optimal vehicle operation in terms of safety, performance, and battery longevity.

As an effective tool, mathematical simulations of the battery can help to obtain a fundamental understanding about how the heat is generated inside the battery, how it

Fig. 4.4.1 High-performance electric sports car 'eVaro' introduced by FVT. (A) Appearance of eVaro. (B) Battery pack used in eVaro.

can be conducted out of the battery during different operating conditions, and how the proper temperature control of a battery can be achieved. To fulfil this aim, a series of modelling studies are developed to estimate the amount of heat generation at the cell level and predict the transient distribution of electrical and thermal behaviour of the battery at different operating conditions with minimum computational effort. Using this information, one can expand the model to a module/pack level in order to design a proper BTMS.

4.5 Battery chemistry

Various battery chemistries have been proposed as the energy source to power electric vehicles since the 1990 California Zero-Emission Vehicle was mandated, which required 10% and 14% of the automobiles sold to be zero emission in 2005 and 2017, respectively. These battery chemistries included improved lead-acid, nickel-cadmium, Ni-MH, and Li-ion batteries, with each of these chemistries having its own advantages and disadvantages. Towards the end of the last century, the competition between battery chemistries was resolved with General Motor's choice of Ni-MH for its EV-1 pure electric vehicles. In the following decade, the technology of the HEV developed by Toyota and Honda matured and gained popularity through its combination of fuel economy, acceptable pricing, and clean safety record. Up to the date of 2011, the leading battery chemistry in these HEVs remained Ni-MH. As the concerns over greenhouse gas emissions and fossil energy shortages grow in the recent years, the development target has shifted from HEV to PHEV, with the eventual target being a purely battery-powered EV. The requirement of a higher energy density in PHEVs and EVs reopens the discussion for automobile battery technologies, giving Li-ion battery chemistry another chance at entering the electric car battery market. In this section, the underlying principles, the current market status, and the future developmental trends of Li-ion battery are discussed.

4.5.1 Basic operation of rechargeable battery

A battery is composed of a positive electrode (holding a higher potential) and a negative electrode (holding a lower potential) with an ion-conductive but electrically insulating electrolyte in between. During charging, the positive electrode is the anode with the reduction reaction, and the negative electrode is the cathode with the oxidation reaction. During discharge, the reaction is reversed, and so, the positive and negative electrodes become cathode and anode electrodes, respectively. As a side note, the positive- and negative-electrode active materials are also conventionally referred to as cathode and anode material, respectively. In a sealed cell, the liquid electrolyte is held in a separator to prevent the direct short between the two electrodes. The separator also serves as a reservoir for extra electrolyte, a space saver allowing for an electrode expansion, an ammonia trap (in Ni-MH battery), and a safety device for preventing shortage due to Li-dendrite formation (in Li-ion battery).

4.5.2 Ni-MH battery operation

A schematic of the Ni-MH rechargeable battery is shown in Fig. 4.5.1. The active material in the negative electrode is metal hydride (MH), a special type of intermetallic alloy that is capable of chemically absorbing and desorbing hydrogen. The most widely used MH in Ni-MH today is the AB_5 alloy with a $CaCu_5$ crystal structure, where A is a mixture of La, Ce, Pr, and Nd and B is composed of Ni, Co, Mn, and Al. The active material in the positive electrode is $Ni(OH)_2$, which is the same chemical used in the Ni—Fe and Ni—Cd rechargeable batteries patented by Thomas Edison more than a hundred years ago. The intrinsic $Ni(OH)_2$ has a poor conductivity; to make up for this shortcoming, coprecipitation of other atoms, formation of conductive network outside the particle, or multilayer coating structure is implemented in the commercial product. The separator is typically made from grafted polyethylene (PE)/ polypropylene (PP) nonwoven fabric. The commonly used electrolyte is a 30 wt%

Fig. 4.5.1 Schematic of the charging operation of a Ni-MH battery.

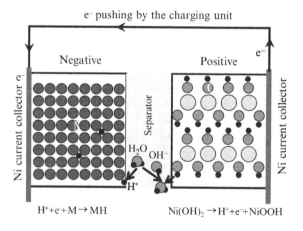

e⁻ pushing by the charging unit

Negative   Positive

$H^+ + e + M \rightarrow MH$ $Ni(OH)_2 \rightarrow H^+ + e^- + NiOOH$

KOH aqueous solution with a pH value of about 14.3. In some special designs for particular applications, certain amounts of NaOH and LiOH are also added into the electrolyte.

During charge, water is split into protons (H^+) and hydroxide ions (OH^-) by the voltage supplied from the charging unit. The proton enters the negative electrode, neutralizes with the electron supplied by the charging unit through the current collector, and hops between adjacent storage sites by the quantum mechanics tunnelling. The voltage is equivalent to the applied hydrogen pressure in a gas-phase reaction and will remain at a near-constant value before protons occupy all of the available sites. OH^- generated by charging will add to the OH^- already present in the KOH electrolyte. On the surface of the positive electrode, some OH^- will recombine with protons coming from the $Ni(OH)_2$ and form water molecules. The complete reaction for charging is as follows:

$$M + Ni(OH)_2 \rightarrow MH + NiOOH \tag{4.5.1}$$

Neither water nor OH^- is consumed; thus, no change to pH value occurs during charge/discharge. The oxidation state of Ni in $Ni(OH)_2$ is 2^+. As protons are consumed at the surface of the positive electrode, more protons are driven out of the bulk from both the voltage and the concentration gradients. Losing one proton increases the oxidation state of Ni to 3^+ in NiOOH. Electrons are collected by Ni form or perforated Ni plate and moved back to the charging unit to complete the circuit.

The whole process is reversed during discharge. In the negative electrode, protons are sent to the electrolyte and recombine with the OH^- as electrons are pushed to the outside load. The electrons re-enter the positive-electrode side of the battery through the outside load and neutralize the protons generated from the water split on the surface of the positive electrode.

4.5.3 Li-ion battery operation

4.5.3.1 Introduction to Li-Ion battery

According to a recent market report published by Transparency Market Research, the global market for Li-ion battery was worth USD 11.70 billion in 2012 and is expected to reach USD 33.11 billion in 2019, growing at a 14.4% compound annual growth rate (CAGR). Fig. 4.5.2 shows a market segment of Li-ion batteries in consumer device vendors, industrial goods manufactures, grid and renewable energy storage segments, and automobile manufacturers. As presented in Fig. 4.5.2, the consumer segment, which accounted for 60.3% of total Li-ion battery revenues in 2013, is expected to see its market share reduced to 23.9% in 2020. This predicted reduction in total market share reflects the expected significant growth in Li-ion battery demand in the automotive and grid and renewable energy storage segments. In Fig. 4.5.2, the global market demand for Li-ion batteries in 2013 is compared with the one predicted in 2020. Despite revenue growth, market share in the industrial segment is expected to decline

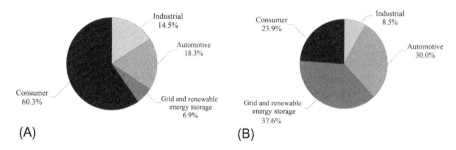

Fig. 4.5.2 Market segment of Li-ion batteries. (A) 2013 and (B) 2020.

Fig. 4.5.3 Schematic of the charging operation of a Li-ion battery.

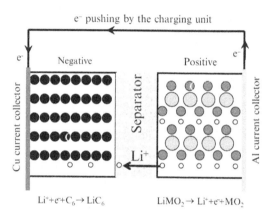

because of the explosive growth in the automotive segment and grid and renewable energy storage segment:

A schematic with two half-cell reactions for the Li-ion battery in charging mode is shown in Fig. 4.5.3.

The complete reaction is

$$C_6 + LiMO_2 \rightarrow LiC_6 + MO_2 \tag{4.5.2}$$

The most commonly used active material in the negative electrode is graphite. During charging, Li-ions, driven by the potential difference supplied by the charging unit, intercalate into the interlayer region of graphite. The arrangement of Li^+ in graphite is coordinated by the surface-electrolyte-interface (SEI) layer, which is formed during the initial activation process. The active material in the positive electrode is a Li-containing metal oxide, which is similar to $Ni(OH)_2$ in the Ni-MH battery but replaces the hydrogen with lithium. During charging, the Li^+ (similar to the H^+ in Ni-MH) hops onto the surface, moves through the electrolyte, and finally arrives at the negative electrode. The oxidation state of the host metal will increase and return electrons to the outside circuitry. During discharge, the process is reversed. Li-ions now move from the intercalation sites in the negative electrode to the electrolyte and then to the original site in the $LiMO_2$ crystal. The commonly used electrolyte is a mixture of organic carbonates such as ethylene carbonate, dimethyl carbonate, and diethyl carbonate containing hexafluorophosphate ($LiPF_6$). The separator is a multilayer structure from PP, which provides oxidation resistance, and PE, which provides a high-speed shutdown in the case of a short.

4.5.3.2 Li-ion cell operation

In Fig. 4.5.4, the internal structure of the battery from cell scale to electrode scale is illustrated. Fig. 4.5.4A and b depicts the unfolded and folded layered structure in the battery. This layered configuration is also referred as 'sandwiched configuration'. In Fig. 4.5.4C, a schematic presentation on charge transfer between electrode pairs during charge and discharge cycle is given; this is also the fundamental scale for thermo-electrochemical battery simulations, since the collective behaviour of these unit cells represents the overall battery behaviour:

Li-ion battery uses a cathode (positive electrode), an anode (negative electrode), and an electrolyte as conductor. The cathode is made of a composite material and

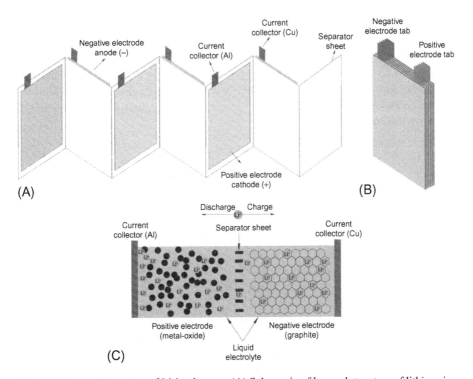

Fig. 4.5.4 Internal structure of Li-ion battery. (A) Schematic of layered structure of lithium-ion battery cells. (B) Folded battery core diagram. (C) Components of unit cell and charge transfer during charge and discharge.

defines the name of the Li-ion battery cell. The electrolyte can be liquid, polymer, or solid. In case of a polymer or solid electrolyte, the electrolyte will act as a separator as well. The separator is porous to enable the transport of lithium ions. It prevents the cell from short-circuiting and protects the cell from thermal runaway. In advanced Li-ion cells, anode (negative electrode) is typically made of carbon-Li intercalation compound, and for this, usually, it is referred as graphite electrode, while the cathode includes metal oxide materials. Indeed, what the cathode is made from determines the cell's capacity as it will be discussed in the next chapter. The critical feature is the rate at which the cathode can intercalate and deintercalate free lithium ions.

The overall chemical reactions inside the Li-ion battery in general form is as follows:

$$6C + LiM_yO_z \rightleftharpoons Li_xC_6 + Li_{1-x}M_yO_z \qquad (4.5.3)$$

where $x \leq 1$, M is the metal in the positive electrode (nickel, cobalt, manganese, or their combination), and O is oxygen. For the charge and discharge processes, the reactions proceed from left to right and right to left, respectively. Charge and discharge reactions in the electrodes are as follows:

$$6C + xLi^+ + xe^- \rightleftharpoons Li_xC_6 \quad \text{(negative electrode)} \qquad (4.5.4a)$$

$$LiM_yO_z \rightleftharpoons Li_{1-x}M_yO_z + xLi^+ + xe^- \quad \text{(positive electrode)} \qquad (4.5.4b)$$

The charge and discharge reactions for rechargeable Li-ion battery are endothermic and exothermic, respectively. If the battery reaction is ideally reversible, the thermodynamic equation under a constant temperature and constant pressure yields the following relation:

$$\Delta G = \Delta H - T\Delta S \qquad (4.5.5)$$

where the residual energy at energy conversion between the enthalpy change ΔH of the battery reaction and the electrical work ($\Delta G = -nF\,V_{oc}$) can be compensated by the heat energy of $T\,\Delta S$. The heat, Q_s, by entropy change, ΔS, is described by the following equation:

$$Q_s = T\Delta S\frac{1}{nF} \qquad (4.5.6a)$$

$$-\frac{\partial \Delta G}{\partial T} = nF\frac{\partial V_{oc}}{\partial T} \qquad (4.5.6b)$$

where T is battery temperature, I is charge/discharge current (defined as positive during charge cycle), F is Faraday constant, V_{oc} is cell voltage for open circuit, and n is the charge number participating to the reaction ($n = 1$ for Li-ion battery). The reaction directions for charge and discharge cycles are opposite to each other; thus, Q_s is endothermic during charge cycle and exothermic during discharge process.

When electric current flows through the cell, the cell voltage, V, deviates from V_{oc} due to overpotential or electrochemical polarization. The total overpotential can be deconvoluted into several parts: the charge transfer or activation overpotential, the ohmic overpotential, and the concentration overpotential. The energy loss by this polarization dissipates as irreversible heat that is exothermic during both charge and discharge cycles. In the following chapters, more discussion will be provided in this regard.

The term 'Li-ion' includes a number of different chemistries; Fig. 4.5.5 shows the chemistries under development in automobile industry. Each chemistry offers a different mix of cost, durability, performance, and safety as discussed in the following.

LCO, $LiCoO_2$, due to its high specific energy, is the most common type of Li-ion batteries in small consumer electronics. The battery consists of a cobalt oxide cathode and a graphite carbon anode. As it is shown in Fig. 4.5.5, the drawback of LCO is relatively short life span, low thermal stability, and limited load capabilities (specific power).

The three-dimensional spinel structure of the cathode in lithium manganese oxide (LMO), $LiMn_2O_4$, provides low internal resistance, high thermal stability, and enhanced safety. Nonetheless, it has limited cycle and calendar life. From Fig. 4.5.5, one can conclude that using LMO in the cathode can roughly offer a

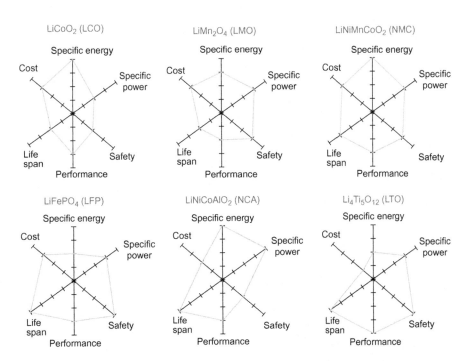

Fig. 4.5.5 Comparison of suitable Li-ions for EV in terms of specific energy, specific power, safety, performance, life span, and cost.

capacity that is one-third lower than LCO-based cathode, but the battery still holds 50% more energy than nickel-based chemistry.

Lithium nickel manganese cobalt oxide (NMC), $LiNiMnCoO_2$, is the most modern manganese-based Li-ion batteries with the cathode combination of nickel, manganese, and cobalt, which offers a unique blend that improves the specific energy, prolongs the life span, and lowers raw material cost due to reduced cobalt content. Fig. 4.5.5 shows that NMC has satisfactory overall performance and excels on specific energy that makes it the preferred candidate for the electric bikes and other electric powertrains.

Lithium iron phosphate (LFP), $LiFePO_4$, as the cathode material for rechargeable lithium batteries, offers acceptable electrochemical performance with low internal resistance. Li-phosphate battery is more tolerant to full charge conditions and is less stressed than other lithium-ion systems if kept at high voltage for a pronged time. It also has excellent safety and long life span but offers moderate specific energy, lower operating voltage, and higher self-discharge rate compared with other Li-ion-based batteries.

Lithium nickel cobalt aluminium oxide (NCA), $LiNiCoAlO_2$, is another type of Li-ion-based battery that shares similarity with NMC in terms of high specific energy and reasonable specific power and a long life span that makes it a candidate for electric vehicles applications. However, high cost and marginal safety remain problematic.

Lithium titanate (LTO), $Li_4Ti_5O_{12}$, that replaces the graphite in the above-mentioned Li-ion batteries, is considered as a promising anode material for applications that require high-rate capability and long cycle life. Due to low nominal cell voltage, they can be fast-charged and deliver a high discharge current. LTO-based batteries provide much better low-temperature performance compared with graphite-based batteries, which makes them excellent in safety. Nonetheless, as it is illustrated in Fig. 4.5.5, low energy density and high cost are the main drawbacks of LTO that needs to be improved.

As summarized in Table 4.5.1, these battery technologies are ranked in terms of key parameters (safety, power and energy density, and life) that are required in vehicular applications.

In general, the batteries can either be of high power density type or high energy density type. Power density provides a good measure on how much energy can be released due to discharge at a given time with regard to kilograms or litres. A high energy density battery is useful in applications where a longer driving distance is required, for example, in EVs and PHEVs, which is intended to be driven on pure electricity for long distances. A high power density battery is useful in an application

Table 4.5.1 **Key parameters of different Li-ion chemistries**

Parameter	Highest performing chemistry	Lowest performing chemistry
Safety	$Li_4Ti_5O_{12}$, $LiFePO_4$	$LiCoO_2$
Power	$LiFePO_4$	$LiCoO_2$
Energy	$LiCoO_2$, $LiNiCoO_2$, $LiNiCoAlO_2$	$LiFePO_4$
Life	$LiFePO_4$, $LiNiCoAlO_2$	$Li_4Ti_5O_{12}$

where a short but intensive power pulse is required, for example, in an ordinary HEVs as the electric motor often only assists the combustion engine in short periods.

4.6 Battery glossary

Cell, module, and packs: EVs and HEVs have a high-voltage battery pack that consists of individual modules and cells organized in series and parallel. A cell is the smallest, packaged form a battery can take and is generally 1–6 V. A module consists of several cells connected in either series or parallel. A battery pack is then assembled by connecting modules together, again either in series or parallel.

C-rates: The charge and discharge current of a battery is measured in C-rate. Most portable batteries are rated at 1C. This means that a 20 Ah battery would provide 20 A for 1 h if discharged at 1C rate. The same battery discharged at 0.5C would provide 10 A (20 × 0.5 = 10) for 2 h. 1C is often referred to as a 1 h discharge; a 0.5C would be 2 h and 0.1C a 10 h discharge.

State-of-charge (SOC): The state of charge refers to the amount of charge in a battery relative to its predefined full and empty states, that is, the amount of charge in amp-hours left in the battery. Manufacturers typically provide voltages that represent when the battery is empty (0% SOC) and full (100% SOC). SOC is generally calculated using current integration to determine the change in battery capacity over time.

Depth of discharge (DOD): It is a measure of how much energy has been withdrawn from a battery and is expressed as a percentage of full capacity. For example, a 100 Ah battery from which 40 Ah has been withdrawn has undergone a 40% DOD. DOD is the inverse of state of charge (SOC). A battery at 60% SOC is also at 40% DOD.

Open-circuit voltage (OCV): The OCV is the voltage when no current is flowing in or out of the battery, and hence, no reactions occur inside the battery. OCV is a function of state of charge and is expected to remain the same during the lifetime of the battery. However, other battery characteristics change with time; for example, capacity is gradually decreasing as a function of the number of charge-discharge cycles.

Terminal voltage (V): The voltage between the battery terminals with load applied. Terminal voltage varies with SOC and discharge/charge current.

Nominal voltage (V): The reported or reference voltage of the battery, also sometimes thought of as the normal voltage of the battery.

Cut-off voltage (V): The minimum allowable voltage. It is this voltage that generally defines the empty state of the battery.

Charge voltage (V): The voltage that the battery is charged to when charged to full capacity. Charging schemes generally consist of a constant-current charging until the battery voltage reaching the charge voltage and then constant-voltage charging, allowing the charge current to taper until it is very small.

Capacity or nominal capacity (Ah for a specific C-rate): The coulometric capacity, the total amp-hours available when battery is discharged at a certain discharge current (specified as a C-rate) from 100% SOC to the cut-off voltage. Capacity is calculated by multiplying the discharge current (in amps) by the discharge time (in hours) and decreases with increasing C-rate.

Specific energy (Wh/kg): The specific energy of a battery is expressed as a nominal energy per unit mass. It is highly dependent on the battery chemistry and packaging.

Energy density (Wh/L): The energy density of a battery is expressed as a nominal energy per unit volume. It is highly dependent on the battery chemistry and packaging.

Power density (W/L): The power density of a battery is expressed as a nominal power per unit volume. It is highly dependent on the battery chemistry and packaging.

Internal resistance: The internal resistance is sometimes considered as the ohmic resistance of the cell, which is the direct voltage change after the application of a current step on a cell in equilibrium. Another definition for the internal resistance is the sum of the ohmic, activation, and diffusion polarization resistances, which is the largest possible voltage drop in the cell. Nevertheless, the result in power dissipation in the form of heat will result due to the complete voltage drop. The voltage drop can be mainly divided as follows:

(1) IR drop is due to the current flowing across the internal resistance of the battery, by ohmic resistance.
(2) Activation polarization refers to the various retarding factors inherent to the kinetics of an electrochemical reaction, like the work function that ions must overcome at the junction between the electrodes and the electrolyte.
(3) Concentration polarization takes into account the resistance faced by the mass transfer (e.g., diffusion) process by which ions are transported across the electrolyte from one electrode to another.

The internal resistance of a battery is dependent on temperature, C-rate, and SOC. Different values for the internal resistance can be found depending on the measurement method. This is caused by the time constants associated with the activation and diffusion polarization resistances; whether the battery electrodes are in equilibrium or not is also important in determining the value of the internal resistance.

4.7   Battery charging methods

The safety, durability, and performance of batteries are highly dependent on how they are charged or discharged. Abuse of a battery can significantly reduce its life and can be dangerous. A current battery management system (BMS) includes both charging and discharging control on-board.

For EV batteries, there are the following common charging methods:

(1) Constant voltage. Constant voltage method charges battery at a constant voltage. This method is suitable for all kinds of batteries and probably the simplest charging scheme. The battery charging current varies along the charging process. The charging current can be large at the initial stage and gradually decreases to zero when the battery is fully charged. The drawback in this method is the requirement of very high power in the early stage of charging, which is not available for most residential and parking structures.
(2) Constant current. In this charging scheme, the charging voltage applied to the battery is controlled to maintain a constant current to the battery. The SOC will increase linearly versus time for a constant-current method. The challenge of this method is how to determine

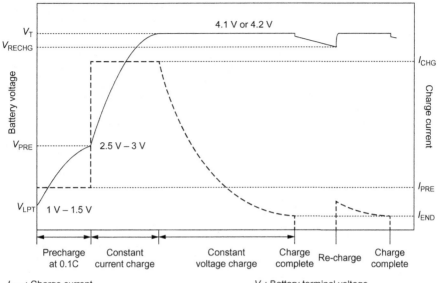

Fig. 4.7.1 Typical Li-ion cell charge profile.

I_{CHG}: Charge current.
  0.5–1C can be considered as fast charge.
I_{PRE}: Precharge current, e.g., 0.1C.
I_{END}: Ending charge current, e.g., 0.02C.

V_T: Battery terminal voltage.
V_{RECHG}: Threshold voltage to start recharge.
V_{PRE}: Voltage when precharge finished.
V_{LPT}: Low protection threshold voltage.

the completeness of a charge with SOC = 100%. The cut-off can be determined by the combination of temperature rise, temperature gradient rise, voltage increase, minus voltage change, and charging time.

(3) The combination of constant-voltage and constant-current methods. During the charging process of a battery, normally both the methods will be used. Fig. 4.7.1 shows a charging profile of a Li-ion cell. At the initial stage, the battery can be precharged at a low, constant current if the cell is not precharged before. Then, it is switched to charge the battery with constant current at a higher value. When the battery voltage (or SOC) reaches a certain threshold point, the charging is changed to constant-voltage charge. Constant-voltage charge can be used to maintain the battery voltage afterwards if the DC charging supply is still available.

For EVs, it is important for batteries to be able to handle random charging due to regenerative braking. As discussed in the previous section, the braking power of regenerative braking can be at the level of hundred kilowatts. Safety limitation has to be applied to guarantee the safe operation of batteries. Mechanical braking is usually used to aid regenerative braking in EVs as a supplementary and safe measure.

It is also critical to know when to stop charging a battery. It would be ideal if the battery SOC can be accurately gauged so that we can stop charging a battery when SOC reaches a preset value (e.g., 100%). As discussed later in the chapter, it has been a very challenging task to accurately estimate SOC. Even if the SOC of a battery can be exactly identified, it is also needed to have some other backup methods to stop charging. The following are some typical methods currently used to stop a charging process:

(1) Timer. It is the most typical stopping method, which can be used for any types of battery. When a preset timer expires, the charging process is stopped.

(2) Temperature cut-off (TCO). The charging will be stopped if the absolute temperature of battery rises to a threshold value.

(3) Delta temperature cut-off (DTCO). When the delta change in battery temperature exceeds the safety value, the charging will be terminated.

(4) Temperature change rate dT/dt. If the temperature change rate is over the safety threshold value, the charging process will be terminated.

(5) Minimum current (I_{min}). When the charging current reaches the lowest limit I_{min}, the charging process stops. This method is normally incorporated with a constant-voltage charging scheme.

(6) Voltage limit. When the battery voltage reaches a threshold value, the charging process will be terminated. This method normally goes together with a constant-current charging method.

(7) Voltage change rate, dV/dt. The charging process stops if the battery voltage does not change versus time or even if it starts to drop (a negative value of dV/dt).

(8) Voltage drop ($-\Delta V$). In Ni-MH battery, upon the completion of the charge process (SOC = 100%), the temperature of the cell starts to increase due to the recombination of hydrogen and hydroxide ions and causes the cell voltage to drop. The charging will be terminated if a preset value of the voltage drop is reached.

4.8   Battery management system

Today's electronic devices have higher mobility and are greener than ever before. Battery advancements are fuelling this progression in a wide range of products from portable power tools to PHEVs and wireless speakers. In recent years, the efficiency of a battery in terms of how much power it can output with respect to size and weight has dramatically improved. Think about how heavy and bulky a car battery is. Its main purpose is to start the car. With recent advancements, we can purchase a lithium-ion battery to jump start a car, and it only weighs a couple pounds and is the size of hand.

The ongoing transformation of battery technology has prompted many newcomers to learn about designing battery management systems. This section provides the BMS architecture, discusses the major functional blocks, and explains the importance of each block to the battery management system.

Battery management systems can be architected using a variety of functional blocks and design techniques. Careful consideration of battery requirements and battery life goals will guide you in determining the right architecture, functional blocks, and related ICs to create your battery management system and charging scheme to optimize battery life.

4.8.1   Building blocks of battery management system

A battery management system can be composed of many functional blocks including cut-off FETs, fuel-gauge monitor, cell voltage monitor, cell voltage balance, real-time clock (RTC), temperature monitors, and state machine. There are many types of

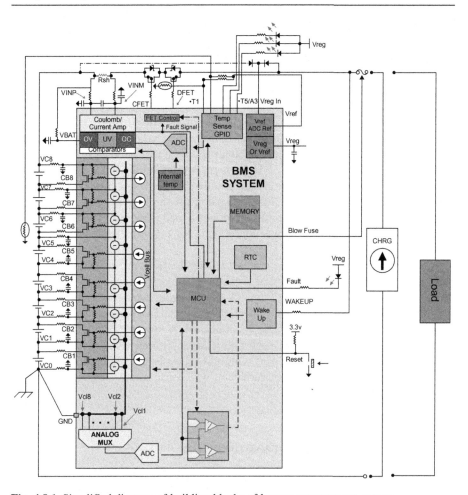

Fig. 4.8.1 Simplified diagram of building blocks of battery management system.

battery management ICs available. The grouping of the functional blocks varies widely from a simple analog front end that offers balancing and monitoring and requires a microcontroller (MCU), to a stand-alone, highly integrated solution that runs autonomously. Now, let us take a look at the purpose and the technology behind each block and the pros and cons of the technology. Fig. 4.8.1 shows a simplified diagram of the building blocks of a battery management system.

4.8.2 Cutoff FETs and FET driver

A FET driver functional block is responsible for the connection and isolation of the battery pack between the load and charger. The behaviour of the FET driver is predicated on measurements from battery cell voltages, current measurements, and real-time detection circuitry. Fig. 4.8.2A and B illustrates two different types of FET connections between the load and charger and the battery pack:

Fig. 4.8.2A requires the least amount of connections to the battery pack and limits the battery pack operating modes to charge, discharge, or sleep. The current flow direction and the behaviour of a specific real-time test determine device's state. For example, Intersil's ISL94203 stand-alone battery pack monitor has a CHMON input that monitors the voltage on the right side of the cut-off FETs. If a charger is connected and the battery pack is isolated from the charger, the current injected towards the battery pack will cause the voltage to rise to the charger's maximum supply voltage. The voltage level at CHMON is tripped letting the BMS device know a charger is present. A load connection is determined by injecting a current into the load to determine if a load is present. If the voltage at the pin does not rise significantly when current is injected, the outcome determines a load is present. The FET driver's DFET is then turned on. The connection scheme for Fig. 4.8.2B allows the battery pack to operate while charging.

FET drivers can be designed to connect to the high side or low side of a battery pack. A high-side connection requires a charge pump driver to activate the NMOS-FETs. Using a high-side driver allows for a solid ground reference for the rest of the circuitry. Low-side FET driver connections are found in some integrated solutions to reduce cost because a charge pump is not needed. A low-side connection does not require high-voltage devices, which consume a larger die area. Using the cut-off FETs on the low side floats the battery pack's ground connection, making it more susceptible to noise injected into the measurement, which can affect the performance of some ICs.

4.8.3   Fuel gauge/current measurements

The fuel-gauge functional block keeps track of the charge entering and exiting the battery pack. Charge is the product of current and time. There are several different techniques that can be used when designing a fuel gauge. A current sense amplifier and an MCU with an integrated low-resolution ADC is one method of measuring the current. The current sense amplifier operates in high common-mode environments and amplifies the signal, enabling higher-resolution measurements. This design technique sacrifices dynamic range. Other techniques are to use a high-resolution ADC or to purchase a costly fuel-gauge IC. Understanding the behaviour of the load in terms of current consumption versus time determines the best type of fuel-gauge design.

The most accurate and cost-efficient solution is to measure the voltage across a sense resistor using a 16-bit or higher ADC with low offset and high common-mode rating. A high-resolution ADC offers a large dynamic range at the expense of speed. If the battery is connected to an erratic load such as an electric vehicle, the slow ADC may miss high-magnitude and high-frequency current spikes that are delivered to the load. For erratic loads, a SAR ADC with perhaps a current sense amplifier front end may be more desirable. Any offset error results in an overall error in the amount of charge in the battery. Measurement errors over time will

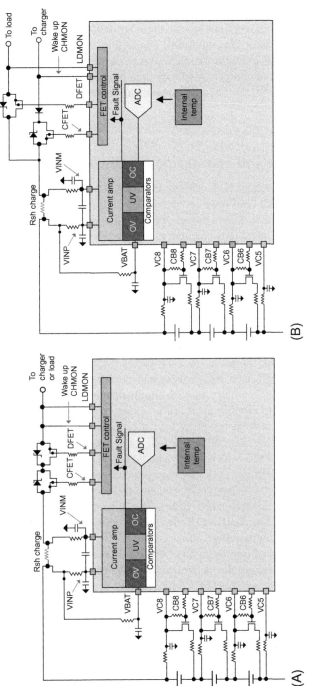

Fig. 4.8.2 Cut-off FET schematic illustrations. (A) Single connection for load and charger. (B) Two-terminal connection that allows for charging and discharging simultaneously.

cause significant battery pack charge status errors. A measurement offset of 50 μV or less with 16-bit resolution is adequate in measuring charge.

With most current measurement blocks, there are analog comparators monitoring for short-circuit and overcurrent conditions. The analog comparator signal is directly connected to FET drivers to minimize latency between the event and isolating the battery pack from the load or charger. A latency time of several tens of microseconds is adequate for most applications, and in most applications, the faster the time to disconnect the battery, the better.

4.8.4 Cell voltage and maximizing battery lifetime

Monitoring the cell voltage of each cell within a battery pack is essential in determining its overall health. All cells have an operating voltage window that charging and discharging should occur to ensure proper operation and battery life. If an application is using a battery with a lithium chemistry, the operating voltage typically ranges between 2.5 and 4.2 V. The voltage range is chemistry-dependent. Operating the battery outside the voltage range significantly reduces the lifetime of the cell and can render the cell useless.

Cells are connected in series and parallel to form a battery pack. A parallel connection increases the current drive of the battery pack, while a series connection increases the overall voltage. Cell voltages are like everything that is manufactured. A cell's performance has a distribution: at time equal zero, the cells charge and discharge rates within a battery pack are the same. As each cell is cycled between charge and discharge, the rate at which each cell charges and discharges changes, resulting in a spread distribution across a battery pack. A simplistic means of determining if a battery pack is charged is to monitor each cell's voltage to a set voltage level. The first cell voltage to reach the voltage limit trips the battery pack charged limit. If the battery pack had a weaker than average cell, this would result in the weakest cell reaching the limit first and the rest of the cells not fully charged. A charging scheme as described does not maximize the battery pack ON time per charge. The charging scheme also reduces the lifetime of the battery pack because more charge and discharge cycles are needed. A weaker cell discharges faster. The same type of occurrence happens on the discharge cycle. The weaker cell trips the discharge limit first, leaving the rest of the cells with charge remaining.

There are two means of improving the ON time of a battery pack per charge. The first one is slowing the charge the weakest cell receives during the charge cycle. This is achieved by connecting a bypass FET with a current-limiting resistor across the cell (see Fig. 4.8.3A). This takes the current from the cell with the highest current resulting in a slowing of charge to the cell, allowing the other cells in the battery pack to catch up. The ultimate goal is to maximize the battery pack's charge capacity, which is achieved by having all the cells reach the fully charged limit simultaneously.

The battery pack can be balanced on the discharge cycle by implementing a charge-displacement scheme. A charge-displacement scheme is achieved by taking charge via inductive coupling or capacitive storage from the alpha cell and injecting the

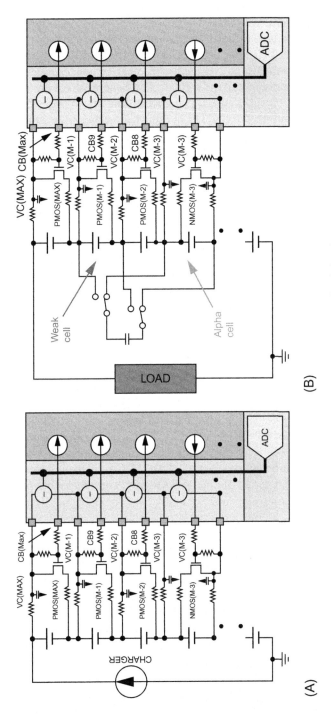

Fig. 4.8.3 Different types of cell balancing. (A) Bypass cell-balancing FETs. (B) Active balancing.

stored charge into the weakest cell. This slows the time it takes the weakest cell to reach the discharge limit. This is known as active balancing (see Fig. 4.8.3B):

Battery packs with one to four batteries in parallel and three or more in series benefit the most from balancing. As the parallel combinations increase per cell, the weak cell's performance is averaged with other cells in parallel. The performance distribution between cells is tighter. The benefit of having more cells in parallel is also a detriment because it is harder to find the weaker cell in a battery pack. A battery pack sitting idle could be burning charge due to the strong cells propping up the weaker cell.

The cell voltage and balancing circuitry receive the harshest treatment from hot-plug events. There is not an OFF button on a battery. Connecting the circuitry to a battery, load, or charger can result in large transients occurring at the inputs of the device. A designer should be aware of the maximum rating of sensitive pins. The maximum voltage rating of a pin is a key specification to determining the likelihood that a transient event will damage the circuitry. The rule of thumb is the higher the voltage rating of a pin, the more robust the part will be in suppressing transients.

An IC manufacturer designing with a high-voltage process ensures that the device is protected from transient events at the expense of design with large geometries. This raises the cost of the device. Other IC manufactures will design with a low-voltage process and stack the devices such that a device never exceeds the process rating. This approach relies upon circuitry such as capacitors, resistors, and diodes to suppress the transient before it reaches the pin. Both manufacturing types require the use of diodes, resistors, and capacitors to dampen transients. Using a high-voltage-rated IC adds further protection against harmful and extraneous signals. Both design approaches will work, but the lower-voltage-rated device may require more tweaking in the development stage to ensure protection against harmful events.

The acquisition time of a voltage cell measurement is dependent on the load behaviour and the number of cells to scan. Erratic behaving loads require fast scan times to monitor a cell's out-of-bound condition. A SAR ADC is often used to achieve quick measurements in a short period of time. A SAR ADC consumes more power and has less resolution.

4.8.5   Temperature monitoring

Today's batteries deliver a lot of current while maintaining a constant voltage, which can lead to a runaway condition that causes the battery to catch fire. The chemicals used to construct a battery are highly volatile, and a battery impaled with the right object can result in the battery catching fire. Not only temperature measurements are used for safety conditions, but also they can be used to determine if it is desirable to charge or discharge a battery.

Temperature sensors monitor each cell for energy storage system (ESS) applications or a grouping of cells for smaller and more portable applications. Thermistors powered by an internal ADC voltage reference are commonly used to monitor each circuit's temperature. The internal voltage reference is used to reduce inaccuracies of the temperature reading versus environmental temperature changes.

4.8.6 State machines or algorithms

Most battery management systems require an MCU or an FPGA to manage information from the sensing circuitry and to make decisions with the received information. In a select few offerings, such as Intersil's ISL94203, the algorithm is encoded, with some programmability, digitally enabling a stand-alone solution with one chip. Stand-alone solutions are also valuable when mated to an MCU because the state machine within the stand-alone can be used to free up MCU clock cycles and memory space.

4.8.7 Other battery management system building blocks

Other BMS functional blocks include battery authentication, a RTC, memory, and daisy chain. The RTC and memory are used for black-box applications where the RTC is used for a time stamp and memory is used for storing data, allowing the user know the battery pack's behaviour prior to a catastrophic event. The battery authentication block prevents the BMS electronics from being connected to a third-party battery pack. The voltage reference/regulator is used to power peripheral circuitry around the BMS system. Finally, daisy-chain circuitry is used to simplify the connection between stacked devices. The daisy-chain block replaces the need for optical couplers or other level-shifting circuitry.

4.9 Battery state of charge estimation

Battery SOC estimation is key component for battery management system. It helps in describing the actual energy level available at the battery. SOC assessment is significant not only for knowing the energy availability of the battery but also for finding the battery lifetime. There are many methods for determining the battery SOC and most of them are based on electrochemical characteristics of the battery and on real-time loading conditions. Real-time loading conditions can be incorporated with battery electrochemical characteristics for the estimation of the battery SOC. The SOC estimation technique should not depend on the battery initial conditions for the avoiding battery relaxing time in real-time loading conditions. Battery cell current, cell voltages, and cell temperature are critical parameters for the estimation of battery SOC. Battery hysteresis behaviour under different real-time loading conditions should be considered under different SOC. In literature, many methods have been reported for the estimation of battery SOC, and most of them are using adaptive extended Kalman filter technique. Battery SOC estimation based on Kalman filter technique needs accurate electrochemical battery models for more accuracy, but it involves complex computation. Therefore, real-time loading conditions can be incorporated with battery electrochemical characteristics for more accurate estimation of the battery SOC and to avoid more complex computation.

4.10   Conclusions

In this chapter, brief overview of the BEV, PHEV, and HEV has been presented. Electrochemical characteristics of the Ni-MH battery, Li-ion battery, and advanced rechargeable battery are discussed. These electrochemical characteristics are useful for determining the battery SOC under dynamic loading conditions of the electric vehicle. Battery chemistry is explained in a detailed manner including an abbreviated modelling approach. Also, the issues of battery-charging method, management and monitoring, and SOC estimation are addressed. The chapter concludes with a discussion on battery cell voltage balancing and temperature monitoring for estimation of battery SOC.

References

Baba, N., Yoshida, H., Nagaoka, M., Okuda, C., & Kawauchi, S. (2014). Numerical simulation of thermal behavior of lithium-ion secondary batteries using the enhanced single particle model. *Journal of Power Sources, 252*, 214–228.

Badwal, S., Giddey, S. S., Munnings, C., Bhatt, A. I., & Hollenkamp, A. F. (2014). Emerging electrochemical energy conversion and storage technologies. *Frontiers in Chemistry, 79(2)*. Available at: http://journal.frontiersin.org/article/10.3389/fchem.2014.00079/full. Accessed 04.10.16.

Baker, D. R., & Verbrugge, M. W. (1999). Temperature and current distribution in thin-film batteries. *Journal of the Electrochemical Society, 146(7)*, 2413–2424.

Bandhauer, T. M., Garimella, S., & Fuller, T. F. (2014). Temperature-dependent electrochemical heat generation in a commercial lithium-ion battery. *Journal of Power Sources, 247*, 618–628.

Cai, L., Dai, Y., Nicholson, M., White, R. E., Jagannathan, K., & Bhatia, G. (2013). Life modeling of a lithium ion cell with a spinel-based cathode. *Journal of Power Sources, 221*, 191–200.

Canadian Automobile Association (2016). *Types of electric vehicles*. Canadian Automobile Association Website. Available at: http://electricvehicles.caa.ca/types-of-electric-vehicles/. Accessed 04.10.16.

Carbone, R. (2011). Electrochemical energy storage. In P. L. Antonucci & V. Antonucci (Eds.), *Energy storage in the emerging era of smart grids* (1st ed., pp. 3–20). Rijeka: IN TECH.

Cho, S., Jeong, H., Han, C., Jin, S., Lim, J. H., & Oh, J. (2012). State-of-charge estimation for lithium-ion batteries under various operating conditions using an equivalent circuit model. *Computers and Chemical Engineering, 41*, 1–9.

Dai, H., Wei, X., Sun, Z., Wang, J., & Gu, W. (2012). Online cell SOC estimation of Li-ion battery packs using a dual time-scale Kalman filtering for EV applications. *Applied Energy, 95*, 227–237.

Dao, T. S., Vyasarayani, C. P., & McPhee, J. (2012). Simplification and order reduction of lithium-ion battery model based on porous-electrode theory. *Journal of Power Sources, 198*, 329–337.

Dhameja, S. (2001). Electric vehicle battery charging. In S. Dhameja (Ed.), *Electric vehicle battery systems* (1st ed., pp. 69–94). Boston: Elsevier.

Doyle, M., Fuller, T. F., & Newman, J. (1993). Modeling of Galvanostatic charge and discharge of the lithium/polymer/insertion cell. *Journal of the Electrochemical Society, 140(6)*, 1526–1533.

Garcia-Valle, R., & Peças Lopes, J. A. (2013). Electric vehicle battery technologies. In K. Young, C. Wang, L. Y. Wang, & K. Strunz (Eds.), *Electric vehicle integration into modern power networks* (1st ed., pp. 15–56). New York: Springer.

Hallaj, S. A., Maleki, H., Hong, J. S., & Selman, J. R. (1999). Thermal modeling and design considerations of lithium-ion batteries. *Journal of Power Sources, 83*(1–2), 1–8.

Hallaj, S. A., & Selman, J. R. (2002). Thermal modeling of secondary lithium batteries for electric vehicle/hybrid electric vehicle applications. *Journal of Power Sources, 110*(2), 341–348.

HybridCars. (2016). New Toyota Prius plug-in hybrid gets solar panel roof, but not in U.S-yet. HybridCars Website. Available at: http://www.hybridcars.com/new-toyota-prius-plug-in-hybrid-gets-solar-panel-roof-but-not-in-u-s-yet/ Accessed 04.10.16.

Intersil Corporation (2015). *Battery management system tutorial*. Intersil Corporation Website. Available at: http://www.intersil.com/content/intersil/en/products/power-management/battery-management-system-tutorial.html. Accessed 04.10.16.

Milligan, R. (2016). *Critical evaluation of the battery electric vehicle for sustainable mobility.* (Ph.D. thesis). Edinburgh Napier University.

National Academy of Engineering. (2014). *Creators of lithium-ion battery and pioneers of innovative education model win engineering's highest honors.* National Academy of Engineering Website. Available at: https://www.nae.edu/Projects/MediaRoom/20095/107830/106261.aspx. Accessed 04.10.16.

Scrosati, B. (2000). Recent advances in lithium ion battery materials. *Electrochimica Acta, 45*(15–16), 2461–2466.

Yazdanpour, M. (2015). *Electro-thermal modeling of lithium-ion batteries.* (Ph.D. thesis) Simon Fraser University.

Next-generation battery-driven light rail vehicles and trains

Koki Ogura
Kawasaki Heavy Industries, Ltd., Kobe, Japan

5.1 Introduction

Electric trains are more energy-efficient and generate fewer CO_2 emissions than other types of transportation systems. Next-generation light rail vehicles (LRVs) have been gaining momentum around the world because of great advancement of their low-floor design and low noise levels as well as passenger- and earth-friendly features.

Kawasaki's SWIMO[1] is an LRV powered by the GIGACELL, Kawasaki's proprietary nickel-metal hydride (Ni-MH) battery, which can operate without overhead power lines. The SWIMO vehicle shown in Fig. 5.1.1 employs a three-car body, three-bogie articulated design to enable smooth curving and flexibility in car combinations.

Fig. 5.1.1 Next-generation low-floor battery-driven LRV 'SWIMO'.

---

[1] SWIMO stands for 'Smooth *Win Mo*ver' because the goal was to realize (WIN) a vehicle (MOVER) with smooth (SMOOTH) boarding and exiting and smooth entry into catenary-free sections. It is the realization of Kawasaki's vision for a vehicle that would provide a smooth riding experience with a seamless transition to catenary-free sections and a win-win green transportation solution via Kawasaki's innovative rail mover technology.

Electric Vehicles: Prospects and Challenges. http://dx.doi.org/10.1016/B978-0-12-803021-9.00005-7

In this chapter, the experimental battery-driven SWIMO vehicle is presented, which includes an impressive technology, concept, and test results. In addition, the examples of battery-driven train and LRV with commercial operations in Japan and France are presented from an innovative point of view.

5.2   Development of a LRV as a means of urban transportation

In 2005, Kawasaki began the development of SWIMO, which is next-generation, battery-driven, low-floor LRV. In November 2007, Kawasaki completed SWIMO verification test at Kawasaki's Harima test track, Japan, and between December 2007 and March 2008, this vehicle successfully completed test runs on actual tracks during winter on lines belonging to the Sapporo City Transportation Bureau, Japan. The development concept for SWIMO has been to make a vehicle that is 'good for people and the environment' with the consideration of the aging of society and environmental concerns.

However, for LRVs to become more commonplace, they must be able to complement the transportation offered by automobiles. For example, if the SWIMO vehicles are operated in areas where there are traffic jams on the roads during rush hour, people will switch to mass transit for the sake of convenience. This will take more cars off the road, essentially increasing the savings in CO_2 emissions attained by SWIMO.

Kawasaki's large, high-performance, Ni-MH battery GIGACELL as shown in Fig. 5.2.1 made a significant contribution towards the development of SWIMO. These batteries were installed in SWIMO vehicles so that they can operate without overhead

Fig. 5.2.1 High-capacity and high-performance nickel-metal hydride battery 'GIGACELL'.

power lines, in other words catenary-free operation by battery power. Detailed explanation about GIGACELL is given in Section 5.4.

5.3 Benefits of catenary-free operation

Vehicles have historically sourced power from overhead power lines. However, over the past decade, catenary-free (nonelectrified) vehicle operations are fast gaining prominence. Catenary-free refers to the removal of the overhead power line equipments from the vehicle system.

Operating vehicles without a catenary, for example, with a power source on board the vehicle by charging the power storage device periodically, can provide various benefits compared with the conventional systems as set out below:

- Reductions in the level of visual intrusion.
- Reduction in the cost of overhead infrastructure.
- Reduction in power usage and CO_2.

The benefits are not common across all catenary-free operating technologies.

Visual intrusion is often an issue for vehicle scheme promoters, particularly where putting a system through a city centre or urban area that does not have a history of vehicle operation. In these instances, there may be very significant opposition to these schemes on the basis of the impact on architectural or cultural landmarks.

The costs associated with designing and getting approval for an overhead system may be very high, and as such, in some recent schemes, the overhead has been dispensed with through the use of catenary-free technology, notably at Seville in Spain where the vehicle system operates through the historic city centre using an on-board super capacitor storage system. In constructing a vehicle system, a major element of cost is the construction and lifetime maintenance of the overhead power line system. Utilizing a catenary-free technology will remove the need for this element of the cost, although the overhead power line equipment is replaced with a different technology. At this stage, in the development of catenary-free operation, it is hard to see the costs of the catenary-free operation being outweighed by the reductions in overhead power line infrastructure costs alone; however, it appears likely that the costs of catenary-free technology will fall in the future and when combined with other potential benefits a positive case may exist. Also, historical bridges have been rebuilt or highways lowered in order to allow vehicles to pass under structures. There are potentially very significant savings in infrastructure modification costs if the vehicle can operate with a reduced height requirement as a result of removing the need for a permanently raised pantograph and overhead power line equipment.

Catenary-free technology, such as on-board battery or super capacitor systems, allows power savings of up to 20% by using energy from regenerative braking. When the vehicle brakes, a vast majority of energy can be returned to the energy storage system rather than burned off in on-board resistors, resulting in energy savings. In the long run, these savings translate into significant power consumption and operating cost reduction for the operators.

5.4   GIGACELL battery

Fig. 5.4.1 shows the structure of GIGACELL battery. The GIGACELL battery is composed of individual battery cells that are connected in series by their cell walls, with the front and rear surfaces becoming positive and negative electrodes, forming the bipolar structure. The thin cell walls provide a large cross-sectional area that minimizes internal resistance and power loss, which occurs when the cells are connected.

Inside each cell, preformed strips of positive and negative electrodes are inserted into the two sides of a pleat-folded separator. Increasing the three-dimensional elements of the preformed strips (height, width, and quantity) can easily expand the capacity of the battery. Therefore, the large capacities of modules can be accomplished by increasing the number of cells connected in the bipolar structure. Also, a heat sink is placed between cells and cooled forcibly by cooling fans to suppress temperature increases.

GIGACELL battery has some of the key advantages as follows:

(1) High scalability: The bipolar 3D design allows for the increase of both the number and capacity of cells in each module. Batteries of extremely large capacities can be made.
(2) Rapid charge and discharge: Low internal resistance enables rapid charge and discharge.
(3) Excellent cycle durability: Designed to withstand frequent cycles of short, rapid charge and discharge.
(4) Environmentally friendly: No lead, mercury, cadmium, or other toxic materials are used.

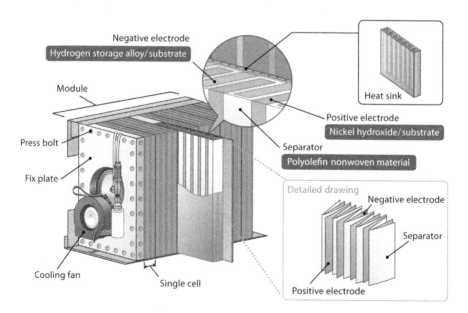

Fig. 5.4.1 Structure of GIGACELL battery.

(5) Simple and safe: Temperature remains relatively low during operation, so there are no restrictions for the installation of the battery. Water-based electrolyte is used, eliminating the risk of fire.

(6) Easy to recycle: No welding is used on the cell cases, so they can be easily disassembled for recycling.

The effect of discharge rate on voltage profile for GIGACELL battery is given in Fig. 5.4.2 (0.2 C discharge, 30 A; 0.5 C discharge, 75 A; 1.0 C discharge, 150 A; 1.5 C discharge, 225 A; 2.0 C discharge, 300 A; 2.5 C discharge, 375 A). There is no significant effect on the shape of the discharge voltage curves for every discharge rates and retain the flat except for the beginning and ending transients.

The specifications of GIGACELL battery module for the SWIMO vehicle are listed in Table 5.4.1.

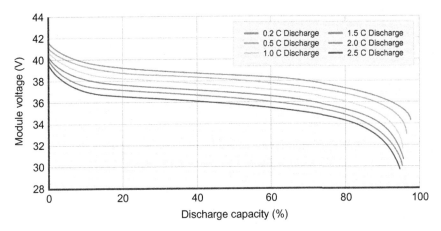

Fig. 5.4.2 Example of GIGACELL discharge characteristics (battery capacity, 150 Ah; ambient temperature, 20°C; 0.2 C × 100% charge).

Table 5.4.1 **Specifications of GIGACELL battery module for SWIMO**

Item	Value
Total number of cell	30
Rated voltage	36 V (1.2 V × 30)
Rated discharge capacity	200 Ah
Energy capacity	7.2 kWh
Outline dimensions	1287 × 218 × 308 mm
Weight	235 kg
Energy density	31 Wh/kg

5.5   Determining on-board battery capacity

Equipping vehicles with many batteries would be effective in terms of the efficient use of regenerative energy and increased degree of freedom for the system. However, when considering factors such as the cost, weight, and space required for those, it is important to find a way to most effectively use of a limited amount of batteries.

Railways are characterized by having predetermined basic operating and running patterns, so basic battery capacity is set by means such as simulations. In gradient sections, securing hill-climbing ability when going up and absorbing regenerative energy when going down must be considered. Rolling stock power consumption is the total of drive load and auxiliary power unit load. Also, decrease in capacity due to deterioration of batteries over time and range used of state of charge (SOC) are also taken into account. The battery-driven SWIMO vehicle is designed to use 16 GIGACELL battery modules in series with a nominal battery voltage of 576 V DC and a total battery capacity of 115.2 kWh. This allows a distance of over 10 km catenary-free operation.

5.6   Overview of SWIMO vehicle

The structure of SWIMO vehicle accommodates a train set length of 15–30 m with three to five car bodies at a width of 2.23–2.5 m on a gauge of 1067–1435 mm. The development of variations intended for vehicle operators both in Japan and overseas was planned.

The experimental SWIMO vehicle is a 15 m type with three car bodies on three bogies (trucks) in an articulated bogie configuration and was the smallest class vehicle among the planned variations. In addition, the vehicle was designed for a minimum curve radius of 14 m and is capable of running on tight curves unique to vehicle routes. Fig. 5.6.1 shows the dimensions of SWIMO, and its main specifications are listed in Table 5.6.1.

Propulsion and braking control units, auxiliary power supplies, air-conditioning units, and other equipment, which are mounted under the floor in ordinary trains, are all mounted on the rooftop to allow passenger compartment floors that are free of steps. The GIGACELL batteries are mounted under the seats in the passenger compartment to use the limited amount space effectively.

The end bogie is a specially constructed direct-mount driving bogie provided with the drive axle at the front position and small-diameter wheels at the rear. The intermediate bogie is a bolsterless bogie with independently rotating wheels without axles. This construction reduces restrictions on floor height, which makes it possible to realize a flat floor 360 mm above the top of rail in most areas of the passenger compartment. The floor plane near the door has been lowered to 330 mm above the top of rail to make it is easier for passengers to board and exit.

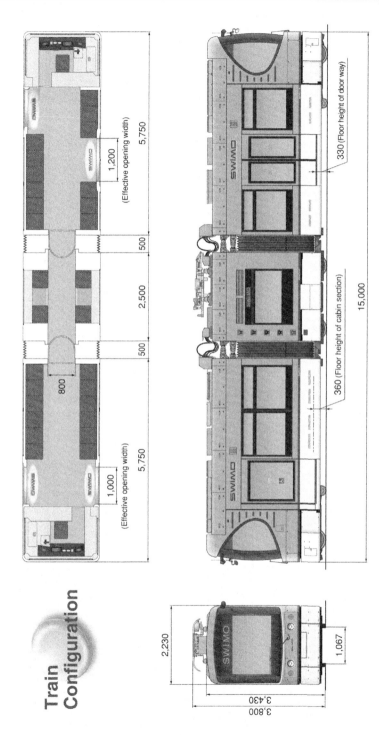

Fig. 5.6.1 SWIMO vehicle dimensions (unit, mm).

Table 5.6.1 SWIMO main specifications

Type of vehicle		Low-floor, battery-driven LRV
Vehicle structure		Three-car body, three-bogie articulated
Operation pattern		One-man operation, rear entrance, and front exit
Power source		600 V DC
Car body dimensions	Length	15,000 mm
	Width	2230 mm
	Height	3800 mm
	Floor height	330 mm (door way)/360 mm (cabin section)
Minimum isle width		800 mm
Tare weight		30 ton
Propulsion battery		Large-capacity nickel-metal hydride battery of not less than 200 Ah capacity
Running gear (low-floor bogies)		Leading bogie, two-axle direct mount; middle bogie, lateral independent wheel set, bolsterless; gauge, 1067 mm
Vehicle performance		Maximum operating speed: 40 km/h (maximum design speed: 50 km/h) Acceleration, 2.5 km/h/s (normal)/3.5 km/h/s (high acceleration) Deceleration, 2.5 km/h/s (max. regenerative deceleration)/3.5 km/h/s (max. normal deceleration)/5.0 km/h/s (max. emergency deceleration)
Passenger capacity (passenger seating capacity)		62 (28)
Charge–discharge control unit		IGBT bidirectional buck-boost converter, 250 kW
Brake equipment		Electric command air brake equipment in combination with regenerative brake
Traction motor		3-phase 50 kW induction motor × 2 units
Propulsion unit		IGBT-VVVF inverter (1C1M) × 2 units
Auxiliary power supply		IGBT-CVCF inverter (18.5 kVA) × 2 units
Current collector		Single-arm type
Air-conditioning unit		rooftop mount 11.63 kW (10,000 kcal/h) × 2
Door system		Electrical plug door
Car body structure		Steel

5.7   Configuration of SWIMO vehicle

The configuration of battery-driven SWIMO vehicle is provided in Fig. 5.7.1. Detailed explanations of each part are presented in the following.

5.7.1   Car structure

The three-car body, three-bogie articulated vehicle (15–18 m class) shown in Fig. 5.7.2A has bogies at both ends and in the middle, allowing for smooth operation even on narrow curves. The vehicle layout can be changed to any configuration.

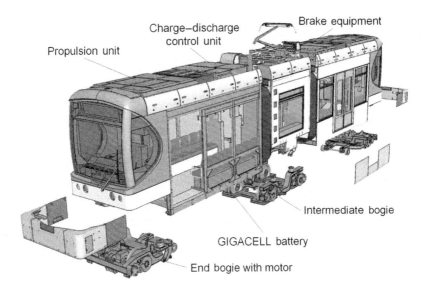

Fig. 5.7.1 Configuration of SWIMO.

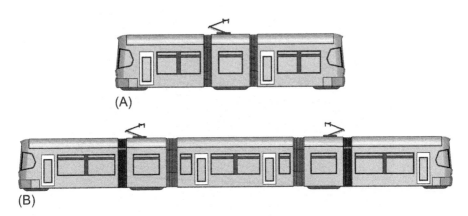

Fig. 5.7.2 Example of body articulated unit: (A) three-car body type (15–18 m class) and (B) five-car body type (20–30 m class).

A five-car body articulated unit (20–30 m class) can be arranged with two bogies at both ends and two bogies in the middle, as provided in Fig. 5.7.2B. The experimental SWIMO vehicle is a 15 m with three-car bodies on three bogies:

5.7.2   Passenger cabin

The floor height in the cabin sections is only 360 mm. The passenger cabins at both ends have completely flat floors, providing maximum flexibility in seating arrangements as shown in Fig. 5.7.3. This was achieved by placing newly developed, extremely compact bogies at both ends and in the middle of the SWIMO vehicle.

5.7.3   Door area

The car floors are only 330 mm off the ground at the door openings. SWIMO's 1200 mm wide sliding double door shown in Fig. 5.7.4 is the widest LRV door in the world.

5.7.4   Operator's cabin

The operator's cabin area is designed to best suite one-man operation with satisfactory operating view and functional equipment layout. Fig. 5.7.5 depicts the main part of the operator's cabin. The video displays at both ends contain images from the cameras at each of the door locations. The display touch screen in the middle can see the battery conditions such as charge or discharge, SOC, voltage, temperature, and error records from battery monitoring system. This screen can also control the air-conditioning, heating, and lighting equipment for the passenger's cabin. The right side of the operator's cabin shows the control handle for accelerating or braking the vehicle.

Fig. 5.7.3 SWIMO passenger cabin.

Fig. 5.7.4 SWIMO door area.

Fig. 5.7.5 SWIMO operator's cabin.

5.7.5 Bogies at both ends

Kawasaki developed compact end bogies designed to fit under the operator's cabin at both ends of the SWIMO in order to make the cabin floor flat. The end bogie shown in Fig. 5.7.6 is developed based on single-axle concept, and the leading axle contains wheels to assist stable operation upon running straight. It also contains a compensating wheel on its second axle to ensure smooth curving. In terms of configuration, the bolster beam type of two-axle truck (with axle) is applied, and the car body is suspended by the direct connection system of the bolster beam and car body. The vehicle adopts a parallel cardan driving device with a three-phase 50 kW induction motor mounted to both end axles by flexible gear couplings (WN couplings). In addition, the flange lubrication system is installed to enhance curving performance and reducing abrasion of parts. A sanding device is also applied in order to prevent wheel slip and skid.

While this low-floor configuration would normally result in protruding wheels (conventional LRV wheels are generally 600 mm in diameter), the SWIMO circumvents this by reducing the size of the wheels on the second axle (the one farther from the operator's cabin) to a diameter of 250 mm, as shown in Fig. 5.7.6. The smaller diameter provides enough room for the flat-floor design and installation of doors over the wheels. The larger wheels have been placed under the operator's cabin along with the motor. While rail vehicles with wheels 360 mm in diameter or smaller have a history of derailing when switching tracks, Kawasaki has come up with a breakthrough development to keep SWIMO running safely on track. Extensive tests have been conducted on its state-of-the-art smaller wheeled bogie to ensure maximum safety.

5.7.6 Intermediate bogie

The intermediate bogie shown in Fig. 5.7.7 is designed as a trailing bogie, independent four-wheel system with no traction motor equipped, thus minimizing restriction on the floor surface of cabin.

Fig. 5.7.6 SWIMO end bogie.

Fig. 5.7.7 SWIMO intermediate bogie.

5.7.7 On-board GIGACELL battery

The GIGACELL battery installed under the seats in Fig. 5.7.8 has been downsized for railway applications and upgraded for larger output. The GIGACELL battery is an extremely safe Ni-MH battery tailored to large-scale applications. It employs a unique structure developed by Kawasaki that prevents the battery from generating excessive heat or igniting, even after rapid charging and recharging a large amount of power.

Fig. 5.7.8 GIGACELL batteries under the seat.

The GIGACELL battery can be charged in 5 min before the SWIMO vehicle has travelled 10 km. That is the same amount of time it takes to turn the LRV around at a terminal.

5.7.8   Roof

Mounting major electric devices on the roof makes SWIMO's low-floor design possible.

5.7.9   Propulsion unit

Roof-mounted propulsion unit (VVVF[2] inverter) is a control device that convert the vehicle's power source to a suitable type of power to drive the traction motors. This inverter converts the incoming DC to AC power and controls the amount of power (voltage and frequency) being supplied in accordance with the vehicle's speed. In addition, this inverter is also capable of regenerating power from the motors when the vehicle decelerates. In this way, this inverter helps the vehicle to smoothly accelerate and brake.

5.7.10   Charge–discharge control unit

The charge–discharge control unit embedded bidirectional DC–DC converter shown in Fig. 5.7.9, developed by Kawasaki, ensures a steady supply of power from the battery and effective use of regenerative power while controlling fluctuations in power consumption from overhead wires. On nonelectrified sections, the control system provides a steady supply of power from the battery when the vehicle is accelerating and returns regenerative energy to the battery for maximum efficiency.

Fig. 5.7.9 Mounted charge–discharge control unit on the roof.

[2]VVVF stands for variable voltage variable frequency.

5.7.11 Braking system

When braking, vehicles use their motors as a generator and return its energy to the overhead power lines. This process, known as regenerative braking, is designed to enhance energy efficiency. Although the regenerative energy is returned to the overhead power lines in Japan railway system, transmission losses occur over long distances, and if there are no other vehicles running nearby to use the regenerative energy, it is wasted. The SWIMO vehicle dramatically enhances energy efficiency by storing all regenerative energy in its on-board GIGACELL batteries and then using it to drive the motors when accelerating.

5.8 Charge–discharge control system

The SWIMO vehicle is equipped with a charge–discharge control unit mounted on the rooftop that controls the charge and discharge for batteries. To ensure stable operation of the vehicle on electrified sections with a large voltage fluctuations in the overhead power lines, a bidirectional DC–DC converter is adopted that allows the batteries to be charged or discharged independently of whether the battery voltage is higher or lower than the overhead power line voltage. The circuit configuration and the component layout of charge–discharge control unit are shown in Figs 5.8.1 and 5.8.2, respectively. Fig. 5.8.3 depicts the appearance of charge–discharge control unit.

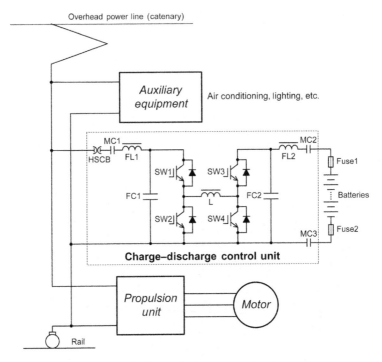

Fig. 5.8.1 Circuit configuration of charge–discharge control unit.

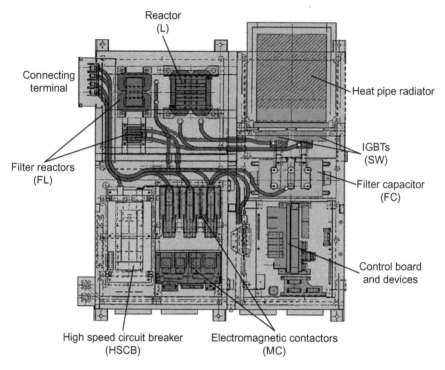

Fig. 5.8.2 Component layout of charge–discharge control unit.

Fig. 5.8.3 Appearance of charge–discharge control unit.

Table 5.8.1 **Specifications of charge–discharge control unit**

Item	Description
Power conversion type	Bidirectional buck–boost converter
Line voltage	600 V DC (nominal)
Battery voltage	576 V DC (nominal)
Switching frequency	4 kHz (PWM)
Maximum output power	250 kW
Battery control	Auto/manual with self-monitoring
Cooling method	Natural cooling
Weight	680 kg
Outline dimensions	$1600 \times 1500 \times 600$ mm
Protective function	Over current, over voltage, low voltage, over heat, abnormal battery condition

Basic specifications of charge–discharge control unit are provided in Table 5.8.1. A charge–discharge control unit, which maximizes the performance of the battery, controls charge and discharge as per SOC to prevent battery deterioration from overcharge or overdischarge. The battery monitoring system that measures battery voltage, battery temperature, and internal pressure is included collaterally. When the battery condition is abnormal, a warning error signal is provided. A self-monitoring function is also included in this unit. The warning signal or a push-button operation in the operator's cab operates the high-speed circuit breaker and electromagnetic contactors to disconnect the batteries and protect the system.

5.9 Bidirectional buck–boost converter

Fig. 5.9.1 shows the circuit configuration of bidirectional buck–boost converter using IGBTs for charge–discharge control unit. In addition, the basic circuit configurations of the buck converter and the boost converter are shown in Fig. 5.9.2A and B for reference.

Operating modes of the bidirectional buck–boost converter are described in Figs 5.9.3 and 5.9.4. This converter can charge and discharge the batteries based on whether the battery voltage or overhead power line voltage (input voltage) is higher.

5.9.1 Battery charge mode

(A) Input voltage $V_{in} \leq$ battery voltage V_{bat}

When the input voltage V_{in} is less than the battery voltage V_{bat}, the switch SW_4 operates in the switching mode to boost the voltage and charge the batteries.

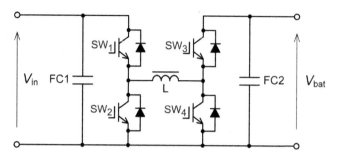

Fig. 5.9.1 Bidirectional buck–boost converter.

Fig. 5.9.2 Basic circuit configurations of buck and boost converter: (A) buck converter and (B) boost converter.

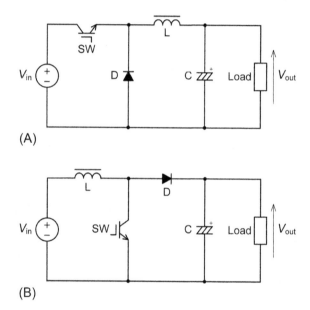

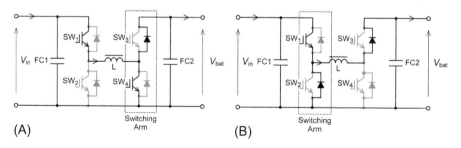

Fig. 5.9.3 Battery charge mode of bidirectional buck–boost converter: (A) $V_{in} \leq V_{bat}$: boost converter and (B) $V_{in} \geq V_{bat}$: buck converter.

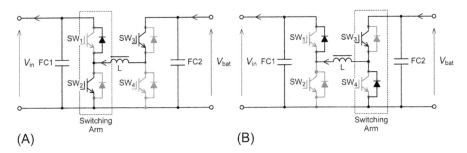

Fig. 5.9.4 Battery discharge mode of bidirectional buck–boost converter: (A) $V_{bat} \leq V_{in}$: boost converter and (B) $V_{bat} \geq V_{in}$: buck converter.

(B) Input voltage $V_{in} \geq$ battery voltage V_{bat}

When the input voltage V_{in} is greater than the battery voltage V_{bat}, the switch SW_1 operates in the switching mode to drop the voltage and charge the batteries.

5.9.2 Battery discharge mode

(A) Battery voltage $V_{bat} \leq$ input voltage V_{in}

When the battery voltage V_{bat} is less than the input voltage V_{in}, the switch SW_2 operates in the switching mode to boost the voltage and discharge the batteries.

(B) Battery voltage $V_{bat} \geq$ input voltage V_{in}

When the battery voltage V_{bat} is greater than the input voltage V_{in}, the switch SW_3 operates in the switching mode to drop the voltage and discharge the batteries.

5.10 Operating mode of battery-driven LRV

5.10.1 Nonelectrified section

On nonelectrified (catenary-free) sections, all power used by the vehicle is supplied with the electric power stored in the batteries. The vehicle operating modes and their power flows under the nonelectrified section are shown in Figs 5.10.1 and 5.10.2, respectively.

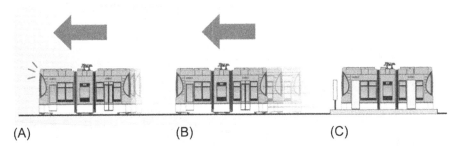

Fig. 5.10.1 Vehicle operating modes under nonelectrified section: (A) braking, (B) accelerating, and (C) waiting/coasting.

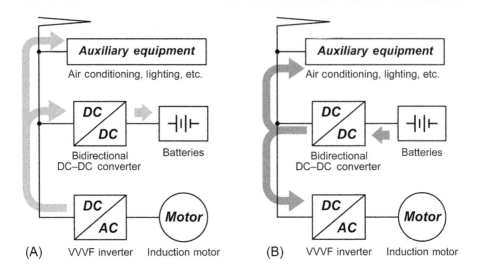

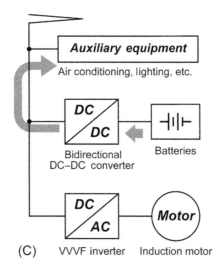

Fig. 5.10.2 Power flows at nonelectrified section: (A) braking, (B) accelerating, and (C) waiting/coasting.

(A) Braking

The energy generated by the regenerative braking is captured by the on-board batteries.

(B) Accelerating

The energy stored by the on-board batteries is discharged. The regenerative energy is used as needed; this accelerating thus reduces total energy requirements and also CO_2 emissions.

(C) Waiting/coasting

The energy stored by the on-board batteries is discharged to the auxiliary equipment such as air conditioning and lighting for passenger cabin.

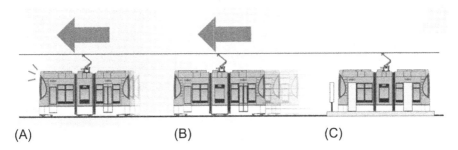

Fig. 5.10.3 Vehicle operating modes under electrified section: (A) braking, (B) accelerating, and (C) waiting/coasting.

5.10.2   Electrified section

On the electrified section, the traction and auxiliary equipment power are supplied directly from the overhead power line via pantograph, while the batteries are charged through the charge–discharge control unit. When the capacity of the batteries rises above a set value or when the overhead power line voltage decreases, the batteries assist accelerating and auxiliary power requirements.

The vehicle operating modes and their power flows under the electrified section are shown in Figs 5.10.3 and 5.10.4, respectively.

(A) Braking
     The energy generated by the regenerative braking is captured by the on-board batteries.
(B) Accelerating
     The accelerating energy is supplied from the overhead power line.
(B') Accelerating at low line voltage
     When the overhead power line voltage is low because of heavy load such as peak/rush hours, the on-board batteries is discharged to assist accelerating.
(C) Waiting/coasting
     The auxiliary equipment such as air conditioning and lighting operates on the overhead power line.

In summary, on the SWIMO vehicle, the regenerative energy generated during braking, whether it is generated on an electrified or a nonelectrified section, is charged into the on-board batteries. Therefore, energy can be used effectively.

5.11   Test runs at revenue service line

After testing the basic performance of the vehicle at Kawasaki's Harima Works test track from October to November 2007 (Fig. 5.11.1), nonservice test runs were conducted on a revenue service line of the Sapporo City Transportation Bureau, Japan. The test runs were conducted over 39 days in the period from December 2007 to March 2008 (Fig. 5.11.2).

The vehicle made 3 or 3.5 round trips per day over an 8.5 km one-way test distance. The test pattern was based on a service operation timetable consisting of mainly

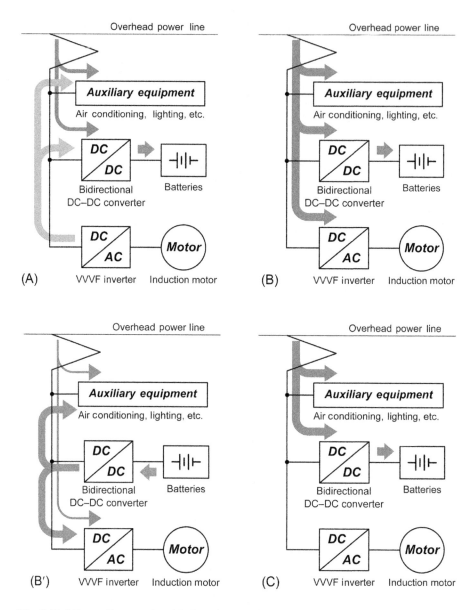

Fig. 5.10.4 Power flows at electrified section: (A) braking, (B) accelerating, (B') accelerating at low line voltage, and (C) waiting/coasting.

running operations with the pantograph raised to come into contact with the overhead power line and running operations with the pantograph folded down so the vehicle had to depend purely on the GIGACELL batteries for power, with both operational modes turned on and off in an alternating sequence.

Fig. 5.11.1 Test run of the SWIMO vehicle on Harima test track.

Fig. 5.11.2 Test runs during winter season in Sapporo.

The total distance travelled in Sapporo was 1842 km, 877 km (about 47%) of which was travelled under battery power.

5.11.1   Test runs on electrified sections (pantograph in use)

The test runs were conducted based on the charge–discharge control method that requires a 70 A constant current charging to be conducted when the SOC falls below 70% with the pantograph in contact with the overhead power line. When the SOC reaches 75%, the constant current charging automatically stops.

When the vehicle is braking, the constant current charging is interrupted, and regenerative power is used to charge the on-board batteries. It was confirmed that cancelled regeneration did not occur on test runs conducted during service operation hours. The SWIMO vehicle is built on a scheme in which regenerative power and battery power are not returned to overhead power line.

It was also confirmed that assist discharging worked normally when the overhead power line voltage dropped in mixed operation with existing vehicles in service, especially during peak/rush hours.

5.11.2 *Battery-powered test runs*

Fig. 5.11.3 provides a result of test runs. In these battery-powered test runs, the vehicle ran with the passenger cabin heating turned on for the entire 8.5 km distance, running in the SOC range of 55%–75%. When the SWIMO vehicle was accelerated at full throttle, the maximum discharge current was reached approximately 350 A. The inverter regeneration ratio (the amount of regenerative energy divided by the amount of accelerated energy) for a 1.5 round-trip test run (25.5 km) reached 36%.

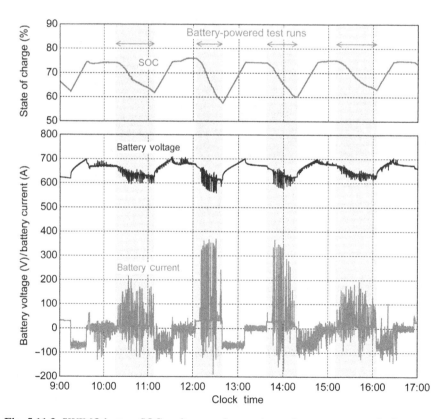

Fig. 5.11.3 SWIMO battery SOC, voltage, and current waveforms at test runs in Sapporo.

Fig. 5.11.4 Outdoor shelf test under low-temperature condition.

The vehicle achieved continuous running over a distance of 37.5 km (corresponding to slightly more than two round trips) without charging in an ordinary service pattern and with the passenger cabin heating turned off. While running without charging, the batteries supplied the total of 64.47 kWh (1.72 kWh/km). It was confirmed that battery-powered running in a cold area poses no serious problems.

5.11.3 Outdoor shelf test under low-temperature condition

Fig. 5.11.4 shows the outdoor shelf test under the low temperature in addition to a snow condition. The SWIMO vehicle was kept at the outdoor yard with no operation. After 5 days, the on-board battery temperature plunged to $-3°C$ (the outside air temperature, $-5°C$); the SWIMO vehicle test run was conducted on a revenue service line. There was no problem to start the vehicle including the electric devices on the roof and also confirmed there was no trouble to run at the low-temperature condition.

5.12 Rapid charging test

Fig. 5.12.1 depicts the photo of the SWIMO vehicle rapid-charging test under a heavy load power supply facility with overhead rigid power line in Kawasaki's Harima test track, Japan. The overhead rigid power line is used for keeping the temperature rise in check at contact point in rapid charging. During rapid charging, the charging time is short because the charging of the SWIMO vehicle in nonelectrified section is conducted as the vehicle turns around at a terminal stop or while the vehicle is stopped at a terminal along its route.

A rapid-charging test was conducted at a charge current of 350 A for the SWIMO vehicle. Fig. 5.12.2 shows the current and voltage waveforms obtained from this test.

Fig. 5.12.1 SWIMO rapid-charging test.

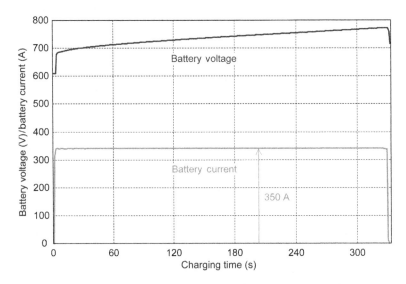

Fig. 5.12.2 SWIMO battery voltage and current waveforms at rapid-charging test.

The on-board battery current was kept constant current at 350 A. As the batteries are charging, the SOC of the on-board batteries rises with the battery voltage.

Rapid charging is accompanied by temperature increases in various parts of a vehicle, which were measured during testing. Fig. 5.12.3 shows the temperature distribution around the pantograph immediately following charging at 350 A. The ambient air temperature during testing was 16°C, while the wind speed was 0.5 m/s, and the maximum pantograph temperature at this time was 30°C. The temperature rise was 14°C,

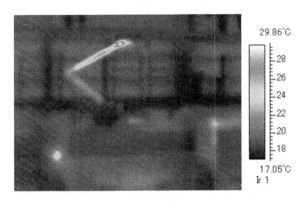

Fig. 5.12.3 Thermography around pantograph after rapid-charging test.

which is not considered to pose problems even in summer. The measured values of the temperature increases do not suggest the presence of any problems in either of the above-mentioned parts and others not mentioned.

During testing, the on-board battery capacity of 31.2 Ah was charged in 5 min 32 s. Converted into a 5 min rate, the amount of charging corresponds to 28.2 Ah (SOC = 14.1%). It was confirmed that, under trial test conditions on the Harima test track, this charge allows 10 km of running operation.

5.13 Battery-driven train with commercial operation in Japan

The East Japan Railway Company (JR East) started commercial operation of the battery-driven train, series EV-E301 named 'ACCUM' in Karasuyama line (Fig. 5.13.1) from March 2014 as a new method towards the reduction of the environmental burden in nonelectric sections. The ACCUM is the Japan's first battery-driven train with commercial operation for nonelectrified section, which was operated by the diesel trains.

The ACCUM train shown in Fig. 5.13.2 has on-board lithium-ion (Li-ion) batteries, enabling it to operate in nonelectrified sections. The train raises its pantograph to receive the electric power for its operation and charging on-board batteries at electrified sections. When the train enters a nonelectrified section, the pantograph is lowered, and the train operates on battery power alone. When the brakes are applied, regenerated energy is used to charge the batteries, which can also be charged at terminal station where rapid-charging facilities have been installed. The introduction of the series EV-E301 has made possible the elimination of exhaust emissions and reductions in CO_2 and noise emission levels from those generated by diesel engines.

5.13.1 Background and objectives

Battery performance had increased tremendously, and the market for those expanded with the popularization of hybrid and electric automobiles. In light of the knowledge gained in the development of diesel hybrid and fuel cell hybrid trains and

Fig. 5.13.1 Map of
Karasuyama line in Japan.

Fig. 5.13.2 Series EV-E301 battery-driven train 'ACCUM'.

advancement in technologies for batteries, feasibility of rolling stock systems for running in nonelectrified sections on electrical energy stored in batteries alone has come into view. JR East thus decided to take on the development of a catenary and battery-powered hybrid train system as a new measure to reduce environmental load in nonelectrified sections.

The objectives of this system are to eliminate exhaust gases from engines and to reduce CO_2 emissions and noise. Ancillary effects that can be expected are ability for trains to run on both electrified and nonelectrified sections, making rolling stock operation more efficient and reduction in maintenance by reducing labour-intensive mechanical parts such as engines and transmissions.

5.13.2 Configuration of catenary and battery-driven hybrid train system

Fig. 5.13.3 shows the configuration of catenary and battery-driven hybrid train system at electrified/nonelectrified sections. In electrified section, the battery-driven train can run just like ordinary electric trains by raising their pantographs, and battery-driven train can charge batteries from overhead power lines when the battery SOC is low.

When moving into nonelectrified section, the pantograph is lowered, and the train runs on power from the batteries alone. When braking, regenerative energy charges to the batteries so as to make effective use of electric energy. Depending on the length of the nonelectrified section, wayside rapid-charging facilities is set up at turn-back station for quick charging of battery-driven train. Setting up charging facilities only in electrified sections (including charging while running in electrified sections) was considered, but having wayside rapid-charging facilities set up at turn-back station is thought to be more reasonable when taking into account system durability and volume/weight of batteries.

The actual operation of the ACCUM train is as follows. The train is operated on the electrified Tohoku Line between Utsunomiya and Hoshakuji station. While the ACCUM train is operated on this line, the on-board batteries are charged from the overhead power line via pantographs. After arriving at Hoshakuji station, this train lowers the pantographs and starts operating with the on-board batteries on the nonelectrified Karasuyama Line (20.4 km). Finally, the ACCUM train reaches the final destination of this line, Karasuyama station.

Fig. 5.13.4 shows the photo of the ACCUM train at Karasuyama station. The ACCUM train charged the on-board batteries at the rapid-charging facility for its return trip. The train raised the pantographs and the short overhead rigid power line constructed above the train. This charge is essential for the battery-driven ACCUM train before a 20.4 km long return trip without electric power supply from the overhead power line.

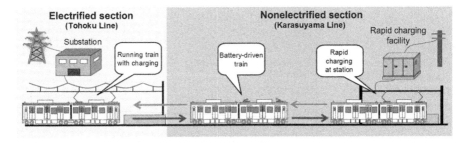

Fig. 5.13.3 Configuration of catenary and battery-driven hybrid train system.

Fig. 5.13.4 Rapid-charging facility at Karasuyama station.

5.13.3   Summary of battery-driven ACCUM train

Table 5.13.1 depicts the main specifications of the ACCUM train. The lithium-ion (Li-ion) battery manufactured by GS Yuasa Corporation is used and installed at the underfloor battery box for this train. The Li-ion battery (GS Yuasa, LIM30H-8A) with specifications and train-mounted battery box are shown in Figs 5.13.5, 5.13.6, and Table 5.13.2, respectively.

Table 5.13.1 Specifications of ACCUM train

Item	Description
Articulation	Two-car body articulated
Train performance	Maximum speed, 100 km/h
	Acceleration, 2.0 km/h/s
	Deceleration, 3.6 km/h/s
Passenger capacity	266 (included seat, 96)
Length	40,000 mm (20,000 mm × 2)
Width	2800 mm
Height	3620 mm
Tare weight	77.9 ton (40.2 + 37.7 ton)
Track gauge	1067 mm
Traction motor	95 kW × 2
Overhead line voltage	1500 V DC (nominal)
Battery voltage	633.6 V DC (nominal)
Battery	Lithium-ion battery, LIM30H-8A 22 in series, 10 in parallel
	Total battery capacity, 190 kWh

Fig. 5.13.5 Lithium-ion battery used for ACCUM train.

Fig. 5.13.6 Train-mounted battery boxes underfloor.

Table 5.13.2 **LIM30H-8A battery specifications**

External dimensions (mm)	W:231 × D:389 × H:147	Weight (kg)	Approximately 20
Nominal voltage (V)	28.8	Nominal voltage (V) per cell	3.6
Nominal capacity (Ah)	30	Operating voltage range (V)	23.2–33.2
Maximum current capacity (A)	600	Continuous energizing current (A)	100
Operating temperature limit (°C)	0–45	Monitoring system	Voltage monitoring for all cells Module temperature monitoring

5.13.4 Operating mode of ACCUM train

5.13.4.1 Nonelectrified section

On nonelectrified (catenary-free) sections, all power used by the train is supplied with the electric power stored in the batteries. The ACCUM's operating power flows under the nonelectrified section are shown in Fig. 5.13.7.

(A) Braking

The energy generated by the regenerative braking is captured by the on-board batteries.

(B) Accelerating

The energy stored by the on-board batteries is discharged. The regenerative energy is used as needed; this accelerating thus reduces total energy requirements and also CO_2 emissions.

(C) Waiting/coasting

The energy stored by the on-board batteries is discharged to the auxiliary equipment such as air conditioning and lighting for passenger cabin.

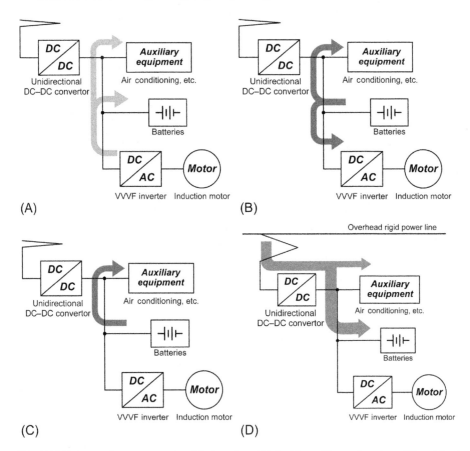

Fig. 5.13.7 Power flows at nonelectrified section: (A) braking, (B) accelerating, (C) waiting/coasting, and (D) rapid charging.

(D) Rapid charging at turn-back station

The ACCUM train charged the on-board batteries at the rapid-charging facility. The train raised the pantographs and the overhead rigid power line constructed above the train. Also, the auxiliary equipment such as air conditioning and lighting operates under the overhead rigid power line.

5.13.4.2 Electrified section

On the electrified section, the traction and auxiliary equipment are fed directly from the pantograph, while the batteries are charged through the charging control unit. The ACCUM's operating power flows under the electrified section is shown in Fig. 5.13.8.

(A) Braking

The energy generated by the regenerative braking is captured by the on-board batteries. If the on-board batteries are full-charged, the regenerative braking energy is returned to the overhead power line to use it by other trains.

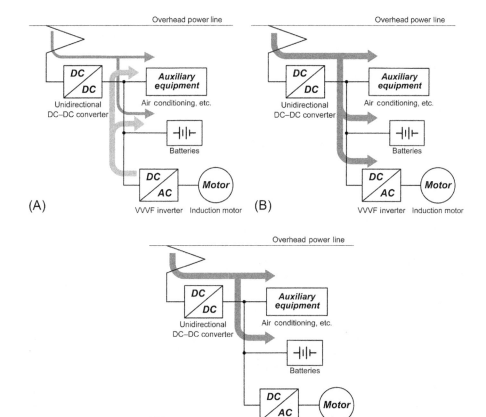

Fig. 5.13.8 Power flows at electrified section: (A) braking, (B) accelerating, and (C) waiting/coasting.

(B) Accelerating

The accelerating energy is supplied from the overhead power line.

(C) Waiting/coasting

The auxiliary equipment such as air conditioning and lighting operates on the overhead power line.

5.13.5 The future

Based on the perspectives of technological innovation and globalization, JR East is committed to the realization of unlimited potential while remaining steadfast in pursuit of eternal missions. Railways are a reputation as an environmentally friendly mode of transportation. However, the automotive industry, one of the main competing modes of transport for railways, has achieved remarkable developments in energy-efficient and environmentally friendly technology in recent years, including hybrid vehicles, electric cars, and fuel cell vehicles. Aiming for the further innovation of railways, JR East aims to embrace open innovation in order to utilize external developmental capabilities and intellectual property to vigorously promote technological innovation. In addressing energy conservation, JR East is moving forth with plans to introduce battery-driven electric train systems for additional operations in selected railway sections to eliminate the use of overhead contact wires. JR East has further plans to build on this technology to develop trains through service between alternating-current (AC) electrified railway lines and nonelectrified segments. This will permit electric train operations without installing overhead power lines.

5.14 Battery-driven LRV with commercial operation in France

The Alstom's Citadis battery-driven LRV shown in Fig. 5.14.1 entered service in Nice, the south of France in November 2007. This is a first modern catenary-free battery-driven vehicle with Ni-MH integrated traction battery systems in the world. This hybrid vehicle for line 1 of Nice's new tramway system shown in Fig. 5.14.2 can switch the source of vehicle power between overhead power line and the on-board batteries for catenary-free operation in the Nice's historic public squares. The city of Nice wanted to keep its two historic town squares (Place Masséna and Place Garibaldi) clear of the overhead power lines. Since the vehicle routes cross these historic town squares, the local authorities asked for a low-impact solution to keep these areas free of overhead power lines. Therefore, Alstom needed an on-board traction battery capable of providing sufficient power for travel over the sections of track places where catenaries are not available.

The system also had to be compact enough for installation in the vehicle roof. Saft developed a fully integrated battery for this new generation of vehicles. The Ni-MH battery systems offer excellent power storage in a compact maintenance-free package and drive the vehicles for around 500 m through each autonomous section. Saft's new generation Ni-MH battery modules provided in Fig. 5.14.3 offered the ideal solution as they have been developed specifically for high-power applications. The system's

Fig. 5.14.1 Battery-driven LRV in Nice, France.

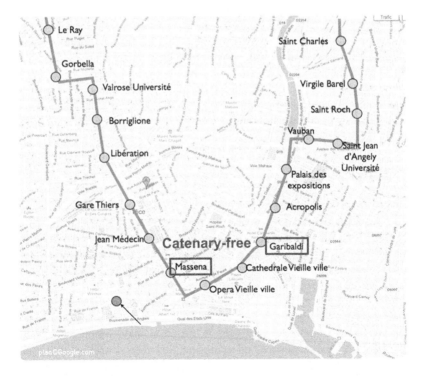

Fig. 5.14.2 Nice Tramway route map.

Fig. 5.14.3 Ni-MH battery system
(Saft, 576 V–34 Ah) used for Citadis
vehicle.

distinguishing technical feature is the use of batteries aboard the vehicle to avoid the necessity of the overhead power line on two historic town sections of the route. The 576 V DC Saft Ni-MH battery system provides approximately 80 kWh of continuous energy. Each battery system incorporates an active cooling device and battery management control for monitoring of battery temperature, voltage, and charging conditions. The whole unit is supplied ready to 'plug and play' in a custom-built tray complete with power and communication capabilities and all necessary safety features. Giving vehicles a range of up to 1 km at a maximum speed of 30 km/h with air conditioning in operation, the switching of power being from either the overhead line or the batteries is activated by the driver, with the pantograph fully lowered when running without catenary. Each vehicle's driver console features visual and audio indications of the need to operate the power changeover sequence. The batteries recharge from the overhead power line supply while in conventional operation. There is no additional external infrastructure needed to operate the vehicles under battery power over the catenary-free sections.

5.15 Conclusions

Test runs on the SWIMO vehicle have provided us with data and know-how needed for actual running operations. In particular, the test runs in Sapporo provided a precious opportunity to demonstrate the feasibility of battery-powered run in cold regions. Additionally, the rapid-charging test provided further data on the operation of vehicles. The total distance travelled amounting to 4663 km (of which battery-powered runs reached 3600 km).

In addition, the examples of battery-driven LRVs and trains with commercial operations were presented from an innovative point of view. Although the technologies currently available for catenary-free vehicle operations have a limited track record, going forward, the market is expected to expand, and costs of these systems are also expected to reduce. Once stabilized, the demand for catenary-free systems is expected to exceed overhead powered systems, due to the multiple advantages offered.

Kawasaki's vision is to realize 'the world's highest standards for operation, safety, and environmental performance'. As technological pioneers, Kawasaki will continue to follow this vision into the future and in efforts to make railway systems highly efficient using our GIGACELL batteries.

Further Reading

Abiko, H. (2012). Development of hybrid railcars and catenary and battery-powered hybrid railcar system. *JR East Technical Review*, *23*, 9–12.

Akiyama, S., Tsutsumi, K., & Matsuki, S. (2008). The development of low floor battery-driven LRV SWIMO. In *The 8th world congress on railway research* (pp. 1–9). Seoul: WCRR.

East Japan Railway Company, (2014). *Accumulator system (ACCUM)*. *JR East Group CSR report* (p. 47).

Global Mass Transit. (2014). *Catenary-free trams: Technology and recent developments*. Global Mass Transit Website. Available at http://www.globalmasstransit.net/archive.php?id=15973 [Accessed 15 September 2016].

Green Car Congress (2005). *Saft providing NiMH batteries to Alstom and Lohr for hybrid tram applications*. Green Car Congress Website. Available at http://www.greencarcongress.com/2005/12/saft_providing_.html [Accessed 15 September 2016].

Griffiths, P. (2012). Technology briefing paper catenary free tram operation. *UK Tram*, 2–4.

GS Yuasa Corporation (2014). *System utilizing industrial-use lithium-ion battery module LIM30H-8A installed in the new EV-E301 series developed by East Japan Railway Company*. GS Yuasa Corporation News Release Website. Available at http://www.gs-yuasa.com/en/newsrelease/article. php?ucode=gs151007580509_154 [Accessed 15 September 2016].

International Railway Journal. (2008). Swimo takes to the rails. *Transit Special*, *8*, 52–54.

Kawasaki Heavy Industries Ltd. (2008). SWIMO's excellence demonstrated. *Newsletters Scope*, *74*, 2–5.

Kawasaki Heavy Industries Ltd. (2009). Inside SWIMO, the next-generation light rail vehicle. *Newsletters Scope*, *78*, 6–7.

Kawasaki Heavy Industries Ltd. (2010). *High-capacity fully sealed nickel-metal hydride battery GIGACELL*. Leaflet.

King, C., Vecia, G., & Thompson, I. (2015). *Innovative technologies for light rail and tram: A European reference resource*: (pp. 1–16). University College London: Sintropher, Briefing Paper 4.

Ogura, K., Yoshida, K., Tsutsumi, K., & Nishimura, K. (2007). The development of low floor battery-driven LRV. In *The 14th jointed railway technology symposium* (pp. 123–126). Tokyo: J-RAIL [in Japanese].

Oku, Y. (2010). Efforts to realize the world's highest standards for operation, safety and environmental performance. *Kawasaki Technical Review*, *170*, 7–12.

Railway Technology. (2007). *Nice Tramway, France*. Railway-technology Website. Available at http://www.railway-technology.com/projects/nice-trams [Accessed 15 September 2016].

Takiguchi, H. (2012). Overview of series EV-E301 catenary and battery-powered hybrid railcar. *JR East Technical Review*, *51*, 45–50 [in Japanese].

Tokyo Railway Labyrinth (2014). *Battery mode operation of the EV-E301 series*. Tokyo Railway Labyrinth Website. Available at https://tokyorailwaylabyrinth.blogspot.jp/2014/06/battery-mode-operation-of-ev-e301-series.html [Accessed 15 September 2016].

Toyo Electric Mfg. Co., Ltd (2016). *Propulsion inverters (VVVF inverter)*. Toyo Denki Seizo Website, Available at https://www.toyodenki.co.jp/en/products/transport/train/vvvf.php [Accessed 15 September 2016].

Yamazaki, H., Akiyama, S., Hirashima, T., Kataoka, M., & Matsuo, K. (2010). Urban transportation that is friendly to people and the environment; SWIMO-X low-floor battery-driven light rail vehicle. *Kawasaki Technical Review, 170*, 21–26.

Sustainable transport, electric vehicle promotional policies, and factors influencing the purchasing decisions of electric vehicles: A case of Slovenia

6

Matjaž Knez
University of Maribor, Celje, Slovenia

6.1 Introduction

Pollution, greenhouse gas emissions (GHG), rising energy demand, and high-energy import dependence present the core of energy problems both in the European Union (the EU) as a whole and in Slovenia. The current energy import dependence in the EU is 50%, while in Slovenia, it is 55% (Government Communication Office, 2009). This dependence, which causes economic, political, and social vulnerability of the EU, must be seen as a challenge and opportunity for sustainable future (Obrecht & Denac, 2013).

The investments in efficient energy and renewable energy sources (RES) are highly important since RES cause little (or no) pollution and enable the use of local resources. In addition, they decrease import dependency and increase the EU competitiveness at the same time. Because 80% of all GHG emissions in the EU and in Slovenia are caused by energy industry (EEA, 2007; Government Communication Office, 2009), the EU intends to lower CO_2 emissions by 20% while increasing the share of RES up to 20% and enhancing EE by 2020. Directive 2009/28/EC within the climate and energy package is mandatory for Slovenia as well. Slovenia's goal is to have 25% of RES in final energy consumption electricity by 2020 (Obrecht & Knez, 2014).

Limited oil reserves and the associated sociopolitical and economic effects are presently the key forces behind the need to develop alternative energy sources and to reduce dependence on imported oil. To reduce the harmful emissions and to make use of finite energy sources more efficient, effective policy measures need to be implemented by the society. One effective approach to attain these objectives is to reduce the use of personal transportation by encouraging the use of bicycles and public transport (Turcksin, Mairesse, & Macharis, 2013).

The transportation sector plays a pivotal role in contemporary societies, consequently, traffic pollution in many cities around the globe causes up to 70% of total carbon emissions (UN HABITAT, 2011). The concept of alternative transportation technologies and alternative energy resources has arisen as a potential long-term

Electric Vehicles: Prospects and Challenges. http://dx.doi.org/10.1016/B978-0-12-803021-9.00006-9

solution for achieving an environmentally friendly future and has become 'an embraced goal' of many countries around the world (Bockarjova & Steg, 2014).

However, most consumers are reluctant to let go of their primary means of transportation, mainly because of current standard of living and strong feelings of independence associated with personal car use (Anderson & Stradling, 2004). It is therefore essential to promote environmentally friendly alternatives. Alternative sustainable technologies as such form one possible solution and changes in driving styles represent another one (Kramberger, Dragan, & Prah, 2014). In this respect, unveiling the consumers' attitudes, preferences, and decision factors towards low-emission vehicles (LEV) is necessary for the formulation of effective policy measures and effective commercialization of LEVs. A report by the National Academy of Sciences (2013) points out that there has been little research to determine which government incentives are most influential in affecting customer decisions, which public education efforts work best, and which kinds of demonstration activities are most helpful. To ensure that investments in LEV adoption are maximized, the federal government could support research on policy effectiveness.

The 1970s oil crisis has had a huge effect on the shaping of initial electric vehicle promotion policies. Since then, various countries have already adopted preliminary plans for electric vehicle promotion with the intention of eventually replacing internal combustion engines vehicles, since their exhausts cause significant environmental pollution (Cowan & Hultén, 1996). Countries such as Japan and the United States have managed to increase electric vehicle sales with the help of effective policies promoting their use. The European Union (EU) and other European countries have followed these trends and put into place their own electric vehicle promotional policies, which in some countries, in particular the EU member states, have been subject to continuous improvement (Fale, 2014).

Low-emission vehicles, especially electric vehicles, hold many promises—from reducing dependence on imported petroleum to decreasing greenhouse gas emissions. However, there are many barriers to their mainstream adoption regardless of incentives and enticing promises to solve difficult problems. The vehicles have some technological limitations, such as restricted electric range and the long time required for battery charging; they cost more than conventional vehicles; and they require appropriate infrastructure for charging the battery (National Academy of Sciences, 2013).

Most potential customers have little knowledge of LEVs and almost no experience with them. Many surveys indicate that they ask many questions, including 'Are these cars powerful enough for freeway driving?', 'Are electric vehicles safe when going through puddles?', 'How much would an electric vehicle add to my home electricity bill?', and 'Are electric vehicles any better for the environment than conventional vehicles?' (Kurani et al., 2009; Turrentine, Garas, Lentz, & Woodjack, 2011, in National Academy of Sciences, 2013). Despite the benefits of hybrid electric vehicles, realization of expected contributions to sustainability ultimately falls on the consumers' willingness to purchase the new technology (Krupa et al., 2014).

Since potential demand, consumers' wishes, preferences, and decision factors are highly important for all car brands and car dealers offering LEVs; these factors have already been studied under very different conceptual frameworks and methodologies, with the main purpose of finding the right leverage to encourage demand for LEVs.

However, understanding of consumer preferences is also important for policy decision-makers since they are responsible for long-term development of the transportation sector and infrastructure related to LEVs. Technological forecasts based on LEV-related research are also crucial for top managers of industry-leading car makers, who are to steer the automotive industry towards a transition to a more sustainable future.

Schaltegger (2008) demonstrates that creating a business case for sustainability requires a good understanding of links between nonmonetary social and environmental activities on the one hand, and business or economic success on the other. The core question, and the basis for any business case for sustainability, is how profit resulting from increased social and environmental activities can be identified and reaped. So, managers need to assess appropriately the economic value generated by innovative environmentally friendly projects.

To assess the economic value created through environmental investments, the Japanese Ministry of Economy, Trade and Industry (2002) established a method for capital investment into environmentally friendly facilities. They recommend comparing alternatives that incorporate not only the economic assessment, such as net present value, but also the environmentally harmful substance reduction benefits such as GHG reduction. Managers then have to make a decision based on both financial value and physical value (Minato, 2011). The first challenge is how to create corporate value from environmental impact reduction. Lyon and Maxwell (1999) argue that corporations can differentiate their products by improving their environmental qualities and thereby charge a higher price; and that green investors may be an increasingly important factor in determining corporate environmental activity. Fairchild (2008) uses a game theory approach to demonstrate that the investment cost and the extent of consumer and investor green awareness affect corporate incentives to make environmental investments. Machlachlan and Gardner (2004) point out some important differences between socially responsible and conventional investors in terms of their beliefs, the importance they ascribe to ethical issues, their investment decision-making style and their perceptions of moral intensity. Kokubu (1999) suggests that green stakeholders, such as green consumers, consider the environmental impacts of purchased products, and green investors appreciate corporate action and corporate policies towards environmental conservation, and will accept additional cost and investment in environmental conservation if this is justified by the reduction of environmental impact.

The implication of studies mentioned above is that value from environmental investments can be created by attracting green consumers who are willing to pay a product price premium and by attracting green investors who are willing to pay a share price premium. Green stakeholders, such as green consumers and green investors, accept the price premium equivalent to the economic value of environmental impact reduction. This aspect can then be included in the investment appraisal calculation. In other words, the social environmental value creation from an environmental investment can be converted into internal corporate value creation (Knez, Muneer, Jereb, & Cullinane, 2014).

Similar but slightly different motivation factors can be identified in demand for LEVs in the private sector, i.e. by households. An interesting study was performed by Ulmer, Huhnke, Bellmer, and Cartmell (2004). They found that costs are more important than environmental benefits when purchasing ethanol-fuelled cars. Golob and Gould (1998) also made a survey about personal vehicle trials and found out that people would choose EV on the basis of very low running costs rather than environmental benefit. Oil price is also an important factor for commercialization of alternative fuels. Popp et al. (2009) found that relative fuel prices are of great importance when choosing a new car. Importance of fuel economy is even higher in case of an increased belief in the ability to positively influence the environment. O'Garra, Mourato, and Pearson (2005), for example, noted that key determinant of acceptability of hydrogen vehicles is high prior awareness of the existence of hydrogen fuel, which is in turn related to gender, age, education, and environmental knowledge. Similar research with similar results was carried out by Thesen and Langhelle (2008).

Johansson-Stenman and Martinsson (2006), on the other hand, write about the importance of status. They suggest that people care more about status than environmental issues when purchasing a new car. Kahn (2007) found that environmentalists are more likely to use public transport, consume less gasoline, and purchase hybrid electric vehicle (HEV) or battery electric vehicle (BEV) rather than a conventional vehicle running on fossil fuel. Flamm (2009), for example, noted that environmental knowledge and proenvironmental attitudes are associated with ownership of more efficient, environmentally friendlier vehicles. Achterberg, Houtman, van Bohemen, and Manevska (2010) performed a study about purchasing habits in the Netherlands and found out that the three most important factors for purchasing new hybrid cars are trust in technology, environmental concern, and sense of need to take care of nature. Van de Velde, Verbeke, Popp, and Van Huylenbroeck (2010) write about the possibility to increase environmental awareness in Belgium by providing information on environmental problems and on cars running on alternative fuels as a possible solution. Because people still desire ranges to be similar to that of conventionally fuelled cars even when their travel diaries indicate that they usually travel only short distances on a daily basis. Kurani, Turrentine, and Sperling (1994, 1996) carried out interactive interviews based on weekly travel diaries. They found that driving range is particularly important, since viable market for EV is within 60–100 mi driving range. Martin et al. (2009), on the other hand, found that driving range of 480 km is acceptable for 90% of respondents in their survey. HEVs are seen as a possible solution for proenvironmental households, having problems with relatively smaller driving range of BEVs. In another survey, Kurani et al. (1996) found out that even if environmental awareness does not lead to purchasing LEVs, it may encourage households to seek out and evaluate EV for purchase considerations. The best valued attribute of EV is the possibility of home recharging. Quiet ride and low maintenance costs are also important. Also very positive for BEV are the findings of Axsen, Kurani, and Burke (2010) that performance requirements of batteries are closer to commercially viable than expected.

When taking into account all features of AFVs, purchase decisions are still mainly driven more by the desired social image of consumers than by environmental issues (Johansson-Stenman & Martinsson, 2006). However, Axen et al. (2010) argued that there are some conditions in consumers' lives convenient for changes in values. He showed that, beside status, 'luminal state to facilitate consideration of new values, alignment with core values, and social network support' could be a turning point in a transition to sustainability-oriented values.

This chapter in first part presents the effects of electric vehicle promotional policies by means of a review of the policies that have been adopted in Slovenia and elsewhere. Internationally, a range of electric vehicle promotional policies have been implemented, including fiscal and other forms of incentives intended to encourage electric vehicle sales. Slovenia has not been so successful, comparing with the leading European Union countries that are successfully promoting electric vehicle use.

The second part of this chapter presents two studies, done in Slovenia; first—the study of customer preferences and opinions about electric vehicles, made in year 2013 and second—the study of a possible sustainable transport solution for a Slovenia town where one of the possible models of integration of RES and EVs is presented.

6.2 Review of policies for promoting the use of electric vehicles around the world

Electric vehicle promotional policies can be divided geographically, according to their adoption on the international, national, or local level (Nilsson, Hillman, & Magnusson, 2012; Ward, 1998). Policies can also be divided from the legislative perspective into internationally binding and nonbinding policies. They can include fiscal and nonfiscal incentives, which encompass a wide range of incentives and measures (May, 2004; Steenberghen & Lopez, 2008). From the end buyer's or user's perspective, these measures can be divided into direct or indirect measures. The use of electric vehicles can be promoted among potential buyers with direct measures, while the development of electric vehicles can be stimulated with indirect measures in order to meet the customer or even the legislative body's needs. Policies can be accepted unilaterally or as an agreement between representatives of the society (Åhman, 2006; Calef & Goble, 2007).

IEA (2013) divides promotional policy measures into those that include fiscal incentives, those that are connected to the research and implementation of infrastructure for electric vehicles, and those that are designed to support the research and development (R&D) of technology related to electric vehicles. Policies for the promotional of electric vehicles were at first intended for manufacturers to develop and produce them, since the necessary technology for their introduction was not yet in place. The incentive for such policies came from government institutions (Fale, 2014). The development and production of electric vehicles have given car manufacturers the possibility of offering electric vehicles on the market. In a parallel move, the initial goals regarding the number and share of electric vehicles on the roads in the near future were also set. These goals would also serve as a useful tool to measure the results of the government incentives, although the initial goals regarding the number of electric vehicles on the road were not met. However, the manufacturers were able to develop the electric vehicles demanded by the government and actually launch them on the roads. We are aware that the demand to first fully develop and produce electric vehicles, all in a very short space of time, was an extremely challenging mission for car manufacturers (Nicholon in Cowan & Hultén, 1996). It is clear that the predictions regarding electric vehicle use must be modified in accordance with technological developments (Iguchi in Åhman, 2006). Technological forecasting is essential, as is the identification of market expectations. These have been taken into account in Europe, where in some countries, there is a clear desire to establish an electric vehicle market. Groups of companies, divided into technology clusters and technology platforms, also took part in this process with financial support from the government. Their goal was to identify the needs and wishes of potential electric vehicle users and to determine if the current technology was able to meet them; however, they discovered that the technology was not far enough advanced to do so (Cowan & Hultén, 1996). Since the initial research in the field of electric vehicles, a great deal of money has been invested in their development, although financial support fell after the initial research was concluded. The development of electric vehicle technology lost its initial impetus, which was especially beneficial for the existing technology, the internal combustion engine, which has been a commanding force in the field of transport vehicles throughout the 20th century. However, investments in electric vehicle technology again increased significantly at the beginning of the 21st century (Fale, 2014).

In recent years, R&D has enjoyed the biggest share of investment, followed by fiscal incentives (financial grants). As presented in Fig. 6.2.1, the smallest share was intended for infrastructure development (IEA, 2013).

An important breakthrough in the introduction of electric vehicles occurred in 1990 when California adopted an act that prescribed the maximum levels of exhaust fumes for new vehicles (Pilkington & Dyerson, 2006). Cowan and Hultén (1996) claim that this is the first example of a government act that promotes the development of low emission vehicles (LEV). This law demanded that car manufacturers produce such vehicles within a certain time frame, although the electric vehicle technology that would best comply with the law, is not yet in place (Calef & Goble, 2007). The legislation adopted in several of the US federal states provided a trigger for technological changes in the field of electric vehicle promotional (Pohl & Yarime, 2012). Despite the evident opposition of the US car industry, it accepted this challenge and began to develop appropriate vehicles. The adoption of this law also allowed smaller car

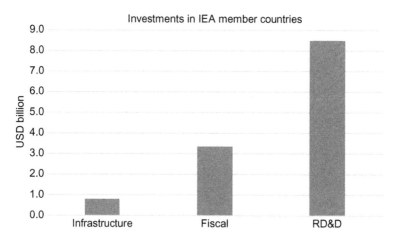

Fig. 6.2.1 Investments in IEA member countries from 2008 to 2012.
Adapted by EVI in IEA (2013). Global EV outlook. Available from: http://www.iea.org/publications/globalevoutlook_2013.pdf Accessed 20.03.14.

manufacturers to enter the electric vehicle market (Calef & Goble, 2007; O'Dell, 2012; Pilkington & Dyerson, 2006), while the authorities have supported the car industry in its goal to develop and to produce these vehicles through fiscal and non-fiscal incentives (Gallagher & Muehlegger, 2011; Ward, 1998). Although companies have managed to develop electric vehicles and present them to the public, the number of electric vehicles has been smaller than expected. The adoption of this law in California has also had an influence on the rest of the world, e.g. Japanese car manufacturers were less opposed to the regulations than their American counterparts. As Japanese carmakers are present on the American market, they adopted strategies in accordance with the law, developing electric vehicles that have met Japanese and European and US standards (Yamaguchi in Åhman, 2006; Pohl & Yarime, 2012; JARI, in Åhman, 2006). These adopted financial incentives offered support for electric vehicles when they hit the market, which had an influence on the growth of electric vehicle sales (JEVA in Åhman, 2006). Carley, Krause, Lane, and Graham (2013) noted that the interest in plug-in electric vehicles is shaped primarily by the consumers' perceptions of electric vehicle disadvantages; financial incentives are therefore welcome in the promotion of further development and demand for electric vehicles.

Developments in the field of electric vehicles has changed the way policies for promoting them are determined; these policies are now being shaped not just by authorities but also with the help of multiple stakeholders such as car companies, educational institutions, government agencies, and local authorities (Watanabe in Åhman, 2006). Goals regarding target numbers of electric vehicles on the roads are not being set by government legislation alone but are accepted as an agreement between working partners. Each partner has taken its own share of responsibility for the introduction of electric vehicles (Calef & Goble, 2007).

However, electric vehicle promotional policies have in reality changed little since their introduction. The only thing that has changed is the level of responsibility shouldered by the authorities and car industry for implementing each of the incentives programs concerning electric vehicles and infrastructure (Nelson & Tanabe, 2013; Toyota, 2013). The preparation and implementation of multilevel policies (where the highest level has an influence on lower levels) is evident in Europe throughout the EU. In countries joining the EU, there is evidence of policies oriented to promote electric vehicles on the EU level that affect policy implementation in the EU member states. This can be attributed to the adopted EU measures in the form of strategies, legislative decisions, and support for the implementation of individual projects (Europa Press Releases, 2013; Elektro crpalke, 2014; European Commission, 2014; Fernandez, Chen, & da Graça Carvalho, 2005). The policies for the promoting of electric vehicle use have been adopted in the EU member states and include fiscal and nonfiscal incentives (France-Diplomatie, 2014; Gass, Schmidt, & Schmid, 2014; Gregorcic, 2011a; Grünig, Witte, Marcellino, Selig, & van Essen, 2011; Hannisdahl, Malvik, & Wensaas, 2013; IEA, 2013; Scientific American, 2011). The effects of those policies are seen in the demand and supply of electric vehicles. Sales of electric vehicles are rising, but the overall share of electric vehicles sold is still extremely low. New models of electric vehicles that customers can actually buy have also influenced the increase in electric vehicle sales.

In the future can be expected a further increase in electric vehicles sales, which should, as forecast by the IEA, reach as high as 5,900,000 vehicles in IEA member states by 2020 (EVI in IEA, 2013).

6.2.1   Review of policies for promoting the use of electric vehicles in Slovenia

The accession of Slovenia to the EU and the signing of the international agreement on environmental goals (Kyoto Protocol) constitute the starting point of policies for the promotion of electric vehicles in Slovenia. Slovenia's accession to the EU in particular is an important milestone, as with this act, Slovenia chose to respect the developmental guidelines on electric vehicles already in place in the EU. Slovenia has modified its legislation by transferring important regulations directly or indirectly connected with the area of electric vehicles promotion into her legislation (European Parliament, 2011).

The first of all electric vehicle types on the Slovenian roads were battery electric vehicles, which were modified cars that previously ran on petrol (internal combustion engine) (Fale, 2014). These cars were modified by individuals who were not supported by the relevant authorities. These converted cars are all individual projects that are now being used mainly for promotional activities (Pecjak, 2014). Much more attention, however, has been given to fuel cell technology (and consequently fuel cell vehicles) than to battery electric vehicles. In 2005, the Slovenian Hydrogen and Fuel Cell Technology Platform (SIHFC) was established by the Chamber of Commerce and Industry alongside the relevant ministry. The SIHFC united companies exploring fuel cells and educational institutions dealing with research projects regarding fuel cells (SIHFC, 2014). The SIHFC also organized the first event, where the technology of electric vehicles, electric vehicles made by car manufacturers (cars that could actually be bought at that time), and electric vehicles converted by individuals were presented to the public (Elaphe, 2009).

One of the largest problems in the transition to alternative fuels is generally the infrastructure itself and, consequently, the accessibility of alternative sources. This is especially important for transport in the business sector where companies cannot allow their cars to be stuck by the side of the road with an empty battery or without hydrogen. It means that for the transition to cleaner technologies, hybrids are the most appropriate choice, since they offer the best possible solution until the infrastructure required by BEV or fuel cell vehicles is established. As an example of the aforementioned, let's mention the case of the New York Police Department, which uses Toyota Prius hybrid vehicles as their patrol cars.

Currently, the infrastructure is better developed for battery electric vehicles. In January 2012, the first fast-charging station for battery electric vehicles was opened in Maribor (Praper, 2012). With the introduction of charging points, there has been an increase in the use of battery electric vehicles, which are mainly used by electricity distribution companies for promotional activities (Avtovizije, 2014). At present, there are over 150 charging points for electric vehicles in Slovenia; however, the number of charging points is still lower than that proposed by the EU Commission (Elektro crpalke, 2014; Gregorcic & Slovenian Press Agency, 2013; Polni.si, 2014).

Potential electric vehicle buyers can apply to an Eco Fund for financial subsidies and loan schemes to help with their purchase. The Eco Fund approves loans for electric vehicles if its emissions do not exceed a predetermined level. These loans are intended specifically for hybrid vehicles, as they do not exceed the strict rules governing emissions (Fale, 2014). Since 2011, the general public and companies have had the chance to receive financial grants for the purchase of electric vehicles that meet predetermined criteria or for customizing any vehicle into an electric one in compliance with the relevant criteria (Eco Fund, 2013c; Official Gazette of the Republic of Slovenia, 2011). Grants are available only for battery electric vehicles and plug-in hybrid vehicles. Unfortunately, these grants remain rather unexploited (Purgar, 2012), as the share of electric vehicles is still relatively low.

The Eco Fund also gives grants to transport companies for the conversion of buses to biogas or compressed natural gas. In 2012, there were no applicants, so no grants were approved or given (Eco Fund, 2013c).

The question must be asked as to whether the public is properly informed about the possibility of receiving grants to purchase electric vehicles (Krause, Carley, Lane, & Graham, 2013). Moreover, Hackbarth and Madlener (2013) revealed that households are willing to pay considerable amounts for greater fuel economy and emission reduction, improved driving range and charging infrastructure and for enjoying vehicle tax exemptions and free parking or bus lane access. However, they are usually insufficiently informed about the public incentives and policies promoting electric vehicles. The possibility of companies receiving grants for electric vehicles can also be very important for their promotion, particularly in large and well-known companies.

Besides the government, which operates electric vehicle promotion policies, local authorities also play a key role, as they are an important partner in international projects, carrying out demonstration projects and adopting goals concerning the introduction of electric vehicles into everyday use (Civitas Elan, 2010; Klajnscak, 2014). The City Municipality of Ljubljana has adopted an electromobility plan in which there are clearly determined measures to encourage the use of electric vehicles in the city (Razpotnik, Loose, Jazbinsek Srsen, Simonovic, & Klancar, 2013). Car dealers, who represent car manufacturers, often do not take part in policies for promoting the use of electric vehicles. Even when they organize special offers for replacing old vehicles with new, cleaner vehicles, a problem arises, as most of the car dealers behind such offers do not sell electric vehicles (Car Dealership Real, 2014; Gregorcic, 2013a).

Past and present electric vehicle promotion policies were adopted as a result of cooperation between public and private institutions, e.g. research institutes and car industry representatives. For this purpose, electric vehicle development strategies were drawn up; financial grant schemes were accepted alongside with other measures and activities. Review of these measures and activities is presented on Table 6.2.1.

Additionally, international project cooperation between public and private sector was also established on this basis. These projects were mainly focused in research and development and cooperation of different EU member states with private sector. Review of Slovenian international project cooperation is presented on Table 6.2.2.

Table 6.2.1 Presentation of activities and measures for promoting the use of electric vehicles in Slovenia

Activity/measure and its time span	Description	Type of included electric vehicle	Competent authority	Interesting details
Loan scheme (available since 2004)	Citizens and legal entities can ask for a loan to buy electric vehicle, which price does not exceed € 40,000	Hybrid vehicles, Battery electric vehicles (since 2009)	Eco Fund of the Republic of Slovenia	In 2009, a new criterion for approving loans came into force: CO_2 levels of electric vehicles cannot exceed 120 g/km (from 2010, 110 g/km); Legal entities were included in loan scheme in 2007
Promotional event CEVELJ (2007–2010, 2012–2013; annual events)	In these events, the public was presented with technology of electric vehicles, electric vehicles made by car manufacturers (cars that could actually be bought at that time) and electric vehicles converted by individuals	All electric vehicles on the market	Various organizations	In 2010, event changed its name from CEVELJ into ECOmeet
Taxation of vehicles (adopted in 2009 by Government of the Republic of Slovenia; came into force in 2010)	Tax rate of vehicles depends of CO_2 emissions of these vehicles (the higher the CO_2 emissions are, higher the tax rate is); Tax rates are also depended on the type of internal combustion engine (diesel engines have higher tax rate than petrol engines if the CO_2 levels are in the same tax class)	All electric vehicles on the market	Financial Administration of the Republic of Slovenia	The lowest tax rate is 0.5% for petrol or LPG vehicles with CO_2 emissions 110 g/km or under and the highest tax rate is 31% for diesel vehicles with CO_2 emissions over 250 g/km
Financial subsidies (available since 2011)	Citizens and legal entities can ask for a financial subsidy to buy electric vehicle or to convert vehicle into electric vehicle; Total sum of money for financial subsidies is € 500,000 per year (€ 200,000 for citizens and € 300,000 for legal entities); Amount of financial subsidies is dependent on type of electric vehicle and on class of	Battery electric vehicles, plug-in hybrid vehicles	Eco Fund of the Republic of Slovenia	Until 2014 price of electric vehicle had to be under € 50,000 (VAT included) if citizens or legal entities wanted to receive financial subsidy; In 2014, there has been a slight change in the amount of financial subsidies

	for vehicles because of the new EU regulation
the vehicle; The lowest financial subsidy is € 1000 and the highest is € 5000; In 2014, financial subsidy for battery electric automobile is € 5000, for plug-in electric automobile € 3000 and for converting automobile into electric automobile € 5000; There are several conditions that citizens or legal entities need to comply with in order to receive and keep subsidy	

Adapted from AMZS (2012). DMV—Motor vehicle tax in 2012. Available from: http://www.amzs.si/si/356/98/default.aspx Accessed 01.12.14 in Slovenian language; Eco Fund of the Republic of Slovenia (2004). Annual report on activities and operations of Eco Fund of the Republic of Slovenia, Public Fund in 2004. Available from: http://www.ekosklad.si/pdf/LetnaPorocila/LP_04_slo.pdf Accessed 17.03.14 in Slovenian language; Eco Fund of the Republic of Slovenia (2006). Annual report on activities and operations of Eco Fund of the Republic of Slovenia, Public Fund in 2005. Available from: http://www.ekosklad.si/pdf/LetnaPorocila/LP_05_slo.pdf Accessed 17.03.14 in Slovenian language; Eco Fund of the Republic of Slovenia (2007). Annual report on activities and operations of Eco Fund of the Republic of Slovenia, Public Fund in 2006. Available from: http://www.ekosklad.si/pdf/LetnaPorocila/LP_06_slo.pdf Accessed 17.03.14 in Slovenian language; Eco Fund of the Republic of Slovenia (2008). Annual report on activities and operations of Eco Fund of the Republic of Slovenia, Public Fund in 2007. Available from: http://www.ekosklad.si/pdf/LetnaPorocila/LP_07_slo.pdf Accessed 17.03.14 in Slovenian language; Eco Fund (2009). Annual report on activities and operations of Eco Fund, Slovenian Environmental Public Fund in 2008. Available from: http://www.ekosklad.si/pdf/LetnaPorocila/LP_08_slo.pdf Accessed 17.03.14 in Slovenian language; Eco Fund (2010). Annual report on activities and operations of Eco Fund, Slovenian Environmental Public Fund in 2009. Available from: http://www.ekosklad.si/pdf/LetnaPorocila/LP_09_slo.pdf Accessed 14.06.14 in Slovenian language; Eco Fund (2011a). Appeal 8SUB-EVOB11. Available from: http://www.ekosklad.si/pdf/doc/8SUB-EVOB11_javni_poziv.pdf Accessed 14.06.14 in Slovenian language; Eco Fund (2011b). Appeal 9SUB-EVPO11. Available from: http://www.ekosklad.si/html/razpisi/main.html Accessed 14.06.14 in Slovenian language; Eco Fund (2011c). Annual report on activities and operations of Eco Fund, Slovenian Environmental Public Fund in 2010. Available from: http://www.ekosklad.si/pdf/LetnaPorocila/LP_10_slo.pdf Accessed 17.03.14 in Slovenian language; Eco Fund (2012a). Appeal 15SUB-EVOB12. Available from: http://www.ekosklad.si/pdf/SUB2012/15SUB-EVOB12_1_JavniPoziv_URL12.pdf Accessed 14.06.14 in Slovenian language; Eco Fund (2012b). Appeal 16SUB-EVPO12. Available from: http://www.ekosklad.si/html/razpisi/main.html Accessed 14.06.14 in Slovenian language; Eco Fund (2012c). Annual report on activities and operations of Eco Fund, Slovenian Environmental Public Fund in 2011. Available from: http://www.ekosklad.si/pdf/LetnaPorocila/LP_11_slo.pdf Accessed 17.03.14 in Slovenian language; Eco Fund (2013a). Appeal 20SUB-EVOB13. Available from: http://www.ekosklad.si/pdf/SUB2013/20SUB-EVOB13_JavniPoziv_url13.pdf Accessed 14.06.14 in Slovenian language; Eco Fund (2013b). Appeal 21SUB-EVPO13. Available from: http://www.ekosklad.si/html/razpisi/main.html Accessed 14.06.14 in Slovenian language; Eco Fund (2013c). Annual report on activities and operations of Eco Fund, Slovenian Environmental Public Fund in 2012. Available from: http://www.ekosklad.si/pdf/LetnaPorocila/LP_12_slo.pdf Accessed 17.03.14 in Slovenian language; Eco Fund (2014a). Appeal 26SUB-EVOB14. Available from: http://www.ekosklad.si/dl/R/14/26SUB-EVOB14_JavniPoziv.pdf Accessed 14.06.14 in Slovenian language; Eco Fund (2014b). Appeal 27SUB-EVPO14. Available from: http://www.ekosklad.si/html/razpisi/main.html Accessed 14.06.14 in Slovenian language; Eco Fund (2014c). Appeal 51OB14. Available from: http://www.ekosklad.si/html/razpisi/main.html Accessed 07.07.14 in Slovenian language; EcoMeet (2010). Programme of the event. Available from: http://ecomeet.si/docs/Program_prireditve/EcoMeetSLO.pdf Accessed 08.08.14 in Slovenian language; EcoMeet (2013). After the event. Available from: http://ecomeet.si/sl/ECOmeet2013/ Accessed 17.06.14 in Slovenian language: Elaphe (2009). Presentation of recent events CEVELJ 1, CEVELJ 2 and CEVELJ 3 in 2007, 2008 and 2009. Available from: http://ecomeet.si/docs/Program_prireditve/PredstavitevDogodkovCeveljlin3.pdf Accessed 17.03.14 in Slovenian language; Government of the Republic of Slovenia (2009). Operational programme for reducing emissions of greenhouse gases until 2012 (OP TGP-1). Available from: http://www.vlada.si/fileadmin/dokumenti/si/projekti/2009/podnebne/op_toplogredni_plini2012_1.pdf Accessed 29.03.14 in Slovenian language; Fale, M. (2014). Review of policies for promoting the use of electric vehicles. Celje, VII, 83 (in Slovenian language).

Table 6.2.2 Presentation of Slovenia's involvement in international projects under the umbrella of EU

Project title	Time span	Field of research	Type of included electric vehicle	Share of funds, contributed by EU	Role of Slovenia	Type of organization from Slovenia (public/private)
MAG–DRIVE	From Oct. 2010 to Sep. 2016	Propulsion system	Battery electric vehicles	71.3%	Coordinator country	Public and private organization
SMARTV2G	From Jun. 2011 to May 2014	Charging of electric vehicles	Battery electric vehicles, Plug–in hybrid vehicles	76.8%	Participant country	Private organizations
EUROLIION	From Feb. 2011 to Jan. 2015	Propulsion system	Battery electric vehicles	71.5%	Participant country	Public organization
CAPIRE	From Dec. 2010 to Nov. 2014	Identifying obstacles for introduction of electric vehicles	Electric vehicles in general	78.9%	Participant country	Private organization
HYSYS	From Dec. 2005 to Nov. 2010	Propulsion system	Fuel cell vehicles	50.6%	Participant country	Public organization

Adapted from European Commission: CORDIS (2011). SMARTV2G. Available from: http://cordis.europa.eu/project/rcn/99306_en.html Accessed 25.06.14: European Commission: CORDIS (2013a). EUROLIION. Available from: http://cordis.europa.eu/project/rcn/109414_en.html Accessed 25.06.14: European Commission: CORDIS (2013b). MAG–DRIVE. Available from: http://cordis.europa.eu/project/rcn/110008_en.html Accessed 25.06.14: European Commission: CORDIS (2013b). HYSYS. Available from: http://cordis.europa.eu/project/rcn/78586_en.html Accessed 25.06.14: European Commission: CORDIS (2014). CAPIRE. Available from: http://cordis.europa.eu/projects/rcn/96977_en.html Accessed 25.06.14: Fale, M. (2014). Review of policies for promoting the use of electric vehicles. Celje, VII, 83 (in Slovenian language).

While the participating partners have defended the success of these activities, the real effects are seen through electric vehicle sales, which have not met the predictions made by car sale companies (Gregorcic, 2011b). The first electric vehicles, which were actually customized vehicles, were registered for the first time in 1993. However, a growth in the number of electric vehicles registered for the first time has been seen only since 2004. The number of first-time registered electric vehicles in Slovenia is presented in Fig. 6.2.2. As is evident from the graph, most of the registered electric vehicles are HEV. BEV started to appear in 2011, and there are still no fuel cell vehicles in Slovenia (Portal NIO, 2012).

If the share of electric vehicles out of all the vehicles registered for the first time in Slovenia, Austria, Germany, and Great Britain, as presented in Fig. 6.2.3, is compared, a huge difference can be noted.

The share of electric vehicles in Slovenia per year does not reach the shares of electric vehicles in other countries where electric vehicle promotion policies are being implemented. The share of electric vehicles in 2013 in Austria, Germany, and Great Britain was more than three times higher than in Slovenia, which means that Slovenian policy is insufficiently promoting electric vehicles and that the policy is not motivating enough to increase electric vehicle demand.

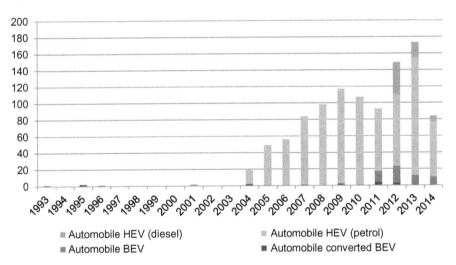

Fig. 6.2.2 Number of electric vehicles (category: automobiles), registered for the first time in Slovenia (per year).
Adapted from Portal NIO (2012). First registered vehicles in 2012, by month. Available from: http://nio.gov.si/nio/data/prvic+registrirana+vozila+v+letu+2012+po+mesecih Accessed 17.03.14 in Slovenian language; Portal NIO (2013). First registered vehicles in 2013, by month. Available from: http://nio.gov.si/nio/data/prvic+registrirana+vozila+v+letu+2013+po +mesecih Accessed 17.03.14 in Slovenian language; Portal NIO (2014). First registered vehicles in 2014, by month. Available from: http://nio.gov.si/nio/data/prvic+registrirana +vozila+v+letu+2014+po+mesecih Accessed 17.03.14 in Slovenian language; Fale, M. (2014). Review of policies for promoting the use of electric vehicles. Celje, VII, 83 (in Slovenian language).

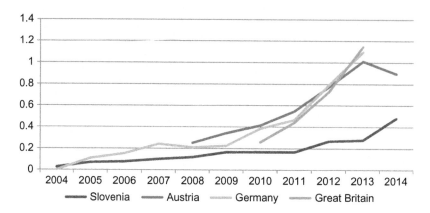

Fig. 6.2.3 Share of electric vehicles in Slovenia, Austria, Germany, and Great Britain (category: automobiles; per year, in %).
Adapted from Portal NIO (2012). First registered vehicles in 2012, by month. Available from: http://nio.gov.si/nio/data/prvic+registrirana+vozila+v+letu+2012+po+mesecih Accessed 17.03.14 in Slovenian language; Portal NIO (2013). First registered vehicles in 2013, by month. Available from: http://nio.gov.si/nio/data/prvic+registrirana+vozila+v+letu+2013+po +mesecih Accessed 17.03.14 in Slovenian language; Portal NIO (2014). First registered vehicles in 2014, by month. Available from: http://nio.gov.si/nio/data/prvic+registrirana +vozila+v+letu+2014+po+mesecih Accessed 17.03.14 in Slovenian language; Statistik Austria (2014). Kraftfahrzeuge-Neuzulassungen. Available from: http://www.statistik.at/web_de/ statistiken/verkehr/strasse/kraftfahrzeuge-_neuzulassungen/index.html Accessed 06.05.14; KBA Umwelt. (2004). Umwelt—Zeitreiche 2004. Available from: http://www.kba.de/cln_031/ nn_191064/DE/Statistik/Fahrzeuge/Neuzulassungen/Umwelt/n__umwelt__z__teil__1.html Accessed 06.05.14; KBA Umwelt. (2013). Umwelt—Zeitreiche 2005 bis 2013. Available from: http://www.kba.de/cln_031/nn_191064/DE/Statistik/Fahrzeuge/Neuzulassungen/Umwelt/n__ umwelt__z__teil__2.html Accessed 06.05.14; Statistical data sets GOV.UK (2014). Table VEH0130. Available from: https://www.gov.uk/government/statistical-data-sets/veh01-vehicles-registered-for-the-first-time Accessed 06.05.14; Fale, M. (2014). Review of policies for promoting the use of electric vehicles. Celje, VII, 83 (in Slovenian language).

6.3 Studies of Slovenia

Further, two studies that were done in Slovenia are presented. Slovenia is a central European country that borders the Adriatic Sea through a small coastal strip. Koper is the main port of the country. The terrain consists of an alpine mountain region adjacent to Italy and Austria and mixed mountains and valleys with numerous rivers to the east. Despite its small size, this eastern Alpine country controls some of Europe's major transit routes. The country is crossed by two TEN-T corridors: Corridor V (Venice-Trieste/Koper-Ljubljana-Budapest-Kiev) and Corridor X (Salzburg-Ljubljana-Zagreb-Belgrade-Thessaloniki) (Fig. 6.3.1).

Fig. 6.3.1 The position of Celje within Slovenia.
Adapted by Kobla.net (2013). The position of Celje within Slovenia. Available from: http://www.kobla.net/lokacija/ Accessed 29.02.13 in Slovenian language.

6.3.1 Study I: Key factors influencing the purchasing decisions of electric vehicles in Slovenia

Here, a study of customer preferences and opinions about electric vehicles is presented. Nowadays the studies of green technologies, especially in the area of green transport, are interesting for policy makers, vehicle producers, customers, and energy. Many stakeholders from public and private sector are investing a lot of effort to identify consumer behaviour for future improvements in development of their green products and strategies. This study supplier (Knez, Jereb, & Obrecht, 2014; Zupan, 2014) is a modification of previously conducted research on customer behaviour on the same topic in Scotland (Borthwick & Carreno, 2012). Study tried to identify the most important parameters of consumer behaviour related to purchase of electric vehicle. To facilitate data analysis, principal component factor analysis was employed to reduce the number of situational variables, which resulted in seven broad factors as presented in Table 6.3.1:

The study was designed to reveal the underlying factors that affect the purchasing habits of people. The results reveal new perspective of purchasers and indicate which factors are the most important for the purchase of an EV.

Two nonfinancial factors are crucial when deciding on a car purchase—(1) 'overall condition and mileage of vehicle (when buying a used car)', and—(2) 'safety

Table 6.3.1 Situational factors of importance in a future vehicle-purchasing decision (Borthwick & Carreno, 2012)

Factors	Attributes
Financial considerations at the time of purchase	• Vehicle price • VAT and other purchase taxes • Value for money
Future financial considerations	• Insurance group for vehicle • Maintenance/repair costs • Warranty (length and coverage) • Biannual/annual VED • Trade-in value
Fuel and performance	• Fuel consumption (miles per gallon/kilometres per litre) • Engine type/size • Fuel type • Fuel economy • Performance/driveability
Exterior design features	• Vehicle make • Model of vehicle • Vehicle size • Style/appearance/colour
Interior design features	• Safety features • Security features • Equipment levels • Entertainment system • Acceleration time
Load space	• Luggage/storage space • Passenger capacity • Body shape
Environmental considerations	• Emissions of CO_2 and other greenhouse gases • Emissions of other air pollutants • Vehicle noise

features'. Other very important factors are vehicle size (exterior), style/appearance/colour, body shape (e.g. hatchback and coupe), and fuel type. Divided results for men and women are presented on Fig. 6.3.2.

The results indicate that there are some differences between male and female population, especially when examining safety features, acceleration, and fuel type. Safety is more important for women; acceleration and fuel type are more important for men. Correlation between education and car features was also studied; however, no significant differences were identified.

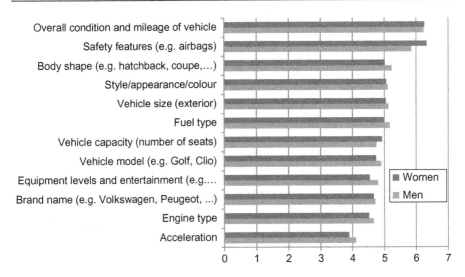

Fig. 6.3.2 Important vehicle performance factors (on a scale from 1 to 7 where 1 means not important and 7 means very important).

When it comes to financial features, most important thing seems to be the total price of the vehicle. Second feature, also very important, is fuel economy. Especially now when gas prices are high and still increasing, information on fuel consumption is crucial. People also put emphasis on repair costs and on the value/money ratio. Two features that are less important are 'trade-in value' (how much money you get when you sell your vehicle) and 'annual road tax'. This makes sense as annual road tax is a relatively smaller expense than, e.g. vehicle price.

A survey by Lane and Banks (2010) identifies fuel economy as one of the most important car purchase factors. Other two most important purchasing factors were size/practicality and vehicle price. When it comes to fuel economy, buyer's primary concern is the size of running cost rather than concern about the environment. Even when a person is green oriented, lower carbon emissions are often taken as a bonus after the main objective of lower running costs has been secured. Carbon emissions and environmental awareness mostly do not have influence on car choice (Anable et al., 2008). Vehicle owners are aware of carbon emissions only if carbon emissions are tied to road taxes, and few car owners can give correct information about carbon emissions for their recently purchased vehicle.

Based on the results from this research, no significant differences between men and women in terms of financial considerations were found. Average grade for individual financial factors was almost the same (Fig. 6.3.3). The respondents were asked about petrol and diesel prices, too. Ten percent of study participants were already seriously thinking about buying a car running on alternative fuels (e.g. biodiesel, bioalcohol, hydrogen, and electricity). If gas prices increased by 30% (to 1.96 EUR/l for petrol and to 1.80 EUR/l for diesel), 58% of respondents would

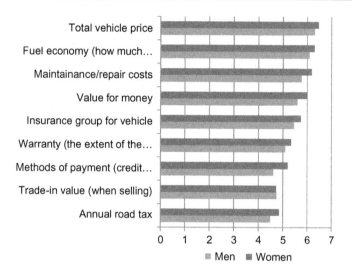

Fig. 6.3.3 Important financial considerations (on a scale from 1 to 7 where 1 means not important and 7 means very important).

start thinking about buying a car that is powered by an alternative fuel (30% increase of petrol prices over several years is not an unlikely situation). The results are presented in Fig. 6.3.4.

In the period from 30 Dec. 2008 to Apr. 2014, petrol prices in Slovenia increased by 83% (from 0.827 to 1.514 EUR/l). It is clearly seen in Fig. 6.3.4 that increasing fuel prices are a motivating factor for people to start considering buying a car running on alternative fuels.

Another key result of this research shows that when people are buying a car, they are more interested in the total price of the car than in different taxes. Two percent of respondents are already thinking about buying an electric car despite their relatively high prices. If the prices of electric cars decreased by 10%, 5% of all research participants

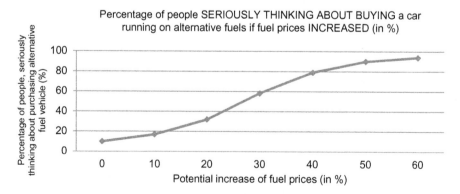

Fig. 6.3.4 Influence of increasing fuel prices on alternative fuel vehicle demand.

would be seriously thinking about purchasing an electric car. If prices decreased by 20%, 25% of all respondents would be seriously thinking about purchasing an electric car; and if prices decreased by 30%, more than one half (59%) of respondents would be seriously thinking about purchasing an electric car. Results are presented in Fig. 6.3.5.

It is recognized that any population is made up of individuals with varying levels of susceptibility towards changing their behaviour (Anable, 2005; Carreno & Welsch, 2009). The influence of taxation and other policy measures upon vehicle purchasing decisions will thus also vary and needs to be accounted for in future policy decisions. K-means cluster analysis was therefore undertaken to identify population segments within the Slovene driver population, resulting in three distinct segments. Based on their response regarding the importance of situational factors and strength of psychological constructs, the following groups were formed:

- Group one—*no-greens* (20% of the total sample)
- Group two—*go-with-the flow-greens* (42% of the total sample)
- Group three—*go-greens* (38% of the total sample)

'No greens' is the group not motivated to buy a LEV in the near future. The information about CO_2 and other emissions is not important to them when buying a car. 'Go-with-the-flow-greens' has a positive opinion about LEVs, but they are still not planning to buy one, like the people in the 'no-green' group. The people in third group 'go-green' are very interested in buying a LEV in the near future. They are aware of their responsibility to reduce environmental impact.

Factors that may influence consumers' willingness to buy environmentally friendly products can be classified into five categories (Laroche, Bergeron, & Barbaro-Forleo, 2001): demographics, knowledge, values, attitudes, and behaviour. Taking into account some of these categories, such as gender, as one of the demographic factors, Fig. 6.3.6 shows that only 11% of all women are in the no-green group, and the remaining 89% are a part of the other two groups. Men are more equally distributed through all three groups (29.4% in no-greens, 39.2% in the go-with-the-flow-greens,

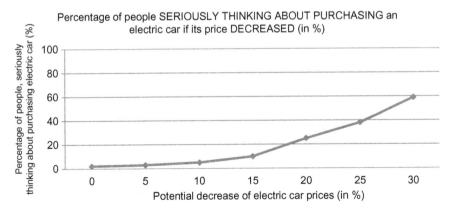

Fig. 6.3.5 Percentage of respondents seriously thinking about purchasing an electric car dependent on potential price decrease of an electric car.

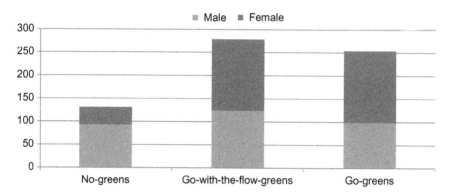

Fig. 6.3.6 Ratio between males and females in different segments of consumers.

and 31.3% in the go-greens). The study shows that there are considerably more men than women in the 'no-green' group, which is consistent with the studies (Banerjee & McKeage, 1994; McIntyre, Meloche, & Lewis, 1993) that found that women tend to be more ecologically conscious than men. Even Borthwick and Carreno (2012) came to a similar conclusion in their study.

Early research identified the green consumer as being younger than average (Anderson & Cunningham, 1972; Berkowitz & Lutterman, 1968; Van Liere & Dunlap, 1981). Surprisingly, this trend has been reversed in the last decade, and several recent studies identified the green consumer as being older than average (Roberts, 1996; Sandahl & Robertson, 1989; Vining & Ebreo, 1990).

Our study also came to the same interesting conclusion as the previously mentioned research that the majority of the no-greens are younger than 44 years. Results are presented in Fig. 6.3.7. Only 11.1% of population above 44 can be defined as no-greens. The ratio of go-greens is increasing with age. It is also very interesting that no people above 60 years who would classify as no-greens were found. Our study also identifies the main reasons for that; they scored high grades on the question about CO_2 and other emissions. Moreover, they are also very attentive about vehicle noise.

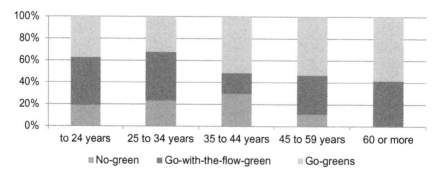

Fig. 6.3.7 Age distribution between different segments of consumers.

Taking into account the 'nonfinancial' factors relevant for future car-purchasing decisions, there are some minor differences between the groups. In general, however, the four most important aspects for every single group are the following:

1. Overall condition and mileage of vehicle (if you buy a used car)
2. Safety features (e.g. airbags)
3. Style/appearance/colour
4. Vehicle size (exterior)

Furthermore, the effect of different push and pull methods on different purchasers was also studied. Table 6.3.2 shows that different push and pull methods would have different impact on different groups. The most influential measure is 'vehicle scrappage scheme'. This 'pull' suggestion would have a strong effect on all three groups.

Borthwick and Carreno (2012) results confirmed greater influence of pull over push measures, and those positive rewards for purchasing an LEV were more influential than those penalising people who choose not to purchase an LEV. The most influential factor in our survey (vehicle scrappage scheme) is also a pull factor. Others include

• motor insurance premiums partly based on carbon emissions (i.e. drivers of higher emission vehicles pay more),
• VAT based on carbon emissions (i.e. buyers of higher emission vehicles would pay more VAT, and buyers of lower emission vehicles would pay less VAT),
• vehicle registration fee based on carbon emissions of the vehicle (i.e. buyers of higher emission vehicles pay more).

6.3.2   Study II: A sustainable transport solution for a Slovenia town

Authorities in Slovenia and other EU member states are confronted with problems of city transportation. Fossil-fuel-based transport poses two chief problems—local and global pollution and dwindling supplies and ever increasing costs. An elegant solution is to gradually replace the present automobile fleet with electric vehicles (EVs). This study explores the economics and practical viability of the provision of solar electricity for the charging of EVs by installation of economic available PV modules. A steep decline in the module, inverter, and installation costs is reported herein. Present estimates indicate that for the prevailing solar climate of Celje—a medium-sized Slovenian town—the cost would be only €2.11cents/kWh of generated solar electricity.

Energy consumption has been growing in the last decade by 1%–2% annually. In order to reach the 20/20/20 objectives, it is necessary to curb and reduce energy consumption. Also, a medium-term reduction of fossil fuel consumption in transport will result in a higher growth of electricity consumption in comparison with other fuels (Obrecht & Knez, 2014). Electricity consumption (in Mtoe) and the share of renewables in electricity production (in %) are presented in Fig. 6.3.8.

Fig. 6.3.8 clearly shows that, on the one hand, energy consumption grew strongly between 2000 and 2007 and has remained almost the same or has even

Table 6.3.2 Different measures that encourage people to purchase a low emission vehicle (Knez, Jereb, & Obrecht, 2014a; Zupan, Jereb, Rosi, & Knez, 2013)

	Average	No-green	Go-with-the-flow-green	Go-green	Difference between average maximum and average minimum
VAT based on carbon emissions (i.e. buyers of higher emission vehicles would pay more VAT)	5.37	4.73	5.35	5.73	1.00
First year rate of road tax derived by a fixed monetary amount (€) per gram of CO_2 (i.e. drivers of low-emission cars pay less)	4.98	4.16	4.93	5.47	1.31
A road user charging scheme based on carbon emissions (i.e. drivers of higher emission vehicles pay more)	5.35	4.66	5.29	5.76	1.10
A vehicle registration fee based on carbon emissions of vehicle (i.e. buyers of higher emission vehicles pay more)	5.36	4.65	5.37	5.71	1.06
Vehicle scrappage scheme with a carbon emissions limit on the replacement vehicle (i.e. you would receive money from the government for getting rid of (scrapping) your old car if you buy a low-emission new car)	5.94	5.28	5.96	6.27	0.99
Annual road tax derived by a fixed monetary amount (€) per gram of CO_2 (i.e. drivers of higher emission vehicles pay more)	5.22	4.50	5.14	5.69	1.46
Parking charges partly based on carbon emissions (i.e. low-emission cars would pay less to park)	4.28	3.64	4.18	4.71	1.07
'Low-emission vehicle lane' (similar to bus lanes, where low emission cars would have separate lanes)	3.66	3.37	3.59	3.88	0.51
Motor insurance premiums partly based on carbon emissions (i.e. drivers of higher emission vehicles pay more)	5.52	4.66	5.21	5.60	0.94

Note: On a scale from 1 to 7 where 1 means not important and 7 means very important.

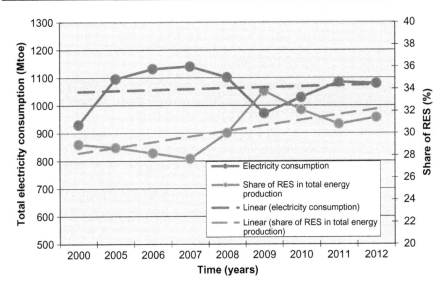

Fig. 6.3.8 Electricity consumption and share of RES in total energy production in Slovenia. Adapted by Eurostat (2014). Solar energy. Available from: http://ec.europa.eu/eurostat/web/environmental-data-centre-on-natural-resources/natural-resources/energy-resources/solar-energy Accessed 03.03.15; Obrecht, M., & Knez, M. (2014). Opportunities for transition to sustainable energy strategy in Slovenia. *Strategic Management, 19*(3), 31–37.

decreased since 2007, which is mostly due to the current economic situation. On the other hand, the share of RES has been fluctuating. Thus, the changes in energy consumption must be considered in planning long-term energy strategy. According to the provisional data, energy dependency of Slovenia in 2010 was 50%, which is slightly more than in 2009 but still relatively low in comparison with other EU member states.

This study is divided into four sections. First, the average mileage of Slovene cars and energy consumption of EVs were calculated, and then the CO_2 emissions of passenger cars in Celje are presented. The third and fourth sections, respectively, present calculation of solar energy availability and the required recharging capacities for proposed EV fleet.

6.3.2.1 Calculating the mileage and energy consumption of EVs in Slovenia

In the first step, the mileage of all vehicles was calculated. At the end of 2010, Slovenia had 1,061,646 registered passenger cars or 518 vehicles per 1000 inhabitants (SURS, 2011b).

The total vehicle mileage (MGE) was calculated in two ways.

$$MGE = N \times D = 23,356,212,000 \text{ vehicle} - km/year$$

where N is number of registered passenger cars and D is the average annual distance travelled by a car in Europe, which is about 22,000 km (ACEA, 2010).

$$MGE2 = Q/FC = 13.192.105.200 \text{ km/year}$$

Q is the amount of gasoline consumption in litres, and FC is average fuel consumption of European cars, which is around 6.5 l/100 km (GCC, 2004). For the purposes of transport, Slovenia spent 651,690,000 kg of gasoline in year 2008 (leaded and unleaded) (MOP, 2009). For calculation purposes, the petrol density of gasoline, which is 760 g/m^3, has been taken into account and thus gets Q in litres, which in our case is 857,486,842 l of gasoline. The average mileage of each car was also calculated that comes out as 47 km/day vehicle.

According to the study prepared by MIT Electric Vehicle Team (MIT, 2008), electric vehicle (EV) consumes 200–300 Wh for a mile on average. In this study, a figure of 200 Wh/km was used, i.e. if all of the above cars in Slovenia were converted to EVs, they would consume 3654 GWh/year. The energy consumed by one electric car would approximately be 3.4 MWh/year or 9.3 kWh/day.

6.3.2.2 CO2 emissions associated with passenger cars in Celje

In 2010, Celje had registered 32,214 road vehicles (SURS, 2010b, 2011b). Fig. 6.3.9 shows the very significant increase in the automobile population, primarily resulting from the membership of Slovenia in EU.

CO_2 emissions of new vehicles registered respectively in the years 2008, 2009, and 2010 have shown a decreasing trend of 159, 156, and 145 g/km. For calculation purposes, the number of cars within Celje, amounting to 32,000 was considered and the average value of CO_2 emissions of cars as 153.3 g/km. This translated to a figure of 84,060 tonnes of CO_2 per annum or 230 tonnes/day.

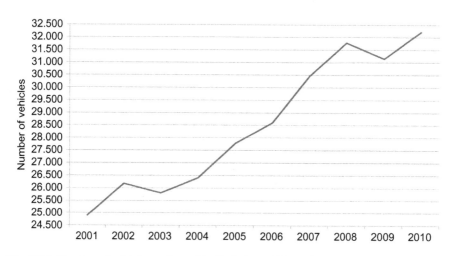

Fig. 6.3.9 Number of vehicles registered in Celje from 2001 to 2010 (SURS, 2010b, 2011b).

6.3.2.3 Solar recharging

Contrary to the current energy strategy, Slovenian energy strategy must be based on a reduced and efficient energy consumption and on the substitution of conventional energy sources with RES. Namely, the central idea of sustainability are circular flows and self-regeneration. The EU energy policy defines sustainability as the development of competitive RES and all other low-carbon sources of energy carriers by reducing energy demand within the EU and by directing the collective efforts to halt climate change and to improve local air quality. Following these three criteria, the construction of thermal power plants is inappropriate. In fact, sustainable development must not be perceived as meeting the needs of the present at the expense of future generations. At a time when we are beginning to realize the global environmental constraints, we still base our development on a quantitative increase in the use of raw materials and energy. We have to move away from restrictive assumptions and change our patterns of thinking with regard to the energy sector and to our everyday lives as this is the only way to a sustainable future (Obrecht & Knez, 2014).

Solar power plants are potentially of interest because they use free energy of the sun, but are not yet highly efficient. Solar power electricity generation has increased rapidly in the EU, reaching 1.5% of net electricity production (Eurostat, 2014); it is mainly generated by photovoltaic panels. As reported by the Eurostat, 'EU electricity generation from solar power increased 20 percent in 2013' (Eurostat, 2014). In rural areas, solar energy may be the ideal source of energy (Painuly, 2001; Razykov et al., 2011). The most important indicator for solar energy production is the amount of energy that is expected at a particular geographical location.

Slovenia measures only 20,256 km^2, but in spite of this its territory into three climate types can be divided: sub Mediterranean, temperate continental, and mountainous (Ogrin, 1996). However, the quantity of energy received due to solar radiation is influenced more by various relief positions than by the different climate types. Average solar radiation in Slovenia is more than 1000 kWh/m^2/annum. The 10-year average of the measured (1993–2003) annual global radiation was between 1053 and 1389 kWh/m^2. Half of Slovenia receives between 1153 and 1261 kWh/m^2.

The construction of solar power plants in Slovenia has shown an extremely rapid growth. In the last years, Slovenia has installed more than 1390 plants that were connected to the grid (Fig. 6.3.10).

Total power plants at the end of 2011 were almost 100 MW, but in 2014, Slovenia's electricity generation was already more than 250 MWh from solar energy, which shows rapid growth.

6.3.2.4 Economics of solar recharging for EV's in Celje

One of the options to reduce CO_2 emissions is the integration of electric vehicles (EV) as far as possible, but only if the EV's would be charged from environmentally friendly sources of energy.

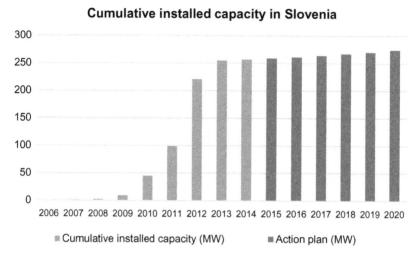

Fig. 6.3.10 Slovenian constructions of solar power plants and the forecast for 2020 (PV Portal, 2015).

The efficiency of solar modules that are available on the market ranges between 8% and 20%. In our study, electric characteristics of monocrystalline silicon photovoltaic modules produced by Bisol Company of Slovenia were used. An average module efficiency of 14% (at temperatures of 25°C and 44°C) was used.

Distribution of electricity in Celje is carried by Elektro Celje Company. Electricity production is still dominated by conventional sources (63%) and 37% from renewable energy sources (Elektro Celje, 2011). The present Carbon footprint for Slovenian electricity is 0.44–0.66 kg CO_2/kWh (Medved, 2003).

As predicted by a study conducted by Purgar (2012), EVs can reach a market share of 10% by 2020. The required electricity that needs to be produced from PV panels to empower the 3200 EVs for Celje's roads by 2020 is shown in Table 6.3.3.

In Celje, there are about 9000 residential buildings and if just one-half of them could be suitable for PV modules installations, it means that at least 3 kW systems could be installed on 4500 homes. This could mean the sum of about 112,000 m^2 of PV modules would generate 14 MW peak power that may be used for charging the battery bank for two- or four-wheel electric vehicles.

The price of the electricity generated by a solar PV system (C_{PV}, €/kWh) is the ratio between annual payment to offset PV installation financing loan and the total annual amount of energy produced (E_{an}). Capital costs can be decomposed in investment:

Table 6.3.3 **Electric energy needed for propulsion of EVs**

Energy demand, MWh	Per day	Per month	Per year
	29.80	907	10,800

panels (I_p), inverter (I_i), installation (I_l), and annual maintenance, M. Note that the replacement costs of the inverter, which has a shorter lifetime than the panels, have to be taken into account. The basic set of equations lead to financing costs.

Annual payment to offset PV installation financing loan,

$$A = P \cdot F_i (1+F_i)^n / [(1+F_i)^n - 1] \tag{6.3.2a}$$

$P =$ Capital costs associated with erection and lifetime maintenance of the PV plant, and n is the payback period, assumed to be 25 years for the present study (life of PV modules).

For a unit square metre of PV module area that has a nominal efficiency of η expressed as a fraction,

$$P, € = 1000\eta \left(I_p + I_l + [L_p/L_i] I_i + nM \right) \tag{6.3.2b}$$

Note that Table 6.3.4, respectively, provide a further explanation for above-used symbols and the present cost of fossil-nuclear-powered electricity. The estimation of the annual-averaged energy generated per square metre of PV module area, E_{an} ought to include the decline of cell efficiency with time (d) and the energy generated per peak Watt installed capacity of the PV modules per year, i.e. kWh/kWp/year. It may easily be shown that

$$E_{an} = 0.5\eta \left[F_0 + F_{25} \right] G_\beta \tag{6.3.2c}$$

G_β is the annual-averaged global irradiation (kWh/m^2) in the plane of (inclined) PV modules. F_0 and F_{25} in Eq. (6.3.2c) are usually provided by module manufacturers and a typical set of values indicate a 97% performance (F_0) for new modules that linearly drops to a figure of 80.2% (F_{25}) after a 25-year use.

The cost equation thus becomes

$$C_{PV}, €/kWh = A/E_{an} \tag{6.3.2d}$$

Using data of Table 6.3.4 and using above Eqs. (6.3.2a)–(6.3.2d) thus obtain for 1 m^2 of PV module area: $P = 364.8$, $A = 23.35$, $E_{an} = 168.34$, and $C_{PV} = 0.139$. Note that in

Table 6.3.4 **Parameters used for the cost equation**

Investment cost for PV panels	I_p	0.6 €/Wp
Investment cost for inverter	I_i	0.2 €/Wp
Investment cost for installation	I_l	0.8 €/Wp
Lifetime for PV panels	L_p	25 years
Inverter lifetime	L_i	10 years
Maintenance costs	M	0.02 €/Wp/year
Interest rate	F_i	4%/year

the above calculations a value of solar irradiation of 1250 kWh/m^2/year for Celje and an efficiency of 15.2% for monocrystalline PV modules was used.

The above estimate of solar electricity (13.9€ cents/kWh) may be compared with the present cost of 14.74€ cents/kWh for fossil-nuclear fuel electricity that is available in Celje.

The present study encompassed the following four tasks:

- Estimate the average mileage of Slovene cars
- Obtain the corresponding CO_2 emissions of passenger cars in Celje, a medium-sized town
- Compute the energy requirements of electric vehicles (EVs)
- Calculate the available solar energy and the required recharging economics for the proposed EVs

It was shown that within Slovenia, there is a strong and robust uptake of solar PV plants with actual installations exceeding the planned capacities by a factor of six. Furthermore, it was noted that the price of PV modules within the past 36 years has dropped by a factor of 104. The present analysis indicates that price of solar electricity that is presently obtainable is 13.9€ cents/kWh. This compares quite favourably with the present cost of 14.74€ cents/kWh (SURS, 2010a) for fossil-nuclear fuel electricity that is available in Celje.

6.4   Conclusions

Some government institutions have implemented a wide range of promotional policies; however, while their reports present the effects of each project, they are simply one piece of the entire puzzle. It is impossible to know where we are on the road to the implementation of electric vehicles. A review of electric vehicle promotion policies is crucial, as this is allows us to efficiently review past activities, determine the current state of affairs and assess the success of individual policies, and pinpoint measures for the wider implementation of electric vehicles. The findings of such research can be used to establish new strategies for electric vehicle use. Stojanovic (2010) claims that this is precisely what Slovenia has been missing.

Based on the number of first-time registered electric vehicles and its share in comparison with other countries, the policies for promoting the use of electric vehicles in Slovenia have been largely unsuccessful, as the share of electric vehicles, e.g. in 2013, has not even reached 1% of all vehicles. Other countries included in our research have already reached this level. Current measures for promoting the use of electric vehicles dictate that the sale of one electric vehicle can be counted as a huge success. In the process of adopting measures, Slovenia must set realistic and measurable goals. Of all the adopted policies for electric vehicle promotion, the effectiveness of the fiscal incentives that were accepted on the market and the promotional activities through which electric vehicles were presented to public and used as a tool to educate interested buyers.

If Slovenia wishes to introduce electric vehicles on a larger scale as soon as possible, the relevant authorities must prepare and adopt a legislative framework for the introduction of electric vehicles with representatives of car manufacturers, energy companies, and local authorities. Past experience has shown us that a combination of legal regulation and financial grants is the best method of ensuring the introduction of electric vehicles and the most effective form of policy for promoting their use. Financial grants and special loan schemes for purchasing electric vehicles have proved to be extremely successful. The subsidy system should also be further enhanced. Financial grants should be based on the difference between the price of the electric vehicle and the price of a comparable vehicle with an internal combustion engine and their environmental performance. These grants should be calculated for each vehicle separately, since huge differences exist between different electric vehicles. Electricity distribution companies and petrol sellers should take responsibility for the establishment of the infrastructure required for certain types of electric vehicles, as they possess the needed know-how, already have a predeveloped infrastructure and will gain most from the growth in electric cars.

Slovenia can start introducing changes through state-owned electricity distribution companies. This should result in energy suppliers following these steps if they wish to stay in touch with the competition and, consequently, developmental trends. Responsibility for the legislative framework falls to government institutions. In this scenario, key is the role of the government (Malyshev, 2009). The second possibility is to allow electric vehicle technology to develop on its own. More attention should also be paid to potential buyers. If they can accept the idea of electric vehicles, they will demand them too. Car manufacturers will have no choice but to respond by fulfilling the buyer's demands. In this situation, government institutions should withdraw fiscal incentives for electric vehicles and other benefits for owning electric vehicles and focus primarily on promoting the development of the technology demanded by the customers (Kahn, 2009). It is important to promote R&D in the car industry, especially in Slovenia, as many companies in this country are part of the global car industry with the know-how to develop and produce a range of electric vehicle parts.

Three groups of people whose opinions about LEVs differ were identified. The biggest group—the 'go-with-the-flow-greens' (42% of the total sample)—have a positive attitude towards LEVs, even if they are not sure about buying one in the near future. The second biggest group—the 'go-greens' (38% of the total sample)—has a very positive attitude towards LEV purchase, and they are planning to buy one in the near future. For the third group—the 'no-greens' (20% of the total sample)—information about vehicle carbon emissions and other environmentally related information is not important when buying a car. They are not planning to purchase a LEV in near future. Members of this group are also the hardest to motivate to purchase a LEV. They would not purchase it even if they had to pay more for a car that is not a LEV.

The main goal of the study was to identify the measures that would increase the people's interest in purchasing LEVs. The measures into 'push' and 'pull' approaches were divided. They have different influence on all three groups. For example, motor insurance premiums partly based on carbon emissions would influence more go-greens than no-greens. People in the no-green group are very difficult to motivate even to start thinking about purchasing a LEV; as a result, a combination of different push and pull factors are suggested.

It is clear that Slovenian government must be aware that single measures do not exist and that they are not effective. If the government wants to increase interest in purchasing LEVs, it should adjust and adopt a variety of different measures, combining both pull and push factors. The most important pull factor is development of special incentives for LEVs. Other relevant measures are VAT based on carbon emissions (buyers with lower emission vehicles would pay less VAT), vehicle registration fee based on carbon emissions of vehicles (buyers of lower emission vehicles pay less), motor insurance premiums partly based on carbon emissions (i.e. drivers of higher emission vehicles pay more), and a road user charging scheme based on carbon emissions (i.e. drivers of higher emission vehicles pay more).

One of the conditions for a faster integration of EVs is the construction of infrastructure for recharging. Slovenia has been here for a step forward. As a partner, Slovenia is participating in a project Central European Green Corridors (CEGC). The project will deploy 115 high-power charging stations in Austria, Croatia, Germany, Slovakia, and Slovenia (26 stations) to create a recharging network with country-wide coverage in Austria, Slovenia, and Slovakia.

The car industry should be also aware that car drivers are more familiar with information about fuel economy than information about a car's environmental influences (e.g. carbon emissions). Most people do not actually know the meaning of 'grams of CO_2 per 100 km'; therefore, our advice to the car industry selling LEVs is that they need to inform people how much money they could save by buying EV.

References

ACEA (2010). The automobile industry pocket guide. Available at: http://www.acea.be/images/uploads/files/2010924_Pocket_Guide_2nd_edition.pdf. Accessed 9 April, 2014.

Achterberg, P., Houtman, D., van Bohemen, S., & Manevska, K. (2010). Unknowing but supportive? Predispositions, knowledge, and support for hydrogen technology in the Netherlands. *International Journal of Hydrogen Energy, 35*, 6075–6083.

Åhman, M. (2006). Government policy and the development of electric vehicles in Japan. *Energy Policy, 34*(4), 433–443.

Anable, J. (2005). 'Complacent car addicts' or 'aspiring environmentalists'? Identifying travel behaviour segments using attitude theory. *Transport Policy, 12*(1), 65–78.

Anable, J., Lane, B., & Banks, N. (2008). Car buyer survey: from 'mpg paradox' to 'mpg mirage'-how car purchasers are missing a trick when choosing new and used cars. Available at: http://www.lowcvp.org.uk/assets/reports/Car_Buyer_Report_2008_Final_Report.pdf. Accessed 8 May 2014.

Anderson, T., Jr., & Cunningham, W. H. (1972). The socially conscious consumer. *Journal of Marketing, 36*(7), 23–31.

Anderson, S., & Stradling, S. G. (2004). Attitudes to car use and modal shift in Scotland. Report of National Centre for Social Research Scotland: Scottish Executive Social Research Available from: http://www.scotland.gov.uk/Publications/2004/03/19062/34290 Accessed 10.05.14.

Avtovizije. (2014). New charging point for electric vehicles in Maribor. Available from: http://www.avtovizije.com/aktualno/reportaze/4002-nova-elektrina-polnilna-postaja-v-mariboru.html Accessed 03.03.14 in Slovenian language.

Axsen, J., Kurani, K. S., & Burke, A. (2010). Are batteries ready for plugin hybrid buyers? *Transport Policy, 17*, 173–182.

Banerjee, B., & McKeage, K. (1994). How green is my value: exploring the relationship between environmentalism and materialism. In C. T. Allen & D. R. John (Eds.), *Vol. 21. Advances in consumer research* (pp. 147–152). Provo, UT: Association for Consumer Research.

Berkowitz, L., & Lutterman, K. G. (1968). The traditional socially responsible personality. *Public Opinion Quarterly, 32,* 169–185.

Bockarjova, M., & Steg, L. (2014). Can Protection Motivation Theory predict pro-environmental behavior? Explaining the adoption of electric vehicles in the Netherlands. *Global Environmental Change, 28,* 276–288.

Borthwick, S., & Carreno, M. (2012). Persuading Scottish drivers to buy low emission cars? The potential role of green taxation measures. In: *8th Annual Scottish transport applications & research conference, Glasgow*: Transport Research Institute, Edinburgh Napier University.

Calef, D., & Goble, R. (2007). The allure of technology: How France and California Promoted electric and hybrid vehicles to reduce urban air pollution. *Policy Sciences, 40*(1), 1–34.

Carley, S., Krause, R. M., Lane, B. W., & Graham, J. D. (2013). Intent to purchase a plug-in electric vehicle: A survey of early impressions in large US cites. *Transportation Research Part D: Transport and Environment, 18,* 39–45.

Carreno, M., & Welsch, J. (2009). MaxSem: Max self regulation model: Applying theory to the design and evaluation of mobility management projects. Available from: http://www.epomm.eu/docs/mmtools/case_studies_TA/MaxSem_applying_theory_to_MM_projects.doc Accessed 01.09.13.

Elektro Celje. (2011). The composition of the primary sources of electricity. Available from: http://www.elektro-celje.si/omrezje/284-sestava-primarnih-virov-el-energije   Accessed 05.05.14 in Slovenian language.

Cowan, R., & Hultén, S. (1996). Escaping Lock-in: The case of the electric vehicle. *Technological Forecasting and Social Change, 53,* 61–79.

Car Dealership Real. (2014). Sales promotion »1.000.000 Euros for ECO subsidies« October. Available from: http://www.avtohisa-real.si/data/slike/dobra_priloznost/OKT_11/pogoji_eko_subvencij_oktober.pdf Accessed 17.03.14 in Slovenian language.

Eco Fund. (2013). Annual report on activities and operations of Eco Fund, Slovenian Environmental Public Fund in 2012. Available from: http://www.ekosklad.si/pdf/LetnaPorocila/LP_12_slo.pdf Accessed 17.03.14 in Slovenian language.

EEA. (2007). EEA, environmental statement. Copenhagen, 2007. Available from: http://www.eea.europa.eu/publications/corporate_document_2007_2 Accessed 29.04.14.

Civitas Elan. (2010). Information about CIVITAS ELAN Project. Available from: http://www.ljubljana.si/file/.../6.-elektromobilnost-18-4-13-konna-verzija.pdf Accessed 25.03.14 in Slovenian language.

Elaphe. (2009). Presentation of recent events CEVELJ 1, CEVELJ 2 and CEVELJ 3 in 2007, 2008 and 2009. Available from: http://ecomeet.si/docs/Program_prireditve/PredstavitevDogodkovCevelj1in2in3.pdf Accessed 17.03.14 in Slovenian language.

Elektro crpalke. (2014). Slovenian portal for searching charging points for electric vehicles. Available from: http://www.elektro-crpalke.si/ Accessed 20.03.14 in Slovenian language.

European Commission. (2014). Green cars initiative. Available from: http://ec.europa.eu/research/transport/road/green_cars/index_en.htm Accessed 20.03.14.

European Parliament. (2011). Electric cars. Available from: http://www.europarl.europa.eu/sides/getDoc.do?pubRef=-//EP//TEXT+TA+P7-TA-2010-0150+0+DOC+XML+V0//SL Accessed 20.03.14 in Slovenian language.

Eurostat. (2014). Solar energy. Available from: http://ec.europa.eu/eurostat/web/environmental-data-centre-on-natural-resources/natural-resources/energy-resources/solar-energy Accessed 03.03.15.

Fairchild, R. (2008). The manufacturing sector's environmental motives: A Game-theoretic analysis. *Journal of Business Ethics*, *79*(3), 333–344.

Fale, M. (2014). Review of policies for promoting the use of electric vehicles. *Celje, VII, 83* [in Slovenian language].

Fernandez, T. R. C., Chen, F., & da Graça Carvalho, M. (2005). "HySociety" in support of European hydrogen projects and EC policy. *International Journal of Hydrogen Energy*, *30*(3), 239–245.

Flamm, B. (2009). The impacts of environmental knowledge and attitudes on vehicle ownership and use. *Transportation Research Part D*, *14*(3), 272–279.

France-Diplomatie. (2014). France is the largest electric vehicle market in Europe. Available from: http://www.diplomatie.gouv.fr/en/coming-to-france/facts-about-france/one-figure-one-fact/article/france-is-the-largest-electric Accessed 22.03.14.

Gallagher, K. S., & Muehlegger, E. (2011). Giving green to get green? Incentives and consumer adoption of hybrid vehicle technology. *Journal of Environmental Economics and Management*, *61*(1), 1–16.

Gass, V., Schmidt, J., & Schmid, E. (2014). Analysis of alternative policy instruments to promote electric vehicles in Austria. *Renewable Energy*, *61*, 96–101.

GCC, Green Car Congress. (2004). Fuel consumption in Europe 47% better than US. Available from: http://www.greencarcongress.com/2004/11/average_fuel_co.html Accessed 18.04.12.

Golob, T. F., & Gould, J. (1998). Projecting use of electric vehicles from household vehicle trials. *Transportation Research Part B: Methodological*, *32*(7), 441–454.

Government Communication Office. (2009). Energy industry. Available from: http://www.evropa.gov.si/si/energetika/ Accessed 20.03.10 in Slovenian language.

Gregorcic, J. (2011a). This year Europeans bought only 5222 electric vehicles. Available from: http://www.siol.net/avtomoto/novice/2011/09/evropski_trg_elektricnih_vozil.aspx Accessed 17.03.14 in Slovenian language.

Gregorcic, J. (2011b). Slovenia confirmed subsidies for purchase of electric vehicles. Available from: http://www.siol.net/avtomoto/novice/2011/09/slovenski_trg_elektricnih_vozi.aspx Accessed 17.03.14 in Slovenian language.

Gregorcic, J. (2013a). When can we expect all-electric Renault cars in Slovenia? Available from: http://www.siol.net/avtomoto/novice/2013/07/renautl_nissan_ze_ev_prihod_v_slovenijo.aspx Accessed 25.03.14 in Slovenian language.

Gregorcic & Slovenian Press Agency. (2013). European Commission: Slovenia should build 3.000 charging points for electric vehicles by 2020. Available from: http://www.siol.net/avtomoto/novice/2013/01/bruselj_slovenija_elektricne_polnilnice.aspx Accessed 25.03.14 in Slovenian language.

Grünig, M., Witte, M., Marcellino, D., Selig, J., & van Essen, H. (2011). An overview of electric vehicles on the market and in development. Available from: http://ec.europa.eu/clima/policies/transport/vehicles/docs/d1_en.pdf Accessed 20.03.14.

Hackbarth, A., & Madlener, R. (2013). Consumer preferences for alternative fuel vehicles: A discrete choice analysis. *Transportation Research Part D: Transport and Environment*, *25*, 5–17.

Hannisdahl, O. H., Malvik, H. V., & Wensaas, G. B. (2013) The future is electric! The EV revolution in Norway—explanations and lessons learned. Available at: http://www.gronnbil.no/getfile.php/FILER/Norway%20-%20lessons%20learned%20from%20a%20global%20EV%20success%20story%20-%20Final.pdf Accessed 22.03.14.

IEA. (2013). Global EV outlook. Available from: http://www.iea.org/publications/globalevoutlook_2013.pdf Accessed 20.03.14.

Japanese Ministry of Economy, Trade and Industry. (2002). *Environmental management accounting workbook. Tokyo: Ministry of Economy, Trade and Industry Japan.*

Johansson-Stenman, O., & Martinsson, P. (2006). Honestly, why are you driving a BMW? *Journal of Economic Behavior and Organization, 60,* 129–146.

Kahn, M. E. (2007). Do greens drive hummers or hybrids? Environmental ideology as a determinant of consumer choice. *Journal of Environmental Economics and Management, 54,* 129–145.

Kahn, M. E. (2009). The green economy. *Foreign Policy, 172,* 34–38.

Klajnscak, T. (2014). Around Maribor with electric car. Available from: http://www.zurnal24. si/mnozicna-izposoja-elektricnih-vozil-clanek-228045 Accessed 29.03.14 in Slovenian language.

Knez, M., Jereb, B., & Obrecht, M. (2014). Factors influencing the purchasing decisions of low emission cars: a study of Slovenia. *Transportation Research Part D: Transport and Environment, 30,* 53–61.

Knez, M., Muneer, T., Jereb, B., & Cullinane, K. (2014). The estimation of a driving cycle for Celje and a comparison to other European cities. *Sustainable Cities and Society, 11,* 56–60.

Kokubu, K. (1999). *Social and environmental accounting.* Tokyo: Chuokeizai-Sha Inc.

Kramberger, T., Dragan, D., & Prah, K. (2014). A heuristic approach to reduce carbon dioxide emissions. *Proceedings of the Institution of Civil Engineers - Transport, 167*(5), 296–305.

Krause, R. M., Carley, S. R., Lane, B. W., & Graham, J. D. (2013). Perception and reality: Public knowledge of plug-in electric vehicles in 21 U.S. cities. *Energy Policy, 63,* 433–440.

Krupa, J. S., Rizzo, D. M., Eppstein, M. J., Brad Lanute, D., Gaalema, D. E., Lakkaraju, K., & Warrender, C. E. (2014). Analysis of a consumer survey on plug-in hybrid electric vehicles. *Transportation Research Part A: Policy and Practice, 64,* 14–31.

Kurani, K., Turrentine, T., & Sperling, D. (1994). Demand for electric vehicles in hybrid households: an exploratory analysis. *Transport Policy, 1*(4), 244–256.

Kurani, K., Turrentine, T., & Sperling, S. (1996). Testing electric vehicle demand in 'hybrid households' using a reflexive survey. *Transportation Research Part D, 1*(2), 131–150.

Kurani, K. S., Axsen, J., Caperello, N., Davies, J., & Stillwater, T. (2009). *Research report UCD-ITS-RR-09-21: Learning from consumers: Plug-in hybrid electric vehicle (PHEV) demonstration and consumer education, outreach and market research program. 2007 Alternative Fuels Incentive program.* Davis, CA: Institute of Transportation Studies, University of California.

Lane, B., & Banks, N. (2010). LowCVP car buyer survey: Improved environmental information for consumers. Available from: http://www.lowcvp.org.uk/assets/reports/LowCVP-Car-Buyer-Survey-2010-Final-Report-03-06-10-.vFINAL.pdf Accessed 25.04.14.

Laroche, M., Bergeron, J., & Barbaro-Forleo, G. (2001). Targeting consumers who are willing to pay more for environmentally friendly products. *Journal of Consumer Marketing, 18*(6), 503–520.

Lyon, T. & Maxwell, J., 1999. Voluntary approaches to environmental regulation: a survey. Available from: http://dx.doi.org/10.2139/ssrn.147888 Accessed 07.03.14.

Machlachlan, J., & Gardner, J. (2004). A comparison of socially responsible and conventional investors. *Journal of Business Ethics, 25*(1), 11–25.

Malyshev, T. (2009). Looking ahead: Energy, climate change and pro-poor responses. *Foresight, 11*(4), 33–50.

Martin, E., Shaheen, S. A., Lipman, T. E., & Lidicker, J. R. (2009). Behavioural response to hydrogen fuel cell vehicles and refuelling: results of California drive clinics. *Internal Journal of Hydrogen Energy, 34*(20), 8670–8680.

May, G. (2004). Europe's automotive sector at the crossroads. *Foresight, 6*(5), 302–312.

McIntyre, R. P., Meloche, M. S., & Lewis, S. L. (1993). National culture as a macro tool for environmental sensitivity segmentation. D. W. Cravens & P. R. Dickson (Eds.), *Vol. 4. AMA summer educators' conference proceedings* (pp. 153–159). Chicago, IL: American Marketing Association.

Medved, S. (2003). *Rational use of materials, space and energy. Faculty of Mechanical Engineering.* Ljubljana: University of Ljubljana.

Minato, N. (2011). *New decision method for environmental capital investment. Environmental management accounting and supply chain management.* London: Springer.

MIT. (2008). Fuel economy numbers for electric vehicles. Available from: http://mit.edu/evt/summary_mpgge.pdf Accessed 22.04.12.

MOP. (2009). *The use of biofuels in the transport sector in the Republic of Slovenia in 2008. Ljubljana: Ministry of Environment.*

National Academy of Sciences. (2013). Overcoming barriers to electric-vehicle deployment: Interim report. Available from: http://gabrielse.physics.harvard.edu/gabrielse/papers/2013/OvercomingBarriersToElectricVehicleDeployment.pdf Accessed 04.12.15.

Nelson, T. D., & Tanabe, M. (2013). Japan continues to offer electric vehicle incentives. Available from: http://www.mondaq.com/unitedstates/x/263904/Renewables/Japan+Continues+To+Offer+Electric+Vehicle+Incentives Accessed 24.03.14.

Nilsson, M., Hillman, K., & Magnusson, T. (2012). How do we govern sustainable innovations? Mapping patterns of governance for biofuels and hybrid-electric vehicle technologies. *Environmental Innovation and Societal Transitions, 3,* 50–66.

O'Garra, T., Mourato, S., & Pearson, P. (2005). Analysing awareness and acceptability of hydrogen vehicles: A London case study. *International Journal of Hydrogen Energy, 30,* 649–659.

Obrecht, M., & Denac, M. (2013). A sustainable energy policy for Slovenia: considering the potential of renewables and investment costs. *Journal of Renewable and Sustainable Energy, 5*(3).

Obrecht, M., & Knez, M. (2014). Opportunities for transition to sustainable energy strategy in Slovenia. *Strategic Management, 19*(3), 31–37.

O'Dell, J. (2012). Will California's zero-emissions mandate alter the car landscape? Available from: http://www.edmunds.com/fuel-economy/will-californias-zero-emissions-mandate-alter-the-car-landscape.html Accessed 24.03.14.

Official Gazette of the Republic of Slovenia. (2011). Proclamation part public auctions. Available from: http://www.uradni-list.si/_pdf/2011/Ra/r2011079.pdf Accessed 30.04.14 in Slovenian language.

Ogrin, D. (1996). Climate types in Slovenia. Geografski vestnik 68. Ljubljana.

Painuly, J. (2001). Barriers to renewable energy penetration; a framework for analysis. *Renewable Energy, 24,* 73–89.

Pecjak, A. (2014). History of electric vehicles in Slovenia. Available from: http://www.ad-pecjak.si/eco/zgodovina_EV_SLO.htm Accessed 17.05.14 in Slovenian language.

Pilkington, A., & Dyerson, R. (2006). Innovation in disruptive regulatory environments: A patent study of electric vehicle technology development. *European Journal of Innovation Management, 9*(1), 79–91.

Pohl, H., & Yarime, M. (2012). Integrating innovation system and management concepts: The development of electric and hybrid electric vehicles in Japan. *Technological Forecasting and Social Change, 79*(8), 1431–1446.

Polni.si. (2014). Portal for searching charging points. Available from: http://polni.si/ Accessed 20.03.14.

Popp, M., Van de Velde, L., Vickery, G., Van Huylenbroeck, G., Verbeke, W., & Dixon, B. (2009). Determinants of consumer interest in fuel economy: Lessons for strengthening the conservation argument. *Biomass and Bioenergy, 33*, 768–778.

Portal NIO. (2012). First registered vehicles in 2012, by month. Available from: http://nio.gov.si/nio/data/prvic+registrirana+vozila+v+letu+2012+po+mesecih Accessed 17.03.14 in Slovenian language.

Praper, A. (2012). Opening of the first fast charging point for electric vehicles. Available at: http://www.avtovizije.com/aktualno/dogodki/4956-otvoritev-prve-hitre-polnilne-postaje-za-elektrina-vozila Accessed 20.03.14 in Slovenian language.

Europa Press Releases. (2013). EU launches clean fuel strategy. Available from: http://europa.eu/rapid/press-release_IP-13-40_sl.htm Accessed 20.03.14.

Purgar, Z. (2012). Ljubljana is overtaking Slovenia in the area of electric mobility. Available from: http://www.delo.si/novice/ljubljana/ljubljana-pri-elektricni-mobilnosti-prehiteva-drzavo.html Accessed 25.03.14 in Slovenian language.

PV Portal. (2015). Solar power plants in Slovenia. Available from: http://pv.fe.uni-lj.si/SEvSLO.aspx Accessed 24.11.15 in Slovenian language.

Razpotnik, I., Loose, N., Jazbinsek Srsen, A., Simonovic, Z., & Klancar, T. (2013). *The plan for sustainable mobility—Strategy for sustainable mobility in City Municipality of Ljubljana.* Ljubljana: City Municipality of Ljubljana, Department of Commercial Activities and Traffic [in Slovenian language].

Razykov, T., Ferekides, C., Morel, D., Stefanakos, E., Ullal, H., & Upadhyaya, H. (2011). Solar photovoltaic electricity: Current status and future prospects. *Solar Energy, 1580*–1608.

Roberts, J. A. (1996). Green consumers in the 1990s: Profile and implications for advertising. *Journal of Business Research, 36*(3), 217–232.

Sandahl, D. M., & Robertson, R. (1989). Social determinants of environmental concern: Specification and test of the model. *Environment and Behavior, 21*(1), 57–81.

Schaltegger, S. (2008). Sustainability and accounting: The question for management managing business cases for sustainability. In: *7th Proceedings of the Australasian conference for social and environmental accounting research (A-CSEAR)*, (pp. 25–33). Adelaide: CAGS.

Scientific American. (2011). Will Germany become the first nation with a hydrogen economy? Available from: http://www.scientificamerican.com/article/will-germany-become-first-nation-with-hydrogen-economy/ Accessed 22.03.14.

SIHFC. (2014). Presentation of SIHFC. Available from: http://www.sihfc.si/prikazi.asp?vsebina=predstavitev%2Fpredstavitev.asp Accessed 24.03.14 in Slovenian language.

Steenberghen, T., & Lopez, E. (2008). Overcoming barriers to the implementation of alternative fuels for road transport in Europe. *Journal of Cleaner Production, 16*(5), 577–590.

Stojanovic, B. (2010). Slovenia still without strategy, some already on the market. Available from: http://www.zelenaslovenija.si/revija-eol-/aktualna-stevilka/logistika/693-slovenija-se-brez-strategije-nekateri-pa-ze-na-trgu-eol-54 Accessed 17.03.14 in Slovenian language.

SURS. (2010). Statistical information. Available from: http://www.stat.si/doc/statinf/03-SI-019-1001.pdf Accessed 24.04.14 in Slovenian language.

SURS. (2010). Registered road motor vehicles and trailers, Slovenia, 2009—final data Statistical Office of the Republic of Slovenia. Ljubljana. Available from: http://www.stat.si/eng/novica_prikazi.aspx?id=3150 Accessed 14.04.12 in Slovenian language.

SURS. (2011b). Registered road motor vehicles and trailers, Slovenia, 2010—final data. Available from: http://www.stat.si/novica_prikazi.aspx?id=3940 Accessed 14.04.12 in Slovenian language.

Thesen, G., & Langhelle, O. (2008). Awareness, acceptability and attitudes towards hydrogen vehicles and filling stations: a greater Stavanger case study and comparisons with London. *International Journal of Hydrogen Energy, 33*, 5859–5867.

Toyota. (2013). Toyota, Nissan, Honda and Mitsubishi to provide financial assistance for electric vehicle charging infrastructure in Japan. Available from: http://www2.toyota.co.jp/en/news/13/11/1112.html Accessed 24.03.14.

Turcksin, L., Mairesse, O., & Macharis, C. (2013). Private household demand for vehicles on alternative fuels and drive trains: A review. *European Transport Research Review, 5*, 149–164.

Turrentine, T. S., Garas, D., Lentz, A., & Woodjack, J. (2011). *The UC Davis MINI E consumer study*. Research Report No. UCD-ITS-RR-11-05 Davis, CA: Institute of Transportation Study, University of California. May. Available from: http://www.its.ucdavis.edu/?page_id=10063&pub_id=1470 Accessed 04.12.15.

Ulmer, J. D., Huhnke, R. L., Bellmer, D. D., & Cartmell, D. D. (2004). Acceptance of ethanol-blended gasoline in Oklahoma. *Biomass and Bioenergy, 27*, 437–444.

UN HABITAT. Global report on human settlement. 2011. Available from: http://mirror.unhabitat.org/downloads/docs/E_Hot_Cities.pdf Accessed 15.01.15.

Van de Velde, L., Verbeke, W., Popp, M., & Van Huylenbroeck, G. (2010). The importance of message framing for providing Information about sustainability and environmental aspects of energy. *Energy Policy, 38*, 5541–5549.

Van Liere, K. D., & Dunlap, R. E. (1981). The social bases of environmental concern: A review of hypotheses, explanations and empirical evidence. *Public Opinion Quarterly, 44*, 181–197.

Vining, J., & Ebreo, A. (1990). What makes a recycler? A comparison of recyclers and non-recyclers. *Environment and Behavior, 22*, 55–73.

Ward, J. (1998). Financing your AFV fleet. The American City & County, Issue 113.3. Available from: http://search.proquest.com.ezproxy.lib.ukm.si/socscijournals/docview/195953689/616443E8BF894C00PQ/507?accountid=28931 Accessed 18.03.14.

Zupan, T. (2014). Factors affecting the decision to purchase environmentally friendly vehicles: Master thesis. Celje.

Zupan, T., Jereb, B., Rosi, B., & Knez, M. (2013). Different measures of low-emission vehicle purchasing. In: *Pre-conference proceedings of the 10th international conference on logistics & sustainable transport 2013, Celje, Slovenia, 13–15 June 2013*, (pp. 270–277). Celje: Faculty of Logistics.

Further Reading

AMZS. (2012). DMV—Motor vehicle tax in 2012. Available from: http://www.amzs.si/si/356/98/default.aspx Accessed 01.12.14 in Slovenian language.

Eco Fund. (2009). Annual report on activities and operations of Eco Fund, Slovenian Environmental Public Fund in 2008. Available from: http://www.ekosklad.si/pdf/LetnaPorocila/LP_08_slo.pdf Accessed 17.03.14 in Slovenian language.

Eco Fund. (2010). Annual report on activities and operations of Eco Fund, Slovenian Environmental Public Fund in 2009. Available from: http://www.ekosklad.si/pdf/LetnaPorocila/LP_09_slo.pdf Accessed 17.03.14 in Slovenian language.

Eco Fund. (2011a). Appeal 8SUB-EVOB11. Available from: http://www.ekosklad.si/pdf/doc/8SUB-EVOB11_javni_poziv.pdf Accessed 14.06.14 in Slovenian language.

Eco Fund. (2011b). Appeal 9SUB-EVPO11. Available from: http://www.ekosklad.si/html/razpisi/main.html Accessed 14.06.14 in Slovenian language.

Eco Fund. (2011c). Annual report on activities and operations of Eco Fund, Slovenian Environmental Public Fund in 2010. Available from: http://www.ekosklad.si/pdf/LetnaPorocila/LP_10_slo.pdf Accessed 17.03.14 in Slovenian language.

Eco Fund. (2012a). Appeal 15SUB-EVOB12. Available from: http://www.ekosklad.si/pdf/SUB2012/15SUB-EVOB12_1_JavniPoziv_URL12.pdf Accessed 14.06.14 in Slovenian language.

Eco Fund. (2012b). Appeal 16SUB-EVPO12. Available from: http://www.ekosklad.si/html/razpisi/main.html Accessed 14.06.14 in Slovenian language.

Eco Fund. (2012c). Annual report on activities and operations of Eco Fund, Slovenian Environmental Public Fund in 2011. Available from: http://www.ekosklad.si/pdf/LetnaPorocila/LP_11_slo.pdf Accessed 17.03.14 in Slovenian language.

Eco Fund (2013a) Appeal 20SUB-EVOB13. Available from: http://www.ekosklad.si/pdf/SUB2013/20SUB-EVOB13_JavniPoziv_url13.pdf Accessed 14.06.14 in Slovenian language.

Eco Fund. (2013b). Appeal 21SUB-EVPO13. Available from: http://www.ekosklad.si/html/razpisi/main.html Accessed 14.06.14 in Slovenian language.

Eco Fund. (2014a). Appeal 26SUB-EVOB14. Available from: http://www.ekosklad.si/dl/R/14/26SUB-EVOB14_JavniPoziv.pdf Accessed 14.06.14 in Slovenian language.

Eco Fund. (2014b). Appeal 27SUB-EVPO14. Available from: http://www.ekosklad.si/html/razpisi/main.html Accessed 14.06.14 in Slovenian language.

Eco Fund. (2014c). Appeal 51OB14. Available from: http://www.ekosklad.si/html/razpisi/main.html Accessed 07.07.14 in Slovenian language.

Eco Fund of the Republic of Slovenia. (2004). Annual report on activities and operations of Eco Fund of the Republic of Slovenia, Public Fund in 2004. Available from: http://www.ekosklad.si/pdf/LetnaPorocila/LP_04_slo.pdf Accessed 17.03.14 in Slovenian language.

Eco Fund of the Republic of Slovenia. (2006). Annual report on activities and operations of Eco Fund of the Republic of Slovenia, Public Fund in 2005. Available from: http://www.ekosklad.si/pdf/LetnaPorocila/LP_05_slo.pdf Accessed 17.03.14 in Slovenian language.

Eco Fund of the Republic of Slovenia. (2007). Annual report on activities and operations of Eco Fund of the Republic of Slovenia, Public Fund in 2006. Available from: http://www.ekosklad.si/pdf/LetnaPorocila/LP_06_slo.pdf Accessed 17.03.14 in Slovenian language.

Eco Fund of the Republic of Slovenia. (2008). Annual report on activities and operations of Eco Fund of the Republic of Slovenia, Public Fund in 2007. Available from: http://www.ekosklad.si/pdf/LetnaPorocila/LP_07_slo.pdf Accessed 17.03.14 in Slovenian language.

EcoMeet. (2010). Programme of the event. Available from: http://ecomeet.si/docs/Program_prireditve/EcoMeetSLO.pdf Accessed 08.08.14 in Slovenian language.

EcoMeet. (2013). After the event. Available from: http://ecomeet.si/sl/ECOmeet2013/ Accessed 17.06.14 in Slovenian language.

European Commission: CORDIS. (2013a). EUROLIION. Available from: http://cordis.europa.eu/project/rcn/109414_en.html Accessed 25.06.14.

European Comission: CORDIS. (2013b). MAG–DRIVE. Available from: http://cordis.europa.eu/project/rcn/110008_en.html Accessed 25.06.14.

European Commission: CORDIS. (2013c). HYSYS. Available from: http://cordis.europa.eu/project/rcn/78586_en.html Accessed 25.06.14.

European Commission: CORDIS. (2011). SMARTV2G. Available from: http://cordis.europa.eu/project/rcn/99306_en.html Accessed 25.06.14.

European Commission: CORDIS. (2014). CAPIRE. Available from: http://cordis.europa.eu/projects/rcn/96977_en.html Accessed 25.06.14.

Government of the Republic of Slovenia. (2009). Operational programme for reducing emissions of greenhouse gases until 2012 (OP TGP-1). Available from: http://www.vlada.si/fileadmin/dokumenti/si/projekti/2009/podnebne/op_toplogredni_plini2012_1.pdf Accessed 29.03.14 in Slovenian language.

Gregorcic, J. (2013b). Fuel cell cars can be from now on refuelled in Slovenia. Available from: http://www.siol.net/avtomoto/zanimivosti/tehnika/2013/09/petrol_vodik_silvan_simcic. aspx Accessed 17.03.14 in Slovenian language.

Kastelec, D., Rakovec, J., & Zaksek, K. (2007). Solar energy in Slovenia. *ZRC SAZU, 76* [in Slovenian language].

KBA Umwelt. (2004). Umwelt—Zeitreiche 2004. Available from: http://www.kba.de/cln_031/nn_191064/DE/Statistik/Fahrzeuge/Neuzulassungen/Umwelt/n__umwelt__z__teil__1. html Accessed 06.05.14.

KBA Umwelt. (2013). Umwelt—Zeitreiche 2005 bis 2013. Available from: http://www.kba.de/cln_031/nn_191064/DE/Statistik/Fahrzeuge/Neuzulassungen/Umwelt/n__umwelt__z__teil__2.html Accessed 06.05.14.

Kobla.net. (2013). The position of Celje within Slovenia. Available from: http://www.kobla. net/lokacija/ Accessed 29.02.13 in Slovenian language.

Portal NIO. (2013). First registered vehicles in 2013, by month. Available from: http://nio.gov. si/nio/data/prvic+registrirana+vozila+v+letu+2013+po+mesecih Accessed 17.03.14 in Slovenian language.

Portal NIO. (2014). First registered vehicles in 2014, by month. Available from: http://nio.gov. si/nio/data/prvic+registrirana+vozila+v+letu+2014+po+mesecih Accessed 17.03.14 in Slovenian language.

Statistical data sets GOV.UK. (2014). Table VEH0130. Available from: https://www.gov.uk/government/statistical-data-sets/veh01-vehicles-registered-for-the-first-time Accessed 06.05.14.

Statistik Austria. (2014). Kraftfahrzeuge-Neuzulassungen. Available from: http://www. statistik.at/web_de/statistiken/verkehr/strasse/kraftfahrzeuge-_neuzulassungen/index. html Accessed 06.05.14.

SunShot. (2012). SunShot Vision Study. U.S Department of Energy. February 2012. Available from: http://www1.eere.energy.gov/solar/pdfs/47927_chapter4.pdf Accessed 19.02.13.

SURS. (2011a). Annual energy statistics, Slovenia, 2010—provisional data. Available from: http://www.stat.si/eng/novica_prikazi.aspx?id=3912 Accessed 20.04.12 in Slovenian language.

The Economist. (2012). Available from: http://www.economist.com/blogs/graphicdetail/2012/12/daily-chart-19 Accessed 21.02.13.

Toyota. (2014). Models. Available from: http://www.toyota.si/ Accessed 20.03.14 in Slovenian language.

Case study for Chile: The electric vehicle penetration in Chile

7

Girard Aymeric[*], Simon François[†]
*Adolfo Ibáñez University, Viña del Mar, Chile, †University of Granada, Granada, Spain

7.1 Introduction

Chile currently has an energy dependency equivalent to 60% (Ministry of Energy, 2014), due to the import of fossil fuels used in power generation. At the same time, the use of these fuels increases levels of air pollution, being the sectors of energy and transport; they contribute the most to harmful gas emissions with 47% and 30%, respectively, of the national emissions of CO_2 equivalent (IEA, 2015). In particular, the transport industry represents 25% of the national potential for energy efficiency according to estimates made by the Chilean Energy Efficiency Agency (Agencia Chilena de Eficiencia Energetica, AChEE) to the year 2020 (Ministry of Energy, 2013), because of the advance new technologies such as electric vehicles (EVs).

The EV market has not yet achieved an important place in Chile but represents an attractive alternative to current transport systems. EVs are not directly dependent on fossil fuels, and their use is responsible for considerably less greenhouse gas (GHG) emission than internal combustion vehicles (ICVs) if powering such vehicles uses renewable energy sources. Moreover, one should not be neglecting the fact that Chile renewable energy resource is huge, particularly the solar energy, thus presenting a great potential for the development of EVs running on clean electricity. However, it is expected that the global vehicle fleet will triple by 2050 and that the use of those types of fuels will remain in today's proportions (WEC, 2013). According to those expectations, the air pollution and energy deficit will grow accordingly. Whereas Chile follows the line of the global forecast, it is time for a change. Since transport is the area with the biggest potential, why not start there?

7.2 Chile energy panorama

7.2.1 Energy consumption

Energy use has a great importance for understanding the energy situation of Chile. It allows us to know the limits and the context of the energy production of the country. The total energy consumption in Chile in 2012 was 35.8 million tonnes of oil equivalent (Mtoe) (U.S. EIA, 2014). The evolution of total primary energy consumption in Chile is presented in Fig. 7.2.1.

In modern history, fossil fuels have always been the main sources of primary energy supply worldwide (BP p.l.c., 2013; IEA, 2012a, 2012b). There are no exceptions in

Electric Vehicles: Prospects and Challenges. http://dx.doi.org/10.1016/B978-0-12-803021-9.00007-0

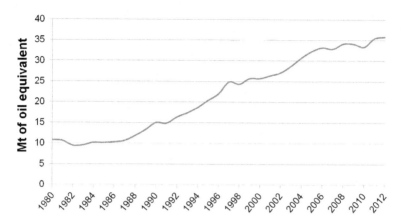

Fig. 7.2.1 Total primary energy consumption in Chile (U.S. EIA, 2014).

Chile; as for most developing countries, pressures from economic and population growth have traditionally forced governments to look at cost-effective options, mainly those exploiting fossil-fuel sources, to cope with the increased demand for electricity from households and industry (Yanine & Sauna, 2013). Fig. 7.2.2 presents the sources of primary energy in 2014.

While economic growth is often linked to increased energy demand, in competitive markets, all producers are interested in lowering the energy consumption per unit of production, while different signs of climate change created an urgent need to reduce emissions from burning fossil fuels. Indeed, burning fossil fuels (oil, gas, and coal) has been identified as the predominant cause of the increase of GHG concentrations in the

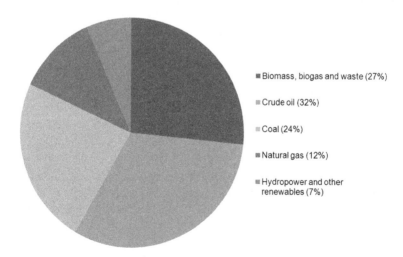

Fig. 7.2.2 Primary energy sources in Chile in 2010 (Ministry of Energy, 2015a).

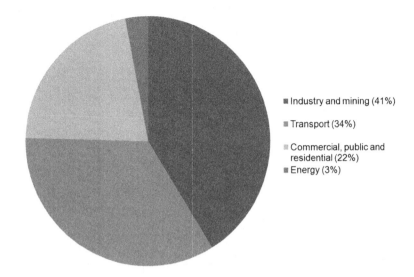

Fig. 7.2.3 Energy consumption per sector (Ministry of Energy, 2015a).

atmosphere. In Chile, the most common source of energy is petroleum, which represents 54% of the final secondary consumption. Virtually, all derivatives are products of the refining of crude oil, which accounted for 96.5% of imports in 2014 (Ministry of Energy, 2015a).

In Chile, a country experiencing rapid industrialization and development but still with a medium income per capita, economic growth, and increased energy consumption is directly related. As shown in Fig. 7.2.3, industry and mining and transport are the sectors with the highest energy demand. These are also the ones that rely mostly on fossil fuels.

Statistics say that there is a link between the GHG growth and the final energy consumed by the population. One of the tasks for the future is to achieve a decoupling between both variables, which would increase competitiveness in a context where economical energy sources will become increasingly scarce.

7.2.2 Environmental loading and CO_2 emissions

According to research, the country has experienced a warming trend ranging from 0.2°C to 1.1°C in the interior regions of the north, centre, and southern, while there has been a cooling of from −0.2°C to −0.5°C in the southern regions of the country, during the period 1901–2005. There has also been a chill in the northern coastal area and south-central Chile of −0.2°C per decade (Garreaud & Falvey, 2009).

Fig. 7.2.4 shows the anomalies or differences between normal and extreme temperatures averaged each year over the period 1961–2010 for the central regions: Valparaíso, the archipelago Juan Fernández, Santiago, Curicó, Chillán, and Concepción. These locations show an increase in the minimum temperatures up to the late 1970s, but later, it does not present a significant increase. In cities of

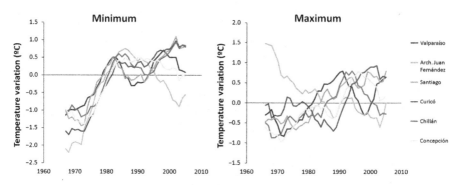

Fig. 7.2.4 Temperature anomalies in central area of Chile, 1961–2010 (MMA, 2012a).

the inner core area, like Santiago, increases in extreme temperatures are recorded in both the minimum and maximum.

Temperature anomalies affect, among other factors, the future availability of water resources, which are essential for human development. Chile has one of the largest and most varied glacial reserves in the world, representing 3.8% of the total world area, excluding Antarctica and Greenland. It also has the largest coverage in South America, with 76% of the glacier area of the continent, estimated at 28,286 km^2. The vast majority of the country's glaciers are experiencing a general trend of mass loss, with rates of linear regression that vary from a few metres annually (especially in the north zone) up to hundreds of metres per year in southern Chile.

One of the main reasons of climate change is airborne emissions of GHG, where the most damaging gases are carbon dioxide (CO_2), methane (CH_4), nitrous oxide (N_2O), hydrofluorocarbon gases (HFCs), perfluorocarbons (PFCs), and sulfur hexafluoride (SF6). While CO_2 has the largest share with 65%, CH_4 and N_2O represent 21 and 14% of the GHG emissions, respectively, and HFCs, PFCs, and SF6 emissions are insignificant. CO_2 emissions are attributed mainly to fossil-fuel burning, cement production, and mining activities. Net emissions from Chile for 2013 were 82 million tonnes (Mt) of CO_2 equivalent approximately (4.7 t per capita and per year) (IEA, 2015; U.S. EIA, 2014); Fig. 7.2.5 shows the trend of total net emissions of CO_2 equivalent for the period between 1980 and 2013. CO_2 emissions in Chile have doubled in 2013 compared with 1995 level. At sector level, the energy industry sector makes a major contribution to and increasingly shapes the values of national emissions, reaching a value of over 38 million tonnes of CO_2 equivalent in 2013 (47% of the total Chilean emissions) (IEA, 2015). The second highest CO_2 emitter is the transport sector with 24.5 Mt. of CO_2 equivalent emitted in 2013 (almost 30% of the national emissions) (IEA, 2015).

7.2.3 Electricity generation

The total electricity consumption of Chile over 1 year period from Nov. 2014 to Oct. 2015 was 71,464 GWh (Generadoras de Chile, 2015). Over that period of time, 60% of Chile's electricity was generated using thermoelectric sources (fossil fuels), specifically oil, coal, and gas, as shown in Fig. 7.2.6. However, Chile is not a fossil

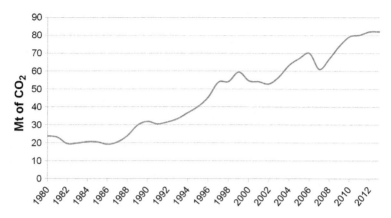

Fig. 7.2.5 CO_2 emissions trend in Chile (IEA, 2015; U.S. EIA, 2014).

energy producer; the country satisfies its internal consumption based mainly on imported fuels (Hanel & Escobar, 2013). The country is dependent on international energy markets in order to secure its needs, which makes Chile vulnerable to supply disruptions and price volatility. The remaining 39% of electricity comes mainly from hydropower (32%), including small and large dams, but their production of electricity varies significantly from 1 year to another. This dependence on the hydrology of a particular period can lead to electricity rationing in dry years. Over that period, 8% of the electricity generated in the country comes from wind (3%), biomass (3%), and solar (2%) sources.

In order to secure energy supply, Chile not only is staying dependent on imported energy but also is going directly against the definition of sustainable development,

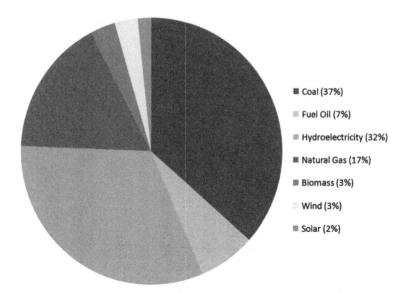

Fig. 7.2.6 Electricity generation in Chile in 2015 (Generadoras de Chile, 2015).

especially since the electricity sector has begun to rely heavily on coal-fired power plants. Up to 3 GW of capacity are being planned to enter the system in the next 3–5 years (Hanel & Escobar, 2013), including the 470 MW Angamos power plant, which has adopted battery storage and seawater cooling tower technologies (Peltier, 2012). However, environmental concerns and local opposition have resulted in delays, cancellations, and court rejections for other coal-fired power projects (Lopez & Ulmer, 2012), and the country is shifting its focus to the expansion of natural gas supplies (Business Monitor International, 2013). Approval has recently been granted to expand regasification capacity at the liquefied natural gas (LNG) plant in Quintero Bay by 50%, from 10 to 15 m^3/day (LNG World News, 2012). As another solution to meet its future electricity needs, Chile has announced that it intends to pursue nuclear power, although it currently has no nuclear power plants (United Press International, 2012). Nuclear energy presents the advantage of not producing carbon emissions when generating electricity, but environmental concerns remain regarding the handling of the wastes it produces. Patterns of the energy sector evolution in Chile can be seen on Fig. 7.2.7 (Reyes, 2013), such as the rapid growth of electricity production since the start of natural gas imports from Argentina in 1998, the variations of hydropower plants production during the years, the recent growth of coal power, and the gas shortage in 2008 replaced by diesel power production.

RE sources for electricity generation in use is mostly hydroelectricity, while wood based biomass and wind account for only 3% for the generation of grid electricity. The construction of five hydroelectric plants with large dams and reservoirs was approved in 2011 on two rivers in the Aysén region (Chilean Patagonia): the Baker and the Pascua, both of which have river basins with the richest biodiversity in the country (Ulloa, 2011). These hydropower plants are supposed to generate approximately 2.75 GW to be incorporated into the SIC network grid. It is expected to flood about 5000 ha of land, currently used for agriculture, recreation, tourism, and biodiversity conservation zones. It is here that some endangered species such as the huemul, a typical Chilean deer, would be affected.

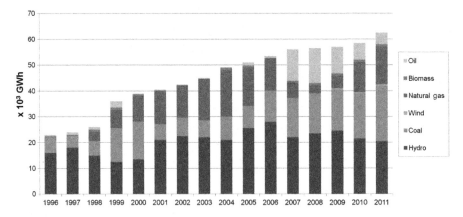

Fig. 7.2.7 Electricity generation in Chile 1990–2010 (Reyes, 2013).

The Chilean electricity market is composed of three independent and private sectors: generation, transmission, and distribution of energy. The government plays a role of regulator in the National Energy Commission (*Comisión Nacional de Energía—CNE*). The CNE is a public organism responsible for analyzing prices, tariffs, and technical standards of energy in Chile and is responsible to make the corresponding analysis in order to coordinate plans and standards for this sector. The Chilean electricity system is divided into four subsystems (see Fig. 7.2.8):

Fig. 7.2.8 Geographic division of electricity production.

- Northern interconnected system (*Sistema Interconectado del Norte Grande*—SING): It represents 28% of total electricity generation. Supplies energy from Arica to the south of Antofagasta and uses mainly thermoelectric, coal, and diesel to generate power.
- Central interconnected system (*Sistema Interconectado Central*—SIC): It represents 71% of the total energy generated, and it is responsible for supplying from the second region to Chiloe. Its main sources of power generation are thermoelectric, hydroelectric, wind, and solar energy.
- Aysén system: It represents 0.3% of the total energy generation and serves the Aysén energy consumption. The electric generation sources for this system are thermoelectric, hydroelectric, and wind power.
- Magallanes system: It supplies energy for Magallanes and Chilean Arctic, and it represents 0.7% of the country's total energy generation, and it does not use renewable sources.

7.2.4   Electricity from renewable sources

Although RE systems are generally more expensive than traditional fossil fuels, they are recognized for reducing environmental and social impacts (European Commission, 2003, 2005). Relying more on RE sources, it aims also at taking more control in the future over the increasing cost of electricity, because, by definition, RE sources are free. Therefore, politicians seriously started to take into consideration solar power as a genuine sustainable solution. Indeed, with some of the highest solar direct normal irradiance (DNI) rates in the world, up to 3300 kWh/m^2, Chile presents a great potential for solar power.

Actually, solar energy development in Chile is small, mostly focusing on water heating applications for the residential sector. The total contribution of solar energy to the primary energy consumption of Chile is negligible (CER, 2013). However, the rise of electricity cost is driving industrial companies to consider self-generation or collaboration on solar projects (Gallego, 2013). Chile's National Energy Commission (CNE) says it is feasible to connect up to 2.2 GW of solar PV plants to the national grid over the next 15 years, in a newly published plan for the expansion of the national transmission system, claiming that the investment price of constructing a solar photovoltaic (PV) power plant is equivalent to or lower than the price of building a coal generation plant (Miller, 2013; Montgomery, 2013; Pekic, 2013).

As recently announced at the PV Insider Latin America conference and by the Deutsche Bank, Chile is currently one of the most exciting markets for PV energy (Gonzalez, 2013). The PV industry in Chile has seen a considerable change from 2013 to 2014 with a growth of 103% of PV plants in operation, while the plants under construction have grown by 184%, according to the PV Insider report 'PV industry roadmap in Chile 2014' (Gonzalez, 2014) (see Table 7.2.1).

The solar thermal power technology (i.e. concentrated solar plant, CSP) presents also a significant potential for covering a large part of the electricity demand (Scanlon, 2012). This well-developed technology is mostly established in the United States and Spain, while in Chile it is still scarcely known. The potential of CSP is greater than PV due to higher capacity factors (CSP plants are able to produce electricity during the night and on cloudy days with the use of thermal storage systems).

Table 7.2.1 Solar PV industry evolution 2013–14 in Chile (Gonzalez, 2013)

		2013	2014	Change
Total capacity (MWp)	Operation	3.7	7.5	103%
	Construction	68.3	244	257%
	Permits approved	3032	4632	53%
	Permitting	1902	2940	55%
Investment (USD million)	Operation	11	19	73%
	Construction	239	602	152%
Energy (MWh/year)	Operation	9250	18,750	103%
	Construction	170,750	485,000	184%

In 2012, Spain-based Ibereolica was one of the first companies to move to Chile and has received environmental approval for its first project, Termosolar Pedro de Valdivia of 360 MW (CSP World, 2012a). The company has also applied for the environmental impact assessment for a second project also in Antofagasta Region, María Elena, a 400 MW CSP complex to use tower technology with molten salt as the heat-transfer fluid and thermal storage medium (CSP World, 2013).

It is only very recently that funding programmes have been approved to develop CSP plants in Chile (CSP World, 2012b, 2012c). For example, in Jan. 2014, Chile's Ministry of Energy and public sector development agency Corporación de Fomento de la Producción (Corfo) awarded Abengoa the contract to build a 110 MW CSP plant in Antofagasta region. It is estimated that the plant would eliminate about 643,000 t of CO_2 emissions per year (Burger, 2014; CSP World, 2014; Parkinson, 2014). The interconnection between the two main grids will directly benefit all these solar plants and boosting the solar industry that will be able to produce electricity in large solar plants in the northern region where radiation is more powerful and supply it where the demand is the strongest in the central region (Galindez, 2014).

Chile's wind resource is also very attractive with many global wind companies currently active in the country. A number of new projects has recently been announced, including Enel Green Power's (formerly Vestas) 90 MW Talinay East project in the Coquimbo region, 250 mi north of Santiago; Pattern Energy's 115 MW Parque Eólico El Arrayán, which the company says will become Chile's largest wind farm when it becomes operational in 2014; and Mainstream Renewable Power's recent USD 1.4 billion joint venture deal with private equity firm Actis to develop around 450 MW of wind projects, to be completed by early 2016 (Bayar, 2013).

The Sustainable Energy Development and Innovation Centre (Centro para la Innovación y Fomento de las Energías Sustentables, CIFES) in Chile says that over 10.3 GW of solar energy projects were approved in 2015 and another 4.1 GW were awaiting an approval. In Dec. 2015, for all renewable energy projects (small hydro, wind, biomass, solar, and geothermal), the CIFES counts 17.8 GW of renewable energy projects approved in Chile and another 6 GW in the planning stage (Fig. 7.2.9) (CIFES, 2015).

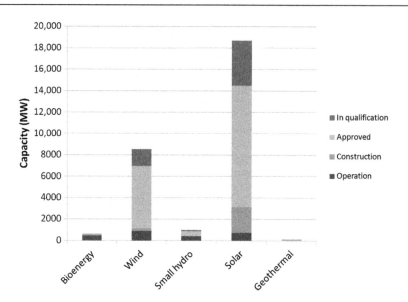

Fig. 7.2.9 State of NCRE projects in Chile (MW) (CIFES, 2015).

Chile's renewable energy capacity soared 40% in 2015 to 2.5 GW (bioenergy 461 MW, small hydropower 397 MW, wind 901 MW, and solar 750 MW), with approximately 11.4% of total capacity in the country grids, according to figures from the CIFES (Fig. 7.2.10) (CER, 2014a; CIFES, 2015; Olivares et al., 2013). NCRE generation was 5.5 TWh in 2015, a 26% increase from 2014 and representing 8% of overall electricity generation (Generadoras de Chile, 2015). In 2015, the installed capacity of NCRE has more than doubled compared with 2013, reaching 2.5 GW in December (CIFES, 2015).

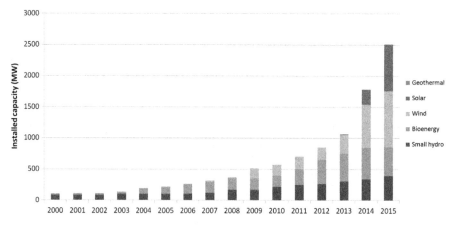

Fig. 7.2.10 NCRE installed capacity evolution. Own elaboration using data from (CER, 2014a; CIFES, 2015; Olivares et al., 2013).

Table 7.2.2 Chilean NCRE potential (CORFO, 2010)

Technology	Potential (MW)	
	Theoretical	Technical
Solar photovoltaic	1000	680
Concentrating solar power	100,000	2200
Geothermal	16,000	2200
Small hydraulic	20,400	4000
Wind	40,000	1900
Biomass	13,600	3300
Total	191,000	14,280

With almost 24 GW in projects in the pipeline approved but not yet built or in qualification, the CER (renamed CIFES in 2015) is launching new funding programmes to help get more of them grid-connected. Among these plans are 1.3 billion pesos (USD 2.3 million) in funding to develop and finance grid-connected projects in the coming year, plus another 2.3 billion pesos (USD 4.1 million) for self-supplying renewable energy systems (CER, 2014b).

CER and German development bank KfW will launch a USD 600,000 programme to help implement more of the projects within that vast pipeline (CER, 2014b). Also in Mar. 2014, the CER is pledging USD 1.8 million to back engineering studies for projects in the preinvestment stage and allocating USD 4.3 million to back smaller projects to develop renewable energy systems for self-consumption; an example is the nation's dairy sector, which seeks to develop more biogas options (Renewable Energy World, 2014).

In order to rapidly achieve energy security and degree of energy independence, Chile is adopting large-scale renewable energy implementation plans, which include strategies for integrating renewable energy systems in a coherent manner designing appropriate energy systems that may be influenced by energy efficiency measures and energy savings, as advised in the scientific literature (Norero & Sauma, 2012; Wilkinson, Smith, Beevers, Tonne, & Oreszczyn, 2007; Yildiz & Güngör, 2009).

Several studies have demonstrated the enormous potential in Chile for NCRE generation and their technical feasibility to implement projects by the year 2025. The results of these studies can be seen in Table 7.2.2 (CORFO, 2010).

7.3 Support mechanisms for renewable energy in Chile

7.3.1 Public policies

Due to the depletion of resources, dependence on imports, and increased admission prices, fossil fuels are no longer an attractive source for energy generation, even without taking into account the environmental impact associated to their use. Moreover, due to the high potential resources for renewable energy generation, Chile has taken notice of the situation and is setting up institutional legislation that forces the electricity generation companies to develop different innovation projects for the diversification of the power grid by nonconventional renewable energy (NCRE),[1] as defined by Chilean law (IEA, 2009). Apart from being a sustainable alternative to fossil fuels, NCREs can also contribute to develop the local economy and provide the country with independence in power generation.

The National Energy Commission (Comisión Nacional de Energía, CNE), body responsible for regulating, preparing, and implementing energy policy, has prepared favorable conditions for electricity generation from NCRE sources, as part of the national energy strategy 2012–30 (Ministry of Energy, 2012). This includes regulation of grid access, integration into the electricity market, and development of expansion strategies and promotional instruments for renewable energies. The public policies that have been developed and implemented in the last decade are presented as follows.

A second priority area of the national energy strategy for the deployment of NCRE is the removal of structural market constraints hindering their expansion in Chile. These include, besides the lack of knowledge about energy resources and their geographic distribution, the lack of experience with planning and approval procedures and with grid connection. Therefore, the Chilean government is financing the GTZ project to investigate the technical and economic energy potential in NCRE sectors and to provide the CNE with advisory services for project planning, approval procedures, and environmental impact studies (CNE, 2009a).

7.3.1.1   Law I ('short law I')

The Ministry of Economy, Development, and Reconstruction, in the year 2004, enacted Law 19.940. Nationally known as the 'Short Law I', the law was implemented to provide consumers with a greater degree of security and quality of supply at a reasonable price. It also provides a modern and more efficient regulatory framework. The law enables small power generation (from 50 kW to 2 MW) to participate in the electricity market. In addition, it includes the partial or total toll exemption for the transmission systems of NCRE with installed capacity smaller than 20 MW (factor of exception), as shown in the Fig. 7.3.1 (CNE, 2009a).

7.3.1.2   Law II ('short law II')

Law 20.018 was enacted in 2005 by the Ministry of Economy, Development, and Reconstruction, mainly because of the uncertainty associated with the availability of natural gas from Argentina. Among the main aspects, the law considers the permission of bidding for long-term contracts by distributing companies and the existence of

---

[1] Nonconventional renewable energy (NCRE) sources include wind power, geothermal energy, solar energy (thermal and photovoltaic), biomass (solid, liquid, and biogas), marine (tides and waves), and small hydraulic energy (<20 MW installed capacity).

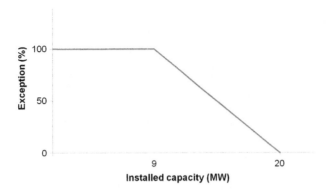

Fig. 7.3.1 Payment exemption of transmission charges to NCRE.

prices higher than the generation transportation rate (not subject to its variation). The law also widens the price adjustment band regulated with respect to free prices, creates a market that allows generating companies to give incentives for clients that consume <2 MW, and stipulates that the lack of supply of Argentinean gas does not constitute a case of force majeure (Ministry of Economy, Development and Reconstruction, 2005).

7.3.1.3 NCRE law

Law 20.257 of 2008, also known as 'NCRE Law' modifies the general law on electric services (Ley General de Servicios Eléctricos) introducing an NCRE quota system. The law requires electricity providing companies, withdrawing electricity to supply their contract commitments, to demonstrate that a certain percentage of their total energy committed was injected in the system by NCRE sources. The energy can be produced by their own plants or by contracting from third parties. This quota came into force at the start of 2010, and until 2014, it will require 5% of electricity to come from NCRE sources. Starting from 2015, the obligation will be increased by 0.5% annually, reaching 10% in 2024. The obligation shown in Fig. 7.3.2 will last for 25 years (2010–34). The law will apply to all agreements executed as of 31 May 2007 (new agreements, renewals, extensions, or similar arrangements). Noncompliance with the law will result in fines per MW not obtained from NCRE sources per year (Ministry of Energy, 2012). On 14th Oct. 2013, the law was reformed and mandates that electric utilities with >200 MW operational capacity should generate 20% of electricity from renewable sources by 2025.

7.3.1.4 Renewable energy centre

In Aug. 2009, the Renewable Energy Centre (Centro de Energías Renovables, CER) institution that consolidates the country's efforts for the development of NCRE was created. The CER acts as a central point of information and support for the promotion of investment and technology transfer.

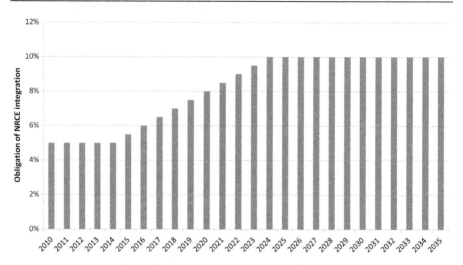

Fig. 7.3.2 Obligations of NCRE integration according to Law 20.257.

The key functions of the CER are the following (CNE, 2009b):

- Study the evolution and development of NCRE technologies and their applicability in Chile, in order to facilitate the elimination of barriers in the materialization of projects.
- Promote and develop an agreement network with centres and institutions, both nationally and internationally, that are promoting innovation in NCRE.
- Serve as a centre of information and orientation for government entities, investors, project developers, and academic researchers.
- Generate natural resources registries for the development of NCRE.
- Watch over the existence of accreditation for the competence of human resources and the certifying of products and services connected to NCRE projects.

7.3.2   Financial and tax incentives for NCRE integration

There are several financial and tax incentives existing in Chile designed to boost the development of renewable energy. In general, financial incentives are channeled through CORFO, a government agency in charge of supporting entrepreneurship, innovation, and competitiveness in the country, which have not necessarily been developed to promote RE, but there are characteristics that coincide with what the instruments require. The mechanisms that can be seized by NCRE technologies are described next.

7.3.2.1   Initiatives of integrated development

These subsidies were created to materialize investment in fixed assets, preferably in parks or technological condominiums, originated by the investment projects. This instrument subsidizes the purchase of critical and/or technological assets in high technology investment projects that extensively promote the development and/or use in the ICTs fields, biotechnology, new materials, and electronics and engineering processes fields. Indeed, projects that apply new production techniques and added value to the natural resources industry in the country, such as renewable generation power plants, are eligible. The subsidy amount could not be higher than 30% of the investment in critical and/or technological assets with a maximum of USD 5000,000. It is addressed to investment projects of USD 2,000,000 minimum (CORFO, 2015a).

7.3.2.2 Technological contracts for innovation

This subsidy supports projects destined to generate innovation in goods, services, commercialization, or organizational methods that have a considerable associated risk. This subsidy is aimed at innovative projects that have the potential to successfully introduce in the market innovations in goods and services and that, at the same time, have the potential to significantly improve the company's performance. Some of the activities that it subsidizes are research for the development of new goods or services, design and construction of prototypes or pilot plants; payment of royalties and patents, and preinvestment studies, among others. The subsidy can amount up to 50% of the total project with a maximum of USD 280,000 (CORFO, 2015b).

7.3.2.3 Tax exemption for extreme zones

There are specific laws in Chile for extreme regions in the country that look to strengthen the development of productive activities in order to stimulate economic growth. These contemplate exceptional tax benefits and subsidies for the installation of services required by the citizens (OLADE-UNIDO, 2011). The zones that benefit from these tax exemptions are the following:

- Regions: Tarapacá, Arica y Parinacota, Aysén, and Magallanes.
- Provinces of Chiloé and Palena.
- Communes of Tocopilla and Isla de Pascua.

Within the described benefits, the country's northern and southern extremes stand out as beneficiaries. Through the tax exemption, the amount of necessary investment to develop renewable energy projects is reduced. This is relevant considering the abundance of solar resources in the North and of water and wind resources in the South.

7.3.3 Prospects for NCRE development

In 2005, the *Invest Chile programme* was launched by the Ministry of Energy and CORFO to support renewable energy projects and finance renewable energy generation nationwide. In the period 2005–09, the programme consisted of a subsidy with a maximum of 50% for studies with a maximum of USD 60,000 and 50% of investment with a maximum of USD 160,000 (IEA, 2014). In this way, the programme supported projects that were trying to generate power based on NCRE with power surplus equal or lower than 20 MW. All kinds of preinvestment studies were financed: prefeasibility and feasibility studies, specialized consultancies necessary to realize the project (prospective studies of energy source, technical and economic, basic engineering, detailed engineering, environmental impact, among others), and studies necessary to evaluate and incorporate projects to the clean development mechanism, among others. In the period 2008–10, the Ministry of Energy transferred USD 2 million to CORFO to continue the programme. After 2010, the applications were received directly without contest. In 2011, CORFO has launched two programmes to subsidize preinvestment studies of NCRE projects, the *support for NCRE development programme* and the *TodoChile programme*. These consisted in a subsidy with a maximum of 50% for studies and 2% of investment with a maximum of USD 60,000 (OLADE-UNIDO, 2011). Since 2012, CORFO has also developed two new contests to finance NCRE projects, the *innovation in renewable energies* that has USD 5 million for subsidies and the *concentrated solar power plant contest*, which has USD 20 million (IEA, 2012c).

Other past CORFO funding programmes such as the *preinvestment programme in NCRE*, the *technological packaging for new businesses*, the *individual business innovation*, and the *innovation projects of fast implementation* in 2011 (OLADE-UNIDO, 2011) also helped to develop NCRE projects in Chile. All the presented funding programmes resulted in a rapid growth of NCRE installed capacity in the recent years (Fig. 7.2.10). Considering the governmental efforts towards cleaner energy development, the projections of installed capacity and NCRE participation are shown in Fig. 7.3.3.

Implementing incentives to increase investments in renewable energy generation combined with a significant increase in gross domestic product (GDP) over the last decades and the stability of external accounts have made Chile a safe and attractive country for investment. According to the Moody's ratings in 2015 (Moody's, 2015), Chile is an 'Aa3' (the 'Aa' categories are the ones with the lowest risks) a long-term country, which means it is the best country in South America in which to invest. Fig. 7.3.4 shows the different Moody's ratings for most countries in Latin America, according to their last publication on the website. This shows how Chile compares with other Latin countries.

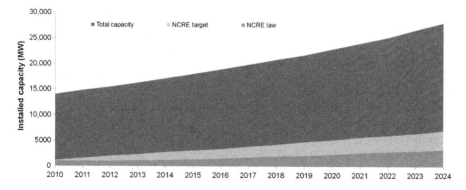

Fig. 7.3.3 Energy capacity projection and NCRE participation in Chile (CORFO, 2010).

Grade	Moody's long term	Moody's short term
Prime	Aaa	P-1
High grade	Aa1	
	Aa2	
	Aa3	
Upper medium grade	A1	
	A2	
	A3	P-2
Lower medium grade	Baa1	
	Baa2	P-3
	Baa3	
Noninvestment grade speculative	Ba1	
	Ba2	
	Ba3	

Fig. 7.3.4 Mood'y country risk ratings for South America (Moody's, 2015).

7.4   Transportation

7.4.1   National context

The environmental effects associated to the use of fossil fuels, combined with their steadily increasing costs (U.S. EIA, 2014), make them unattractive to developed countries that have managed to achieve 'nondependence' of these resources thanks to technological advances in the last era. Chile, on the other hand, imports 74% of all fuels that it uses, therefore, is dependent on energy from other countries, because 60% of the electricity generated in Chile comes from methods that use fossil fuels, while 57% of imported fuels are used in transportation (Ministry of Energy, 2015a), a figure with high reduction potential. Indeed, 99.9% of the vehicle fleet in Chile is dependent on gasoline or diesel, as shown in Fig. 7.4.1 (INE, 2014).

The total number of motorized vehicles in circulation in 2014 is 4,468,450 including light- and medium-duty vehicles (motorcycles, cars, four-wheel drives, vans, and minibuses), buses, and trucks (INE, 2014). Most of the vehicle fleet in Chile is located in the metropolitan and central regions (Fig. 7.4.2), with a number of over 3.35 million vehicles in 2014 in these two regions, which represent 74% of the total fleet in the country located on only 15% of the country surface area (INE, 2014). Fig. 7.4.3 shows the vehicle distribution per type.

Electric transportation has gained an interesting degree of prominence in international systems, especially when compared with other types of transportation, due to its characteristics of greater efficiency and performance (Silva, Neves de Melo, Trovao, Pereirinha, & Jorge, 2013). In Chile, different models of electric

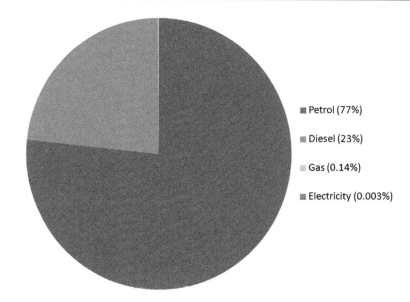

Fig. 7.4.1 Vehicle distribution per combustible type in Chile (INE, 2014).

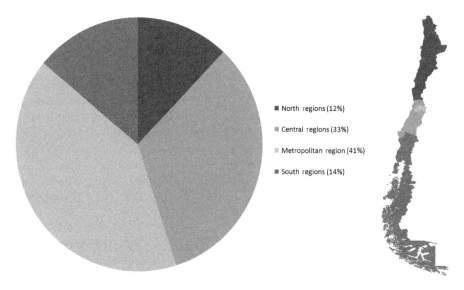

Fig. 7.4.2 Regional distribution of vehicles per zone (ANAC, 2015).

and hybrid cars have been integrated to the automobile market in the past decade, but without achieving significant penetration. This is probably due to their high acquisition costs compared with the conventional cars. Indeed, the price for an EV in Chile in 2015 can easily triple the one of a conventional ICV with similar characteristics (Ibarra, 2014).

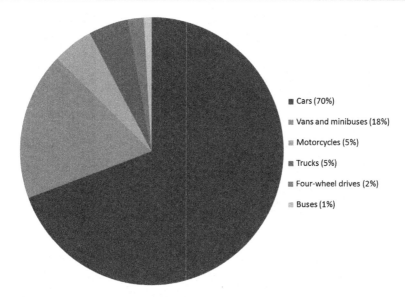

Fig. 7.4.3 Vehicle distribution per type in Chile (INE, 2014).

7.4.2 Air pollution and awareness

The phenomenon of air pollution in Santiago dates from colonial times and has been present in the last 50 years. The relationship between city (population) and pollution has led to an endemic and inherent issue. The factors that have determined that air pollution settles in Santiago as a stable part of the landscape (Fig. 7.4.4) can be grouped into two categories, natural or anthropogenic. Among the former, the geographical location of the capital plays an important role. Indeed, the city of Santiago is located in the basin of the Maipo River enclosed by mountain ranges, which impede a fluid circulation of particulate pollutants. On the other hand, the explosive development of the city along with the intensification of the public and private transport in the region results in increasing levels of pollutant concentrations.

Air pollution is usually considered as smog, particulate materials, or chemical molecules introduced into the atmosphere and causing damage, diseases, or death to living organisms. The transport sector is the second major contributor to the atmospheric pollutant emissions in Chile generating about 25% of the pollutant emissions (the first being the energy sector with 45% of the national atmospheric pollutant emissions) (MMA, 2012a). The main direct emissions caused by traffic vehicles are carbon monoxide (CO), nitrogen oxides (NOx), volatile organic compounds (VOC), and particulate matter (PM). It is estimated that in the metropolitan region (MR), ICVs are responsible for 81% of the CO emissions, 76% of the NOx emissions, and 42% of the emissions of PM10 and PM2.5[2] (USACH, 2014).

During the first half of the 20th century, various specialists conducted research about the problems linked with air pollution. For example, it was found that population

---

[2]PM10 and PM2.5 refer to the concentration in air of particles with diameter below 10 and 2.5 μm.

Fig. 7.4.4 Smog over Santiago de Chile (Photo by Jason Vargo).

growth, with an increase of >100% between 1940 and 1960, and the extension of the urban area of the city, directly affect the increase in particulate pollutants in the capital, which were at this time mostly due to home fireplaces (Morales Segura, 2006). However, it was in the second half of the century that it became a matter of ongoing concern, and public policies were designed to address the problems of air pollution, firstly with the Decree n° 144 of the Health Ministry regarding the pollutant emissions in 1961, with the creation of the National Commission of ambient decontamination in 1970 and with the Sanitary Code launched in 1972 (Riveros Requena, 1997).

In the 1990s, media and public opinion pressure led to the protection of the environment and its resources and became permanent part of government policies. A concrete reflection of that was the enactment in 1994 of 19.300 Law or Basic Environmental Law, thereby creating environmental institutions. A series of measures linked to this new body of law sought to cushion the effects of pollution. Among them, in Jun. 1996, the Santiago MR was officially declared a saturated zone for four atmospheric pollutants: total suspended particulates (TSP), breathable particulate matter (PM10), carbon monoxide (CO), and ozone (O$_3$).[3] Also that year, the Special Commission on Decontamination of the Metropolitan Region was created. This commission developed the Prevention and Atmospheric Decontamination Plan for the Metropolitan Region (PPDA) in 1997 and the Urban Transport Plan for Santiago (SUTPs) in 2000. All these actions had reduced levels of air pollution in the capital in a remarkable way, from 37 preemergency alerts[4] in 1997 to only two in 2004 (Morales Segura, 2006). However, as air pollution is directly linked with economic growth, the trend has been reversed, and pollutant levels have constantly been increasing in the recent years. From only two in 2004, eight preemergency alerts were registered in Santiago in 2008 (INE, 2011), and the figure rose to 17 alerts in 2014 (El Dínamo, 2015).

---

[3] Decree No. 131, 12 Jun. 1996.

[4] The emergency alert on air pollution is set when the particulate material exceed 240 mg per cubic metre.

7.4.3 Public transport infrastructure in the metropolitan region

The public transport system in Santiago includes city buses, subways, and taxis. The entire public transportation system is regulated, controlled, and monitored by the metropolitan transportation directory (*Directorio Metropolitano de Transporte*), which is a governmental body.

The bus transport system ('Transantiago') is composed of 6500 buses operated by seven different companies, but the entire system can be utilized with the same travel card called 'bip!', and a multitude of charging points can be found across the city. The 'transantiago' bus system is used by about 6.2 million people over the 32 communities of Santiago and covers a geographic area of about 680 km^2. Within one working day, more than three million transactions are made in 'Transantiago' buses. The seven companies in charge of covering different areas of the city are distinguished by different colours as shown in Fig. 7.4.5 (Transantiago, 2014).

Fig. 7.4.5 Transantiago fleet coverage (Transantiago, 2014).

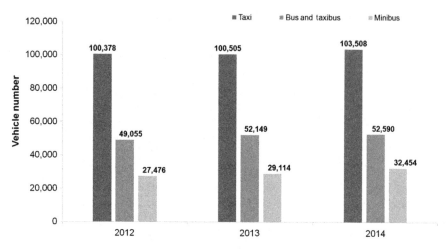

Fig. 7.4.6 Public transport per type of vehicle in the MR (INE, 2012, 2013, 2014).

The Santiago's subway is a major component of the public transport system in Santiago. The subway fleet is composed of 1090 wagon trains travelling in five lines and a total of 108 stations covering different districts of the city. Most metro stations are connected with Transantiago bus stops or other intercity modal stations. The Santiago's subway registers over two million ticket transactions on a daily basis.

The Chilean public transport sector holds in total about 190,000 vehicles in 2014, divided between taxis, minibuses, and buses, of which 71,000 are counted in circulation in the MR (Fig. 7.4.6) (INE, 2012, 2013, 2014). The number of taxis and minibuses in 2014 and buses in 2013 are the highest ever registered in the MR.

7.4.4   Transport public policy

The structure of the national transport policy in Chile is based on a strategic development objective, which main working areas are ensuring people mobility through infrastructure improvement in the same time as considering social and economic development (MTT, 2013a). To achieve the goal of infrastructure, the plan contains different working areas, focusing efforts on capacity, quality of service, efficiency, and operational continuity. The country has adopted investment plans for road, rail, port and airport constructions, or refurbishments in order to meet the long-term needs. As for the social development, efforts aim at improving the access to public transport and the efficiency of urban transport. The public policies for transport that have been developed are presented as follows.

7.4.4.1   Law 20.378

In 2009, a Public Transport Subsidy Law was approved by the government (Law 20.378). This law introduced a permanent and a transitory subsidy for public transport of over USD 500 million for Santiago and about the same amount for the rest of the country to finance transportation initiatives, connectivity, and development of regional transportation. In 2013, almost 4200 public transport services (bus and rail) in the country have received subsidies; over 3000 regional buses have been able to propose a reduced fare; about 500 ferry or plane services were developed to give access to public transport to >370,000 people in isolated areas; 631 free school buses were financed to allow 42,000 children from low revenue families to go to school; about 2260 regional buses were replaced; and about 6300 new stop stations were built (MTT, 2013b).

7.4.4.2   2025 Santiago maestro plan

In 2014, the Ministry of Transport and Telecommunications (MTT) launched the 2025 Santiago maestro plan (MTT, 2014). Considering that the number of vehicles in Santiago is expecting to double between 2012 and 2025, the plan intends to give guidance on the management and strategic infrastructure investments in order to develop an urban transport system that will be able to satisfy the needs for population and goods mobility on the long term. The 2025 maestro plan offers propositions to develop coordinated projects with the objective of establishing an efficient, equitable, sustainable, and secure transport system in Santiago.

7.4.4.3   Zero emissions mobility program

Since 2012, new policies are being developed, such as the *zero-emission mobility programme* ('Programa de Movilidad Cero Emisiones') launched by the Ministry of Environment (Chilectra, 2012). This programme, which includes several initiatives such as the PM2.5 decontamination plan and the public transport improvement plan, is aiming to promote electric transport, giving greater emphasis to environmental and health impacts caused by pollution. In the programme, the incorporation of EVs comes with the purpose of reducing polluting particles emissions and thereby substantially improving air quality in the region. However, government efforts are focusing at the moment on the public transport sector giving subsidies for electric taxi renewals. The purpose of the incentive is to bring EVs purchase price to the same competition level with conventional ICVs.

7.4.4.4   Emission standards

Emission standards for light- and heavy-duty vehicles have been in place since the early 1990s. While the standards are based on US and EU emission regulations, they are not necessarily equivalent. Dual standards often exist, allowing new engines to meet either US or EU standards. Durability requirements were added in 2012 for light and medium vehicles with Decree 29 (MMA, 2012b) and for buses and trucks with Decree 4 (MMA, 2012c). The emission limits are based on the date that application is first made to register the vehicle in the national vehicle registry and the geographic region in which it operates. The application date for registration, as opposed to the vehicle model year, is presumably used to control emissions from imported used vehicles. To legally operate an on-road vehicle in Chile, a coloured sticker must be attached to the vehicle. The colour of the sticker determines the region of the country in which a vehicle may operate; rules for issuing stickers depend on the vehicle class.

Due to severe pollution problems, many vehicle emission standards for the Santiago MR are more stringent and/or were introduced earlier than those for the rest of the country. The Decree N° 66/10 issued in 2010, which revise, reformulate, and update the prevention and atmospheric decontamination plan for the MR (PPDA), required a number of programmes to be established in the MR by 2011 to accelerate the uptake of cleaner vehicles. These programmes include the implementation of a low-emission zone for heavy vehicles and a voluntary truck scrappage programme (SEGPRES, 2010).

7.5   The role of EVs in the Chilean transport market

EVs are transport methods that use an electric motor as the engine. Electric motors are more energy efficient converting virtually all of their fuel energy into usable power compared with traditional internal combustion engines, which are <20% efficient. Moreover, EV motors present more simplistic mechanical characteristics than internal combustion engines, and so their maintenance is more straightforward. There are no gaseous emissions produced while driving EVs, and they are more silent than ICVs. Also, other problem pollutants such as oil, transmission fluid, and radiator fluid are eliminated in EVs, and the only hydrocarbon-based substance used is the grease that lubricates bearings. However, EVs present two main disadvantages, namely, an elevated purchase price and low autonomy, which represent a significant barrier to the EV market development. The key features of EVs that can influence their integration in the vehicle market in Chile are presented as follows.

7.5.1   GHG emissions—EVs vs. ICVs

There are several EVs available in the market at the moment, and so, benchmark indicators of autonomy, consumption, or purchase price depend on model and specifications of each vehicle. The comparison of GHG emissions from the electrical 2016 model of the Nissan Leaf with battery capacity of 24 kWh to average gasoline and diesel cars (Table 7.5.1) shows a lower value of GHG emissions for the electric model compared with conventional diesel and gasoline cars when the EV is charged from both national grids. According to the CNE, the emission factors of the national grids can be assumed as 0.77 and 0.38 kg of CO_2 equivalents per kilowatt hour of electricity generated ($kgCO_{2eq}$/kWh) for the SING and SIC networks, respectively (Ministry of Energy, 2015b). The difference between the two networks is due to the difference of primary energy sources used in the electricity generation mix. It is worth mentioning that these figures tend to lower with the progressive integration of RE electricity into national grids as shown in Section 7.2.4. The emissions of the 2016 Nissan Leaf when charged from the SIC network found in Table 7.5.1 remain under most of the GHG emissions from small advanced diesel engine car (EPA, 2014, 2016; Janic, 2014; Ministry of Energy, 2015b).

Table 7.5.1 **Passenger car GHG emission comparison**

	Average consumption	Emission factors		CO_{2eq} emissions (gCO_{2eq}/km)
2016 Nissan Leaf	0.186 kWh/km[a] (EPA, 2016)	SING	0.77 $kgCO_{2eq}$/kWh (Ministry of Energy, 2015b)	143
		SIC	0.39 $kgCO_{2eq}$/kWh (Ministry of Energy, 2015b)	73
Average diesel car	0.058 L/km (Janic, 2014)	2.69 $kgCO_{2eq}$/L (EPA, 2014)		156
Average gasoline car	0.073 L/km (Janic, 2014)	2.35 $kgCO_{2eq}$/L (EPA, 2014)		172

[a]Based on 45% highway, 55% city driving test from EPA (EPA, 2016), while manufacturer displays 0.15 kWh/km under CEPE/ONU n° 101 test.

One of the main advantages of EVs over ICVs is that their batteries can be charged from renewable electricity, and in this case, the corresponding CO_2-equivalent emissions would fall to a much lower value. For example, assuming the Nissan Leaf being charged by solar PV panels, in which CO_2 emission can be considered of 0.03 $kgCO_{2eq}$/kWh,[5] the corresponding CO_2-equivalent emissions would be 6.6 gCO_{2eq}/km.

7.5.2 *Market penetration*

The electricity-fuelled vehicle is not a recent development. In fact, the EV has been around for over 100 years, since the German engineer Andreas Flocken built the first battery four-wheeled electric automobile in 1888 (Anderson & Anderson, 2012). The high cost, low top speed, and short range of battery-EVs, compared with internal combustion engine vehicles, led to focus efforts on technological developments on the ICVs, although EVs have continued to be used in the form of electric trains.

It is only very recently, at the beginning of the 21st century that interest in EVs has increased due to growing concern over the problems associated with hydrocarbon-fuelled vehicles, including damage to the environment caused by their emissions. As a result, great technological progress has been achieved regarding EVs, and several models have entered the market place since 2010, when the first EV was introduced in Chile.

[5] According to the latest life-cycle analyses, which measure the environmental impact of solar PV panels from production to decommission, GHG emissions have come down to around 30 g of CO_2 equivalents per kilowatt hour generated (gCO_{2eq}/kWh), compared with 40–50 gCO_{2eq}/kWh 10 years ago (Fthenakis et al., 2011; NREL, 2013).

Although projections for the EV penetration in Chile were very positive at first, indicating that 200,000 EVs would be in circulation by 2020 in Chile, in 2013 that projection was re-evaluated to 70,000 according to the National Chilean Automobile Association (Asociación Nacional Automotriz de Chile, ANAC) (Lazcano, 2013). Actually, the popularity and market penetration of EV in Chile can be qualified as very poor, although they present great efficiency and performance characteristics. In 2014, only 0.003% of the vehicle fleet in Chile is electricity-fueled, which amount to a total of 136 vehicles (98 in the MR) (INE, 2014). The low penetration of EVs in the transport market can be explained by two factors, that is, the high battery cost that involves high EV cost and low availability of charging points detailed as follows.

7.5.3 Market development barriers

While ICVs can store diesel or gasoline in a 50 L tank and have autonomy approximately for 500–1000 km, EVs need to store electricity in batteries, which provide much lower autonomy (about 100–200 km) and take a lot of space and weight, and the cost of which is high. Li-ion batteries usually account for 50% of the cost of an EV (Nykvist & Nilsson, 2015; Yuzawa, Archambault, Burgstaller, Yang, & Kurian, 2015). However, the literature reveals that Li-ion battery costs are coming down (Catenacci, Verdolini, Bosetti, & Fiorese, 2013; Gerssen-Gondelach & Faaij, 2012; Weiss et al., 2012). According to Nykvist and Nilsson (2015), the cost of battery packs used by market-leading EV manufacturers have declined by 8% annually between 2007 and 2014, from above USD 1000 per kWh to around USD 300 per kWh, and that learning rate in 2015 remained between 6 and 9%. Although technological advances play a significant role, growing economies of scale are the key driver of cost reductions. According to Yuzawa et al. (2015), costs could be reduced 30%–40% to an expected value of USD 150 per kWh following the startup in 2017 of Tesla Gigafactory of consumer batteries (50 GWh factory capacity) and below USD 100 per kWh by 2020. On the other hand, a number of EV makers are planning to cut auto battery costs 30%–40% by 2020 compared with 2015 levels via energy density increases achieved through cathode material innovation and scale, and between 2015 and 2020, analyst forecast battery weights to fall by over 50%, and the battery EV range is expected to increase by over 70%, due to technological advances (Goldman Sachs, 2015).

Just as gasoline and diesel stations, the presence of electricity charging points in sufficient quantity and wisely distributed across specific areas can be an effective solution to respond to the issue of EV low autonomy, although the infrastructure required would mean a certain challenge to the Chilean authorities. Due to low autonomy issues and a relatively long charging time of batteries, EV usage is limited to local travels only, until significant technological advances are made in electricity storage systems. However, in an urban environment, where transport is necessary only within a local area such as the MR, EVs can perfectly fit the requirements, probably as a second vehicle for particular or for public city buses transportation. In this case, the increase of charging points can overcome the market barrier of low autonomy EVs.

7.5.4   Financial and tax incentives for EV

Considering that the acquisition price for an electric car in Chile is between CLP\$ 20 and 50 million (between USD 30,000 and 75,000), such vehicles cannot compete equally with conventional cars, the price of which can be half that of a similarly equipped EV (El Mercurio Inversiones, 2015). At the moment, no incentive plan exists in Chile to boost the marketing of EVs in the private sector. However, new measures were published on the 5th Jan. 2016 by the Environment Ministry in the Resolution N° 1.260 of the PPDA, including a new incentive plan strategy to be implemented by 2017 for the purchase of EVs (MMA, 2016).

Regarding the public transport system, the Ministry of Environment launched in late 2014 the PM2.5 decontamination plan. The plan measures the introduction of electric buses in the public transportation system of Santiago. The first objective, which would serve as a pilot programme, is to incorporate 100 electric buses to the fleet by the end of 2016 (CMM, 2014). The Mario Molina Centre (CMM) final report (CMM, 2014) presents a 20-year economic analysis of the different alternatives for public transportation, including the total costs of a Euro5 diesel bus, a battery-electric bus, a trolleybus, and a hybrid bus, as shown in Fig. 7.5.1. Costs were evaluated on a 20-year life cycle, assuming 5% discount rate, 7% loan interest rate, 2% annual increase of diesel cost, 1% annual increase of electricity cost, and 10-year battery life. Results shown that trolleybuses and battery buses could compete economically with advanced diesel buses depending on the contracts made with buses companies.[6]

Chilectra, one of the main electricity distribution companies, would indeed directly benefit from the increase of EVs, and so, it has been constantly supporting EV

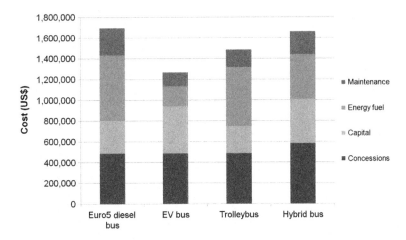

Fig. 7.5.1 Economic analysis over 20 years of different public bus types.

---

[6]Concession contracts are awarded in the form of incentives to the bus companies of the Transantiago network to promote public transportation.

initiatives. Together with a research group of the Mario Molina Centre (CMM), they have predicted that by 2022, 35% of the public transport system will be electricity-fuelled, integrating 1500 battery buses and 775 trolleybuses. Projections also indicated that by 2020, the national car fleet would be composed by 10% of EV and that 20% of the new vehicle sales in 2020 be electricity-fuelled (Mujica Carvajal, 2014).

As of Oct. 2014, there were 21 electric taxis in Santiago, which are part of the public transport improvement plan initiated by the Ministry of Transport and Telecommunications (Martínez Gaete, 2014). Since Nov. 2015, the same programme has proposed a new subsidy of CL$ 6 million (over USD 8500) for taxi drivers who wish to buy an electric taxi. The expectations of this programme are to aim for 70 electric taxis by the end of 2016 (Reyes, 2015).

By the end of 2015, Chilectra and Copec completed the construction of 12 public charging stations in the city of Santiago. These charging points are located not >10 km from each other in the communities of Santiago, Vitacura, Las Condes, Huechuraba, Pudahuel, and Providencia, in order to ensure normal circulation of EVs within the MR area. By this initiative, it is intended to increase the use of EVs. However, although such infrastructure developments provide support to actual EV users, it does not promote directly the purchase of EVs (Paleo, 2015).

7.5.5 Sustainable development

The notion of sustainability is defined as the '*development that meets the needs of the present without compromising the ability of future generations to meet their needs*' (World Commission on Environment and Development, 1987). Sustainable development requires the reconciliation of environmental, social equity, and economic demands referred to as the three E's of sustainability. In the diagram shown in Fig. 7.5.2 (Adams, 2006), the 'environment' region of the diagram refers to the conservation of natural resources and the reduction of impacts on ecosystems, 'ethical' refers to the protection of the communities' health and the education and empowerment of populations to participate in the process, and 'economic' region relates to cost. Economic feasibility is required if sustainability is to remain viable

Fig. 7.5.2 The three E's of sustainability.

in the long term. For example, the generation of incentives for sustainable practices (such as tax credits for solar panels or EVs and feed-in tariffs for electricity generation from renewable sources) is one means of making important sustainability issues economically viable and accessible.

EVs represent an attractive alternative that can make an important contribution towards Chile's sustainable transport development, particularly in the MR. The Chilean government and financing agencies, through tax credit and incentives, have the capacity to generate an economic market for EV in the MR, which can help car owners to shift from conventional vehicles to more efficient and cleaner ones. The uptake of the EV market, combined with the high penetration of NCRE into the Chilean electricity generation mix, has the potential to significantly reduce air pollution in Santiago, thus improving the health and quality of life. At the same time, the intensification of an EV economy can lead to job creation, thus fitting perfectly the concept of sustainable development.

7.6   Lithium mining in Chile

The extraction and refinement of lithium has become one of the leading challenges for new technologies, powered by the increasing demands of lithium-ion batteries used in mobile phones, laptops, and electric cars. The key raw material for lithium-ion battery manufacturing is lithium carbonate (Li_2CO_3), which can be extracted from hard rock minerals (generally spodumene, petalite, and lepidolite) or from lithium brine bodies in salt lakes or salars (Vikström, Davidsson, & Höök, 2013). Chile is the first producer of lithium in the world with 33% of the global production in 2013 (USGS, 2015). The Chilean lithium is extracted exclusively from brine in the Atacama salar, in the extreme north of the Chile, where the lithium reserves are estimated to represent 31% of the world reserves (Fox Davies Capital, 2013).

The brine is formed in underground basins where water that has leached the lithium from the surrounding rock is trapped. As shown in the diagram (Fig. 7.6.1), the process of extracting the lithium involves pumping the brines into a series of evaporation ponds to crystallize other salts, leaving lithium-rich liquor. This liquor is further processed to remove impurities before conversion to either lithium carbonate or lithium chloride for further upgrading to lithium hydroxide, which can also be used in battery manufacturing (Pavlovic, 2014). The Atacama region is one of the best places in the world to produce lithium from brine, due to three main factors. The brine has high lithium and potassium contents, which allow the coproduction of potash (potassium salts), secondly, the costs of production are lower than in other places because the brine has low magnesium content, and thirdly, the climate conditions in the region (low rainfall and high solar irradiation) are ideal for a fast evaporation process (Garcés Millas, 2002).

Lithium is considered more environmentally friendly, in comparison with existing nickel-metal hydride or lead-acid battery manufacturing technologies, due to a recovery process having virtually no waste when mining. Once the lithium is recovered, the chemicals can be recycled, for example, the production of by-product compounds such as potash and boron. Moreover, lithium extraction from

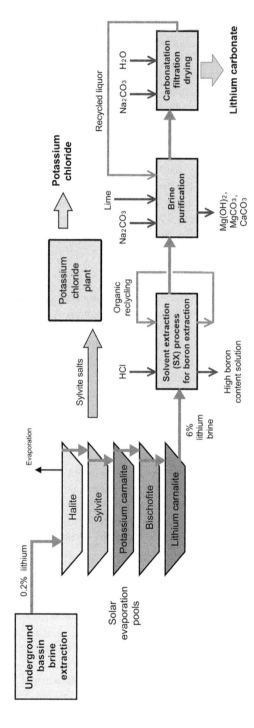

Fig. 7.6.1 Lithium carbonate production process (Pavlovic Zuvic, 2014).

brines has low environmental impact, as demonstrated by Stamp, Lang, and Wäger (2012), and most of its carbon footprint is due to product transportation.

In Chile, lithium is regarded as an issue of national security for being a 'material of nuclear interest', due to its potential in nuclear fusion processes. For this reason, despite huge investments in mining operations in the Atacama region, neither company owns Chile's lithium reserves. Under the 1979 Decree law Nr. 2.886, lithium reserves are exclusively property of the state, and lithium is controlled and regulated by Chile's nuclear energy commission (CChEN), governmental body created in 1965 under the Law Nr. 16.319. Since 1983, the law on mining concessions Nr.18.097 and the Chile Mining Code provide that the exploitation of lithium can only be performed directly by the state of Chile or its companies and could not be the object of mining concessions, except for those validly granted before the introduction of 1979 Decree law Nr. 2.886. According to the mining code, lithium concessions can only be handled outside the general system for granting mining concessions, by means of administrative concessions or special operation contracts (Mining Ministry, 2015).

Chile lithium production in 2013 was estimated to 11,200 t[7] (USGS, 2015), equivalent to nearly 60,000 t LCE. The US Geological Survey reports an approximately 7.5 million tonnes of lithium reserves in Chile (USGS, 2015), which equals to 670 years of supply at the 2013 rate of demand. The worldwide exportations of Chilean lithium amount to USD 226 million in 2013 (Central Bank of Chile, 2015) and the sector counts about 1000 employees in Chile (Bussi, 2012).

In 2013 alone, the world consumption of lithium increased by 10% (Fox Davies Capital, 2013), and with few alternatives for lithium within portable devices, the industry is set for continuing strong demand. Considering that in the coming years lithium-ion batteries may be adopted by the global automobile industry in a progressive transition to EVs, battery demand for lithium in hybrid and EVs is expected to increase annually by over 27% until 2025, according to signumBOX estimation (Table 7.6.1) (Fox Davies Capital, 2013). The lithium demand in lithium carbonate

Table 7.6.1 Global lithium demand projections (Fox Davies Capital, 2013)

Application	2011 (tonnes LCE)	2025 (tonnes LCE)	Annual growth rate 2011–25
Batteries for portable devices	30,416	111,176	9.70%
Batteries for grid	500	7500	21.30%
Batteries for hybrid and EVs	6967	204,901	27.30%
Other applications	91,400	174,994	4.70%
Total lithium demand	129,283	498,571	10.10%

[7]Lithium metal content. Conversion factor 5.323 for lithium carbonate equivalent (LCE).

equivalent (LCE) in all types of batteries combined is estimated to represent around 65% of the total consumption in 2025.

Possible development of Bolivia's extensive reserves, uncertainties about the technology for electric car batteries, increasing competition from China, and a potentially evolving lithium recycling market could change the long-term picture for lithium demand (Speirs, Contestabile, Houari, & Gross, 2014). However, in the short term, it appears that Chile's position is secure, and companies operating in Chile appear quite capable of meeting the challenges of increasing global demand.

7.7 Case study

This section introduces a comparison of the fuel costs given between the Nissan V-16 vehicles (ICV taxis of Santiago) and equivalent EVs. Then, a technical analysis is done on the performance potential of a solar power charging station for electricity-fuelled taxi vehicles in Santiago.

7.7.1 Fuel costs comparison

In order to compare the cost per km of city driving for a Nissan V-16 and an EV, the price of gasoline must be compared with that of electricity. However, the comparison is not straightforward since the two variables are not expressed in the same units: the price of gasoline corresponds to a quantity in $/L, while the electricity price is in $/kWh. To allow comparison, the cost to fill up the tank of the ICE and the batteries of an EV can be calculated from Eqs (7.7.1), (7.7.2), respectively:

$$C_{ICV} = \frac{V_{tank} \times P_g}{D_{ICV}} \tag{7.7.1}$$

where C_{ICV} is the fuel cost per km of the Nissan V-16 vehicle (USD/km), V_{tank} is the tank volume capacity (litres), P_g is the price of gasoline (USD/L), and D_{ICV} is the distance range that can be travelled with a full gasoline tank.

$$C_{EV} = \frac{E_b \times P_e}{D_{EV} \times \eta_t} \tag{7.7.2}$$

where C_{EV} is the fuel cost per km of the battery EV (USD/km), E_b is the electrical energy required to fill up the battery (kWh), P_e is the electricity price (USD/kWh), D_{EV} is the distance range that can be travelled with a full battery charge, and η_t is the total well-to-wheel efficiency between the battery and the wheel transmission (%), which can be calculated using Eq. (7.7.3):

$$\eta_t = \eta_m \times \eta_i \times \eta_b \times \eta_c \tag{7.7.3}$$

where η_m is the electrical motor and drivetrain efficiency (%), η_i is the inverter efficiency (%), η_b is the battery efficiency (%), and η_c is the charger efficiency (%).

Table 7.7.1 Nissan V-16 and EV parameter assumptions

	T_{cap}	50 l	(Nissan, 2015a)
	D_{ICV}	450 km in urban area	(Nissan, 2015a)
Gasoline Nissan V-16	P_l	1.14 USD/l	(Econsult, 2016)
24 kWh battery EV (based on the 2016 Nissan Leaf)	Battery type	Lithium-ion	(Nissan, 2015b)
	E_b	36 kWh	(Gallardo Lozano, 2012)
	D_{EV}	170 km in urban area	(Nissan, 2015b)
	η_m	90%	(Markowitz, 2012)
	η_i	95%	(Markowitz, 2012)
	η_b	90%	(Markowitz, 2012)
	η_c	95%	(Markowitz, 2012)
	P_e	0.16 USD/kWh	(Chilectra, 2016)

The analysis takes into account that the Nissan V-16 is able to drive 450 km on one full gasoline tank and that the EV is equipped with a 24 kWh battery capacity able to drive 170 km on one full charge. The parameters shown in Table 7.7.1 are the other assumptions also taken in account in the calculations (Chilectra, 2016; Econsult, 2016; Gallardo Lozano, 2012; Markowitz, 2012; Nissan, 2015a, 2015b).

Results show that the fuel cost for driving the Nissan V-16 is 0.127 USD/km and 0.046 USD/km for driving the EV. The fuel cost is more than half cheaper to drive the EV compared with the conventional gasoline Nissan V-16.

7.7.2 Performance of a solar power charging station

In the following, the performance of a solar PV power station installed in Santa Isabel supermarket's parking is investigated. Many Santiago's taxis are based close to metro station 'Grecia', on Santa Isabel parking lot, which is a frequented place because of its shopping mall, and therefore conducive to improvements in public transport. A possibility would be to instal solar PV panels on the roof of that large and flat building. Below is an investigation on the amount of electricity that can be generated from a solar PV station occupying the rooftop area of the supermarket building.

In Fig. 7.7.1 (Google earth, 2015), the red circle area shows the taxis parking lot and the five available roof areas of the supermarket's building (noted 1–5) available to carry the solar PV modules. The area shown in the red rectangle is unavailable as it is occupied by the building's air treatment facilities.

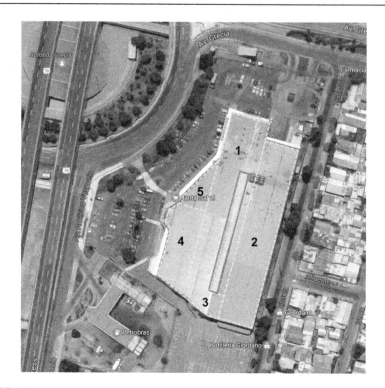

Fig. 7.7.1 View of Santa Isabel's roof (Google earth, 2015).

The total flat area including the five zones is 6930 m². It is considered that 20% of the surface is left unoccupied in order to allow passage between panels for maintenance purposes. Assuming the PV modules to be installed at 45° inclination and orientated towards the north, the useful PV module area is 7817 m².

At this location, it assumed a global horizontal irradiation (GHI) of 1961 kWh/m²/year and a monthly distribution of daily GHI as shown in Table 7.7.2 (NREL, 2015). Assumptions account that 60% of the GHI can be harvested by the north orientated, 45° inclined panels, and PV module efficiency of 16%. The average daily electricity that can be generated by the solar PV station is shown in Table 7.7.2.

Assuming that the daily distance travelled by taxis in Santiago is 130 km in average (Martínez Gaete, 2014), the daily electricity energy required to charge battery EV taxis from the grid would be 31.2 kWh. Table 7.7.2 presents the number of battery EVs, which can be charged daily by the PV solar station. On average, the solar station could charge daily 129 battery EV taxis.

7.8 Conclusion

Over the past 10 years, CO_2 emissions in Chile have increased dramatically, reaching values close to 82 million tonnes of CO_2 equivalent in 2012 (U.S. EIA, 2014). One of the main reasons of this rise is the expansion of the public transport sector. This

Table 7.7.2 **Santa Isabel's solar station performance**

	Daily GHI (kWh/m^2)	Daily electricity generated (kWh)	Number of battery EV taxis charged per day
Jan.	8.57	6431	206
Feb.	7.563	5676	182
Mar.	6.171	4631	148
Apr.	4.326	3246	104
May.	2.8	2101	67
Jun.	2.436	1828	59
Jul.	2.571	1929	62
Aug.	3.348	2512	81
Sep.	4.495	3373	108
Oct.	6.031	4526	145
Nov.	7.671	5757	185
Dec.	8.483	6366	204
Average	5.372	4031	129

chapter discloses the improvements that can be implemented to reduce the CO_2 emissions of this sector in Chile. It highlights the comparison between the conventional ICV taxi to the EV version. To this end, the CO_2 emissions from the two compared technologies were found, obtaining emissions of 143 gCO_{2e}/km and 73 gCO_{2e}/km for the EV charged from the SING and SIC grid, respectively. For the average gasoline and diesel ICVs, we obtained emissions of 156 and 172 gCO_{2e}/km, respectively.

Then, another assessment was concerned with the economic dimension of the cost per kilometre travelled by the Nissan V-16 ICV compared with an equivalent size EV. It was obtained a value of USD0.127/km for the ICV and USD0.046/km for the EV model. Regarding GHG emissions for the electrical version, two types of emissions are relevant. There are the emissions emanating from charging the EV on the Chilean grid and those emanating from charging the EV on a solar charging station, which led to emissions of close to zero. Thus, EVs do not particularly help to reduce CO_2 emissions when their batteries are powered from the Chilean electricity grid. This is because the grid contains energy derived from burning fossil fuels. However, when charged with PV power, EVs show a significant drop in emissions, down to 6.6 gCO_{2e}/km. The case study investigates on the possibility of installing a rooftop solar PV power station at the Santa Isabel supermarket's parking. It is found that on average, 129 battery EVs could be charged daily considering the solar station characteristics. However, further studies should focus on how to sustain solar power during radiation poor periods in winter.

Though obstacles remain, the Chilean government has made an important progress in laying out a policy framework for greater future EV adoption. In the short and medium run, expanding incentives for EV adoption and pilot programmes could serve to jumpstart demand. Fiscal incentives would likely be most effective in promoting

Fig. 7.8.1 The three pillar of success for EV.

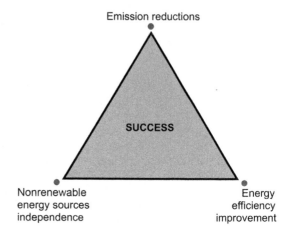

EV adoption by lowering upfront costs. Alternatively, the government can consider options like feebates, establishing a fee on high-emitting vehicles and in turn providing a rebate for low-emitting ones. The three pillars for the success of EV adoption in Chile are presented in Fig. 7.8.1.

The slow adoption of EVs in Chile largely reflects their high upfront costs and a lack of incentives to reduce them. Widespread EV adoption will require bolder incentives in the short-, medium-, and long-term commitments from public and private actors designed to link e-mobility with renewable power generation initiatives in order to lessen the country's reliance on fossil fuels and reduce emissions.

Acknowledgements

This work was supported in Chile by the projects CONICYT/FONDAP/15110019 (SERC-CHILE); the UAI Earth Research Centre; and engineers Daniel Kincade, Eva Ossandón, and Gregoire Cholat-Namy and in the United States by the University of California Berkeley with the project SWITCH Chile and engineer Patrica Hidalgo.

References

Adams, W. M. (2006). *The future of sustainability: Re-thinking environment and development in the twenty-first century*. Switzerland: The World Conservacion Union IUCN.

ANAC (2015). *Mercado Automotor-Junio 2015*. Santiago: Asociación Nacional Automotriz de Chile (ANAC).

Anderson, C. D., & Anderson, J. (2012). *Electric and hybrid cars: a history*. Jefferson, NC: McFarland and Company.

Bayar, T. (2013). *Chile Ups renewable energy target*. Nashua, NH: Renewable Energy World.

BP p.l.c. (2013). *BP statistical review of world energy*. London: BP Petrol.

Burger, A. (2014). *South America's largest solar thermal power plant to be built in Chile*. San Francisco, CA: Triple Pundi.

Business Monitor International (2013). *Chile oil and gas report Q1 2013*. Market Research.

Bussi, M. I. (2012). *El cuento del Litio en Chile*. Le Monde Diplomatique.

Catenacci, M., Verdolini, E., Bosetti, V., & Fiorese, G. (2013). Going electric: Expert survey on the future of battery technologies for electric vehicles. *Energy Policy, 61*, 403–413.

Central Bank of Chile (2015). *Statistical synthesis of Chile 2010–2014*. Santiago: Central Bank of Chile.

CER (2013). *CER Report November 2013, NCRE Project Status in Chile*. Santiago: Renewable Energy Center (Centro de Energías Renovables, CER).

CER (2014a). *CER Report September 2014*. Santiago: Renewable Energy Center (Centro de Energías Renovables, CER).

CER (2014b). *Centro de Energías Renovables detalló plan de fomento ERNC 2014*. Santiago: Renewable Energy Center (Centro de Energías Renovables, CER).

Chilectra (2012). *Chilectra y Parque Arauco inauguran primera electrolinera de Las Condes*. Santiago: Chilectra.

Chilectra (2016). *Tarifas de suministro eléctrico para clientes sujetos a regulación de precios*. Chilectra.

CIFES (2015). *CIFES Report (December 2015)*. Santiago: Sustainable Energy Development and Innovation Center (Centro para la Innovación y Fomento de las Energías Sustentables, CIFES).

CMM (2014). *Propuesta de regulaciones para la reducción del MP2,5, sus precursores y contaminantes que afecten al cambio climático, para las distintas fuentes estacionarias de la Región Metropolitana*. Santiago: Mario Molina Center (Centro Mario Moilna, CMM).

CNE (2009a). *Las Energías Renovables No Convencionales en el Mercado Eléctrico Chileno*. Santiago: National Energy Commission (Comisión Nacional de Energía, CNE).

CNE (2009b). *Chile crea Centro de Energías Renovables*. Santiago: National Energy Commission (Comisión Nacional de Energía, CNE).

CORFO (2010). *Con énfasis en la Industria Auxiliar se realizó el V Foro Internacional de Inversiones en ERNC*. Santiago: CORFO.

CORFO (2015a). *Iniciativas de Fomento Integradas–IFI: Apoyo a Proyectos de Inversión Tecnológica*. Santiago: CORFO.

CORFO (2015b). *Contratos Tecnológicos para la Innovación*. Santiago: CORFO.

CSP World (2012a). *Termosolar Pedro de Valdivia receives environmental approval from Chilean authorities*. CSP World.

CSP World (2012b). *Chile, an emerging market for Concentrated Solar Power, receives $200 million support from World Bank*. CSP World.

CSP World (2012c). *Chile receives $67 M from Climate Investments Funds to develop CSP projects*. CSP World.

CSP World (2013). *Chile's attractiveness for CSP companies keeps growing*. CSP World.

CSP World (2014). *Abengoa wins first CSP tender in Chile, will build a 110 MW tower plant to generate around the clock*. CSP World.

Econsult (2016). *Informes Económicos Combustibles*. Econsult.

El Dínamo (2015). *Revisa acá el calendario de restricción vehicular en Santiago para el 2015*. El Dínamo.

El Mercurio Inversiones (2015). *Cuál es el costo real de un auto eléctrico*. El Mercurio Inversiones-Edición N° 8.

EPA (2014). *Greenhouse gas emissions from a typical passenger vehicle. EPA-420-F-14-040a*. United States Environmental Protection Agency (EPA), Office of Transportation and Air Quality.

EPA (2016). *EPA fuel economy*. United States: Environmental Protection Agency (EPA).

European Commission (2003). *External costs: Research results on socio-environmental damages due to electricity and transport*. Brussels: Directorate-General for Research, Directorate J-Energy.

European Commission (2005). *Extern E—Externalities of energy: methodology 2005 update.* Brussels: Directorate-General for Research, Sustainable Energy Systems.

Fox Davies Capital (2013). *The lithium market.* FDC.

Fthenakis, V., Kim, H. C., Frischknecht, R., Raugei, M., Sinha, P., & Stucki, M. (2011). *Life cycle inventories and life cycle assessment of photovoltaic systems.* International Energy Agency (IEA), PVPS, Task 12, Report T12-02:2011.

Galindez, M. (2014). *New grid inter-connectivity in Chile to boost solar industry.* Nashua, NH: Renewable Energy World.

Gallardo Lozano, J. (2012). Electric vehicle battery charger for smart grids. *Electric Power Systems Research, 90*, 18–29.

Gallego, B. (2013). *CSP and mining in Chile.* Nashua, NH: Renewable Energy World.

Garcés Millas, I. (2002). *Litio y derivados: La industria del litio en Chile.* Antofagasta: Universidad de Antofagasta.

Garreaud, R., & Falvey, M. (2009). Regional cooling in a warming world: Recent temperature trends in the SE Pacific and along the west coast of subtropical South America (1979–2006). *Journal of Geophysical Research, 114*, D04102. http://dx.doi.org/10.1029/2008JD010519.

Generadoras de Chile, A. G. (2015). *Boletín del Mercado Eléctrico, Sector generacón, December 2015.* Santiago: Dirección de Estudios y Contenidos.

Gerssen-Gondelach, S. J., & Faaij, A. P. C. (2012). Performance of batteries for electric vehicles on short and longer term. *Journal of Power Sources, 212*, 111–129.

Goldman Sachs (2015). *The low carbon economy—GS SUSTAIN—Equity investor's guide to a low carbon world, 2015–25.* Manhattan, NY: Goldman Sachs.

Gonzalez, B. (2013). *Profiting from PV in Chile.* PV Insider.

Gonzalez B. (2014). PV plants under construction in Chile grow 184% from last year. PV Insider.

Google earth. (2015). Google earth.

Hanel, M., & Escobar, R. (2013). Influence of solar energy resource assessment uncertainty in the levelized electricity cost of concentrated solar power plants in Chile. *Renewable Energy, 49*, 96–100.

Ibarra, A. (2014). *Los usuarios de autos eléctricos alaban sus virtudes, pero se quejan de su alto precio*: (p. A10). El Mercurio.

IEA (2009). *Chile—Energy policy review.* Paris: International Energy Agency (IEA), OECD/IEA.

IEA (2012a). *Key world energy statistics.* Paris: International Energy Agency (IEA), OECD/IEA.

IEA (2012b). *CO₂ emissions from fuel combustion highlights (2012 edition).* Paris: International Energy Agency (IEA), OECD/IEA.

IEA (2012c). *Support for non-conventional renewable energy development programme.* International Energy Agency (IEA),OECD/IEA.

IEA (2014). *Invest Chile project.* International Energy Agency (IEA), OECD/IEA.

IEA (2015). *CO₂ emissions from fuel combustion highlights (2015 edition).* Paris: International Energy Agency (IEA), OECD/IEA.

INE (2011). *Compendio Estdísticos 2011.* Santiago: Statistics National Institute (Instituto Nacional de Estadisticas, INE).

INE (2012). *Parque de Vehículos en circulación 2012.* Santiago: Energy National Institute (Instituto Nacional de Estadística, INE).

INE (2013). *Parque de Vehículos en circulación 2013.* Santiago: Energy National Institute (Instituto Nacional de Estadística, INE).

INE (2014). *Parque de Vehículos en circulación 2014.* Santiago: Energy National Institute (Instituto Nacional de Estadística, INE).

Janic, M. (2014). *Advanced transport systems: Analysis, modeling, and evaluation of performances.* London: Springer 978-1-4471-6287-2.

Lazcano, P. (2013). *La lenta marcha de los autos eléctricos*: (p. 34). La Tercera.

LNG World News (2012). *Enagas completes GNL Quintero stake acquisition.* Chile: LNG World News.

Lopez, E., & Ulmer, A. (2012). *Chile top court rejects $5 bln Castilla power project.* Reuters.

Markowitz, M. (2012). *Wells to wheels: Electric car efficiency.* Energy Matters.

Martínez Gaete, C. (2014). *Los taxis eléctricos llegan a Santiago en octubre!.* Plataforma Urbana.

Miller, A. (2013). *Chile proving to be a hot market for solar.* Renewable Energy World.

Mining Ministry. (2015). *Lithium: An energy source, an opportunity for Chile—Final report* Santiago: Mining Ministry (Ministerio de Minería), National Lithium Comission (Comisión Nacional del Litio)

Ministry of Economy, Development and Reconstruction (2005). *Ley 20.018.* Santiago: Ministerio de Economia, Fomento y Reconstrucción (Ministry of Economy, Development and Reconstruction.

Ministry of Energy (2012). *National energy strategy 2012–2030.* Ministry of Energy (Ministerio de Energía): Santiago.

Ministry of Energy (2013). *Plan de Accion de Eficiencia Energetica 2020.* Ministry of Energy (Ministerio de Energía): Santiago.

Ministry of Energy (2014). *Importación y producción.* Ministry of Energy (Ministerio de Energía): Santiago.

Ministry of Energy (2015a). *National energy balance 2014.* Santiago: Energía Abierta, Commission National of Energy (Comisión Nacional de Energía, CNE), Ministry of Energy (Ministerio de Energía).

Ministry of Energy (2015b). *Factores de Emisión SIC—SING.* Santiago: Energía Abierta, Commission National of Energy (Comisión Nacional de Energía, CNE), Ministry of Energy (Ministerio de Energía).

MMA (2012a). *Official environment status report 2011.* Santiago, Chile: Minister of Environment (Ministerio del Medio Ambiente, MMA).

MMA (2012b). *Decreto N°29 modifíca el decreto N° 211, de 1991, del MTT, que establece las Normas sobre emisiones de vehículos motorizados livianos.* Santiago: Ministry of Environment (Ministerio del Medio Ambiente, MMA).

MMA (2012c). *Decreto N°4 que modifica el Decreto N° 55 de 1994, del MTT, que establece las normas de emisión aplicables a vehículos motorizados pesados.* Santiago: Ministry of Environment (Ministerio del Medio Ambiente, MMA).

MMA (2016). *Anteproyecto Del Plan de Prevención y Descontaminación Atmosférica Para La Región Metropolitana de Santiago.* Santiago: Ministry of the Environment (Ministerio del Medio Ambiente, MMA).

Montgomery, J. (2013). *Latin America report: Chile's road to solar grid parity.* Renewable Energy World.

Moody's (2015). *Country Economy: Sovereigns Ratings List.* NY: Moody's. Available at: http://countryeconomy.com/ratings/moodys.

Morales Segura, R. (2006). *Contaminación atmosférica urbana: episodios críticos de contaminación ambiental en la ciudad de Santiago.* Santiago: Universidad de Chile, Centro de Química Ambiental.

MTT (2013a). *Política Nacional de Transportes.* Santiago: Ministry of Transport and Telecommunications(Ministerio de Transportes y Telecomunicaciones, MTT).

MTT (2013b). *División De Transporte Público Regional.* Santiago: Ministry of Transport and Telecommunications (Ministerio de Transportes y Telecomunicaciones, MTT).

MTT (2014). *Maestro de Transporte 2025 Santiago.* Santiago: Ministry of Transport and Telecommunications(Ministerio de Transportes y Telecomunicaciones, MTT).

Mujica Carvajal, J. I. (2014). *Recomendaciones para la introducción progresiva de la tecnología de propulsión eléctrica en el transporte público de Santiago de Chile.* Santiago: Universidad de Chile.

Nissan (2015a). *Nissan Sentra V16—Especificaciones técnicas.* Switzerland: Nissan.

Nissan (2015b). *Nissan LEAF MY'16—Especificaciones técnicas.* Switzerland: Nissan.

Norero, J., & Sauma, E. (2012). Ex-ante assessment of the implementation of an energy efficiency certificate scheme in Chile. *Journal of Energy Engineering—ASCE, 138,* 63–72.

NREL (2013). *Crystalline silicon and thin film photovoltaic results—Life cycle assessment harmonization.* National Renewable Energy Laboratory (NREL).

NREL. (2015). Solar and wind energy resource assessment (SWERA), NREL, Retrieved from https://maps.nrel.gov/swera/#/ (22.09.15)

Nykvist, B., & Nilsson, M. (2015). Rapidly falling costs of battery packs for electric vehicles. *Nature Climate Change.* http://dx.doi.org/10.1038/NCLIMATE2564.

OLADE-UNIDO (2011). *Chile—Final report—Component 3: Financial mechanism.* Observatory of Renewable Energy in Latin America and The Caribbean.

Olivares, A., et al. (2013). *Guía de Gestion, Aspectos claves en el desarollo de proyectos ERNC.* Santiago: Renewable Energy Center (Centro de Energías Renovables, CER).

Paleo D. (2015). Venta de autos híbridos creció 150% en ocho años: hoy hay 1.830 unidades en Chile, Plataforma Urbana.

Parkinson, G. (2014). *Abengoa to build 110 MW solar tower storage plant in Chile.* Renewable Energy World.

Pavlovic, P. (2014). La industria del litio en Chile. *Ingenieros,* 30–35.

Pavlovic Zuvic, P. (2014). La industria del litio en Chile. *Ingenieros,* 30–35.

Pekic V. (2013). Chile prepares to add 2.2 GW of solar power generation. PV Magazine

Peltier, R. (2012). *Plant of the year: AES Gener's Angamos power plant earns POWER's highest honor.* Power Mag.

Renewable Energy World (2014). *Latin America report: How Chile is shepherding its renewable energy expansion.* renewable Energy World.

Reyes, J. R. (2013). *Technology assessment for embarking countries.* Vienna: International Center Chilean Nuclear Energy Commission (CCHEN).

Reyes, P. C. (2015). *Gobierno ofrece hasta $ 6 millones para que taxistas opten por vehículos eléctricos:* (p. 26). La Tercera.

Riveros Requena, C. (1997). *El problema de la contaminación atmosférica en Santiago de Chile: 1960–1972.* Santiago: Pontificia Universidad Católica de Chile.

Scanlon, B. (2012). *Thermal storage gets more solar on the grid.* Renewable Energy World.

SEGPRES (2010). *Decreto N° 66 que Revisa, Reformula y Actualiza el Plan de Prevención y Descontaminación Atmosférica para la Región Metropolitana (PPDA).* Santiago: Ministry of the General Secretary of the Presidency (Ministerio Secretaría General de la Presidencia, SEGPRES).

Silva MA, Neves de Melo H, Trovao JP, Pereirinha PG, Jorge HM. (2013). Hybrid topologies comparison for electric vehicles with multiple energy storage systems. EVS27—International battery, hybrid and fuel cell electric vehicle symposium. Barcelona: EVS27.

Speirs, J., Contestabile, M., Houari, Y., & Gross, R. (2014). The future of lithium availability for electric vehicle batteries. *Renewable and Sustainable Energy Reviews, 35,* 183–193.

Stamp, A., Lang, D. J., & Wäger, P. A. (2012). *Journal of Cleaner Production*, *23*, 104–112Environmental impacts of a transition towards e-mobility: the present and future role of lithium carbonate production.

Transantiago. (2014). Transantiago.

U.S. EIA (2014). *International energy statistics*. Washington, DC: U.S. Energy Information Administration.

Ulloa, G. (2011). *Autoridades aprueban proyecto hidroeléctrico HidroAysén*. Chile: Biobio.

United Press International (2012). *Chile thinking again of nuclear power use*. Washington, DC: United Press International.

USACH (2014). *Informe Final—Estudio Actualización y sistematización del inventario de emisiones de contaminantes atmosféricos en la Región Metropolitana*. Santiago: Santiago de Chile University (USACH)—Physic Department.

USGS (2015). *Mineral commodity summaries 2015*. Reston, VA: US Geological Survey (USGS).

Vikström, H., Davidsson, S., & Höök, M. (2013). Lithium availability and future production outlooks. *Applied Energy*, *110*, 252–266.

WEC (2013). *World Energy Scenarios: Composing energy futures to 2050*. London: World Energy Council (WEC).

Weiss, M., Patel, M. K., Junginger, M., Perujo, A., Bonnel, P., & van Grootveld, G. (2012). On the electrification of road transport—Learning rates and price forecasts for hybrid-electric and battery-electric vehicles. *Energy Policy*, *48*, 374–393.

Wilkinson, P., Smith, K. R., Beevers, S., Tonne, C., & Oreszczyn, T. (2007). Energy, energy efficiency, and the built environment. *Lancet*, *307*, 1175–1187.

World Commission on Environment and Development (1987). *Our common future, report of the world commission on environment and development*. Oxford: Oxford University Press Published as Annex to UN General Assembly document A/42/427.

Yanine, F. F., & Sauna, E. E. (2013). Review of grid-tie micro-generation systems without energy storage: Towards a new approach to sustainable hybrid energy systems linked to energy efficiency. *Renewable and Sustainable Energy Reviews*, *26*, 60–95.

Yildiz, A., & Güngör, A. (2009). Energy and exergy analyses of space heating in buildings. *Applied Energy*, *86*, 1939–1948.

Yuzawa, K., Archambault, P., Burgstaller, S., Yang, Y., & Kurian, A. (2015). *Cars 2025: Vol. 2, Solving CO_2: Engines, batteries and fuel cells*. Goldman Sachs.

Electric vehicles: Case study for Spain

8

Eulalia Jadraque Gago
University of Granada, Granada, Spain

8.1 Introduction

The 28 countries that make up the European Union (EU 28) depend largely on oil imports for mobility and transport (European Commission, 2013). The European Union imports 53% of the energy consumed within its borders, of which 43% were of oil, 35% of natural gas, and 22% of solid fuels such as coal. The cost of energy imports rose in 2013 to around 400,000 million euros, equivalent to over a fifth of the total imports of the European Union. Nuclear energy accounted for 11.5% and renewable energy 12% of the energy sector (Eurostat, 2013).

In 2013, the transport sector was the sector with the highest consumption of energy in the European Union, accounting for 31.6% of the final consumption of energy. This percentage reached 39.4% in Spain. Furthermore, this sector is responsible for a quarter of all CO_2 emissions (European Commission, 2010; Sánchez-Braza, Cansino, & Lerma, 2014), with road emissions being at the forefront, creating over 91% of the emissions. In Spain, road traffic is responsible for 24%, 38%, and 24% of CO_2, NOx, and $PM_{2.5}$ emissions, respectively (MMA, 2012). These emissions not only damage the atmosphere but also are harmful to health (possibly causing cardiovascular and/or respiratory diseases). People exposed to high levels of noise generated by traffic suffer not only sleep loss but also their ability to communicate can suffer, which, in the case of children, can affect their learning process (Stead, 2008).

These negative impacts of road traffic have been some of the main preoccupations in recent policies on sustainable transport. In Spain, in recent years, steps taken to mitigate climate change caused by road traffic have been taken through the use of different strategies aimed at saving energy or improving energy efficiency. The Ministry of Industry, Energy, and Tourism presented the Energy Saving and Efficiency Strategy in 2004, which led to three consecutive action plans. The current Energy Saving and Efficiency Plan 2011–2020 aims to reduce the total nationwide consumption of energy by 20% before 2020; a third of this must come from savings in the road transport sector.

Electric vehicles are gaining greater global interest as a viable component in the search for alternatives to sustainable mobility. Electric vehicles could reduce the consumption of fossil fuels, greenhouse gas emissions, and other contaminants.

For EU-28, an electric mobility promotional strategy has been included in the European Green Cars' Initiative, which is part of the European Economic Recovery Plan.

Electric Vehicles: Prospects and Challenges. http://dx.doi.org/10.1016/B978-0-12-803021-9.00008-2

8.2   Current situation of energy use and environmental loading

8.2.1   Background information

The energy demand in Spain has grown alongside its economy. It is worth noting the impact of hydrocarbons on the Spanish energy supply and the way in which periods of high inflation have affected investments in energy, especially nuclear energy (Menéndez Pérez, Feijoo Lorenzo, & Cámara Rascón, 2006). The oil crisis of the seventies meant not only a return to coal and problematic development of nuclear energy but also led to high inflation and low economic growth. A substantial increase in energy demand following the 2008 economic crisis affected oil prices, although these recovered in 2012.

During the last two decades, the Spanish economy has undergone a significant rising trend. The gross domestic product (GDP) rose 61.7% between 1990 and 2010, and the population increased by 18.1%, leading to a 30.7% increase in CO_2 emissions during this period (IEA, 2012).

The Spanish energy model shows signs of unsustainability that are similar to those of the global model: an uncontrolled growth in demand and CO_2 emissions, alongside a high dependency on fossil fuels (CCEIM, 2011).

The Spanish economy has concentrated its activities in steel, cement, and brick, products associated with the construction sector. This leads to a belief that the Spanish economy has based itself on those sectors that have an unsustainable energy model. Construction, alongside low-cost tourism, has enabled spectacular economic growth, whilst proportioning a high demand for energy. Meanwhile, most of the advanced economies of the EU 15 have specialized in value-added activities, which reduce energy intensity and emissions (Girard, Gago, Ordoñez, & Muneer, 2016).

In Spain, fossil fuels accounted for 77% of the primary energy sources used in 2011 (44.9% oil, 22.1% natural gas, and 9.9% coal) (MITyC, 2011). Nuclear energy accounted for 11.5% and renewable energy 12% of the energy sector. In terms of energy production, renewable energy in Spain made up 31.8% of the total in 2012, mainly wind and hydraulic energy (REE, 2013). Spain is the second country in Europe after Germany with the largest installed wind power capacity, with 22,796 MW installed at the beginning of 2013 (GWEC, 2013).

8.2.2   Energy and environmental challenges

The analysis of the current energy production model and its comparison with the evolution of energy consumption clearly reflects the limits of conventional energy resources and the progressive deterioration of the environment (Jefferson, 2006; Xiaowu & Ben, 2005). The dramatic increase in energy consumption witnessed in the European Union (EU) since the industrial revolution has led to a dependency on overseas energy sources that could reach 70% of the total energy consumed by 2020 (Kusku, 2010). Fortunately, there is an increasing social awareness of the negative impact on the environment brought about by the use of fossil fuels and also of the environmental problems relating to the use of nuclear energy. This new social awareness to environmental problems and their possible solutions will have a direct effect on the way in which we rise to the challenges posed by energy in the future. This without a doubt will mean a reduction in the consumption of energy from fossil fuels, alongside an increase in the production of renewable energy or energy generated from natural resources (Jäger-Waldau, 2007).

In order to promote the use of renewable energies and make more efficient use of energy, Europe, has developed various directives, these included Directive 2009/28/EC of the European Parliament and of the Council of 23 April 2009 on the promotion of the use of energy from renewable sources (BOE, 2009a). This Directive sets amongst its general objectives that a minimum of 20% of the final consumption of energy in the European Union (EU) come from renewable sources and a minimum of 10% of energy consumed in the transport sector in each member state should also come from renewable sources, by the year 2020 (Lechón et al., 2006). To achieve this aim, the directive establishes targets for each of the member states to be reached by 2020, and a minimum indicative trajectory in the run up to that year. In Spain, the target is that renewable energy sources make up at least 20% of final consumption of energy by the year 2020, as in the rest of Europe, alongside a minimum contribution of 10% of renewable energy sources in the transport sector by the same year. These targets have been laid down in Act 2/2011, of Mar. 4, on sustainable economy (BOE, 2011).

Along these lines, the Renewable Energy Action Plan 2011–20 offers a series of proposals that aim to meet the European requirements, reach the national targets by 2020, and achieve, in line with the methodology outlined in the Directive 2009/28/EC, a gross final consumption of energy from renewable sources equivalent to 20.8% of the total final consumption of energy, and a final consumption of renewable energy that will make up 11.3% of the gross final consumption of energy in transport (IDAE, 2011b). Below is a summary table that includes both mandatory targets and the indicative path of the energy quotas from renewable energy sources in the gross final consumption as presented in Directive 2009/28/EC. Table 8.2.1 also shows the level of compliance to these targets, taking into account the predictions for gross final consumption of energy coming from renewable energy sources, based on the application of different initiatives proposed in the Renewable Energy Action Plan 2011–20.

Regarding the calculation of the degree of compliance with the renewable energy targets in transport, in row C.3 of the table, the calculation method for the numerator is included (IDAE, 2011b). Row C.2 refers to the biofuels mentioned in Article 21, Appendix 2 of the Directive 2009/28/EC, which alludes to those biofuels obtained from waste, residues, nonfood cellulosic materials, and lignocellulosic materials. Row D represents the sum of the total gross final consumption of energy from renewable sources. Row F meanwhile shows the gross final consumption of energy in all sectors of energy consumption, which is used as a denominator to calculate the compliance with the indicative trajectory and the mandatory goal of 20%. It is worth noting that this consumption has been corrected in some specific years, in compliance

Table 8.2.1 Global targets of the Renewable Energy Action Plan 2011 and level of compliance to these mandatory and indicative targets set out in Directive 2009/28/CE.

ktoe	2005	2010	2011	2012	2013	2014	2015	2016	2017	2018	2019	2020
A. Gross final consumption of electricity from renewable sources	4624	7323	7860	8340	8791	9212	9586	9982	10,547	11,064	11,669	12,455
B. Gross final consumption of energy from renewable sources for heating and refrigeration	3547	3933	3992	4034	4109	4181	4404	4651	4834	5013	5152	5357
C. Gross final consumption of energy from renewable sources in the transport sector	245	1538	2174	2331	2363	2418	2500	2586	2702	2826	2965	3216
C.1. Gross final consumption of electricity from renewable sources in the road transport sector	0	0	0	0	5	11	21	34	49	67	90	122
C.2. Consumption of biofuels from article 21.1	0	5	15	45	75	105	142	167	193	177	199	252
C.3. Subtotal of renewables in compliance with the target for transport: (C) + (2.5 − 1)·(C.1) + (2 − 1)·(C.2)	245	1543	2189	2376	2446	2540	2674	2804	2969	3104	3299	3651
D. Total consumption of energy from renewable sources (avoiding double	8302	12,698	13,901	14,533	15,081	15,613	16,261	16,953	17,776	18,547	19,366	20,525

counting of renewable electricity in transport)													
E. Gross final consumption of energy in transport	32,431	30,872		30,946	31,373	31,433	31,714	32,208	32,397	32,476	32,468	32,357	32,301
F. Gross final consumption of energy in heating and refrigeration, electricity and transport	101,719	96,382		96,381	96,413	96,573	96,955	97,486	97,843	98,028	98,198	98,328	98,443
Transport targets (%)													
Minimum mandatory target by 2020		5.0											10.0
Level of compliance with the mandatory target by 2020 (C.3/E)		5.0											11.3
Sources of renewable energies in the transport sector (%) (Directive method)	0.8			7.1	7.6	7.8	8.0	8.3	8.7	9.1	9.6	10.2	11.3
Global targets (%)													
Indicative trajectory (average for each 2-year period) and minimum mandatory target by 2020				11.0		12.1		13.8		16.0		20.0	
Level of compliance with the indicative trajectory and the minimum mandatory target by 2020	8.2	13.2		14.7		15.9		17.0		18.5		19.7	20.8

Prepared by the author with data from Institute for Energy Diversification and Savings IDAE (Instituto para la Diversificación y Ahorro de la Energía) (2011b). Plan de Energías Renovables 2011-20. [Online] Available at: http://www.idae.es/index.php/mod.documentos/mem.descarga?file=/documentos_11227_per_2011-2020_def_93c624ab.pdf Accessed 07.04.16.

Table 8.2.2 Targets of the Renewable Energy Action Plan 2011–20 in the transport sector.

ktoe	2005	2010	2011	2012	2013	2014	2015	2016	2017	2018	2019	2020
Bioethanol/bio-ETBE	113	226	232	281	281	290	301	300	325	350	375	400
Of which biofuels of Article 21.2[a]	0	0	0	0	0	0	7	7	7	19	19	52
Biodiesel	24	1217	1816	1878	1900	1930	1970	2020	2070	2120	2170	2313
Of which biofuels of Article 21.2[a]	0	5	15	45	75	105	135	160	186	158	180	200
Electricity from renewable sources	107	96	126	172	182	198	229	266	307	356	420	503
Of which road transport	0	0	0	0	5	11	21	34	49	67	90	122
Of which nonroad transport	107	96	126	172	176	187	207	232	258	289	330	381
Total biofuels	137	1442	2048	2159	2181	2220	2271	2320	2395	2470	2545	2713
Total renewables energies in transport	245	1538	2174	2331	2363	2418	2500	2586	2702	2826	2965	3216

[a]Article 21, Appendix 2 of the Directive 2009/28/EC—biofuels obtained from waste, residues, nonfood cellulosic materials, and lignocellulosic materials.
Prepared by the author with data from Institute for Energy Diversification and Savings IDAE (Instituto para la Diversificación y Ahorro de la Energía) (2011b). Plan de Energías Renovables 2011-20. [Online] Available at: http://www.idae.es/index.php/mod.documentos/mem.descarga?file =/documentos_11227_per_2011-2020_def_93c624ab.pdf Accessed 07.04.16.

with Article 5, Appendix 6 of the Directive 2009/28/EC, which stipulates that the quantity of energy consumed in aviation in a given year shall be deemed as not exceeding 6.18% of the gross final consumption of energy in that same year. The lower part of the table shows the levels of compliance to mandatory targets set for 2020 regarding renewable energy in the transport sector and in the gross final consumption and the indicative trajectory set out in Directive 2009/28/EC.

Table 8.2.2 offers a breakdown of all renewable energy sources used in the transport sector. It is worth noting the appearance of targets for electric vehicles that came into play before the middle of the decade.

8.3 Automobile population and its contribution to CO$_2$ emissions and impact on air quality

Transportation is a core activity for the wellbeing and freedom of citizens, whilst transportation of freight is a decisive factor in the competitiveness and economic growth of any nation. Therefore, mobility and economic growth tend to vary consistently. Along these lines, changes in the gross domestic product (GDP) are often reflected in the volume of traffic and the level of activity in the transport sector (Annema & De Jong, 2011; Dargay, Gately, & Sommer, 2007). Thus, in 2014, Spain witnessed a recovery in mobility in line with the growth of its GDP. According to the latest revision of the national accounts in Spain (Dec. 2015), the GDP grew +1.4%, inland freight transport (measured in tonnes according to data taken from the permanent survey of carriage of goods by road (EPTMC in Spanish), amongst others), grew by +5% and international transport grew similarly by 5%. Inland passenger transportation (measured in passenger-kilometre), on the other hand, witnessed a decrease of −2% (data taken from the Directorate General of Roads (DGC in Spanish), amongst others), although these data could well be conditioned by the changes in methodology made by the directorate general regarding ways to estimate transportation by measuring heavy vehicle capacity, a measure that especially effects buses (MFOM, 2015a; Fig. 8.3.1).

It is also important to note that the data relating to mobility show Spain as a country with a high intensity of mobility (defined as mobility in relation to GDP) when compared with other European countries and especially compared with the four big European economies (Germany, France, the United Kingdom, and Italy).

Regarding passenger transport, the relationship between the mobility of people in passenger-kilometre and the Spanish GDP is similar to that of the other EU 28 countries, with a slight downward trend witnessed in all countries, as can be seen in the graph below (Fig. 8.3.2).

It is in the transportation of freight where Spain stands out as a country with intense mobility. That is to say that for every unit of GDP, a lot of physical movement of goods is required (Fig. 8.3.3).

As stated in the Annual Report 2014 of the Spanish car manufacturer's association (Anfac), the total number of vehicles in Spain in 2014 was 27.76 million, 0.5% more than in the previous year. To break this down into vehicle type, family cars made up a total of 22.11 million, a rise of 0.4% compared with the 22.02 million on the road in

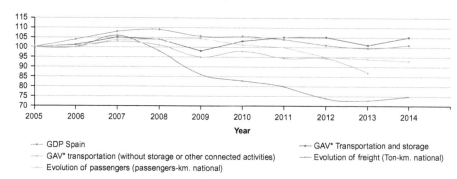

Fig. 8.3.1 Evolution of the gross domestic product (GDP), gross-added value of 'transportation and storage' and inland mobility of passengers and freight. 2005–14. *GVA—Gross-value added. GAV is the measure of the value of goods and services produced in an area, industry or sector of an economy. In national accounts, GVA is output minus intermediate consumption; it is a balancing item of the national accounts' production account.
Prepared by the author with data from Ministry of Development MFOM (Ministerio de Fomento). (2015a). Observatorio del Transporte y la Logística en España. Informe Anual 2015 [Online] Available at: http://observatoriotransporte.fomento.es/NR/rdonlyres/0AE839CF-9E00-46F3-A27C-88B14AC37715/136237/INFORME_OTLE_2015.pdf Accessed 19.03.16.

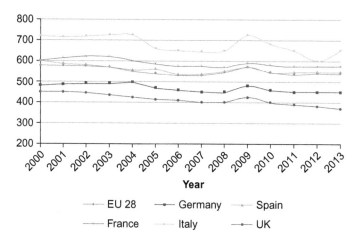

Fig. 8.3.2 Intensity of passenger transport in relation to GDP (passengers km/1000 constant euros). Spain and the main European countries. 2000–13.
Prepared by the author with data from Ministry of Development MFOM (Ministerio de Fomento). (2015a). Observatorio del Transporte y la Logística en España. Informe Anual 2015 [Online] Available at: http://observatoriotransporte.fomento.es/NR/rdonlyres/0AE839CF-9E00-46F3-A27C-88B14AC37715/136237/INFORME_OTLE_2015.pdf Accessed 19.03.16.

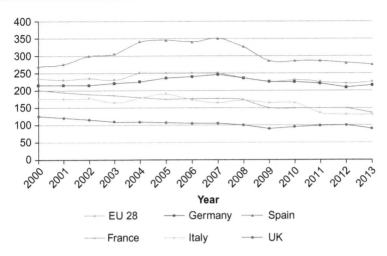

Fig. 8.3.3 Intensity of freight transport in relation to GDP (t/km/1000 constant euros). Spain and the main European countries. 2000–13.
Prepared by the author with data from Ministry of Development MFOM (Ministerio de Fomento). (2015a). Observatorio del Transporte y la Logística en España. Informe Anual 2015 [Online] Available at: http://observatoriotransporte.fomento.es/NR/rdonlyres/0AE839CF-9E00-46F3-A27C-88B14AC37715/136237/INFORME_OTLE_2015.pdf Accessed 19.03.16.

2013. There were 4.9 million trucks in circulation in 2014, 0.3% more than in 2013. As for buses and coaches, in 2014, there were a total of 59,677, a reduction of 0.4% compared with 2013, and the number of tractors grew 1.8% to 186,112 in 2014 (ANFAC, 2014).

This report by Anfac also shows that at the close of 2014, there were 473 passenger cars per 1000 inhabitants. This figure is at its highest since 2008. Notwithstanding this growth, the number of passenger cars per inhabitant in Spain is below the European average, which, in 2012, was at 487 vehicles per 1000 inhabitants, with some countries such as Luxembourg (663/1000) and Italy (621/1000) going way over this mark (ANFAC, 2014).

In 2013, the transport sector was the sector with the highest consumption of energy in the European Union, accounting for 31.6% of the final energy consumed. This percentage was higher in Spain (39.4%) (see Figs. 8.3.4 and 8.3.5).

If we observe the data on final consumption of energy in the transport sector, we can see that road transport is responsible for over 90% of the final consumption of energy in the transport sector, with over a million terajoules (TJ) consumed in the year 2013 (see Table 8.3.1).

The following graph shows that, as with energy consumption, greenhouse gas emissions in Spain are above the European average (27.5% compared with 23.7%) (Kahn et al., 2007). These emissions have a negative effect on air quality as they emit particulate matters (PM_{10} and $PM_{2.5}$), NOx, HC, and CO, and can cause related health

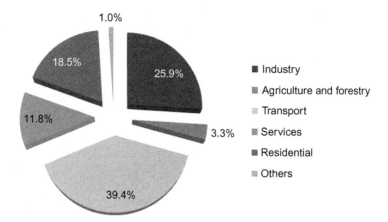

Fig. 8.3.4 Final consumption of energy in the transport sector compared with other sectors. Spain. 2013.
Prepared by the author with data from Ministry of Development MFOM (Ministerio de Fomento). (2015a). Observatorio del Transporte y la Logística en España. Informe Anual 2015 [Online] Available at: http://observatoriotransporte.fomento.es/NR/rdonlyres/0AE839CF-9E00-46F3-A27C-88B14AC37715/136237/INFORME_OTLE_2015.pdf Accessed 19.03.16.

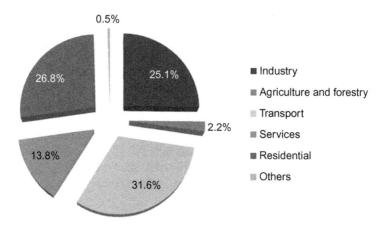

Fig. 8.3.5 Final consumption of energy in the transport sector compared with other sectors. European Union. 2013.
Prepared by the author with data from Ministry of Development MFOM (Ministerio de Fomento). (2015a). Observatorio del Transporte y la Logística en España. Informe Anual 2015 [Online] Available at: http://observatoriotransporte.fomento.es/NR/rdonlyres/0AE839CF-9E00-46F3-A27C-88B14AC37715/136237/INFORME_OTLE_2015.pdf Accessed 19.03.16.

problems, especially in urban areas (Figs. 8.3.6 and 8.3.7; European Commission, 2010; Lumbreras, Borge, Guijarro, Lopez, & Rodríguez, 2014).

Road transport is the leading cause of emissions, responsible for over 91% of the total (see Fig. 8.3.8).

Table 8.3.1 **National final consumption of energy by mode of transport (TJ). Period from 2010 to 2013.**

	2010	2011	2012	2013
Road	121,233.3	1,165,038.2	1,106,743.9	1,059,196.0
Rail	16,658.6	17,769.9	17,103.3	16,610.0
Aviation	53,053.8	50,412.9	43,148.5	35,915.8
Sea transportation	44,535.4	34,850.5	36,221.0	21,256.6
Total national transport	1,326,580.8	1,268,071.5	1,203,216.7	1,132,978.4

Prepared by the author with data from Ministry of Development MFOM (Ministerio de Fomento). (2015a). Observatorio del Transporte y la Logística en España. Informe Anual 2015 [Online] Available at: http://observatoriotransporte. fomento.es/NR/rdonlyres/0AE839CF-9E00-46F3-A27C-88B14AC37715/136237/INFORME_OTLE_2015.pdf Accessed 19.03.16.

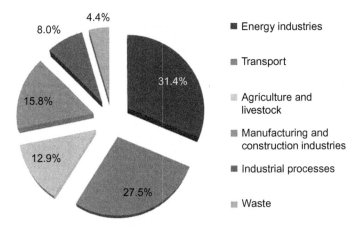

Fig. 8.3.6 Greenhouse gas emissions from transport compared with other sectors. Spain. 2012. Prepared by the author with data from Ministry of Development MFOM (Ministerio de Fomento). (2015a). Observatorio del Transporte y la Logística en España. Informe Anual 2015 [Online] Available at: http://observatoriotransporte.fomento.es/NR/rdonlyres/0AE839CF-9E00-46F3-A27C-88B14AC37715/136237/INFORME_OTLE_2015.pdf Accessed 19.03.16.

Within the area of road passenger transport, passenger cars make up around 85% of the movement on the road, and cause 90% of road transport emissions; buses are responsible for 5% of emissions (MFOM, 2015a). In Spain, road traffic causes 24%, 38%, and 24% of CO_2, NOx, and $PM_{2.5}$ emissions, respectively (MMA, 2012).

With the aim of reducing polluting emissions, systems of evaluation based on models that evaluate the different strategies in place to reduce emissions have been developed (Steenhof & McInnis, 2008; Winiwarter & Schmiak, 2005). These road transport models are born of formulations ranging from average speed up to those that define different traffic situations and more realistic driving patterns (Borge et al., 2012).

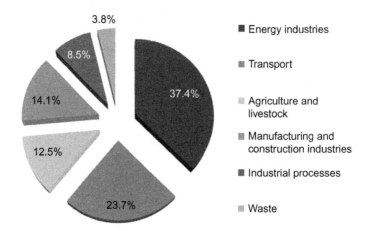

Fig. 8.3.7 Greenhouse gas emissions from transport compared with other sectors. European Union. 2012.
Prepared by the author with data from Ministry of Development MFOM (Ministerio de Fomento). (2015a). Observatorio del Transporte y la Logística en España. Informe Anual 2015 [Online] Available at: http://observatoriotransporte.fomento.es/NR/rdonlyres/0AE839CF-9E00-46F3-A27C-88B14AC37715/136237/INFORME_OTLE_2015.pdf Accessed 19.03.16.

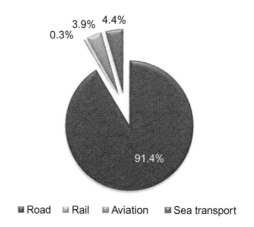

Fig. 8.3.8 Greenhouse gas emissions from transport in Spain, according to mode of transport. 2011.
Prepared by the author with data from Ministry of Development MFOM (Ministerio de Fomento) (2015c). Observatorio del Transporte y la Logística en España. Informe Anual 2013 [Online] Available at: http://observatoriotransporte.fomento.es/NR/rdonlyres/775612E3-0CDE-43C3-8A6A-13853CCF3919/127878/INFORMEOTLE2013v02.pdf Accessed 19.05.16.

An exhaustive literature review has been carried out by Smit, Ntziachristos, and Boulter (2010), in which the authors present a meta-analysis of 50 studies dealing with the validation of the different types of traffic emission models. A more detailed discussion of its complexities, advantages, and disadvantages is presented in Smit, Dia, and Morawska (2009).

In Europe, the main model used to estimate emissions from road transport on a national or regional scale is that developed by the European Environment Agency (EEA), a software called COPERT 4 (Ntziachristos, Gkatzoflias, Kouridis, & Samaras, 2009). This model can estimate fuel consumption and vehicle emissions over a specific area and is currently integrated into the EMEP/EEA methodology for calculating emissions (EEA, 2009).

Lumbreras et al. (2014) applied a model that they had developed themselves called EmiTRANS, to quantify the effects of the different policies and measures in place to reduce emissions generated by road transport in Spain from the year of publication (2005) until 2020. A detailed description of the model can be found in Lumbreras et al. (2008), where a combined scenario including previous assumptions on biofuel promotion, higher technological penetration and lower mobility was included to show the results of a 'green' scenario. It was called maximum feasible reduction (MFR) to reflect the hypothetical situation of the maximum emission reduction. Figs. 8.3.9–8.3.11 show the results of CO_2, NO_x, and $PM_{2.5}$ emissions:

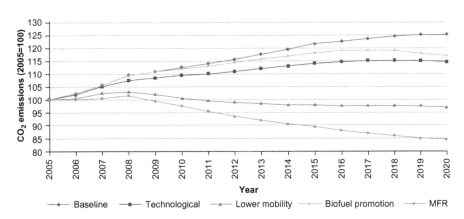

Fig. 8.3.9 CO_2 emission projections for road transport in Spain relative to 2005 (2005 value corresponds to 100).
Prepared by the author with data from Lumbreras, J., Rodríguez, E., López, J. M., Guijarro, A., Villimar, R., & Rodríguez, J. (2008). *Herramienta integral para el cálculo de emisiones atmosféricas del transporte por carretera. Escuela Técnica Superior de Ingenieros Industriales* (1st ed.). Madrid: Universidad Politécnica de Madrid.

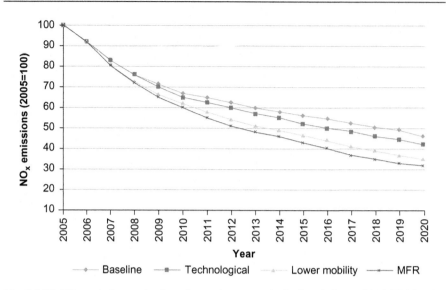

Fig. 8.3.10 NO$_x$ emission projections for road transport in Spain relative to 2005 (2005 value corresponds to 100).
Prepared by the author with data from Lumbreras, J., Rodríguez, E., López, J. M., Guijarro, A., Villimar, R., & Rodríguez, J. (2008). *Herramienta integral para el cálculo de emisiones atmosféricas del transporte por carretera. Escuela Técnica Superior de Ingenieros Industriales* (1st ed.). Madrid: Universidad Politécnica de Madrid.

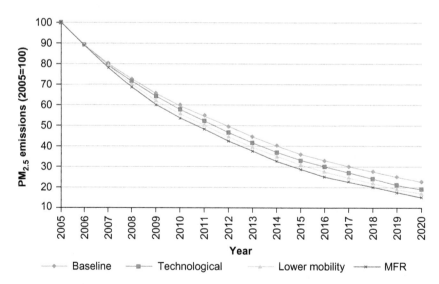

Fig. 8.3.11 PM$_{2.5}$ emission projections for road transport in Spain relative to 2005 (2005 value corresponds to 100).
Prepared by the author with data from Lumbreras, J., Rodríguez, E., López, J. M., Guijarro, A., Villimar, R., & Rodríguez, J. (2008). *Herramienta integral para el cálculo de emisiones atmosféricas del transporte por carretera. Escuela Técnica Superior de Ingenieros Industriales* (1st ed.). Madrid: Universidad Politécnica de Madrid.

For the period 2006–20, according to the scenarios analysed, reducing passenger and freight mobility are the most effective measures in reducing CO_2 emissions and air quality pollutants, whilst vehicle scrapping systems are likewise efficient but only in reducing air quality pollutants. Mobility measures are the most influential for reducing CO_2 emissions, decreasing highway speed, and downsizing policies also have a considerable effect. Fleet renewal, however, brings about quite limited reductions. As far as NO_x, is concerned, the implementation of scrapping systems to renew the passenger car fleet increasing Euro 5 vehicles is the best way to reduce emissions in accordance with the emission factors outlined in COPERT. Mobility measures, highway speed decrease, and petrol/diesel ratio changes are less effective although they also have material effects. Finally, to reduce $PM_{2.5}$, increasing petrol passenger car percentage yields relevant reductions. Old vehicle substitution, mobility measures and, to a lower extent, decreasing highway speed were found effective in cutting down $PM_{2.5}$ emissions.

8.4   Public transport infrastructure

8.4.1   System of responsibilities

In Spain, the legislative framework for public transport is structured on three levels that correspond to the way in which the government is organized: local, autonomous, and national (Martín Urbano, Ruiz Rúa, & Sánchez Gutiérrez, 2012).

- Local or municipal level—the local authorities have jurisdiction over urban transport services within their municipality. In Spain, there are 8180 municipalities.
- Regional or autonomous regional level—the autonomous governments have jurisdiction over two or more municipalities present in their autonomous region. In Spain, there are 17 autonomous regions, eight of which have only one province.
- National or state level—the Ministry of Development holds jurisdiction over public transport services between two or more autonomous regions.

8.4.2   Funding

The four models of funding most commonly identified with urban transport are (i) public budget funding models; (ii) off-budget public funding from business enterprises (such as RENFE and ADIF in the case of rail transport); (iii) private funding models, usually through subsidies; and (iv) mixed funding models where usually both the public and the private sectors participate. In the Spanish case, besides those mentioned for rail transport, to understand urban and metropolitan transport funding, the authorities are of fundamental importance, given that, as well as having responsibilities in this area, they also have say in the management of collective transport services. Urban transport funding in Spain is conditioned by its jurisdictional scope, by its fare system and by the legislation covering public transport funding.

The state sectorial legislation on passenger transportation (LOTT) is split into three categories: (i) urban transport, run exclusively on urban land, which joins different areas of any one municipality; (ii) interurban transport, which does not meet the previous conditions; and (iii) metropolitan transport.

The authority figure in public transport, which the Spanish Law of Sustainable Mobility claims should be a public body, responsible for planning and managing the public transport system in any given metropolitan area, must also be taken into account. The role of this figure is fundamental as far as funding is concerned, as they can channel the funds awarded by the administration to the urban and metropolitan transport operators.

Spanish urban transport is a cornerstone in improving the mobility of the nation's citizens. To improve mobility, the development of the public transport system is crucial. Public transport depends greatly on support and funding, and its development has only ever been possible with strong support from public funds.

In Spain, there are three programmes set aside in the general state budget for the funding of the Spanish public transport system:

- Programme 912C, dedicated to providing funds for the provision of urban public transport to local corporations
- Programme 513 B of subsidies and support for land transport, given over to funding part of the current expenditure and buying the equipment necessary for the operation of metropolitan transport
- Programme 513 A. Rail transport infrastructures. A programme dedicated to funding the construction of new railway infrastructures: metro, tram, and regional railways

8.4.3 Productive efficiency

Notwithstanding the significant variety of cases born from the diverse areas and modes of transport, in general, local public transport services depend highly on public subsidies, which prove to be less effective in rail services, especially tram and light rail, when compared with buses, and according to service area, suburban services tend to be less efficient than urban services (OMM 2008–2010).

8.4.4 Service use

In 2010, the total number of trips taken on Spanish local public transport was around 3100 million. Of these, approximately 51% corresponds to (urban and suburban) bus services and the other 49% to suburban railways (OMM, 2012).

8.5 Private transport sector

Private transport refers to all transport carried out by an individual, of their own accord, whether for personal needs or as a complementary activity carried out by an individual alongside a company's or an establishment's main activities, always when directly linked to the proper development of said activities (MFOM, 2011).

Private transport, especially on the road, is the cause of varied and significant energy problems amongst developed economies such as dependency on foreign oil imports, local contamination, and the rise of greenhouse gas emissions (Loureiro, Labandeira, & Hanemann, 2013).

In Spain, there are 245 complementary private transport fleets with over 25 heavy vehicles, totalling 15,170 vehicles of this type, according to a study by the Ministry of Development (todotransporte.com, 2003). The main uses for these vehicles are construction, with 48.46%, collection and treatment of waste, 21.54%, food, 20.77%, and the rest, different activities.

8.6   Spanish infrastructure policy

Spain is the country that has seen the highest growth in infrastructure (both interurban and long distance) in recent years. However, demand has not kept pace with the increase in supply (Albalate, Bel, & Fageda, 2015), with a 15-year plan (2005–20) called PEIT in Spanish (strategic infrastructure and transport plan) that forecasted a 249,000 million euro investment in transport infrastructure (MFOM, 2005), being reduced to 138,000 million due to the changes witnessed in the Spanish economic situation (see Fig. 8.6.1). The original Plan PEIT was replaced by the 12-year PITVI (Infrastructure, Transport, and Housing Plan, 2012–24) (MFOM, 2015b).

Starting with the overland networks, without a doubt, the factor that has characterized Spanish infrastructure policies since the beginning of 2000 has been the growth of

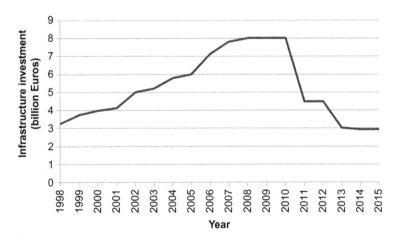

Fig. 8.6.1 Spanish state investment in infrastructure. Data on investments in infrastructures made by the Ministries of Development and the Environment.
Prepared by the author with data from Ministry of Development MFOM (Ministerio de Fomento). (2015b). PITVI: Plan de Infraestructuras, Transporte y Vivienda. [Online] Available at: http://www.fomento.gob.es/MFOM/LANG_CASTELLANO/PLANES/PITVI/ PITVI_DOCU/ Accessed 15.04.16.

the high-speed rail network, with new lines dedicated almost exclusively to passenger transport. For this reason, investment in the rail sector during 2014 rose to 3213.73 million euros. The highest investor was ADIF (Spain's railway infrastructure administrator), investing 2192.95 million euros in high-speed trains and 318.9 million in the conventional network. Following this were the Directorate Generals of the Autonomous Regions, with investments of 216.87 million euros. Other investors included the state-owned land transport infrastructure company (SEITT) that invested 190.81 million euros, RENFE Operadora (the Spanish national rail network operators), with an investment of 106.12 million euros, the Directorate General of Railways, which invested 99.94 million euros and the narrow-gauge railways (FEVE) of the Autonomous Regions and of private companies, which invested 88.14 million euros (MFOM, 2014).

Notwithstanding these investments, the length of railway in Spain fell 0.5% compared with 2013, reaching 16,233 km in 2014, 15,279 of which belonged to ADIF (Spain's railway infrastructure administrator), 871 km to the Autonomous Regions, and the rest (83 km) to private companies. The length of electrified railways reached 9910 km of which 9288 belonged to ADIF, 650 to the Railways of the Autonomous Regions, and 32 to private companies. This means that Spain ranks second in terms of total number of working network kilometres and first in terms of number of kilometres per million inhabitants (50% more than France). However, an analysis of the demand shows that the number of trips taken on the Spanish rail network stood at 16.8 million, compared with 110 million trips taken in France (Albalate et al., 2015).

Spain also has the most extensive road network of all countries in the European Union. The total length of the interurban road network on 31 Dec. 2014 was 116,284 km, 923 km more than the previous year. This network is made up of toll roads, toll-free motorways, A roads, dual carriageways, and country roads. Neither urban roads and streets nor forestry or farm tracks or roads are included. Aside from the lengths of principle roads mentioned above, there are 3438 intersections with a total length of 4339 km.

The investment made by the Directorate General of Roads of the Ministry of Development fell by 10.4% compared with the previous year. The total investments made in the road network in 2014 fell by 8.2% (MFOM, 2014). Furthermore, as with the railways, the fact that Spain has one of the most extensive road networks in Europe does not mean that the volume of traffic is similarly large. According to Albalate et al. (2015), the number of millions of passengers/km for each km of motorway in Italy was 4.3 times higher than in Spain. In Germany, it was 2.6 times higher and, in France, 2.8 times.

As for investments in aeronautical infrastructures made by Aena, S.A. and ENAIRE, these saw an overall decrease of 38.1% in 2014 compared with the previous year. Investments in airports fell 37.5%, and investments in air traffic were down 41.3%.

As far as air traffic is concerned, in 2014, Spanish airports registered a total of 194.95 million passengers on commercial flights (4.6% more than in 2013). Both domestic and international traffic was up 2% and 5.7%, respectively, and there were

1.59 million flights, 3.2% more than the previous year. A total of 684,905 tonnes of freight were transported, 7.2% more than in 2013 (MFOM, 2014). Fageda and González-Aregall (2014) analysed the link between number of passengers and surface area of Spanish airport terminals and compared their results with a sample of international airports in the year 2010. The average number of passengers per metre squared of terminal in Spain was 71, whilst the international average was 101. Furthermore, 16 of the Spanish airports had an average number of passengers per metre squared of below 50.

Finally, in 2014, the investments made by the entire state-owned port system in maritime transport infrastructures, works, and port facilities dropped by 9.5% compared with the previous year, reaching a total value of 421.90 million euros. Between 2000 and 2009, investments in Spain in this field increased to 21,988 million euros, more than double those made in Italy, more than triple the German investments and six times those of France. However, the fleet registered under the Spanish flag underwent a negative evolution in 2014 in terms of registered tonnage of the total passenger vessels, dropping from 2.47 million GT in 2013 to 2.21 million GT. The Special Register of Ships of the Canary Islands also recorded losses in the GT of the aforementioned vessels, from 2.44 million GT in 2013 to 2.18 million GT.

The total port freight traffic (not including provisions, fishing, and local traffic) increased by 5.1% compared with the previous year. This rise can be seen in the three main groups of freight: liquid bulk, solid bulk, and general cargo (Fageda & González-Aregall, 2014).

8.7 The role of electric vehicles

Sustainable mobility is a reality that is becoming more and more visible in urban areas. In February, alternative fuel vehicles already made up 2.6% of the market as a whole. In this regard, vehicles that use alternative energy for fuel will play an important part in the future of our cities, given that they can be driven without damaging the environment. Spain is one of the powerhouses for the production of four-wheeled electric vehicles in the world—with some of the top automotive brands such as Nissan, Peugeot, and Citroën y Renault and with factories in Catalonia, Galicia, Castilla León, and Murcia (FENERCOM, 2015). Of the 41 makes of car produced in Spain, five are electric and one hybrid (ANFAC, 2014).

Electric vehicles are essential for a sustainable transport model as, due to their efficiency, they reduce local CO_2 gas emissions (85%–90%) as opposed to traditional vehicles (90%), and can also be recharged using renewable energy (Newton, 2009). The Integral Strategy for the Promotion of the Electric Vehicle 2010–2014 highlighted the need to promote the development and use of electric vehicles in Spain. Within the framework of this strategy, the direct concession of subsidies for the purchase of electric vehicles in 2014 (the MOVELE programme) was brought to light (BOE, 2015b). This programme had its continuity in 2015 with MOVELE 2015 (BOE, 2015b). For the year 2016, the MOVEA plan (plan to promote mobility with alternative energy vehicles) came into play (BOE, 2015a). This plan regulates the grants on offer for alternative energy vehicles, but does not modify the amount indicated in the MOVELE 2015 in place for electric vehicles, and includes grants for the installation of public electric car charging points. The size of the grant varies depending on the vehicle category (for passenger cars and motorbikes, a limit was set), type of fuel used, and, in some cases, the maximum authorized mass (MAM) or on a vehicle's autonomy when powered exclusively by electricity. The grants range from 200 euros for bikes up to 20,000 euros for some category M3 vehicles. The sum total of the grants is 16.6 million euros, of which 13.3 million is set aside for electric mobility (vehicles and charging points).

In Spain, electric vehicles increased by 213.58% in February 2016, with 254 vehicles sold. During the first 2 months of the year, a total of 496 vehicles was reached, an increase of 204.29%. That is to say, around a quarter of the total number of cars sold in the whole of the previous year were sold in just 2 months (ANFAC, 2014).

In the case of hybrid vehicles, the trend is similar. In Feb. 2016, there was an increase of 66.72%, a total of 2084 vehicles. Between January and February, a total of 4211 vehicles were registered, an increase of 54.76% (ANFAC, 2014).

An electric vehicle has a storage capacity of between 15 and 30 kWh, giving it an autonomy of between 150 and 200 km. Given these limitations, the best use for electric vehicles still seems to be daily use, over short distances. The Spanish electricity system (generation, transmission, distribution, and marketing) supplies power to the vehicles in trickle charge mode,[1] from an unsupervised access for low penetration, with automatic algorithms to reduce simultaneous charging for medium penetrations, to control systems for high penetrations (FAEN, 2016). Rapid charging calls for the development of specific infrastructures, and this is the method used in charging stations or in the car parks of shopping centres, hospitals, etc. In Mar. 2014, Spain had reached third place in Europe with the implementation of 108 CHAdeMO[2] electric car charging points (CHADEMO, 2016).

The advanced load control and the bidirectional vehicle to grid (V2G) services come under smart grid technologies, whose aim is global optimization of the resources of electrical systems. This can be achieved by allowing large amounts of distributive and renewable energy to enter the system, and the clients to actively participate in the running of the system (FENERCOM, 2015).

The most common and the most widely developed way to recharge an electric car, not only in Spain but also around the world, is through the conductive method. This is when the car is connected to an electrical outlet using a cable, whether a normal household socket or a specific electric car charging point. Currently, on the market are models such as the Tesla S, whose battery has been designed to enable recharging in 90 s (the idea is the exchange of used batteries by new ones in the recharge centres

---

[1]Trickle charging is the most suited given that, by using the network already in place, it has the lowest possible impact on the charging infrastructures. To feed average daily mobility needs, trickle charging at night, during peak periods, or periods of high local availability of renewable energy, is best.

[2]The protocol for recharging electric car batteries that supply up to 62.5 kW of high-voltage direct current through a connector for recharging electric vehicles.

of the company), but this option seems more viable in two-wheeled vehicles (scooters and bicycles), where vehicles with a removable battery already exist.

Inductive or wireless charging, through magnetic induction or microwaves, is one of the most promising markets for powering electric cars, and Spain boasts a number of technology companies in Madrid who have developed highly efficient systems for recharging vehicles using microwaves. In this regard, in Spain, it is worth noting the Victoria project in Malaga, where Endesa (the largest electric utility company in Spain) is developing a system to recharge a moving electric bus without cables.

8.8   Renewable energy charging of electric cars

8.8.1   Renewable energy production in Spain

The importance of the renewable energy sector for the Spanish economy has grown steadily, and its contribution will continue to grow over the coming years. In constant terms, and based on 2010 values, the direct contribution of the renewable energy sector and gross domestic product of Spain (GDP) have shown a positive development, accumulating a growth of approximately 56.7% between 2005 and 2009. According to the predictions made by the International Monetary Fund about the growth of the Spanish gross domestic product in the following years (until 2015) and supposing a yearly growth of 2.5% between 2016 and 2020, the direct contribution of the Spanish renewable energy industry will make up 0.88% of the GDP in 2015 and 1.03% by 2020 (see Fig. 8.8.1; IDAE, 2011a).

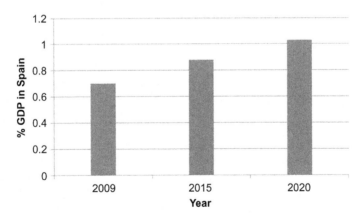

Fig. 8.8.1 Percentage of the renewable energy sector over the total of the Spanish economy. Prepared by the author with data from Institute for Energy Diversification and Savings IDAE (Instituto para la Diversificación y Ahorro de la Energía) (2011a). Impacto económico de las energías renovables en el sistema productivo español. Estudio Técnico PER 2011–2020. [Online] Available at: http://www.idae.es/uploads/documentos/ documentos_11227_e3_impacto_economico_4666bcd2.pdf Accessed 19.04.16.

In Spain, renewable energy sources cover 44% of the annual demand for electricity, with a potential to supply around 75% of this demand during progressively more hours per year (FENERCOM, 2015). With the aim of integrating the maximum production of renewable energy possible into the electricity system, whilst maintaining levels of quality and safety in the supply, halfway through 2006, the Spanish electricity network set up the Centre for the Control of Renewable Energy (CECRE), a pioneering project and world leader in monitoring and controlling renewable energies (REE, 2016).

Amongst the types of renewable energy, solar power has the highest potential. In terms of installed electric energy, Spain has a number of terawatts (TW) of power from solar energy. Wind energy is in second place, with an estimated potential of around 340 GW. Hydroelectric power in Spain is also very high, with approximately 33 GW of potential, although this has mostly already been developed. Other energies, wave and geothermal, have a potential of around 50 and 20 GW, respectively (IDAE, 2011b).

8.8.2 The legislation regulating renewable energy in Spain

The Electricity Sector Act 54/1997 (BOE, 1997), was passed on 27 Nov. Its main aim was to regulate the activities directed towards the supply of electrical energy, and it includes the special regime previously approved in the Royal Decree 2366/94. This Act guarantees access to the network of facilities under the special regime and also introduces an economic regime and production plan developed after the act in successive Royal Decrees (Royal Decree 2818/1998, of 23 Dec., 436/2004, of 12 Mar., and 661/2007, of 25 May). The Act also awards power to each Autonomous Region for the development of the legislation and regulation, and for the implementation of the state's basic regulations for electricity.

Royal Decree 1955/2000 (BOE, 2000), passed on 1 Dec., regulates the procedures for approval of production facilities and electricity transmission and distribution networks when their use affects an area larger than an Autonomous Region, when the power to be installed is greater than 50 MW, or when transportation or distribution will be outside the territory of the Autonomous Region. In these cases, the competent authority is the General Policy Directorate of Energy and Mines, part of the Ministry of Industry, Tourism, and Trade.

Royal Decree 842/2002 (BOE, 2002), passed on 2 Aug., approves the low-voltage electrotechnical regulation and its complementary technical instructions (CTI) BT 01–BT 51, which are to be applied to all renewable energy generation facilities connected to low voltages. The use of geothermal resources is specifically regulated in the Mines Act 22/1973 passed on 21 Jul. (modified by Act 54/1980 of 5 Nov.). The responsibility for developing the legislation and enacting the basic state Acts on mining falls to the Autonomous Regions.

Royal Decree 661/2007 (BOE, 2007b), passed on 25 May, which regulates the activity of electrical energy production under a special regime, is developed in the Electricity Sector Act 54/1997 and establishes a legal and economic regime for the power generation and cogeneration facilities that use renewable energy and waste as the raw materials for power generation, with the specific aim of establishing a stable, predictable system that guarantees sufficient profitability in the production of electrical energy under the special regime.

Royal Decree 1028/2007 (BOE, 2007a), passed on 20 Jul., establishes the administrative procedures to follow when applying for authorization to install power generation facilities in territorial waters.

Royal Decree 1578/2008 (BOE, 2008), passed on 26 Sep., sets out a new economic regime for photovoltaic facilities and creates a prepay register for this technology (PREGO) that affects the facilities entered permanently into the administrative register of production facilities under special regime (RIPRED) since 8 Sep. 2008. This new framework is based on a system of increased quotas and lower tariffs and is described in more detail in Appendix 4.9.1.

Royal Decree Act 6/2009 (BOE, 2009b), passed on 30 Apr., establishes a preallocation register for facilities under the special regime. Entry in the preallocation of rewards register will be a requisite for granting the right to the economic regime laid out in Royal Decree 661/2007, of 25 May, regulating the production of electricity under the special regime.

Royal Decree 1565/2010 (BOE, 2010a), passed on 19 Nov., regulates and modifies certain aspects relating to the production of electric energy under the special regime. The act sets out the technical requisites for a facility to be considered eligible for substantial modifications in order to enable electrical energy production from cogeneration or wind power. The decree also modifies the payment method for reactive energy, establishes the conditions for experimental wind technology facilities, and, in the third additional provision, innovative solar thermal power plants will have the right to additional remuneration alongside the market remuneration, through a tender process of up to 80 MW.

Royal Decree 1614/2010 (BOE, 2010b), passed on 7 Dec., regulates and modifies certain aspects of the production of electrical energy through thermal solar and wind power technologies. This decree sets a limit on the equivalent working hours with rights to an equivalent premium and a decline in the premium for wind energy plants.

Royal Decree Act 14/2010 (BOE, 2010c), passed on 23 Dec., defines the urgent measures needed to correct the rate deficit in the electricity sector. This regulation limits the equivalent working hours for photovoltaic plants with rights to a premium economic regime. The decree also sets two restrictions, one temporary, in place until 31 Dec. 2013, for those facilities covered by the economic regime set out in Royal Decree 661/2007, and the other permanent, restricting all other plants covered by the economic regime defied in Royal Decree 1578/2008 and Royal Decree 661/2007 of 1 Jan. 2014.

Act 2/2011 (BOE, 2011), passed on 4 Mar., incorporates elements of support frameworks for renewable energy that must be present to ensure the sustainability of future growth in the sector, including stability, flexibility, progressive internalization of costs, and prioritization in the incorporation of these facilities including technological innovations, production efficiency optimization, transportation, and distribution, leading to increased management capacity by reducing emissions of greenhouse gases, thus ensuring adequacy and stability in energy.

Royal Decree Act 1/2012 (BOE, 2012), passed on 27 Jan., led to the suspension of preallocation of rewards procedures and suppressed economic initiatives for new electricity production using cogeneration, renewable energy and waste. It also eliminated incentives for the construction of these facilities, in order to avoid additional costs created by the energy system.

Royal Decree Act 9/2013 (BOE, 2013), passed on 12 Jul., authorizes the government to approve a new legal and economic framework for existing production facilities producing energy from renewable sources, cogeneration and waste. The framework sets a tariff that allows the renewable energy, cogeneration, and waste plants to cover costs and sets them up to be as competitive on the market as plants that rely on other technologies and to make a reasonable yield. Another relevant aspect of this act is that the government can determine the right to remuneration for facilities producing electricity from cogeneration or using nonconsumable or nonhydro renewable energy and biomass, biofuels or agricultural waste, livestock or services, as prime materials, even if the electricity production facility has an installed capacity greater than 50 MW.

8.8.3   Renewable energy charging of electric vehicles in Spain

When evaluating the ecological value of electric cars, the core consideration must be the amount of electricity that they consume. If this electricity came from renewable sources (solar, wind, and hydraulic), then these cars really could be considered ecological, given that an electric car whose battery has been charged with the current power generation park, where gas dominates, emits 56 g of CO_2/km into the atmosphere. This is obviously much lower than the 180 g CO_2/km emitted by a conventional fuel vehicle but is still an emission of harmful gases that continue to contribute to global warming.

Some of the better developed kinds of renewable energy currently available (wind and solar power) are by their nature unpredictable and only available intermittently and at certain times, regardless of whether they are needed or not. That is to say they are sources of energy that cannot be controlled by demand as is the case with fossil fuels or with other kinds of renewables such as biomass, hydraulic, or thermo solar. Although this fact may seem problematic, it actually means that renewable energies and electric cars complement each other magnificently. For example, as seen previously, Spain has an enormous wind energy park that produces an important amount of the energy consumed in the country, making wind energy the third most important energy source in Spain in 2015, producing 47,704 GWh, covering 19.4% of the total energy demand (AEE, 2016). A large amount of this energy is produced at night when the demand is low. Given that wind energy is hard, in some cases impossible, to store, the aero generators tend to be disconnected. By promoting mass charging of electric vehicles at night time, this energy could be used during the day in the running of said vehicles.

Along these lines in Spain, since 2011, the University of Alcala in Madrid has been running a solar charging station. That is a charging station for electric cars powered by solar panels with a capacity to recharge four cars and five bicycles. The solar charging station has an annual energy potential of 5600 kWh with 15 solar panels and a total of nine power points. Of the four power points set aside for cars, which can also charge motorbikes and other electric vehicles, two are smart chargers, meaning the charging process can be controlled remotely (e.g. from a mobile phone). The other two are conventional (UAH, 2016). Other examples include the following:

- Merlyn model solar charging station for smart charging in the Plaza del Milenio, Valladolid, has eight smart charging points.
- The Centre for Innovation in Companies, part of the Chamber of Commerce of Lanzarote in the Canary Islands, also boasts a solar charging station since 2012. Designed by the company Suntelco, it has a total of three photovoltaic panels. The system sends any extra energy to the Chamber of Commerce building.

8.9 Conclusion

There is a clear understanding of the difficulties raised by transportation and its links to problems with the energy structure, the climate change and the production model, and the global consequences that the expansion of the sector and midterm predictions will have over the planet's unsustainability. Greenhouse gas emissions are affecting the climate all over the world, with increasing average temperatures causing changes in the ice melting, causing sea levels to rise, and changing rainfall levels alongside average temperatures in certain parts of the world. This will all lead to increasingly negative results, and even catastrophic situations for many areas of our planet, particularly coastal areas, if the necessary measures are not taken.

The transport sector has the second highest impact on climate change of all sectors, and is also in second place for increases in greenhouse gas emissions on a global scale. Both the documents that have been developed (plans and strategies) and the action taken up to the present time, including the lines of research carried out on infrastructures, modes of transport, and social behaviour, do not allow sufficient conclusions to be reached regarding the situation of the planet, the energy problematic, environmental sustainability and its effect on global change set out in Horizon 2020. Thus far, there have been liberalization processes of transport systems, public investments fostering a boom in the number of vehicles and its associated model of society, and a drive towards a territorial model that is incompatible with oil and limited, expensive energy sources, and much less with environmental sustainability and the fight against climate change.

In Spain, the actions taken so far have not been sufficient to reverse the dynamic of greenhouse gas emissions, to limit the generation of external costs, or to encourage collaboration with the wellbeing of its citizens in the field of transport, in line with the international compromises reached in Horizon 2020.

In this context, the measures aimed at improving the technological performance of vehicles and fuel in order to increase energy efficiency in transport used for people and for freight, as well as those measures put in place to fine tune vehicles with dual energy sources (hybrids) or alternative energy sources (electric, gas, second-generation biofuels, etc.), go a long way to reducing consumption, emissions, and energy dependency.

Along these lines, renewable energy sources are being developed and launched to replace the generation of electricity from fossil fuels, and to reduce greenhouse gas emissions, along with other pollutants such as nitrogen oxides (NOx) and sulphur dioxide (SO_2). In Spain, renewable energy sources cover 44% of the annual demand for electricity, with a potential to supply around 75% of this demand during progressively more hours per year. Amongst types of renewable energy, solar power has the highest potential. In terms of electrical energy facilities, Spain boasts a number of terawatts (TW) of power from solar energy. Wind energy is in second place, with an estimated potential of around 340 W.

Finally, a key aspect in evaluating the value of electric cars stems from the source of the power used. For this reason, a combination of electric vehicles and renewable energy has the potential to significantly reduce global dependency on fossil fuels, resulting in a decrease in greenhouse gas emissions.

References

AEE (Asoiacion Empresarial Eolica) (2016). http://www.aeeolica.org/es/sobre-la-eolica/la-eolica-en-espana/ Accessed 20.04.16.

Albalate, D., Bel, G., & Fageda, X. (2015). When supply travels far beyond demand: Causes of oversupply in Spain's transport infrastructure. *Transp. Policy*, *41*, 80–89.

ANFAC (Asociación Española de Fabricantes de Automóviles y Camiones) (2014). Memoria anual 2014 [online] Available at: http://www.anfac.com/documents/tmp/MemoriaANFAC 2014.pdf Accessed 18.03.16.

Annema, J. A., & De Jong, M. (2011). The history of the transport future—Evaluating Dutch transport scenarios of the past. *Transp. Rev.*, *31*(3), 341–356.

BOE (Boletín Oficial del Estado) (1997). LEY 54/1997, de 27 noviembre. del Sector Eiectrico. [Online] Available at: https://www.boe.es/boe/dias/1997/11/28/pdfs/A35097-35126.pdf Accessed 07.04.16.

BOE (Boletín Oficial del Estado) (2000). REAL DECRETO 1955/2000, de 1 de diciembre, por el que se regulan las actividades de transporte, distribución, comercialización, suministro y procedimientos de autorización de instalaciones de energía eléctrica. [Online] Available at: https://www.endesaclientes.com/static/iberia/RD-1955-2000.pdf Accessed 07.04.16.

BOE (Boletín Oficial del Estado) (2002). REAL DECRETO 842/2002, de 2 de agosto, por el que se aprueba el Reglamento electrotécnico para baja tensión [Online]. Available at: https://www.boe.es/boe/dias/2002/09/18/pdfs/A33084-33086.pdf Accessed 07.04.16.

BOE (Boletín Oficial del Estado) (2007a). REAL DECRETO 1028/2007, de 20 de julio, por el que se establece el procedimiento administrativo para la tramitación de las solicitudes de autorización de instalaciones de generación eléctrica en el mar territorial. [Online] Available at: http://www.boe.es/boe/dias/2007/08/01/pdfs/A33171-33179.pdf Accessed 07.04.16.

BOE (Boletín Oficial del Estado) (2007b). REAL DECRETO 661/2007, de 25 de mayo, por el que se regula la actividad de producción de energía eléctrica en régimen especial. [Online] Available at: https://www.boe.es/boe/dias/2007/05/26/pdfs/A22846-22886.pdf Accessed 07.04.16.

BOE (Boletín Oficial del Estado) (2008). REAL DECRETO 1578/2008, de 26 de septiembre, de retribución de la actividad de producción de energía eléctrica mediante tecnología solar fotovoltaica para instalaciones posteriores a la fecha límite de mantenimiento de la retribución del Real Decreto 661/2007, de 25 de mayo, para dicha tecnología. [Online] Available at: http://www.boe.es/boe/dias/2008/09/27/pdfs/A39117-39125.pdf Accessed 07.04.16.

BOE (Boletín Oficial del Estado) (2009a). Directiva 2009/28/CE del Parlamento Europeo y del Consejo de 23 de abril de 2009 relativa al fomento del uso de energía procedente de fuentes renovables y por la que se modifican y se derogan las Directivas 2001/77/CE y 2003/30/CE (Texto pertinente a efectos del EEE). [Online] Available at: https://www.boe.es/doue/ 2009/140/L00016-00062.pdf Accessed 07.04.16.

BOE (Boletín Oficial del Estado) (2009b). Real Decreto-ley 6/2009, de 30 de abril, por el que se adoptan determinadas medidas en el sector energético y se aprueba el bono social. [Online] Available at: https://www.boe.es/boe/dias/2009/05/07/pdfs/BOE-A-2009-7581 Accessed 07.04.16.

BOE (Boletín Oficial del Estado) (2010a). Real Decreto 1565/2010, de 19 de noviembre, por el que se regulan y modifican determinados aspectos relativos a la actividad de producción de energía eléctrica en régimen especial. [Online] Available at: http://www.boe.es/boe/dias/ 2010/11/23/pdfs/BOE-A-2010-17976.pdf Accessed 20.04.16.

BOE (Boletín Oficial del Estado) (2010b). Real Decreto 1614/2010, de 7 de diciembre, por el que se regulan y modifican determinados aspectos relativos a la actividad de producción de energía eléctrica a partir de tecnologías solar termoeléctrica y eólica. [Online] Available at: http://www.boe.es/boe/dias/2010/12/08/pdfs/BOE-A-2010-18915.pdf Accessed 07.04.16.

BOE (Boletín Oficial del Estado) (2010c). Real Decreto-ley 14/2010, de 23 de diciembre, por el que se establecen medidas urgentes para la corrección del déficit tarifario del sector eléctrico. [Online] Available at: https://www.boe.es/boe/dias/2010/12/24/pdfs/BOE-A-2010-19757.pdf Accessed 07.04.16.

BOE (Boletín Oficial del Estado) (2011). Ley 2/2011, de 4 de marzo, de Economía Sostenible. [Online] Available at: https://www.boe.es/buscar/pdf/2011/BOE-A-2011-4117-cons olidado.pdf Accessed 07.04.16.

BOE (Boletín Oficial del Estado) (2012). Real Decreto-ley 1/2012, de 27 de enero, por el que se procede a la suspensión de los procedimientos de preasignación de retribución y a la supresión de los incentivos económicos para nuevas instalaciones de producción de energía eléctrica a partir de cogeneración, fuentes de energía renovables y residuos. [Online] Available at: http://www.boe.es/boe/dias/2012/01/28/pdfs/BOE-A-2012-1310.pdf Accessed 07.04.16.

BOE (Boletín Oficial del Estado) (2013). Real Decreto-ley 9/2013, de 12 de julio, por el que se adoptan medidas urgentes para garantizar la estabilidad financiera del sistema eléctrico. [Online] Available at: http://www.boe.es/boe/dias/2013/07/13/pdfs/BOE-A-2013-7705. pdf Accessed 07.04.16.

BOE (Boletín Oficial del Estado) (2015a). Real Decreto 1078/2015, de 27 de noviembre, por el que se regula la concesión directa de ayudas para la adquisición de vehículos de energías alternativas, y para la implantación de puntos de recarga de vehículos eléctricos en 2016 MOVEA. [Online], Available at: https://www.boe.es/boe/dias/2015/11/28/pdfs/BOE-A-2015-12900.pdf Accessed 18.04.16.

BOE (Boletín Oficial del Estado) (2015b). Real Decreto 287/2015, de 17 de abril, por el que se regula la concesión directa de subvenciones para la adquisición de vehículos eléctricos en 2015 (Programa MOVELE 2015). [Online] Available at: http://www.idae.es/uploads/ documentos/documentos_Boletin_CD_62-15_MOVELE_2015_180bacb0.pdf Accessed 18.04.16.

Borge, R., de Miguel, I., de la Paz, D., Lumbreras, J., Pérez, J., & Rodríguez, E. (2012). Comparison of road traffic emission models in Madrid (Spain). *Atmos. Environ., 62*, 461–471.

CCEIM (Centro Complutense de Estudios e Información Medioambiental) (2011). Cambio Global España 2020/50. Energía, economía y sociedad. [Online] Available at: http://pendientedemigracion.ucm.es/info/fgu/descargas/cceim/programa_energia_2020_ 2050.pdf Accessed 09.02.16.

CHADEMO Association (2016). http://chademo.com/pdf/interface.pdf Accessed 27.04.16.

Dargay, J., Gately, D., & Sommer, M. (2007). Vehicle ownership and income growth, worldwide: 1960–2030. *Energy J.*, *28*(4), 143–190.

EEA (European Environment Agency) (2009). EMEP/EEA air pollution emission inventory handbook-2009. EEA Technical Report 9/2009. [Online] Available at: http://www.eea. europa.eu/publications/emep-eea-emission-inventory-guidebook-2009 Accessed 19.03.16.

European Commission (2010). A European strategy on clean and energy efficient vehicles. Communication from the Commission to the European Parliament, the Council and the European Economic and Social Committee of April 28. Brussels.

European Commission (2013). Clean power for transport: A European alternative fuels strategy. Communication from the Commission to the European Parliament, the Council, the European Economic and Social Committee and the Committee of the Regions of January 24. Brussels.

Eurostat (2013). *Statistics.* Brussels: European Commission epp.eurostat.ec.europa.eu/ Accessed 25.05.16.

FAEN (Fundación Asturiana de la Energía) (2016). http://www.faen.es/batterie/Recarga_ vehiculo_electrico.pdf Accessed 27.04.16.

Fageda, X., & González-Aregall, M. (2014). Port charges in Spain: The roles of regulation and market forces. *Int. J. Shipp. Transp. Logist.*, *6*(2), 152–171.

FENERCOM (Fundación de la Energía de la Comunidad de Madrid) (2015) http:// www.fenercom.com/pdf/publicaciones/Guia-del-Vehiculo-Electrico-II-fenercom-2015.pdf Accessed 27.04.16.

Girard, A., Gago, E. J., Ordoñez, J., & Muneer, T. (2016). Spain's energy outlook: A review of PV potential and energy export. *Renew. Energy*, *86*, 703–715.

GWEC (Global Wind Energy Council) (2013). Global wind statistics 2012. GWEC, pp. 1-4. [Online] Available at: http://www.gwec.net/publications/global-wind-report-2/.

IDAE (Instituto para la Diversificación y Ahorro de la Energía) (2011a). Impacto económico de las energías renovables en el sistema productivo español. Estudio Técnico PER 2011–2020. [Online] Available at: http://www.idae.es/uploads/documentos/docu mentos_11227_e3_impacto_economico_4666bcd2.pdf Accessed 19.04.16.

IDAE (Instituto para la Diversificación y Ahorro de la Energía) (2011b). Plan de Energías Renovables 2011–20. [Online] Available at: http://www.idae.es/index.php/mod. documentos/mem.descarga?file=/documentos_11227_per_2011-2020_def_93c624ab.pdf Accessed 07.04.16.

IEA (International Energy Agency) (2012). CO_2 emissions from fuel combustion highlights. In *IEA statistics* (2012 ed., pp. 1–138). Paris: IEA/OECD.

Jäger-Waldau, A. (2007). Photovoltaics and renewable energies in Europe. *Renew. Sustain. Energy Rev.*, *11*, 1414–1437.

Jefferson, M. (2006). Sustainable energy development: Performance and prospects. *Renew. Energy*, *31*, 571–582.

Kahn, S., Kobayashi, S., Beuthe, M., Gasca, J., Greene, D., Lee, D. S., et al. (2007). Transport and its infrastructure. In B. Metz, O. R. Davison, P. R. Bosch, R. Dave, & L. A. Meyer et al. *Climate change 2007: mitigation. Contribution of working group III to the fourth assessment report of the intergovernmental panel on climate change* (pp. 1–386). New York: Cambridge University Press. [Online] Available at: http://www.ipcc.ch/pdf/assessment-report/ar4/wg3/ar4-wg3-chapter5.pdf Accessed 19.04.16.

Kusku, E. (2010). Enforceability of a common energy supply security policy in the EU: An intergovermentalist assessment. *Cauc. Rev. Int. Aff.*, *4*(2), 145–148.

Lechón, Y., Cabal, H., de la Rúa, C., Lago, C., Izquierdo, L., Sáez, RM. and San Miguel, M. (2006). ACV de combustibles alternativos para el transporte. Fase II: ACV comparativo del biodiésel y del diesel [LCA of alternative fuels for transport. Phase II: Comparative LCA of biodiesel and diese. Ministerio de Medio Ambiente (MMA)]. Madrid. [In Spanish]. [Online] Available at: http://rdgroups.ciemat.es/documents/10907/12207/Analisis2_p8.pdf/ccb89843-ba82-4845-a995-5ce97063953c Accessed 19.04.16.

Loureiro, M. L., Labandeira, X., & Hanemann, M. (2013). Transport and low-carbon fuel: A study of public preferences in Spain. *Energy Economics*, *40*, S126–S133.

Lumbreras, J., Borge, R., Guijarro, A., Lopez, J. M., & Rodríguez, M. E. (2014). A methodology to compute emission projections from road transport (EmiTRANS). *Technological Forecasting & Social Change*, *81*, 165–176.

Lumbreras, J., Rodríguez, E., López, J. M., Guijarro, A., Villimar, R., & Rodríguez, J. (2008). *Herramienta integral para el cálculo de emisiones atmosféricas del transporte por carretera*. *Escuela Técnica Superior de Ingenieros Industriales* (1st ed.). Madrid: Universidad Politécnica de Madrid.

Martín Urbano, P., Ruiz Rúa, A., & Sánchez Gutiérrez, J. I. (2012). El sistema de transporte público en España: una perspectiva interregional. *Cuadernos de Economía*, *31*(58), 195–228.

Menéndez Pérez, E., Feijoo Lorenzo, A., & Cámara Rascón, A. (2006). Energy scenarios in Spain (Escenarios energéticos en España). In *Claridad* (pp. 51–71). Madrid: Comisión Ejecutiva Confederal de UGT.

MFOM (Ministerio de Fomento) (2005). PEIT: Plan estratégico de infraestructuras y transporte 2005–2020 [Online]. Available at: http://www.fomento.gob.es/MFOM/LANG_CASTELLANO/_ESPECIALES/PEIT/ Accessed 15.04.16.

MFOM (Ministerio de Fomento) (2011). Legislación del Transporte por Carretera. [Online] Available at: http://www.fomento.gob.es/MFOM.CP.Web/handlers/pdfhandler.ashx?idpub=TTW001 Accessed 15.04.16.

MFOM (Ministerio de Fomento) (2014). Los Transportes y las Infraestructuras. Informe Anual 2014. [Online] Available at: http://www.fomento.gob.es/MFOM.CP.Web/handlers/pdfhandler.ashx?idpub=BTW027 Accessed 15.04.16.

MFOM (Ministerio de Fomento) (2015a). Observatorio del Transporte y la Logística en España. Informe Anual 2015 [Online] Available at: http://observatoriotransporte.fomento.es/NR/rdonlyres/0AE839CF-9E00-46F3-A27C-88B14AC37715/136237/INFORME_OTLE_2015.pdf Accessed 19.03.16.

MFOM (Ministerio de Fomento) (2015b). PITVI: Plan de Infraestructuras, Transporte y Vivienda. [Online] Available at: http://www.fomento.gob.es/MFOM/LANG_CASTELLANO/PLANES/PITVI/PITVI_DOCU/ Accessed 15.04.16.

MITyC (Ministry of Industry, Tourism and Trade (Minetur)). State Secretary Of Energy (2011). La energía en España, 2011. [Online] Available at: http://www.minetur.gob.es/energia/es-ES/Paginas/index.aspx Accessed 19.09.15.

MMA (Spanish Ministry of Environment) (2012). CORINAIR Spain 1990–2010 inventories of emissions of pollutants into the atmosphere (Spanish Atmospheric Emission Inventory). Prepared by AED for the Secretariat-General for the Environment, Directorate-General for Environmental Quality and Assessment.

Newton, T. (2009). *How cars work:* (1st ed., pp. 1–96). California: Black Apple Press.

Ntziachristos, L., Gkatzoflias, D., Kouridis, C., & Samaras, Z. (2009). COPERT: a European road transport emission inventory model. In I. N. Athanasiadis, P. A. Mitkas, A. E. Rizzoli, & J. Marx Gómez (Eds.), *Information Technologies in Environmental Engineering. Environmental science and engineering* (pp. 491–504). Berlin, Heidelberg: Springer-Verlag.

OMM (Observatorio de Movilidad Metropolitana) (2008–2010). *Informe del Observatorio de la Movilidad Metropolitana*. Madrid: Ministerio de Medio Ambiente y Medio Rural y Marino.

OMM (Observatorio de Movilidad Metropolitana) (2012). *Informe del Observatorio de la Movilidad Metropolitana*. Madrid: Ministerio de Medio Ambiente y Medio Rural y Marino.

REE (Red Eléctrica de España) (2013). The Spanish electricity system 2012 (El Sistema Eléctrico Español, Informe 2012). [Online] Available at: http://www.ree.es/sites/default/files/downloadable/inf_sis_elec_ree_2012_v2.pdf Accessed 19.09.15.

REE (Red Electrica Española) (2016). http://www.ree.es/es/actividades/operacion-del-sistema-electrico/centro-de-control-de-energias-renovables# Accessed 27.04.16.

Sánchez-Braza, A., Cansino, J. M., & Lerma, E. (2014). Main drivers for local tax incentives to promote electric vehicles: The Spanish case. *Transport Policy, 36*, 1–9.

Smit, R., Dia, H. and Morawska, L. (2009). Road traffic emission and fuel consumption modelling: Trends, new developments and future challenges. In: Demidov S., Bonnet J., (Eds), Traffic related air pollution and internal combustion engines. USA: New York; Nova Publishers. [Online] Available at: https://www.novapublishers.com/catalog/product_info.php?products_id=9546 Accessed 19.03.2016.

Smit, R., Ntziachristos, L., & Boulter, P. (2010). Validation of road vehicle and traffic emission models—A review and meta-analysis. *Atmos. Environ., 44*, 2943–2953.

Stead, D. (2008). Institutional aspects of integrating transport, environment and health policies. *Transport Policy, 15*(3), 139–148.

Steenhof, P. A., & McInnis, B. C. (2008). A comparison of alternative technologies to de-carbonize Canada's passenger transportation sector. *Technol. Forecast. Soc. Chang., 75*, 1260–1278.

todotransporte.com (2003). http://www.todotransporte.com/en-espana-existen-245-flotas-de-transporte-privado-integradas-por-mas-de-25-vehiculos-pesados/ Accessed 18.04.16.

UAH (Universidad de Álcala) (2016). http://www.uah.es/ Accessed 20.04.16.

Winiwarter, W., & Schmiak, G. (2005). Environmental software systems for emission inventories. *Environ. Modell. Softw., 20*(12), 1469–1477.

Xiaowu, W., & Ben, H. (2005). Energy analysis of domestic-scale solar water heaters. *Renew. Sustain. Energy Rev., 9*, 638–645.

Further Reading

MFOM (Ministerio de Fomento) (2015c). Observatorio del Transporte y la Logística en España. Informe Anual 2013 [Online] Available at: http://observatoriotransporte.fomento.es/NR/rdonlyres/775612E3-0CDE-43C3-8A6A-13853CCF3919/127878/INFORMEOTLE2013v02.pdf Accessed 19.05.16.

The scenario of electric vehicles in Norway

9

Mohan Lal Kolhe, T.P. Chathuri Madusha
University of Agder, Kristiansand, Norway

9.1 Introduction

The efficient transportation facility is one of the major impacts for the economic and social development of a country. It needs to be practical, economic, and environmentally friendly conditions. Among the vehicle fleets in the world, electrical vehicles (EVs) are the cleaner vehicle technology that minimizes the production of conventional vehicles' tailpipe emissions that are harmful to the environment (Jong, Åhman, Jacobs, & Dumit, 2009).

At present, Norway is the forefront in using EVs and expanding its EV market without hesitations due to incentives provided by the Norwegian government. The number of EVs usage is increasing as the result of multiple economic incentives such as exemption from value-added tax (VAT) and permission to use transit lanes. (Aasness & Odeck, 2015). Today, Oslo has the world's highest number of electric vehicles per inhabitant (http://www.avere.org/www/newsMgr. php?action=view&frmNewsId=611&section=&type=&SGLSESSID=tqiice0pmj dclt7l4q0s3s1o27). The renewable energy source, that is, hydropower, provides major power in Norway. Hydropower is still the mainstay for producing electricity in low cost and will continue to play vital role in the future. Electrical propulsion is the most energy efficient process in transport with reduction of climate gas emissions and noises from the road transport sector. In this chapter, the importance of EVs to the country, the evolution process, incentives, and current situation of EV market in Norway are presented. The current situation of EV market has been outlined with the EV users, charging infrastructures, and EV models.

9.1.1 Importance of electrical vehicle technology

The EVs perform an important role in moving towards sustainable mobility. Battery electrical vehicle (BEV) is a vehicle that uses an electric motor for propulsion powered by rechargeable battery packs rather than a gasoline engine. This type of vehicles stores all its energy in rechargeable batteries that are powered from a power grid. Most BEVs have a theoretical range of 160–240 km between recharges, but the practical range is small.

Electric Vehicles: Prospects and Challenges. http://dx.doi.org/10.1016/B978-0-12-803021-9.00009-4

However, these vehicles may be powered by gasoline engines and by electrical motors, leading it to take part of the hybrid technology. This type of hybrid electrical vehicles (HEV) runs partly on electricity and partly on some other fuel (gas or diesel). Vehicles that can be plugged in to charge their batteries are called plug-in hybrid electric vehicles (PHEVs). EVs and PHEVs are clearly much better for the sustainability than using ICE and HEV (Lozano, 2012).

Overall, EV has more advantages than the disadvantages. The main advantage is the reduction of CO_2 emission that leads to a reduction in global warming. Further, recharging battery pack for long distances is the main issue in which it needs to charge batteries for 4–8 h depending on the kind of EV types and charging facilities. By increasing the number of charging stations and fast charging facilities in the country, EVs penetration can be increased further (Argueta, 2010). Moreover, EV users have a great benefit to charge using any electricity outlet, including outlets in home garage, on the walls of workplaces, etc. rather than finding a gas or fuel station as ICE vehicle users.

9.1.2 The characteristic of Norway and EV evolvement

Norway is one of the five Nordic nations situated in European continent. As northerly country, Norway has colder climate with an annual average temperature of 3.8°C at the country's capital, Oslo. Due to its air quality and other factors, it is officially one of the best places to live in the world. Since 1960, exports of oil and gas have become very important elements of the economy of Norway. Thus, the oil producing industry is not part of the ICE regime in Norway (Tietge, Mock, Lutsey, & Campestri, 2016). Through the economic view, Norway stands out with the highest GDP per capita with the economic safety around its petroleum reserves. Norway's electricity production is 98% from hydropower power plants.

Norway is expected to reduce 50% in CO_2 emissions by 2030 and zero emissions in 2050. However, as per the Paris Agreement 2016, Norway plans to bring forward its target for reaching carbon neutrality to 2030, two decades sooner than its current 2050 goal. These geographic, economic, and renewable energy backgrounds imply the importance for the electrification of passenger transport in Norway. The high penetration of EV usage has increased with the benefits of policies and incentives introduced by Norwegian transport sector (see Section 9.1.3).

Norway has the highest EV market penetration per capita in the world, and it has the largest plug-in electric market share of new car sales. The market share of EVs is 12.5% of new vehicle sales in 2014, and it is increased to 20% in the first quarter of 2015 (OFVAS, 2015). Most other countries have less than 1% market share as shown in Fig. 9.1.1.

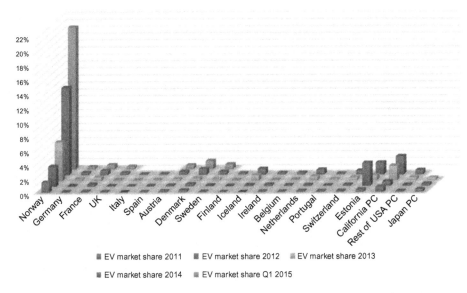

Fig. 9.1.1 The percentage of EV market shares in Europe, the United States, and Japan. *Source*: Figenbaum and Kolbenstvedt, 2015 (TØI Report 1420/2015) (Figenbaum, E., Kolbenstvedt, M. (2015b). *Pathways to electromobility—Perspectives based on Norwegian experiences.* Oslo,Norwa: Institute of Transport Economics (TØI), TØI Report 1420/2015).

9.1.3   EV regional diffusion in Norway

Norway has a large area with population of approximately 5 million people. The population density is $14/km^2$ is the second lowest in Europe. Due to the low population density and long coastlines of Norway, its public transport is less developed than in many other European countries, especially outside the major cities. The alternative for public transport is the usage of private cars. The Norwegian transportation sector has various taxes for registration (e.g. annual tax on vehicle ownership). However, numerous incentives have been introduced to reduce the taxes on EVs for increasing the EVs use.

Among the 428 municipalities in Norway, geographical diffusion of the share of EVs in total fleet has enlarged from small areas into larger areas (Fig. 9.1.2). The expansion process of EV market has primarily based within each municipality, and the number of municipalities without BEVs has gone down dramatically over the years due to the incentives and policies introduced by the Norwegian government. At present, Norway is the worlds fastest EV diffusion country in 2015 with 2% of fleet.

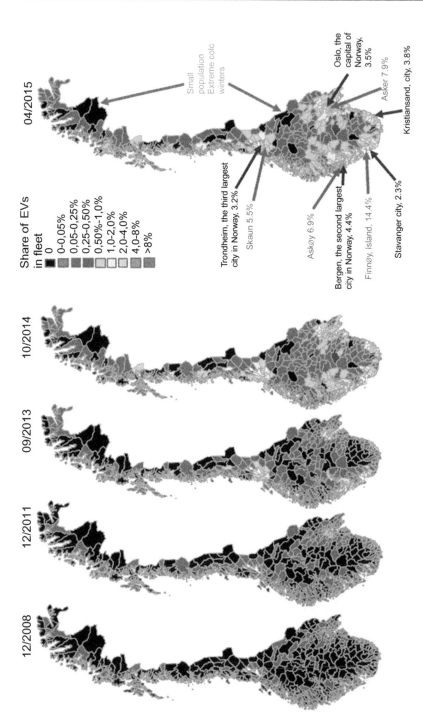

Fig. 9.1.2 Geographical diffusion of BEVs in Norway's 428 municipalities 2008–15. Share of BEV in total fleet in different points in time. BEV and total fleet data from the Norwegian Public Roads Administration and the EV association, OFVAS 2015, Statistics Norway 2015. *Source:* Figenbaum and Kolbenstvedt, 2015 (TØI Report 1422/2015) (Figenbaum, E., & Kolbenstvedt, M. (2015a). *Competitive Electric Town Transport Main results from COMPETT—An Electromobility+ project.* Oslo: Institute of Transport Economics, Norwegian Centre for Transport Research, TØI Report 1422/2015).

9.2 Electric vehicle evolution in Norway

In Norway, the process of being a forefront EV user country has been started since 1970s. In fact, there is currently EV 'fever' in Norway. The global vehicle manufacturers are providing new EV models for the e-mobility. In gradual development of EV evolution in Norway, five main phases, concept development, testing, early market, market introduction, and finally from 2012 entering the market expansion phase, have been conceptualized.

9.2.1 The concept development phase (1970–90)

In the concept development phase from 1970 to 1999, the development of prototypes of EVs was focused. Some private enterprises such as Bakelittfabrikken (forerunner of Think), Strømmens Værksted, and ABB received the financial support from the Research Council of Norway for their involvement in developing the EV prototypes and their propulsion systems. In 1989, ABB Battery Drives in Vestby outside Oslo developed the first VW Golf Citystromer electric cars; however, it was never reached to the market.

9.2.2 The testing phase (1990–99)

The EV testing with different technologies and incentives were considered in this phase from the beginning of the 1990s until 1999. The initial attempts in commercializing the Norwegian made EVs (e.g. 'Think car vehicle') were launched and tested in this phase. It had a great advantage in Norway due to Norwegian electricity at lower prices. The high tax on registration of vehicles made impossible to buy an electric car in Norway at the early time of this phase. The Norwegian EV association, Norstart, was established, and it had introduced first EV incentives, exemption from registration tax and the annual vehicles license fee and exemption from toll road charges. Later in this phase, the EV users had been benefitted with the free parking facilities of municipalities parking slots and reduction in the imposed benefit tax on corporate EVs. After the introduction of the exemption of registration tax, 'Kewet electric vehicles' were imported from Denmark. At the end of this phase, the first production version of EV, that is, Think City had started; however, the company went bankrupt before the production set to the market.

9.2.3 The early market phase (1999–2009)

The Ford Motor Company bought the Think city vehicle in 1999 and later developed an EV model, which has been better suited for the US market, targeting on cost reduction and quality improvement. Further, Kollega Bil has bought the assets from the Danish bankrupted producer and established the production of the Kewet vehicle in Norway. The Norwegian Kewet producer faced an expired type approval, and they introduced and registered the two types as L7e with a simplified approval procedure. However, Ford Motor Company pulled out of Think in 2003 due to the insufficient technology development. Then, EVs demand was filled by imported French EVs during 1998–2002. At the end of this stage, Norwegian investors bought Think and launched the new model in 2009.

Powerful and attractive measures were established during the early market phase. Some incentives such as exemption from VAT (25% in Norway) were launched from 2001. Further, it's granted the access to bus lanes in the larger Oslo region from 2003 (permanent and nationwide from 2005). In this phase, EVs have been able to have their own number plate starting with EL, and it has helped the administration on making free parking incentives easier. These caused for new evolution to buy EVs despite the fact that they were not in advanced stage or comfortable as other cars.

9.2.4   The market introduction phase (2009–12)

The new Think and Pure Mobility (known as Buddy or Kewet) car models were launched from 2009. However, the electric vehicle production in Norway faded away and Think and Pure Mobility went bankrupt in 2011. It was over in 2012 after Tata sold the company to a battery manufacturer. Pure Mobility has restarted to production of Buddy again, now operating under the name Buddy Electric AS. Further, in 2010/11 big car manufacturers such as Mitsubishi, Peugeot, Citroën, and Nissan launched their EV vehicles to EV market in Norway.

The EV market has expanded rapidly from 2010, and the EV association in Norway has been recognized as an important organization during this period. The association started supporting their members to get the information on charging facilities, recruited new EV drivers, and other dissemination EV activities through the internet user forum. Further, the government organization 'Transnova' that established in 2009 has started doing testing and expansion of new technologies for establishing the EV charging stations and to reduce CO_2 emissions from the transportation sector.

9.2.5   The market expansion phase (from 2012)

The first PHEVs were started launching in to the market at the end of 2012 with lower registration tax than traditional hybrid vehicles. However, those PHEVs were not able to obtain EV incentives. The competition between global vehicle manufacturers such as Mitsubishi, Peugeot, Citroen, Nissan, Tesla, Renault, Mia, Toyota, Opel, Volvo, Fisker, BMW, VW, Audi, Smart, Daimler, and Ford have been increased for expanding their EV products. The car dealers started marketing their EVs by offering different businesses models such as Nissan offered free of charging loan for 20 days and the first 3 years of EV ownership on the beginning of this period. During this time, only one small company, Buddy has been manufacturing EVs in Norway. The new financial incentives for EVs are continuing until the end of 2017 according to the climate policy settlement in Norwegian Parliament 2012 (Figenbaum & Kolbenstvedt, 2013).

It is likely that the EV market expansion in Norway continues with the attractive incentives, charging infrastructure facilities, and knowledge about the technology, and also, the vehicle manufacturers are continuously launching new EVs. Currently, the Norwegian EV market share of the new car sales has been the highest in the world.

The EV evaluation process in Norway with incentives has been outlined as Fig. 9.2.1.

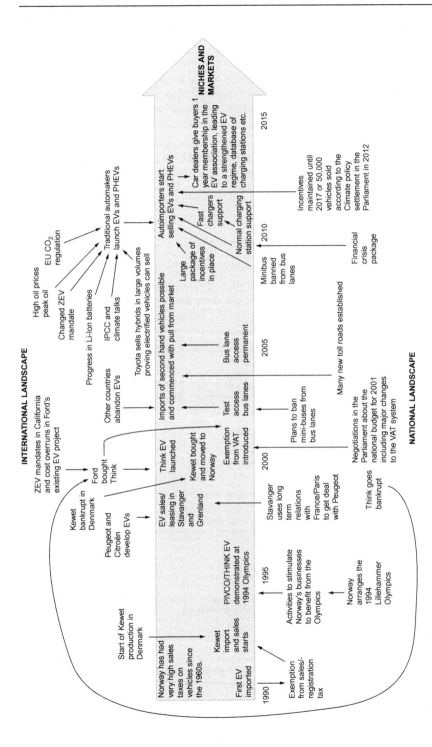

Fig. 9.2.1 The electrical vehicle evolution process in Norway.

Source: Figenbaum and Kolbenstvedt, 2015 (TØI Report 1422/2015) (Figenbaum, E., & Kolbenstvedt, M. (2015a). *Competitive Electric Town Transport Main results from COMPETT—An Electromobility+ project.* Oslo: Institute of Transport Economics, Norwegian Centre for Transport Research, TØI Report 1422/2015).

9.3   EV policies and incentives

Norwegian Ministry of Transport and Communication has gradually developed generous incentives to continue the development of EV market throughout the country over the years. These economical benefited incentives are motivating and encourage the use of EVs in Norway. The results of EU research project competitive electric town transport (COMPETT) have summarized the EV policies in 2015 (Norwegian Center for Transport Research, 2015). Table 9.3.1 summarizes the national incentives and policies in Norway; Figenbaum and Kolbenstvedt (2013, 2015a, 2015b). This section has provided EV policies and incentives that the Norwegian government has planned for future adjustments of incentives.

9.3.1   Exemption from registration tax

The registration taxes mainly depend on vehicle weight and emissions of CO_2 and NO_x. The conventional vehicles have high taxes in Norway. EVs have been exempted from the registration tax charge since 1990, and it became permanent in 1996. In practice, most EVs are free from the registration tax. Some plug-in hybrid vehicles have low CO_2 emissions that achieve a large deduction in vehicle emission fee of registration tax than the conventional vehicles. The exemption of registration tax incentive on

Table 9.3.1 **National incentives, policies, and initiatives in Norway**

Incentives	Year
Fiscal incentives reduction of purchase price	
Exemption from registration tax	1990/1996
Exemption from VAT	2001
Reduced annual vehicle license fee	1996/2004
Reduced company car tax	2000
Direct subsidies to users reducing usage costs and range challenges	
Free toll roads	1997
Exemption from paying car ferry fees	2009
Financial support for charging stations	2009
Financial support for fast charge stations	2011
Reduction of time costs and giving relative advantages	
Access to bus lanes	2003/2005
Free parking	1999
Free charging	

Source: Figenbaum and Kolbenstvedt, 2015 (TØI Report 1422/2015) (Figenbaum, E., & Kolbenstvedt, M. (2015a). *Competitive Electric Town Transport Main results from COMPETT—An Electromobility+ project*. Oslo: Institute of Transport Economics, Norwegian Centre for Transport Research, TØI Report 1422/2015).

EVs will be continued until 2020 to achieve the Norwegian climate goals for 2020 and 2030 (Figenbaum & Kolbenstvedt, 2015a). For ICE vehicles, the registration tax will be tuned further towards reducing the emissions from the vehicles.

9.3.2 Exemption from VAT

The additional VAT of 25% of the purchase value is added to all goods and services sold in Norway, and for the conventional vehicles, the tax is added to sales without the registration tax. EVs have been fully exempted from VAT since 2001, and it has increased the number of EV consumers and discouraged the usage of conventional vehicles. PHEV has no more benefits like the pure EVs such as paying full VAT. However, they allow to get a deduction of 26% in the component of purchase tax. There are high taxes on high-emission cars and lower taxes on low- and zero-emission cars (Torper, 2015). Further, the exemption from VAT will be continued till the end of 2017, and it will be considered to replace a subsidy scheme that will initially be at the level of the value of the VAT exemption. A maximum subsidy per car may be set, but without leading to large price increases for any EV models. In the future, this incentive may be ramped down.

9.3.3 Reduced annual vehicle license fee

The annual license fee is applied on three rates for all private vehicles registered in the vehicle register. It is valid from Jan. 1 until the end of the year. EV had a total exemption from the fee in the 1996–2004 period. The lowest rate previously covered a third-party injury fee, a state fee that was intended to cover the costs during the vehicle accidents. Moreover, half rate of ICE vehicles could be introduced from 1 Jan. 2018 and full rate from 2020.

9.3.4 Reduced company car tax

The company cars can be used for private driving, and the user must pay a fee for taking this benefit. The reduced company car tax for electric vehicles was introduced in 2000, and it has been imposed that EV benefit taxation has been halved in relation to other regular vehicles. According to future incentive decision taken from political agreement on May 2015, this incentive may be removed. Also, 13% of all workers receive an allowance for using their own vehicle at work. EV drivers receive an allowance of 0.52 €/km, while drivers of regular vehicles gets 0.5 €/km (Figenbaum & Kolbenstvedt, 2013).

9.3.5 Free toll roads

In 1997, EVs were exempted from toll roads in Norway. Tolls are used to improve the public transport and expand the road capacity. This is a considerable financial benefit to the EV owners and easy to drive in rush cities such as Oslo. The effect of free toll roads has lost the large income to road toll companies. According to new policies, the government will appraise the environmental effects of introducing differentiated fees for toll roads (main roads and toll rings around cities).

9.3.6 Exemption from paying car ferry fees

EVs have been exempted from paying car ferry fees on Norwegian highway ferries since 2009. However, passengers in the vehicle still have to buy a ticket. This incentive is similar manner to toll roads saving money for those using car ferries frequently, especially the coastal region of Norway. Based on the environmental characteristics of ferries, the Norwegian government has encouraged the low rate for zero emission vehicles to introducing such incentives.

9.3.7 Access to bus lanes

Since 2003, EVs have been allowed to access bus lanes on selected test roads (in the greater Oslo region), and it became permanent from 2005. From 2009, minibuses were banned from the bus lanes. It directly affected to the sale of EVs in the Asker municipality due to the time delay on the E18 during rush hour congestion (Figenbaum & Kolbenstvedt, 2013). The free access to public transport lanes in larger cities of Norway is very efficient, in advantage and helpful for time savings, especially during busy hours. The time saving is also equivalent to money savings as economic savings. In future, local authorities will handle this process with the possibilities of restrictions.

9.3.8 Free public parking with free charging

EVs have been given free public parking access since 1999. However, it has been allowed in some places since 1993. Further, the number of electric charging stations and charging points has been increased with the large share of government support and financial support for charging stations by many municipalities. The effect of these policies has lost their income to the state, municipalities, and road toll companies. Thus, in many free parking places in Norway, EVs can charge for free. In 2013, plug-in hybrids were given access to the charging stations.

The free public parking with free charging facilities in some cities are more attractive and economically benefits of using EVs. In addition, the benefit for EV users is to save time looking for a space that depends on the number of spaces available. According to the political agreement in 2015, local authorities will be given the authority to decide whether this incentive continues in their district.

These policies have been consistently increasing the usage of pure EVs in Norway, indicating the compatibility with social needs. The purchase incentives have been effectively speeding up to motivation to be a BEV user. The BEV fleet in Norway with incentives shows the rapid expansion process from 1997 to 2014 (see Fig. 9.3.1).

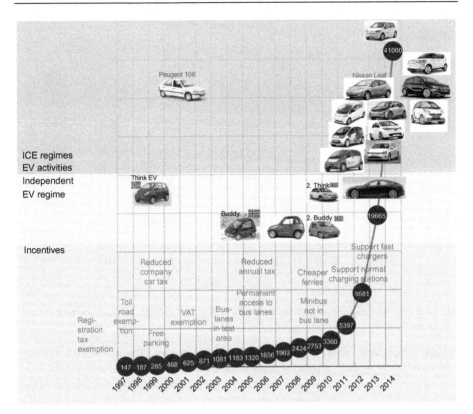

Fig. 9.3.1 The expansion of the BEV fleet in Norway with the incentives agreed by Norwegian Ministry of Transport and Communication and makes and models.
Source: Figenbaum and Kolbenstvedt, 2015 (TØI Report 1422/2015) (Figenbaum, E., & Kolbenstvedt, M. (2015a). *Competitive Electric Town Transport Main results from COMPETT—An Electromobility + project*. Oslo: Institute of Transport Economics, Norwegian Centre for Transport Research, TØI Report 1422/2015).

9.4 Current situation of EV market

The substantial incentive package for EV is creating day by day increment in the Norwegian EV market. It is leading towards for progress to zero emission EVs. According to the EVS29, International Battery, Hybrid and Fuel Cell Electric Vehicle symposium, the Norwegian EV market share is stabilizing on 20%, and the most of other countries are holding around 1% of market share. It is approximately 5 years of advancement in Norwegian EV market rather than the most of other countries (Haugneland, Bu, & Hauge, 2016).

The Norwegian EV association involves lot of programs for the motivation of EV usage. Since 2012, the consumer requirements and facilities have been surveyed for future decision-makings in government and the industrial sector as well as globally. The increment of EV share in Norway has become the largest European plug-in

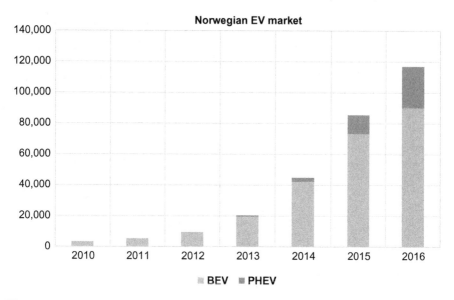

Fig. 9.4.1 The development in purchase of EVs in Norway (EAFO updated 31/08/2016).

electric vehicle and fourth largest in the world EV market. The Norwegian government have motivated the people for using EVs to achieve environmental friendly form of transport, while simultaneously reducing the usage of oil. The contribution of the Norwegian EV owners by sharing the customer experiences and opinions plays the major role in developing policies towards the end target.

In Norway, the growth of the electrical car market is almost exponential and from 2011 to 2014, EV market has been doubled and in 2015, EV had a 22% market share (Fig. 9.4.1). Also, the market share for plug-in hybrids (PHEVs) is growing.

According to Fig. 9.4.1, the current situation at the end of Aug. 2016 shows that close to 90,000 EVs and about 25,000 PHEVs have been registered in Norway. It is predictive that there is no longer use petrol and diesel for road transportation by 2050, and the proportion of zero emission EVs may increase to 100% in 2050 (Engen et al., 2015a; Engen, Amundsen, Bratland, & Barkved, 2015b).

9.5 The typical EV owners

Most of the Norwegian EVs users belong to multivehicle households, and it has been observed that almost all EV owners (87%) will continue to buy EVs in the future (Figenbaum, Fearnley, Pfaffenbichler, & Hjorthol, 2015). Owners of both EVs and ICE vehicles use their EV more for daily traveling than ICE vehicle due to the compatibility of usage. The motivations to buy EVs are related to economy, incentives, environment, and easiness of monitoring with low noise and comfort.

The policy makers are continuously trying for developing the motivation for EVs for achieving the zero and low emission cars instead of high-emission ICE cars. In the Norwegian car tax system, 'polluter pays principle' is applied. EV owners enjoy local incentives such as access to bus lanes, free public parking exemption from toll roads, and reduced rates on coastal main road ferries. Moreover, the EV has more benefit due to low maintenance compared with ICE vehicles.

The Institute of Transport Economics (TØI-Transportøkonomisk Institutt) has surveyed the attitudes towards EVs among a sample of nationwide vehicle owners in each year. In 2016, the 3111 EV owners, 2065 PHEV owners, and 3080 ICEV owners are considered for analyzing their behaviors. The perception of general characteristics of EVs among different vehicle owing groups is shown in Fig. 9.5.1, per TØI 2016

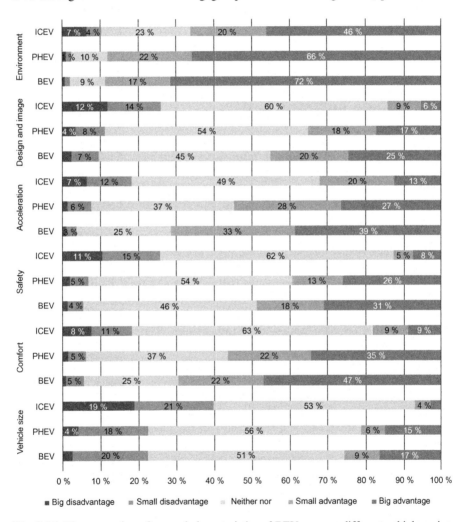

Fig. 9.5.1 The perception of general characteristics of BEVs among different vehicle owing groups. $n_{BEV} = 3111$, $n_{PHEV} = 2065$, and $n_{ICEV} = 3080$, *Norwegian PEV consumer survey*. *Source*: Figenbaum and Kolbenstvedt, 2015 (TØI Report 1492/2016) (Figenbaum, E., & Kolbenstvedt, M. (2016). *Learning from Norwegian Battery Electric and Plug-in Hybrid Vehicle users—Results from a survey of vehicle owners*. Oslo: Institute of Transport Economics, Norwegian Centre for Transport Research, TØI Report 1492/2016).

survey report. The environment, comfort and acceleration are rated as the big general advantage by EV and PHEV owners than ICE vehicle owners. Further, design and image and safety are rated above average by EV and PHEV owners and slightly below average by ICE vehicle owners (Figenbaum & Kolbenstvedt, 2016).

As per the TØI 2016 survey report, the operating cost is high in ICE vehicle owners compared with EV owners and little less than the PHEV owners. Resale value has not been important parameter to EV and PHEV owners, but ICE vehicle owners are still cautious about that. The fiscal incentives reduction of purchase price in Norway makes EVs more competitive in purchase price. Among these advantages, the limited range is the biggest disadvantage to EV owners. This can be solved through well-managed charging infrastructure facilities in the country.

9.6   Charging infrastructure

The improvements in EV technologies are helping in having large variety of EV models in the market. The charging infrastructures have advanced with the increment of number of EVs to build a sustainable transport facility in country. Infrastructure facilities that facilitate normal and fast charging are essential to the EV users, and it needs to be facilitated the access at public and private, own house, or apartment buildings. For longer distance trip, a well-organized charging facility must be in place. Even EV users are charging at home and manage without charging daily, fast charging is very important option that they can use when needed.

Normal charging at public parking spaces is usually free. Because of the e Tesla scheme of lifetime free fast charging, the half of the Tesla users did not pay for fast charging. Although fast charging was sometimes free of charge a few years ago (Fearnley, Pfaffenbichler, Figenbaum, & Jellinek, 2015) and now, they are also willing to pay a higher price for the service of fast charging. The world's most extensive network of charging stations of EVs has developed in Norway, and the locations of charging stations are listed on publicity available database by Norwegian electric vehicle association (NOBIL). NOBIL is the quality assurance database with the information about charging stations, what kind of power outlet is available at the station, access information, images, etc. This database has provided correct and reliable online information, and it can be transferred to the navigation systems in the vehicles to access the closest charging station information and directions. It will be continuously expanded as new charging stations are built in Norway. The current situation in 2016, Norway has reached to 6600 normal charging points and 900 fast charging points (EAFO, 2016). As shown in Fig. 9.6.1, the rapid increment of EVs has led to a continuous increase in charging facilities.

According to European Clean Power for Transport directive recommendations, there should be one public available charging point for every 10 electric cars by

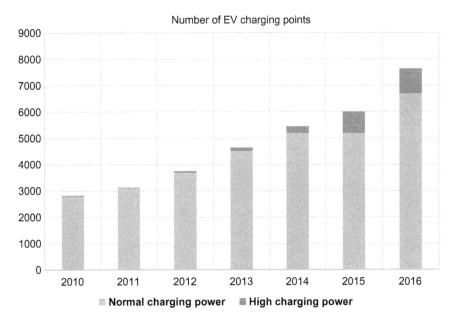

Fig. 9.6.1 The publicity accessible EV charging infrastructure facilities. *Source*: EAFO (European Alternative Fuels Observatory).

2020. The exponential increment of EV usage in Norway shares from 22% in 2015 to 30% in 2020, the Norwegian EV population may reach to 250,000 by 2020. It shows that there should be around 25,000 public charging points available by 2020. Hence, the Norwegian government have launched a program to finance the establishment of at least two multistandard fast charging stations every 50 km on all main roads in Norway by 2017 (Haugneland et al., 2016).

9.7 The models in Norway EV market

The sales of EVs in Norway are increasing exponentially over last few years due to the attractive policies for purchasing and using of EVs. Due to that, various types of EV models are available in the market. With the new EV models and the emerging battery technologies, longer distance travel is becoming possible. The number of new EVs sold in the Norwegian market with EV models is given in Fig. 9.7.1.

Overall, the registration of top 10 EVs from 2011 to 2015 are given in Table 9.7.1. The Nissan Leaf and Kia Soul EV are with the largest share of registration (almost 70%).

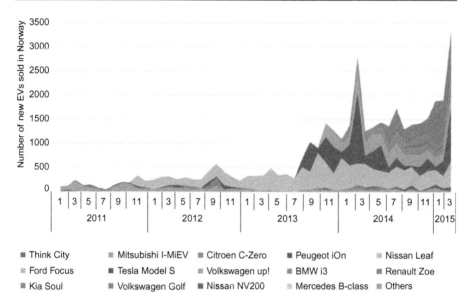

Fig. 9.7.1 The EV sales of models sold per year and month in Norway market.
Source: Figenbaum and Kolbenstvedt, 2015 (TØI Report 1422/2015) (Figenbaum, E., & Kolbenstvedt, M. (2015a). *Competitive Electric Town Transport Main results from COMPETT—An Electromobility + project*. Oslo: Institute of Transport Economics, Norwegian Centre for Transport Research, TØI Report 1422/2015).

The EV market in Norway has been increasing due to lower prices and incentives and a better selection of vehicles that matches user requirements. There have been available mini, small, compact, and large/luxury EV models in Norway to satisfy the selection of user needs since 2010.

9.7.1   Price of EVs in Norway

The sale price statistics of EVs in Norway from 1998 to 2015 has been given in Fig. 9.7.2 (according to the COMPETT project). It has indicated that the price reduction of EVs in 2001 is due to VAT exemption on EVs during that year. After battery technology shift to the Li-ion technology, the new Think city vehicle was available in the market in 2009, and its price has gone down rapidly as shown in Fig. 9.7.2. The PHEV was introduced to the Norway EV market in 2012, and until then, the PHEV prices have remained high. The exemption of VAT and other taxes to EVs has been reduced the market prices.

9.8   The electric vehicle industry in Norway

The EV evolution in Norway has started since 1970s, and currently, it has become large EV consumption country (see Section 9.2). The expansion of EV market base on the availability of different EV models with their requirements. In 1991, the electric

Table 9.7.1 Registration of top 10 plug-in electric vehicles by model in Norway (2011–15) (Bekker, 2013; Moberg, n.d.-a, n.d.-b)

Model		Total registration	Market share	2015	2014	2013	2012	2011
				New only		Includes new and used imports		
Nissan Leaf	Combined	21,231	25.2%	5277	7013	6073	2487	381
	Only new	15,245		3189	4781	4604	2298	373
	Used imports	5986	51.4%	2088	2232	1469	189	8
Volkswagen e-Golf		10,961	13.0%	8943	2018			
Tesla Model S		10,064	11.9%	4039	4040	1985		
Volkswagen e-Up!		5056	6.0%	1507	2971	578		
BMW i3		4494	5.3%	2403	2040	51		
Mitsubishi Outlander P-HEV		4363	5.2%	2875	1485	3		
Kia Soul EV	Combined	3355	4.0%	2064	445			
	Only new	1311		866	445			
	Used imports	2044	17.5%	2044	NA			
Mitsubishi i-MiEV		3077	3.6%	490	413	453	671	1050
Renault Zoe		2071	2.5%	1634	433	4		
Volkswagen Golf GTE		2000	2.4%	2000				

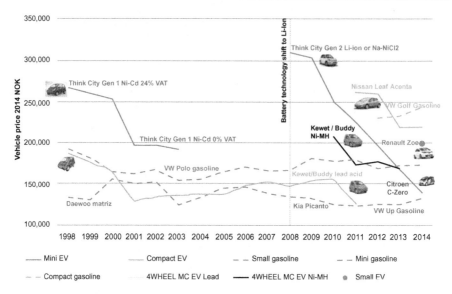

Fig. 9.7.2 The sales price statics of BEVs in 2015 NOK (without all taxes according to the Norwegian incentives) compared with ICE vehicles (including registration taxes and VAT). 2014 currency rate, 1€ = 8.35 NOK.
Source: Figenbaum and Kolbenstvedt, 2015 (TØI Report 1420/2015) (Figenbaum, E., Kolbenstvedt, M. (2015b). *Pathways to electromobility—Perspectives based on Norwegian experiences*. Oslo,Norwa: Institute of Transport Economics (TØI), TØI Report 1420/2015).

vehicle industry in Norway was started manufacturing the electric vehicle 'Think' as the Eureka project partially funded by the Research Council of Norway and EU funds, with contributors and companies in Norway. The Think City electric vehicle was planned to launch for the end of 1998, but it went bankrupt the same year. The Ford Motor Company (the United States) was purchased the company and developed and launched the Think City vehicle in 1999. Think had a patent for the molded plastic body plates that manufactured locally at the factory in Aurskog. In the middle of 2002, the company was put up for sale and Indian businessman purchased the company, but Think went into bankrupt again. In 2006, Norwegian investors purchased and launched the vehicles in 2008 with the greatly delay due to financial problems. The original Norwegian electric-car maker, Think was in serious financial difficulties, and the company again went into bankrupt for the fourth time in 20 years. Hence, in 2011, a Russian inventor bought, and the company was moved to Germany. A total of 2500 Think vehicles had been manufactured at Oslo-based production facility (Figenbaum & Kolbenstvedt, 2013).Other smaller companies were also involved the EV production industry in Norway. Kollega Bill-Elbil Norge-Buddy Electric, the company has had many names and produced Kewet and Buddy development project and gone into bankruptcy in several times. Recently, it has started up again and registered the Buddy as a four-wheel motor cycle for local transport in urban areas. Norwegian vehicle parts production industries, Kongsberg Automotive, and Eltek supplied the parts to 'Think' in the initial phase. Eltek has a contract to deliver chargers to Volvo's V60 plug-in hybrid, while Kongsberg Automotive delivers, among other things, gear selectors and other components for several electric vehicles being marketed by the large car manufacturers (Figenbaum & Kolbenstvedt, 2013). Mostly in Norway, different companies have started EV imports and sales. Since 2011, the established vehicle manufacturers have completely taken over the EV market in Norway (described in Section 9.7).

9.9   Government impacts of increasing electromobility

On the focus of electromobility, EVs convert the main impact of transportation sector in Norway rather than using ICE vehicles. The environmental friendliness of EVs and attractive incentives are clearly emerging as an important feature when considering a purchase. EVs are much cheaper to operate than ICE vehicles in Norway due to low-priced electricity. Further, the incentives on taxes and attractive exemptions are leading to impact on the EV market.

9.9.1   Environmental impacts

The EVs are powered by electricity, and there is no contribution to tailpipe emissions such as IECs. The effect of local air quality will not be large until EVs are a main part of the vehicle fleet in Norway. COMPETT has studied and measured the noise impact of EVs and concluded that the noise of EVs are 2–5 dB lower than of ICE vehicles at speeds below 30 km/h, depending on traffic conditions (Figenbaum, 2015). The noise deduction is very small compared with the IEC, and there is no need to impact on policies to reduce the noise effect in cities. The sustainable and cleaner energy in transportation has attracted a worthy attention in Norwegian government and EV consumers are encouraged with new policies and tax exemptions. The reduction of CO_2 emissions is the main advantage of EVs rather than using ICE vehicles. The two broad political agreements (2008 and 2012) on climate in Norwegian parliament have agreed to protect existing incentives until 2018 or until 50,000 EVs reached. It is reported that the average emission in 2015 is 99 g/CO_2/km and the most important goal from climate agreement is 85 g/CO_2/km in 2020 (Berger-Røsland (Governing Mayor of Oslo), 2015).

9.9.2   Economic impacts

The incentives are increased the usage of EVs to become a profitable satisfaction consumer. The large manufacturing price reduction in Norway market has been seen in since 2008, developing a competitive market to BEVs with the ICE vehicles (Figenbaum, 2015). Time and money savings due to the incentives such as reduction of taxes, free access to parking, and bus lanes as mention in Table 9.3.1 contribute the economic advantage to EV consumer side.

Moreover, all these EV incentives will influence the government budgets either directly or indirectly. The budget cost-effectiveness of the incentives refers to the

Table 9.9.1 **Government cost, market impact, and cost per BEV of incentives in 2020**

BEV policy	Effect, number of BEVs	Budget effect ('cost'), NOK millions	Cost per BEV ('cost-effectiveness'), NOK
VAT exemption only	10,102	527	52,143
Road charges only (toll roads or similar)	2949	186	63,021
Free parking only	1882	171	90,719
Annual tax only	4240	82	19,305
Purchase tax only	20,101	1514	75,332
VAT and purchase tax	35,700	3340	93,546
Bus lane access only	18,255	55	3025
All incentives combined	77,335	6563	84,861

NOK and number of BEVs, 2014 currency rate. 1 Euro: 8.35 NOK.
Source: Figenbaum and Kolbenstvedt, 2015 (TØI Report 1422/2015) (Figenbaum, E., & Kolbenstvedt, M. (2015a).
Competitive Electric Town Transport Main results from COMPETT—An Electromobility+ project. Oslo: Institute of Transport Economics, Norwegian Centre for Transport Research, TØI Report 1422/2015).

effect on the market take up relative to public budget costs. The Table 9.9.1 indicates the budget effect and budget cost effectiveness of the incentives.

The effective incentives should be improved, and ineffective incentives should be removed to enhance the quality. In addition, purchase incentives on EVs result in large loss of budget tax income, and it can be eliminated by increasing taxes on IEC vehicles and fossil fuels and by reducing incentives when sales mature and production costs go down.

9.10   Norwegian transport in future: a vision for low carbon society

The Energy and Climate Dialog in Norway in 2014 described a vision for Norway as a low-carbon society in 2050. The framework conditions and incentives are focused to have almost free greenhouse gas emissions, with efficient energy use and a high proportion of renewables. The petrol and diesel ICE vehicles are no longer use in 2050, and the number of personal car has fallen sharply to achieve the low carbon in society. The dialog meetings took place between Nov. 2013 and Feb. 2015 were summarized that the society in 2050 is fully organized around walking, cycling, and using of electric bicycles and other small, electric, single-person vehicles. In addition, public transport and self-driving taxis have become so cheap and available. The improved technologies of batteries in EVs have also become an alternative for long distances and short charging times. Direct emissions of greenhouse gases and other environmentally harmful gases have been eliminated, since Norway has become the first country in the world to have all cars in electrified, and the proportion of zero emission vehicles has increased to 100% (Engen et al., 2015a, 2015b). Moreover, in the Paris agreement 2016, Norway plans to bring forward its target for reaching carbon neutrality to 2030.

9.11 Summary and conclusion

Norway has the highest EVs per capita. In Norway, it is estimated to have more than 1 million EVs (nearly 28% EV market share) by the end of 2016. The Norwegian EV market share in 2016 has reached to 100 times compared with 2010. The development of electric mobility across all transport modes is necessary to meet the sustainable targets. The EVs are well suited for reducing the greenhouse gas emissions and improving the air quality and noise level. Norway has ratified the Paris agreement on climate change in Jun. 2016 to bring forward the target for reaching carbon neutrality by 2030, two decades sooner than the current 2015 goal.

   The government of Norway has motivated the EV usage for substantial reduction in the GHG emissions from its transportation sector. The clean and environmentally friendly electricity generation in Norway creates low-cost electricity, and it has encouraged EV users to become economically smart and environment friendly. The main reason for Norway's success for increasing penetration of the EVs is due to the combination of financial and convenience incentives policies of the Norwegian government for EV owners. Since 1990, Norwegian EV policies and incentives have been developed to make attractive EV market and to make the best place on the world market in terms of the number of EVs to the population. Economic incentives such as exemption from vehicle taxes (VAT and registration tax) have secured the possibility to sell EV competitively. Most of the 27 electric cars entering the Norwegian EV market are BEVs, and it is due to the Norwegian government incentives of the large exemption from registration taxes and VAT to BEVs over PHEVs. The successful nonfiscal incentives, free parking and charging and access for EVs to bus lanes has begun to progress the EV usage in country. Modifying and introducing new incentives triggered reduction on EV prices and inflamed the EV purchasing rate in Norway. The well functional charging schemes have introduced to increase vehicle range and charging challenges reported in EV owners. Thus, the scenario of EVs in Norway remains on top as among one of the world's most important countries to follow within e-mobility.

Bibliography

Aasness, M. A., & Odeck, J. (2015). The increase of electrical vehicle usage in Norway—Incentives and adverse effects. *European Transport Research Review*, 7–34.

Argueta, R. (2010). *A technical research report: The electric vehicle.* CA: University of California Santa Barbara, College of Engineering.

Bekker, H., 2013. 2012 (Full Year) Norway: Best-selling electric car models. Available at: http://www.best-selling-cars.com/electric/2012-full-year-norway-best-selling-electric-car-models/ Accessed 21.02.17.

S. Berger-Røsland (Governing Mayor of Oslo). *Breakthrough for electrical vehicles.* Oslo: s.n. 2015.

EAFO (European Alternative Fuels Observatory), 2016. Available at: http://www.eafo.eu/content/norway

Engen, S., et al. (Eds.), (2015). *Norway 2050: A vision for a low-carbon society.* Oslo: The Bellona Foundation.

Engen, S., Amundsen, T., Bratland, A., & Barkved, L. (Eds.), (2015). *Norway 2050: a vision for a low-carbon society.* Oslo: The Bellona Foundation.

Fearnley, N., Pfaffenbichler, P., Figenbaum, E., & Jellinek, R. (2015). *E-vehicle policies and incentives: assessment and recommendations.* Oslo: Compett deliverable D5.1 TØI report 1421/2015.

Figenbaum, E., 2015. *COMPETT—COMPetitive Electric Town Transport.* Oslo, s.n.

Figenbaum, E., Fearnley, N., Pfaffenbichler, P., & Hjorthol, R. (2015). Increasing the competitiveness of e-vehicles in Europe. *European Transport Research Review, 2015*, 7–28.

Figenbaum, E., & Kolbenstvedt, M. (2013). *Electromobility in Norway–Experiences and opportunities with electric vehicles.* Oslo: TØI Report.

Figenbaum, E., & Kolbenstvedt, M. (2016). *Learning from Norwegian battery electric and plug-in hybrid vehicle users—Results from a survey of vehicle owners.* Oslo: Institute of Transport Economics, Norwegian Centre for Transport Research. TØI Report 1492/2016.

Figenbaum, E., & Kolbenstvedt, M. (2015a). *Competitive Electric Town Transport Main results from COMPETT—An Electromobility+ project.* Oslo: Institute of Transport Economics, Norwegian Centre for Transport Research. TØI Report 1422/2015.

Figenbaum, E., & Kolbenstvedt, M. (2015b). *Pathways to electromobility—Perspectives based on Norwegian experiences, 1420/2015.* Oslo, Norway: Institute of Transport Economics (TØI). TØI Report.

Haugneland, P., Bu, C., & Hauge, E. (2016). *The Norwegian EV success continues.* Montréal, QC: EVS29 Symposium, Norwegian Electric Vehicle Association.

Jong, R. d., Åhman, M., Jacobs, R., & Dumit, E. (2009). *Hybrid electric vehicles: And overview of current technology and its application in developing and transitional countries.* Kenya: UNEP.

Lozano, A. P., 2012. Intelligent energy management of electric vehicles in distribution systems. s.l.: Aalborg University.

Moberg, K., n.d.-a *44 prosent av oss vil ha ladbar bil (44 percent of us will have rechargeable car).* Available at: http://www.dinside.no/motor/44-prosent-av-oss-vil-ha-ladbar-bil/61186084 Accessed 10.03.16.

Moberg, K., n.d.-b *Bilsalget i 2015: Tidenes bronseplass (Car sales in 2015: Ages bronze).* Available at: http://www.dinside.no/motor/bilsalget-i-2015-tidenes-bronseplass/60992664 Accessed 05.03.16.

Norwegan Center for Transport Research TEI. (2015). *Environment and Climate. Electromobility conference in Oslo June* Available at: https://www.toi.no/electromobility-conference-in-oslo-june-11-12-2015/category1546.html.

OFVAS (2015). Brand, segment and EV sales statistics from OFVAS, Available at: http://www.ofvas.no/bilsalget/category404.html

Tietge, U., Mock, P., Lutsey, N., & Campestri, A. (2016). *Comparison of leading electric vehicle policy and deployment in europe.* Berlin: International Council on Clean Transportation.

Torper, P. A. (2015). *Norwegian electromobility policy for 2020.* Oslo: s.n.

Further Reading

European Alternative Fuels Observatory (EAFO). http://www.eafo.eu/content/norway
    Accessed on 2016/08/31.
Ministry of Climate and Environment, Government of Norway, Norway has ratified the Paris
    Agreement' date 2016-06-21. https://www.regjeringen.no/en/aktuelt/norge-har-ratifisert-
    parisavtalen/id2505365/.
New in Norway, New Edition 2017, The Norwegian Directorate of Integration and Diversity.
    ISBN: 978-82-8246-166-5.
The Norwegian Electric Vehicle Association (Norsk Elbilforening) (2016). The Norwegian
    Charging Station Database for Electromobility (www.nobil.no).

A case study for Northern Europe 10

Aisling Doyle, Tariq Muneer
Edinburgh Napier University, Edinburgh, Scotland, United Kingdom

10.1   Introduction

This chapter will present a review of the automobile industry in Northern Europe focusing on the United Kingdom as a case study. To fully understand where the electric vehicle stands in this industry, it is important to consider the factors that led to its return to the vehicle market. The passenger vehicle was invented in 1801 by Richard Trevithick and was powered by steam. However, the steam-powered passenger carriage never out-performed horse-powered carriages, and it wasn't until 1834 when Thomas Davenport invented the first direct-current electric motor that the vehicle industry was introduced to its first game changer. In the years that followed the Scotsman, Robert Anderson invented the first battery-powered electric vehicle but the fall of the electric automobile was a result of three major reasons. The range of the battery was limited, it was a technology that was not very flexible to compete with Henry Ford's production line in 1909 and so did not benefit from mass production, and finally in 1911, Charles Kettering's internal combustion vehicle starter motor was the ultimate blow that led to the extinction of the electric vehicle in the 1930s.

So why has the electric vehicle found its place back in the automobile market? This chapter will discuss the various energetic, environmental, and economic factors on a global and national scale (United Kingdom) to give some insight to why the electric automobile is competing with the conventional internal combustion as a solution to factors aforementioned.

10.2   A review of energy demand

10.2.1   Global energy consumption

Energy is needed to carry out day-to-day activities in both developed and developing countries. Activities may include agricultural and fishery, domestic and industrial, and electricity production and transport. Fig. 10.2.1 shows the global exponential growth of energy consumption since 1971. The cause for this growth may be linked to a few factors, some of which may be loosely linked to the growth in population across the globe. Scientists report that extreme weather condition is an effect of climate change. Heating and cooling of buildings requires a large quantity of energy all year-round, thus resulting in increased energy usage. The freedom people have acquired to travel is also a major factor that has contributed to the propulsion of this trend.

Electric Vehicles: Prospects and Challenges. http://dx.doi.org/10.1016/B978-0-12-803021-9.00010-0

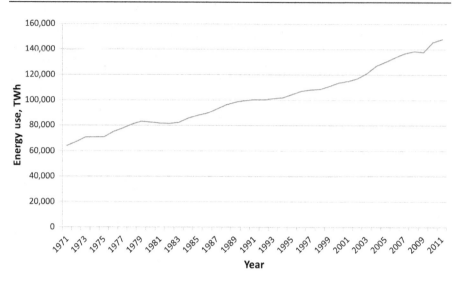

Fig. 10.2.1 Global energy use (World Bank, 2014a).

10.2.2 European Union's energy consumption

The EU has seen a general increase in energy consumption. As this continues to rise, legislations have been implemented to stabilize this demand to a manageable level. Fig. 10.2.2 shows that unlike the global demand that is continually increasing, the EU trend illustrates periods when the demand decreased from the previous year. The most recent decline in demand is between the period 2007 and 2009 that saw

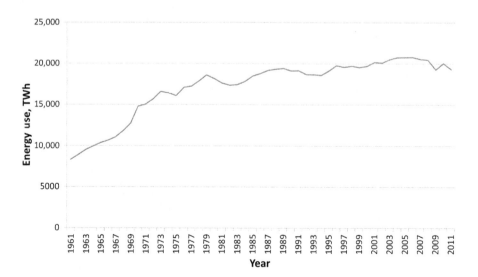

Fig. 10.2.2 European Union energy use (World Bank, 2014a).

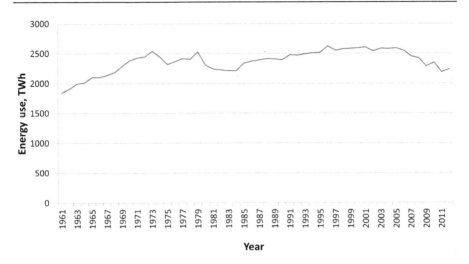

Fig. 10.2.3 UK energy use (World Bank, 2014a).

a decrease in energy demand 1400 TWh in 2009 than in 2006. That represents a 7% decrease in energy usage for that period in the EU. However, one may also comment that the economic recession experienced throughout the EU member states in this time frame was a contributor to this decline in energy use. As economies recover, this is reflected in the amount energy consumed.

10.2.3   UK energy consumption

Fig. 10.2.3 shows the energy usage trend for the United Kingdom. From 2002, the trend begins to decrease with a few exceptions. The largest drop in demand was seen in 2011 as a result of the increase in energy usage experienced in 2010. This increase could be a result of the harsh weather conditions experienced throughout the United Kingdom in 2010 during the winter months.

To put each region into context, Fig. 10.2.4 shows the global, EU, and UK energy usage trends since 1961. This log-scaled graph can allow us to put the units into global perspective and see how they have increased with each other comparably.

10.2.4   Energy consumption by sector

In 2011, global energy consumption reached 153,569 TWh, as seen in Fig. 10.2.5 (U.S. EIA, 2015). The sector that consumed the most energy in 2011 was the industrial sector followed by the transport and then the residential sector.

This trend should not be considered a general template of energy distribution for every country as the economic structure of a particular country, be it developed or developing, will vary greatly. What can be taken from the information presented is that worldwide transport is one of the major energy consumers and the sustainability

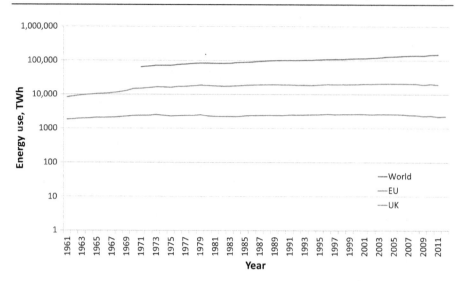

Fig. 10.2.4 Global, European Union, and UK energy use (World Bank, 2014a).

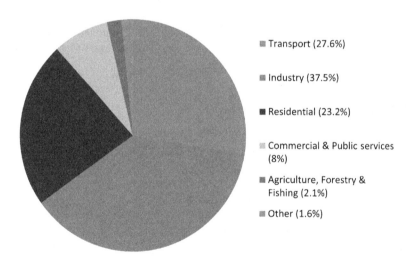

Fig. 10.2.5 Global energy consumption by sector 2011 (IEA, 2013).

of a robust transport network is essential should energy demand increase as mentioned in Section 10.2.1. Furthermore, it would be highly desirable to sever the link between the latterly mentioned energy usage and carbon emission. This factor goes a long way in the support of electric propulsion as electricity may now be sourced from sustainable means at a monetary cost that is dropping sharply.

Looking at Europe closer, Fig. 10.2.6 shows that similarly the three major energy consumers are industrial, transport, and residential. However, unlike globally where

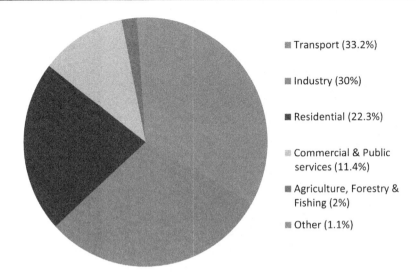

Fig. 10.2.6 EU27 share of final energy consumption by sector in 2011 (IEA, 2013).

industry was the prominent energy consumer, the transport sector utilizes the most energy in the EU27 countries.

The United Kingdom follows similar trends to that of European countries with the transport sector utilizing the most energy of any other sector as seen in Fig. 10.2.7. In 2013, over 17 TWh of energy was consumed in the United Kingdom, predominantly in the transport and domestic sectors.

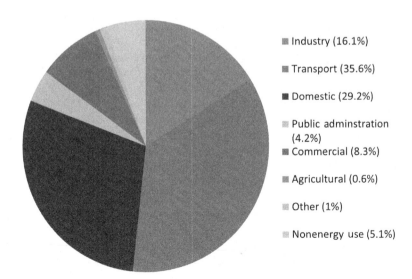

Fig. 10.2.7 UK share of energy consumption by sector in 2013 (DECC, 2014).

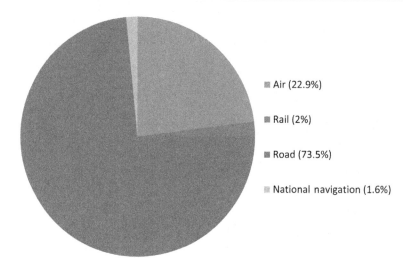

Air (22.9%)

Rail (2%)

Road (73.5%)

National navigation (1.6%)

Fig. 10.2.8 Energy distribution in the UK transport sector in 2013 (DECC, 2014).

10.2.5   Energy used in the transport sector

As stated by the Department of Energy and Climate Change (DECC), the transport sector consumes 36% of the total energy consumed in the United Kingdom. To break down this energy consumption further within the transport sector, Fig. 10.2.8 illustrates that most of the energy is consumed by on-road vehicles.

10.2.6   The composition of on road vehicles

The amount of energy consumed by road vehicles has increased both globally and in the United Kingdom as seen in Fig. 10.2.9. People's freedom to travel more and travel longer distances has resulted in increased energy consumption. Fig. 10.2.10 shows how energy consumption from on-road vehicles in the United Kingdom has increased since the 1960s. A dramatic increase in the consumption of energy was experienced in the United Kingdom until more recently. After 2007, there is a drop in energy consumption. This is a result of improvements in vehicle energy efficiency. Further improvements by automobile manufacturers to meet government legislation will see this trend continue to fall as on-road vehicles become more energy efficient.

Fig. 10.2.11 shows how the energy used in the transport sector has varied since 1961 as a percentage of total energy consumption. Across the globe, transport has always required around 14%–15% of total energy supply. Both the EU and UK values show that the percentage of energy required for the transport sector continues.

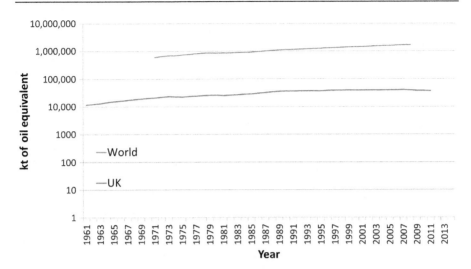

Fig. 10.2.9 Energy consumption by road transport globally and in the United Kingdom (World Bank, 2014c). Note: 1 ktoe = 11.63 GWh.

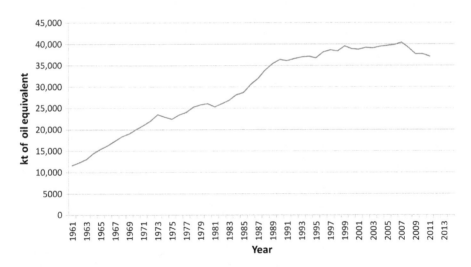

Fig. 10.2.10 Energy consumption by road transport in the United Kingdom (World Bank, 2014c). Note: 1 ktoe = 11.63 GWh.

The transport sector can consume up to 20% of total energy usage in the United Kingdom. Since 1970, the United Kingdom has required more energy in the transport sector than energy used in the EU transport sector as a percentage of total energy usage.

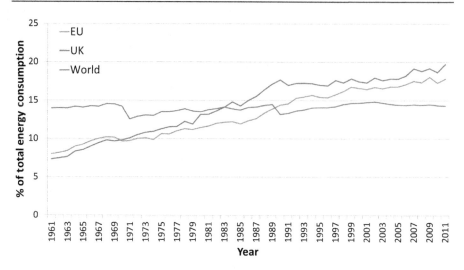

Fig. 10.2.11 Energy consumption in the transport sector as a percentage of total energy consumption (World Bank, 2014c).

10.3   Environmental review

10.3.1   General considerations

Climate change is a bone of contention in the scientific community and the media. Sustainable energy is a topical conversation that has caught the attention of all stakeholders, especially governments who recognize that it is essential to address the issue in order to reduce further damage to the environment. Anthropogenic climate change activities and how they can be managed to mitigate further pollution are to the fore of these discussions. This section will present material to chronologically relate these issues and effects relating to climate change and environmental sustainability.

Fig. 10.3.1 shows the anomalous behaviour of global temperature change. This figure illustrates the influence of the latter part of the industrial revolution on the earth's ambient temperature when significant carbon loading of the planet had ensued. Fig. 10.3.2 shows the exponential growth in carbon dioxide (CO_2) concentration in the atmosphere. It can be said that Fig. 10.3.3 is loosely connected to the latter presented. As the human population is globally and nationally increasing, it can be assumed that human activity will subsequently increase and have an undesirable effect on CO_2 emissions and global temperature trends.

10.3.2   CO₂ emissions

10.3.2.1   Global CO₂ emissions by sector

The automobile industry is primarily dependent on the burning of fossil fuels, and the release of these CO_2 emissions will continue to contribute to the adverse effects on global temperature. This vicious cycle will continue unless an infrastructure is put

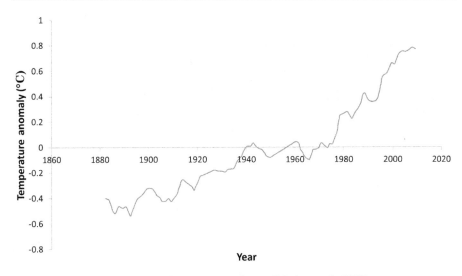

Fig. 10.3.1 Chronology of global temperature change (Morice et al., 2012).

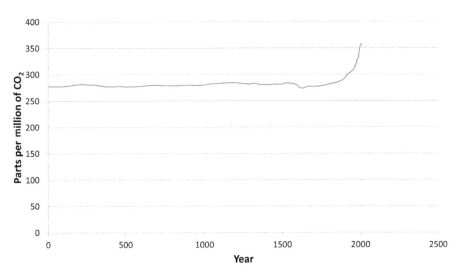

Fig. 10.3.2 Chronology of global atmospheric CO_2 concentration (IPCC, 2007).

in place to stabilize temperature before they reach unbearably high levels. There is a strong correlation between energy demand and CO_2 emissions. As discussed previously, energy demands are rapidly increasing and so CO_2 levels will increase respectively unless action is taken to limit these greenhouse gases (GHG). Fig. 10.3.4 illustrates the contribution of each economic sector to CO_2 emissions. The power sector responsible for the production of electricity and heat is the main contributor of this harmful gas. In the year 2012, transport was responsible for 23% of global CO_2

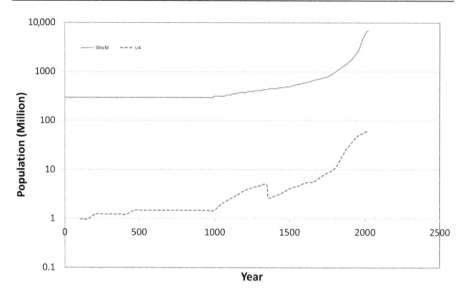

Fig. 10.3.3 Human population increase (Emmott, 2013; World Bank, 2015b).

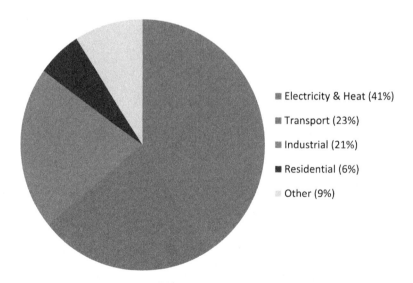

Fig. 10.3.4 World CO_2 emissions by sector in 2012 (IEA, 2014a, 2014b).

emissions; see Fig. 10.3.4 (IEA, 2014a, 2014b). Similar to energy consumption, it cannot be assumed that these figures are adaptable for every country worldwide as different countries with different economic structures will have different CO_2 levels depending on anthropogenic activities in that specific country. However, it is clear that on a global scale, transport is a significant contributor with emission levels higher than that from industrial and residential activities.

10.3.2.2 European CO_2 emissions by sector

Looking at the EU levels of CO_2 by sector published by the World Bank (2014d) for the year 2011 (see Fig. 10.3.5), transport is the second highest contributor of CO_2 emissions with the energy sector being the main culprit.

10.3.2.3 UK CO_2 emissions by sector

Looking at more localized figures, the DECC published data illustrating that almost 27% of CO_2 emissions in the United Kingdom are from the transport sector (see Fig. 10.3.6). It is a common trend that at a larger or global scale right to a national scale

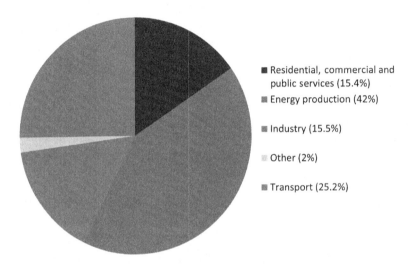

Fig. 10.3.5 CO_2 emissions by sector for EU in 2011 (World Bank, 2014b).

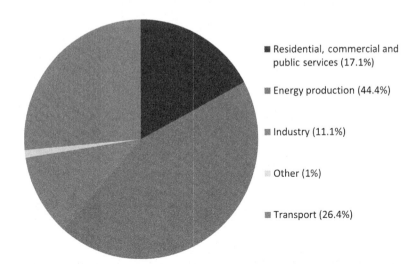

Fig. 10.3.6 CO_2 emissions by sector in the United Kingdom in 2014 (DECC, 2014).

in the United Kingdom, the main activities influencing CO_2 levels are consistently the energy sector followed by transport. In order to manage these harmful gases to a safe and sustainable level, it is essential that the focus of research, government policies, and developments are on the energy production, transport, and industrial sectors.

10.3.3 CO_2 and its relationship with rise in sea levels

With rising ambient temperature, it is no surprise ocean temperatures will rise respectively. Levitus et al. (2012) presented the thermal loading of the sea with 1980 as its baseline. The heat content has dramatically increased since the 1980s and continues to rise. Fig. 10.3.7 should be read in conjunction with Fig. 10.3.8 that demonstrates a sharp decline of solubility of CO_2 in sea water should ocean temperatures increase. CO_2 can be stored in three different mediums: the atmosphere, ocean, and land biospheres. The gas can move easily from ocean to atmosphere or likewise between the atmosphere and land biosphere. The ocean is the main regulator of CO_2 as approximately 93% of CO_2 can be found in ocean waters (Water Encyclopedia, 2015). When dissolved in water, the greenhouse gas can no longer trap heat. However, should ocean temperatures continue to rise, more CO_2 will be released to the atmosphere. This will have a knock-on effect that will trap heat and increase ambient temperature causing ocean temperatures to rise, which reduces the oceans ability to act as a proficient sequester of the harmful gas.

This characteristic should be a major consideration in nuclear power waste heat management. Nuclear power is subject to criticism in the scientific world as to whether it should be characterized as a renewable source of energy or not. A considerable amount of heat is produced as a by-product of nuclear energy. Once the process of cooling has taken place, this water is safely released into the ocean. Poor management of nuclear power plants and their waste heat could result in elevated levels of unnecessary CO_2 emissions being released into the atmosphere. From Figs. 10.2.10 to 10.3.8, the linear relationship between 0°C and 20°C suggests that

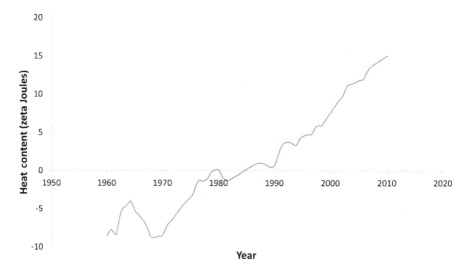

Fig. 10.3.7 Thermal loading of sea. Note: 1 zetta joule $= 10^{21}$ joules (Levitus et al., 2012).

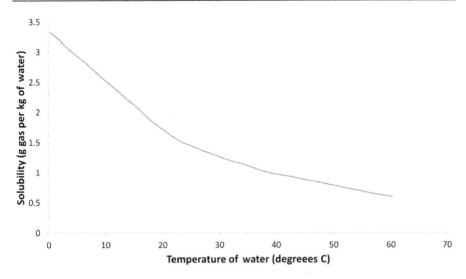

Fig. 10.3.8 Solubility plot for CO_2 in water (Engineering Toolbox, 2015).

should the ocean temperature increase by 1°C, this would result in 0.0835 g of CO_2 per kg of water to be released into the atmosphere. With 1335×10^{12} kg of water in oceans worldwide (Eakins & Sharman, 2010), should ocean temperatures increase by 1°C globally, 1.11×10^{14} g of CO_2 will be released into the atmosphere. On Mar. 5, 2015, CO_2 in the atmosphere was 400.14 ppm globally in the atmosphere (ESRL, 2015). The total mean mass of the atmosphere is given as 5.15×1018 kg (Trenberth & Smith, 2005). A ratio of 0.054 was calculated to represent the increase of CO_2 emissions for every degree of temperature increase.

10.3.4   Effects of aerosols on climate

To date, the focus of most government policies and strategies is to reduce CO_2 emissions that contribute to climate change. A study by Unger et al. (2009) presented a very interesting review on climate change attributes. This study discussed the symbiotic relationship that exists with GHG and aerosols. GHG are harmful gases that contribute to climate change and aerosols are on the contrary cooling the atmosphere. The latter study organizes these emissions and displays them in terms of their economic sectors. The emissions include ozone, sulphate, nitrate, black carbon, organic carbon, aerosol indirect effect, methane, nitrous oxide, and carbon dioxide. This study aids policy makers to consider a more holistic approach when considering the mitigation of harmful gases for short-term planning for year 2020 based on the year 2000 values. Unger et al. acknowledge that the impacts these potential policies will have on mitigating such emissions may take decades to be evident due to the long lifetime of various gases. The data presented in Fig. 10.3.9 are in relation to the sectors net value for radiative forcing for the near 2020 predicted values. Radiative forcing measures the amount of gas or any other forcing agent that throws the energy balance in the atmosphere and whereby contributes to climate change.

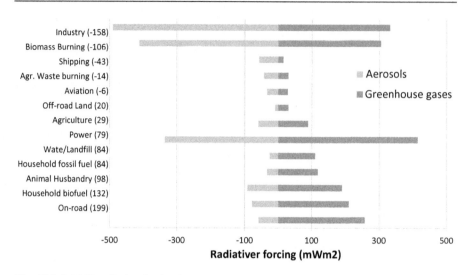

Fig. 10.3.9 2020 radiative forcing by sector (Unger et al., 2010a).

The above study does not solely focus on carbon dioxide, the dominant contributor to climate change, as most often it is the focus of environmental impact studies but it includes all factors that contribute to increasing earth temperatures and also cooling agents. As mentioned previously, the power sector is the main culprit of CO_2 emissions. Fig. 10.3.9 illustrates that although the power sector emits the most GHG, its aerosol production is reasonably high to counteract the harmful effects of GHG alone. Unger et al. predict that in 2020, the power sector will be only the sixth main contributor to radiative forcing with 79 mW m^{-2} of radiative force being produced. This figure allows effective policy making to address the sectors most in need of attention to successfully mitigate appropriate forcing factors to combat the complex issue of global temperatures rising.

Unger et al. illustrate that on-road pollution requires the most attention to mitigate harmful gases in 2020. As it stands in 2020, on-road emissions have a ratio of 1:4 in terms of cooling and warming agents, respectively. The power sector's warming agents are only 1.2 times greater than its cooling agents (Unger et al., 2010b). One might say that the electric vehicle is not a true solution to climate change as it is only shifting emissions from the transport sector to the energy production or power sector. However, the potential that the power sector has to counteract the warming effects with aerosol production is far greater than that of the transport sector. This being said, it is vital that the power sector continue to make strides towards introducing more and more renewable energy to the power sectors network to reduce harmful emissions.

Fig. 10.3.10 gives a projection of how three major economic sector's radiative forces are trending towards the year 2100. This figure shows the evolution of the total radiative forces from the sum of all agents and how they change with time. Since 2000, on-road emissions have always had a surplus of warming agents over aerosols.

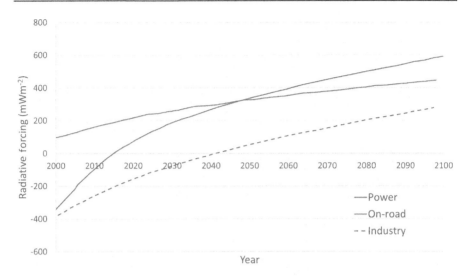

Fig. 10.3.10 The future evolution of radiative forcing in the economic sectors: power, on-road, and industry (Unger et al., 2010b).

By 2020, the power sector has just moved out of the negative (or predominately aerosol production) into a surplus of warming agents. Similar trends are seen in the industry sector moving into increasing levels of warming forces. Although the on-road sector is to have a radiative forcing of almost 200 mWm^{-2}, the highest of any other economic sector, the power sector is soon to surpass and become the most dominant radiative forcing contributor by 2050 (Unger et al., 2010b).

10.3.5 CO_2 emissions within the transport sector

The *Railway Handbook 2014* breaks down the transport sector further into navigation, aviation, rail, road, and other modes of transport in relation to CO_2 emissions. According to this document published by the IEA, road transport is by far the largest contributor of CO_2 emissions with almost 74% and 72% contribution on a global and EU27 scale, respectively; see Figs. 10.3.11 and 10.3.12, respectively.

10.3.6 CO_2 in the automobile industry

The information presented in Section 10.3.5 illustrates the importance of developing the transport sector in order to reduce the energy demand and dependency on fossil fuels but also in order to reduce CO_2 emissions. In particular, the on-road transport is of particular interest as this sector is the mode of preference chosen as mentioned in Section 10.3.5. In Fig. 10.3.13, improvements in the on-road sector focusing on private automobiles and taxis in the United Kingdom can be seen. Environmental awareness created by the scientific community and affirmed by the public and media

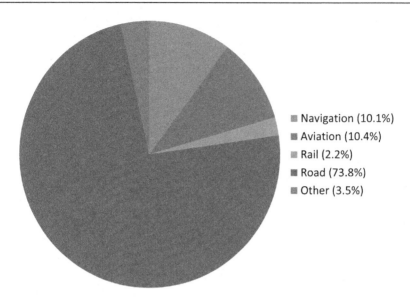

Fig. 10.3.11 Global CO_2 emissions within the transport sector in 2011 (IEA, 2014b).

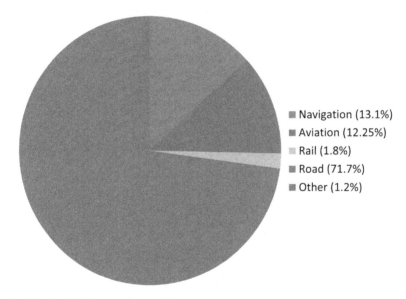

Fig. 10.3.12 EU 27 CO_2 emissions within the transport sector in 2011 (IEA, 2014b).

has influenced the manufacturers of automobiles to consider the energy and environmental efficiency of their product at the planning and production stages. Their improvements are illustrated in Fig. 10.3.14 that shows the g/km of CO_2 emitted by a new car in 2012 is around 130 g/km, a 50 g/km reduction from that of a new car in 2000.

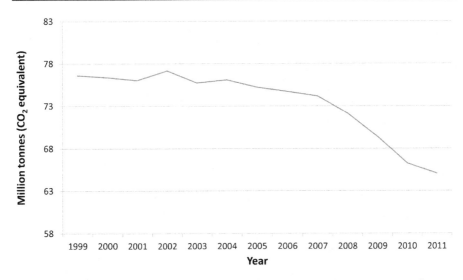

Fig. 10.3.13 CO_2 emissions from car and taxis in the United Kingdom (Department for Transport, 2013a).

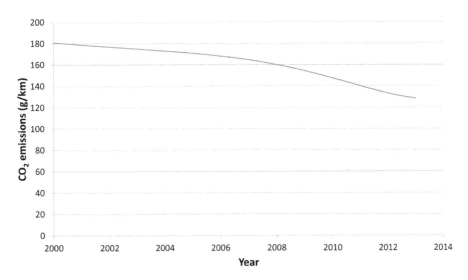

Fig. 10.3.14 Average new car CO_2 emissions (Department for Transport, 2013).

10.4 Economic review

10.4.1 A review of automobile usage

Figs. 10.4.1 and 10.4.2 can be viewed together to show the trend in modern lifestyle and how the automobile is a part of the daily routine rather than a luxury. The number of cars registered on the roads has increased throughout the world and also in the

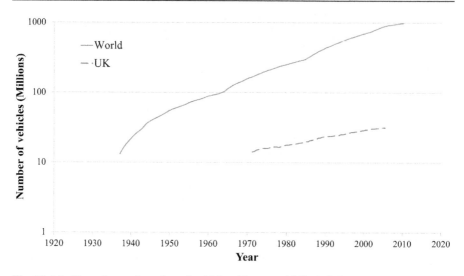

Fig. 10.4.1 Chronology of number of vehicles (Emmott, 2013; Leibling, 2008).

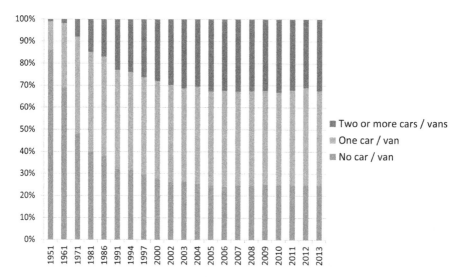

Fig. 10.4.2 Household car availability in England from 1951 to 2013 (Department for Transport, 2016).

United Kingdom. A study carried out by the UK Department for Transport shows the nature of automobile use has dramatically changed since 1951. Fig. 10.4.2 illustrates how in 1951 around 85% of households in England didn't have access to a car and virtually almost no one had access to two or more cars or vans. In 2013, almost the reverse is seen as around 75% of households in England have access to an automobile. Over 30% of households have access to two or more automobiles in 2013.

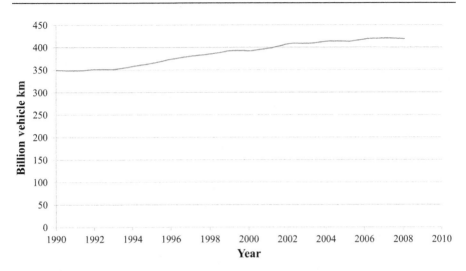

Fig. 10.4.3 Total usage of automobiles in the United Kingdom (UK Committee on Climate Change, 2010).

Vehicles have become a way or life or 'essential' in today's society as the number of vehicles registered on the road continues to rise along with more households gaining access to a vehicle.

Along with the number of vehicles on the road. it is important to look at the usage of vehicle in the United Kingdom. The Committee of Climate Change (CCC) has investigated how many kilometres are undertaken in the United Kingdom by automobiles. Fig. 10.4.3 illustrates the increase of kilometres travelled on United Kingdom roads. Between 1990 and 2008, the distance travelled in the United Kingdom increased by 20% from 350 billion to 418 billion vehicle kilometres. As the distance travelled increases, it is interesting to compare this figure with the environmental figures aforementioned. In Section 10.3, Fig. 10.3.13 illustrates the that decline in CO_2 emissions in recent years from private cars and taxis demonstrates the success of research and development of the automobile industry to work towards effective efficiency while battling with increased usage of the automobile but also promoting the product.

10.4.2   Cost of conventional fuel

Fossil fuel is a topical conversation in the media and scientific community as its depletion is forcing alternative power sources to flourish. Data presented by the World Bank demonstrate that the strain on diesel and gas resources is reflected in their market prices. The increase in prices has been experienced across the globe and presented in Fig. 10.4.4 In the United Kingdom, fuel prices have increased from under 1 US dollar in 1998 to around 2.2 US dollars in 15 years. The automobile that has become an essential part of one's daily routine is becoming more and more expensive. The fossil fuel market is very unstable, and should these trends continue, it can be said that the automobile may become too expensive or even unreliable for some households.

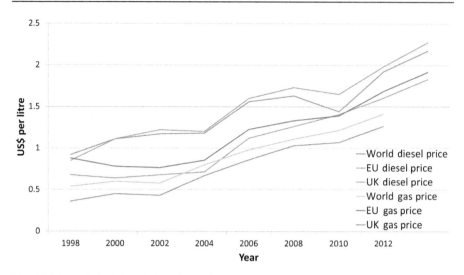

Fig. 10.4.4 Global, EU, and UK prices for diesel and gas (World Bank, 2014d).

10.4.3   The composition of electricity generation

The integrity of the EV evolution lies with the distribution network and its composition. As discussed in Chapter 2, a well-to-wheel analysis of the EV illustrates that depending on the source from which the EV's battery is charged will define if the vehicle is truly 'zero emission'.

Fig. 10.4.5 illustrated the gradual increasing trend of electricity production in the United Kingdom from 1990 to 1999. CO_2 emissions from the same period saw a

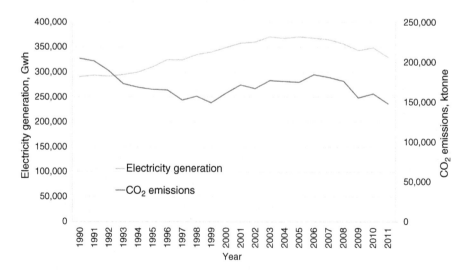

Fig. 10.4.5 Electricity generation and CO_2 emissions in the United Kingdom (DEFRA, 2013).

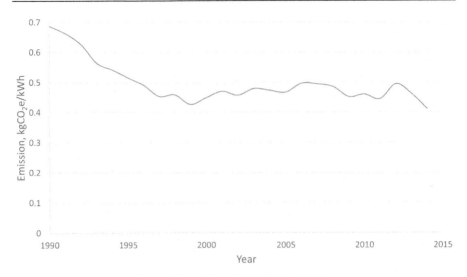

Fig. 10.4.6 CO_2 reduction in UK electricity grid (DEFRA, 2016).

dramatic fall despite the strain to produce more electricity. From 1999 to 2006, CO_2 emission in the electricity sector increased as generation increased. As generation gradually reduced in 2008, the CO_2 emissions from the period 2008 to 2009 saw a dramatic fall. In the anticipation that EV's will penetrate the transport market, the distribution system must be prepared for an increase in electricity production. However, it is essential to be aware that CO_2 emissions will naturally increase should the distribution network stay in its current position and so the composition of electricity generation should be to the fore of the planning stage of any EV project. Fig. 10.4.6 illustrates the reduction of the United Kingdom's carbon intensity electricity as a result of mixed energy sources.

The structure of electricity generation as it stands is shown in Fig. 10.4.7. The world is overall still heavily dependent on coal as a source of energy. On a more national level, Scotland has made significant moves towards lowering emissions in their distribution network. Nuclear and renewables contribute towards 33% and 27% of electricity generation, respectively, in Scotland. Scotland is leading by example in Northern Europe to cleaning up its electricity, and should others follow, a global robust and greener network can be established to support the EV industry worldwide.

10.4.4   Alternative means to fuel automobiles

The EV has found its place again to bridge the concern of an expensive and unpredictable future of the conventional internal combustion vehicle. The EV's battery requires recharging, and the cost of electricity should be considered when comparing the increase in fuel prices. Electricity prices have increased in recent years and continue to rise; see Fig. 10.4.8. This factor will not gain the confidence of the new

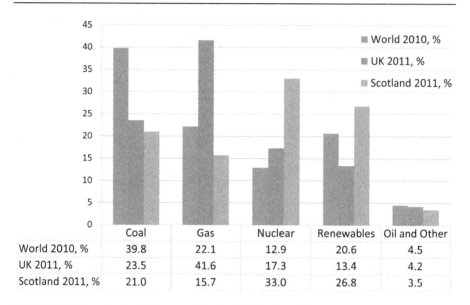

	Coal	Gas	Nuclear	Renewables	Oil and Other
World 2010, %	39.8	22.1	12.9	20.6	4.5
UK 2011, %	23.5	41.6	17.3	13.4	4.2
Scotland 2011, %	21.0	15.7	33.0	26.8	3.5

Fig. 10.4.7 Share of electricity generation by fuel for the United Kingdom and Scotland and globally (Hemingway & Michaels, 2012; OECD, 2013).

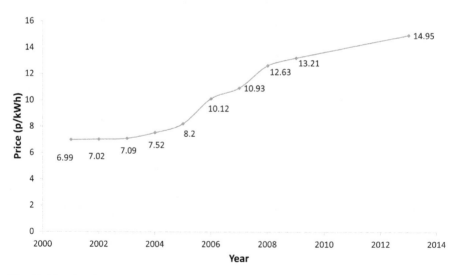

Fig. 10.4.8 Electricity domestic price in the United Kingdom (Bolton, 2014; AMDEA, 2015).

consumers of EVs. However, recharging by renewable means elevating this concern and provides a true green mode of transport. With technology moving at a rapid pace, technologies such as hydro, wind, solar, biomass, and others are becoming more integrated into the distribution network and provide a robust source of power. With public awareness, these technologies are becoming more accepted, and also mass production

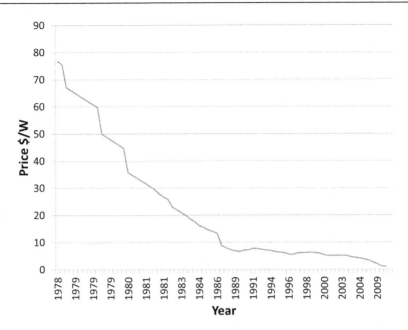

Fig. 10.4.9 Historical price of crystalline silicon photovoltaic cells (Bloomberg New Energy Finance, 2014).

will see a reduction in technology prices, and the use of these alternative energy sources will become more frequent in distribution systems. Solar technology has seen improvements in recent years and its installation costs have dramatically fallen. In order to provide a sustainable transport network, it is vital that technologies such as those aforementioned are supported to develop further.

Fig. 10.4.9 illustrates the historical prices of the PV cell. Bloomberg New Energy Finance shows from 2012 onwards, the cells cost less than one dollar. The fall in cost of the of crystalline silicon photovoltaic cells reduces the overall cost of solar technology to an attractive and affordable means to charge the electric car. Other forms of renewable energy sources are experiencing similar fall in price as with further development and improved public acceptance, the technologies can reap the rewards of mass production.

10.4.5   The electric vehicle in today's UK market

EV acceptance was not immediate and is still struggling to find its place in this niche market. However, over the last few years, improvements have been seen as the number of ultra-low emission vehicle (ULEV) on UK roads has increased with around 15,000 vehicles registered in the United Kingdom by 2014's third quarter, which can be seen in Figs. 10.4.10 and 10.4.11. The Department for Transport defines an ULEV as a vehicle whose tailpipe emissions are significantly lower than that of a conventional vehicle. These vehicles include electric, hybrid, and fuel cell vehicles or a vehicle

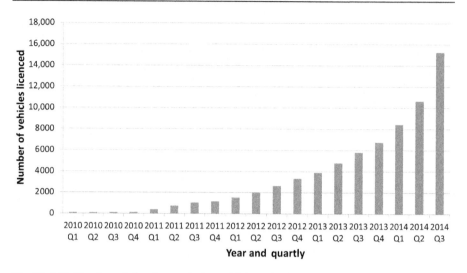

Fig. 10.4.10 Number of ultra-low emission vehicles (ULEV) licenced in the United Kingdom (Department for Transport, 2016).

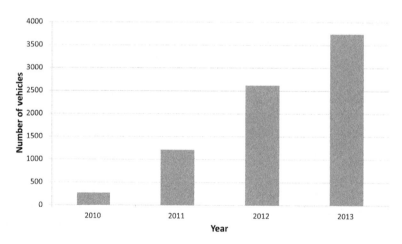

Fig. 10.4.11 Ultra-low emission vehicles (ULEV) registered for the first time in the United Kingdom (Department for Transport, 2016).

whose tailpipe CO_2 emissions are below 75 g/km. It should be noted that although these figures show promising trends, they are still relatively small in respect to the total number of vehicles deployed on UK roads. Barriers still hinder a confident penetration of these vehicles into the market. Through further research and enforcing government policy, these actions will overcome barriers and then we will see the ultimate revenge of the electric vehicle.

Scotland is making its way onto the EV map with over 1100 EVs on Scottish roads in 2014. Edinburgh has the highest EV usage in Scotland, mostly due to its successful supporting infrastructure. A survey carried out by RAC Foundation showed that only 60% of the charging points in Glasgow and Dundee were used in Aug. 2014 versus all 38 of Edinburgh's charging stations being used in that period. Low usage of these charging stations contributes to negative public perception of the EVs; however, the capital is showing great confidence that EVs will play a part in Edinburgh's sustainable transport strategies (The Scotsman, 2015).

10.5 Case Studies in Northern Europe

10.5.1 Introduction

The following section will discuss various case studies in Northern Europe that have been implemented to encourage the uptake of low emission vehicles. The TEV Project, a concept suggested by an American inventor, is discussed in this section as a solution to our transport needs of the future. Improvements to the transport sector and support of the EV market are presented on a small and larger scale, from small businesses like taxi companies going electric and car clubs to government incentives nationwide to support and address peoples initial financial concerns. It is the authors' hope that an enthusiasm from the industry, government, and all stakeholders involved is portrayed through these case studies to instil confidence in the reader that the EV is the transport of the future and can be a reliable, efficient,and attractive alternative to the conventional vehicle.

10.5.2 Ultra-low emission zones, London, UK

Various projects are running throughout Northern Europe to encourage 'cleaner' modes of transport. The low emission zone (LEZ) has been running in the greater London region since 2008. This project runs 24 h a day, 7 days a week, and 365 days a year. The aim of this project is to encourage vehicles with high emissions driving into the centre of the city, mainly heavy diesel vehicles, to review its environmental impact and adopt a more sustainable means for transporting goods.

This charge is a separate charge to the congestion fee that is also running in the city. If a vehicle complies with the congestion requirements and is exempt, charges may still be incurred in regard to the LEZ charge. Vehicles that do not comply with emission standards set out for the LEZ will pay £100 to £200 daily depending on the vehicle type.

Cameras are in operation within the zones. The camera feeds the registration number to a database that gives details of the vehicle and its environmental emission value to calculate an appropriate charge. This database will also give details on whether the vehicle has paid its fee. Parked vehicles within the zone are exempt from charges. Signs are also provided to avoid the LEZ and a detour route is suggested.

The success of this project will be shown in the improved quality of air and the positive effect on people's health in London. Further improvements are constantly being made to this LEZ project with the support of the city's mayor to tighten emission restrictions in the greater London region more and more to implement an ultra-low emission zone (ULEZ).

Range anxiety is an issue that hinders the widespread adoption of the EV. Urban Foresight predict that the concern will shift from 'How far can a vehicle go?' to 'Where can a vehicle go?' (Urban Foresight, 2014). With an ULEZ to be in operation in London by 2020, this will restrict the majority of conventional vehicles to travel into the city without paying a considerable tariff.

The mayor of London along with Transport for London sets out goals in the 2013 Transport Emissions Roadmap. Here, the goal to achieve the world's largest ULEZ is outlined. The roadmap states that some vehicles will need to meet a CO_2 emission level of 35–75 g/km to be exempt from the emission charges. As of the 7th of Sep. 2020, all light-duty and heavy-duty vehicles will be subject to emission charges of £12.50 and £100 plus, respectively (European Comission, 2015). LEZs are implemented in over 200 cities throughout Europe, Norway, Sweden, and the Netherlands, to name a few participating countries. These environmental zones will have various emission restrictions depending on the strategy in place in that city.

10.5.3 Breaking the boundaries by Tesla in Northern Europe

A Californian company founded in 2003 provided the game-changing technology that has pushed the limitations of the EV. Elon Musk, CEO of Tesla Motors, has a claim that sustainable transport is available globally. The Tesla Model S is eliminating any concerns in regard to range with its 85 kWh battery reaching 312 miles in ideal conditions (Tesla, 2015b). The evolution of charging times for the Renault Zoe EV is seen below in Table 10.5.1; however, Tesla's innovative idea of the battery swap allows the driver to have an unlimited range by swapping the vehicles battery for a battery with 100% charge in 90 s. This facility is only available in the state of California, but

Table 10.5.1 Evolution of battery charging times for the Renault Zoe and Tesla's battery swap facility

Charging times					
Charger type	Phases	Current (A)	Voltage (V)	Power (kW)	Charge time
Very slow	1	10	230	2.3	9.5 h
Slow	1	16	230	3.7	6.0 h
Fast	1	32	230	7.4	3.0 h
AC-rapid	3	32	230	22	1.0 h
DC-rapid	3	63	230	43	0.5 h
Battery swap[a]	–	–	–	–	90 s

[a]Tesla's battery swap service (only available in California).

should Tesla Motors continue to grow at its rate in Northern Europe, these facilities may soon become available to Tesla users throughout Europe. The Tesla's super charger will charge the cars battery to half its capacity in 20 m. In 2014, the Tesla was the most sold vehicle in Norway, beating its next biggest competitors the Volkswagen Golf and the Nissan Leaf (Gas2, 2015).

It is without question that with the luxury of Tesla car, it will not be affordable to all customers. Development and innovation is required across the automobile industry as a whole to bridge these gaps that exist and follow Tesla's ethos to provide the appropriate infrastructure to allow the EV an equal platform to compete with its rivals.

10.5.4   Battery right sizing in the Netherlands

The most cost-intensive aspect of the EV is its battery. The battery can be up to 30% of the overall cost. For this reason, it is important that the appropriate battery size is selected for custom use. A larger battery than required will result in unnecessary increase in costs and vehicle weight, so for smaller mileages, a smaller battery is required. Amsterdam Airport Schiphol has adapted a bus system to service the airport with correct battery sizing to the fore of the project to optimize the projects economic value.

The Netherlands airport is the first ever airport to adopt electric buses to transport its passengers to and from the aircraft more sustainably. In 2013, 35 electric buses were purchased from Chinese manufacturer BYD. The short journey lengths and low speed requirements are ideal for the technology, and with the smaller usage demand, the battery is respectively smaller. At that moment, the bus can travel 250 km on a single charge. The battery is subject to right sizing, and as it is, the capacity of the battery is achieving far more than what is required. With a smaller battery, this will reduce the cost of the project for the airport making it a successful means of passenger transport and set an example for future airports or businesses to consider electric as an alternative option.

A document published by Urban Foresight (2014) suggests that this battery right sizing technique may also be applied to existing vehicles. Should an EV user wishing to travel a distance that is constrained by the capacity of its existing battery, this technique will allow the battery to be swapped with a battery of adequate capacity to allow the driver to travel a greater distance (Urban Foresight, 2014).

10.5.5   Government initiatives

Governments in Northern Europe are making plans to promote modern transport systems to encourage economic growth. Low emission vehicles are being introduced to improve the countries environmental status along with maintaining public mobility needs. These environmentally friendly technologies help to combat the issue of climate change without penalizing the motorist. To ensure the success of the EV, governments have introduced frameworks or strategies to achieve their visions of a sustainable transport system in their concerning countries.

Focusing on the United Kingdom, for example, the Office for Low Emission Vehicles has published 'The Plug-In Vehicle Infrastructure Strategy' that outlines the government's aims to achieve successful utilization of the EV throughout Britain. The drivers for change are climate change, the growth of the 'green' community, energy security, decarbonizing of the electricity system, and air quality. Europe saw a 37% increase in the electric car market in 2014; however, EV still only makes up 0.6% of the total new registered vehicles (ENDS Europe, 2015). The United Kingdom has seen successful adoption trends in the EV market over recent years, mainly due to the £5000 incentive for the customer to go electric. Incentives have not had as much of an impact in countries such as the Netherlands, even with tax exemptions witnessing a drop in EV's sales of 42% in 2014 compared with the previous year's figures (ENDS Europe, 2015). In Ireland, similar government funding exists. Consumers can avail of the €5000 grant when buying an electric car and €4000 trade in scheme should they trade in their conventional petrol or diesel vehicle; see Fig. 10.5.1 (IEA, 2014a). This €9000 grant and free installation of a domestic 3 kW slow charger are attractive incentives for the early up-takers. At that moment, both in the United Kingdom and Ireland, the public can use the recharging infrastructure free of charge. Once the EV successfully penetrates the automobile market, there will be a fee incurred to charge a vehicle at charging stations.

In Apr. 2015, France is to launch a 'scrappage scheme'. French diesel vehicle drivers who choose to replace their ICV with a plug-in hybrid or electric vehicle can qualify for a bonus of up to €10,000. The Minister of Ecology, Sustainable Development and Energy stated that in France, the country has taken the move towards sustainable transport and ultimately 'the fight against air pollution' has been too slow with 60% of the public living in areas of poor air quality. Around half of the French people live in the most polluted areas of the country, and the €10,000 incentive will be available alongside existing grants.

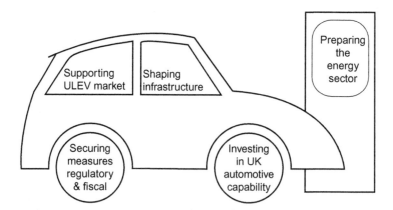

Fig. 10.5.1 UK governments objectives to drive forward the ULEV industry (IEA, 2014a).

10.5.6   Connecting Europe through TEN-T projects

The European Commission introduced a new transport infrastructure policy as of Jan. 2014. Throughout Europe, the Trans-European Transport Network (TEN-T) projects are in place to connect all of Europe's member states and allow easy mobility for passengers and goods. Through this project, a competitive and resource-efficient transport network is to be established. Adequate road networks are vital for the economies of all European member states to ensure that goods are transported fast, safely, efficiently, flexibly, and cheaply. The European Commission states that 44% of all goods in Europe are transported by road. People primarily travel by on-road transport that caters for 73% of all passenger traffic in Europe. In terms of the economic benefit of road transport, 5 million people across the EU are employed by the road transport sector, producing nearly 2% of the EU's GDP (European Commission, 2012).

There have been many drivers for the TEN-T project as mentioned in the European Commission's document. These drivers for change include congestion on European roads, which costs the EU economy the equivalent of 1% GDP that is more than budgeted. Others are the need for a safer and more secure means to transport goods and people, the instability of fuel prices of petrol and diesel, air and noise pollution, and the increased awareness of transport carbon footprint.

The need to move towards more reliable, efficient, and effective modes of transport is essential to ensure the success of TEN-T. The project will depend on new technologies in the automobile market such as hybrid and electric vehicles to address the issues the EU currently face. The EU acknowledges that on-road transport is the main culprit to many of the aforementioned problems where 45.9% of goods and 73.7% of passengers travelled by on-road transport in 2010 in EU countries, the most popular mode of transport both in the goods and passenger transport sectors. The European Commission supports research and development of more efficient engines with electric being the preferred power source for short journeys, methane and hydrogen for middle distances, and liquefied gas for longer journeys. With the TEN-T project introducing a trans-European network with improved infrastructure and implementing policies to ensure fair competition within the transport community, Europe is moving towards a more sustainable transport network.

10.5.7   Norway leading the way for other European countries

With a population of 5 million people, Norway is the world leader for EVs and Oslo the EV capital of the world (World Bank, 2015a). Norway has moved towards a more sustainable economy due to its nationwide acceptance and implementation of renewable energy. Norway's statistic office reported energy production in 2012 hit an all-time high as over 147 TWh of energy was produced, of which a net value of 17 TWh was exported. Norway has an abundant supply of hydro energy that accounts for 97% of Norwegians energy production along with wind and thermal energy accounting for the other 3% of energy production.

Not only is Norway leading the way for European countries but also sets an example worldwide of how the electric vehicles can be successfully integrated and ultimately dominate the transport sector as a truly clean new technology. There are many incentives for the EV customer in Norway: they are not subject to toll charges and have access to municipal parking and access to bus lanes. They also have no purchasing taxes on a new EV, and the EV consumer has 25% VAT reduction (Elbilforening, 2012). There are many exciting projects happening in Norway as the world watches in the hope of following suit.

Grønn Bil, which translates as 'Green Car' in English, is a project running to encourage people to purchase an electric vehicle works with the Norwegian Association of Local and Regional Authorities (KS) and the NGO ZERO. Grønn Bil's goal is to get 200,000 electric vehicles on the road by 2020. The project is owned by Energy Norway, a nonprofit organization that represents around 270 companies who are involved in the EV market from production to trading. Transnova, a government-run project, funds Grønn Bil that is a project that works towards reducing CO_2 footprint of road transport in Norway (Energy Norway, 2015). To achieve this, the project will work with stakeholders such as automobile manufacturers, infrastructure providers, policy makers, and the potential customers to work together to successfully penetrate the automobile market.

The Zero Emission Resource Organization (ZERO) is an independent climate foundation whose aim is to work towards reducing anthropogenic emissions without jeopardizing global energy demands. The ZERO rally is an annual event where low emission vehicles, plug-in electric vehicles, hydrogen, bio-fuels, and others set out and travel various routes. These routes have included the Green Highway, the longest electric car highway in the world between Sweden and Norway. The Green Highway is working towards establishing a fossil-fuel-free highway through mid-Scandinavia that will not only result in environmental benefits but also strengthen the region's economy and competition. Events such as the ZERO rally is a great way to engage with the public in terms of education and also addressing any misconception or perceptions of the limitations that previously existed with these new technologies (ZERO, 2015).

Another innovative project in Norway was the awarded ship of the year 2014, the ZeroCat battery-electric ferry. This ferry that was developed in association with Norled and built by Fjellstrand is lightweight and made of aluminium that halves energy consumption in comparison with convention fossil-fuelled ferries. The ferry can accommodate 120 cars and 360 passengers. This zero-emission and minimum noise ferry is the world's first electrified car ferry. The project is the result of a government-based competition encouraged by the Minister of Transport and communications in 2011, and the ZeroCat was scheduled to enter service on the 1st of Jan. 2015. Propulsion energy for this ferry is supplied by on board batteries battery recharged by smart grid components on the national power-grid. The nature of the Norwegian energy generation means the ZeroCat has shown the way towards decarbonizing the marine transport sector by electrical means. The ferry will travel 5.7 km across the Sognefjord that is located between Lavik and Oppedal, two Norwegian villages, making approximately 34 trips per day. The on-board 1 MWh lithium-polymer battery is charged within 10 m, the amount of time the ferry is to be stationed at the port before making its return journey (SMM, 2014).

The Taxi Trondheim project that ran for 2 years in Norway was implemented to see if battery-powered vehicles would prove a good alternative to conventional petrol- or diesel-powered taxis. The taxi operators Trøndertaxi and Stjørdal taxi collaborated together to answer these questions if the EV could overcome its limitations in a commercial environment with the support of the utility company NTE and their three fast chargers. The project worked closely with Nissan. Six Nissan Leafs were deployed for the project and their data were logged and shared to analyse the energy usage of the taxis and potential benefits. The success of the project was evident early on with the public perception towards the new electrified taxi service being very positive. The taxis were found to be very efficient around the city with an annual average of 33,000 km travelled by each vehicle. However, the project did uncover that the electrified fleet of taxis were not profitable at the end of the project. The outcome of the Taxi Trondheim project was that although taxi drivers adapted well to the battery-powered alternative by acquiring the appropriate driving skills to optimize the performance of the new taxi, for an electrified taxi fleet to be successful, it requires appropriate supporting infrastructure and a smarter trip management plan on the operators behalf (EV Norway, 2015).

The Norwegian Charging Station Database for Electromobility (NOBIL) is a public-owned database network for charging stations coordinated by the government body Transnova and the Norwegian Electric Vehicle Association, which allows the public to access standardized data free of charge so that services can be built and/or improved. The database encourages dialogue between all EV users to share high-quality information by building a robust tool for EV owners to gather and share information on appropriate infrastructure facilities available to maximize knowledge on the technology in the country. The database was launched in June of 2010 by the Minister of Transport and Communications. A state-funded domestic charger is required to provide data, and according to the Norwegian Electric Vehicle Association, this covers around 80% of charging stations in Norway. Should an individual desire information from the database, both Norwegian and other residents of the Nordic countries can do so should the individual share their EV information to strengthen NOBIL's database. Information is uploaded daily. This tool has not only been used for EV owners but has been used to improve public perception, for policy makers to see real-life results and potential infrastructure installers in order to get an overview of the countries existing infrastructure (Elbilforening, 2012). This project led to another project called EVR MAP that is developing a wider database for Nordic countries.

10.5.8   Ecotricity's electric highway in the UK

Ecotricity, the world's first green electricity company based in the United Kingdom, has supported EV uptake recognizing that the infrastructure for electrified vehicles to travel long distances did not exist. The company has highlighted the problem that people will not buy the EV as the supporting infrastructure that exists in the United Kingdom is not sufficient enough to allow people the same freedom of travel that is provided by the conventional ICV. With very little uptake of the electric vehicle, stakeholders are cautious whether to invest in the installation or improvement of the infrastructure. To bridge the limitation of battery range and ultimately break this vicious cycle, Ecotricity launched the first ever electric highway in the United Kingdom in 2011. A national network of charging points allows the EV driver to travel the length and breadth of the country and not have to worry about running out of charge. The company has installed 22 kW AC fast chargers, 50 kW DC chargers, and 50 kW DC/43 kW AC. The electric highway is free to the public and drivers can register through the Ecotricity website. Registered members are then issued a swipe card to utilize the charging points on the electric highway. The company's plan is to maintain the free service as to encourage and support the revolution of the electric vehicle (Ecotricity, 2015).

10.5.9 Electric taxis in Dundee, Scotland

Scotland as a whole has shown great movements towards supporting EV uptake. A survey was carried out across the United Kingdom with 433 councils participating. The question posed to the councils was:

Under the Freedom of Information Act 2000 I seek the following information: The number of road legal electric powered vehicles the council currently owns, manages or leases.

The survey showed that in the top 5 councils that have adopted EVs, 80% of them were Scottish. Dundee is a community that has embraced electric transport over the last few years with 38 EVs in Dundee City Council (the highest EV council adopter in the United Kingdom) and South Lanarkshire Council following with 24 vehicles (Intelligent Car Leasing, 2015).

Entrepreneur David Young saw an opportunity for his taxi business that would incorporate EVs and continue the growth of EV adoption in Dundee. David Young was the CAO of Dundee Private Taxi Hire (also known as 203,020) that later merged with Tele-Taxis and has been known as a single entity since 2012. Young's side of the company has developed a fleet of 100 vehicles, of which 30 are electric Nissan Leafs. The business decision was made to move towards EV after extensive research into infrastructure, battery technology, and best practice. When asked, had the public noticed the change from conventional vehicles to electric, Young answered that they did not notice a difference, the people still got the service they required delivered in an efficient, reliable, and professional manner. The project that is still in its early stages has yet to be fully advertised to the people of Dundee, but plans are in place to allow the customer, should they want an EV to provide their service, they will soon be allowed to request a Leaf but any advertisement through local media has been received very positive with the public.

The taxi company has invested approximately £500,000 with the support of Transport Scotland, the Energy Saving Trust (EST), and Dundee City Council. The electric fleet is supported by five rapid chargers and ten 7 kW, 32 Amp fast chargers installed by APT. The company is satisfied with the new direction the business has taken and has had a positive experience in terms of driving, the supporting infrastructure, and savings on maintenance cost that has incurred as a result of moving electric. Should an EV exhaust its range, the company has a pickup service in place; however, with the support of Dundee City Council and their 44 charging points, range is not an issue for the vehicles. Because of taxis high-mileage business nature, Dundee University members working with the taxi company claim that with 100 taxis going electric, it will mitigate 1200 t of carbon emissions being released into Dundee's environment, improving the air quality (The Courier, 2014). When asked what David Young's vision is for his company, he replied: 'If I can show people that if a taxi can do it, I can do it and people will see the benefits of the Electric car which will benefit and help Dundee's environment'. The taxi has become somewhat of an icon in cities around the world; New York and their yellow taxis or London and their black cabs, Young wants Dundee to be put on the map as the 'Electric Taxi City'. Dundee is becoming a template to encourage others to do the same within the United Kingdom and Europe of how EVs can effectively take the place of the conventional vehicle. The future of the Dundee taxi company has yet to be written, but an all-electric fleet is the ultimate goal for both business reasons and should it help the environment and health of the city's residents, it would have been worth the business risk.

10.5.10 The electrification of buses in Italy and Scotland

Public transport plays a huge role in urban dwellers daily lives. People travel more frequently and further than ever before due to the success of public road transport. Buses serve for business and recreational purposes on a daily basis in fast-moving cities and towns. Urban foresight acknowledged that buses operate almost 5–10 times more than the average passenger vehicle, and so are key contributors to CO_2 emissions in cities around the globe.

Rome has one of the largest electrified bus fleets in Europe. The Italian capital serves 945 million passengers per year (Agenzia, 2015), and the electric bus fleet has been running since 1989 (Eltis, 2015). The bus network incorporates 60 all-electric mini-buses that serve five different routes around Rome. The bus's battery swapping system allows a reliable and emission-free service. These mini-buses can easily manoeuvre around the small narrow streets of this historic city. These buses also comply with the cities LEZ and noise regulation policies.

Edinburgh is another example of where low emission buses are being implemented. In 2011, Transport Scotland worked with the cities bus operator Lothian Buses to introduce 15 new Alexander-Dennis Ltd. diesel-electric hybrid buses to route 10 in Edinburgh. They were the first ever double-decker bus of their kind in Scotland. The project was funded by Scottish Green Bus Fund. The number 10 route was chosen because of the high portion of car commuters that utilizes the service and the route also passes through two air quality management areas. To make the service more appealing and encourage road users to use public transport, commuters can avail of the free Wi-Fi on board and audio–visual units have been installed to display stops along the route. Transport Scotland has reported the project to be a success. The number of passengers to avail of the service has increased, and the bus operator has experienced an improvement in fuel saving of 59% resulting in 600 tonnes of carbon emissions being mitigated from the cities air per year. The Falkirk-based Scottish company, Alexander-Dennis Ltd., has also enhanced the regions economy by becoming one of Europe's leading suppliers of low-carbon hybrid-electric buses, supplying the region and environs with around 900 jobs and a further 2000 across its wider supply chain network (Transport Scotland, 2013).

10.5.11   Car Sharing Clubs, UK and France

Scotland has a very successful network of commercial nonprofit car clubs. The car clubs are pay-as-you-drive services for both urban and rural communities in the country. Registered members can hire a car 24 h a day, 7 days a week. The car club covers the cost of owning, operating, and maintaining the car as well as covering insurance, tax, and fuel costs. Car club members pay an annual subscription and then an hourly rate for renting the car; this charge will cover the mileage undertaken and wear and tear of the vehicle. The car club is an affordable way for drivers to experience driving an EV. Co-wheels who operate the club provide its members with EVs in Aberdeen and introduced one EV in Dundee, and Transport Scotland continues to support the electrification of car clubs (Transport Scotland, 2013). A car club based in London allows electric car access to its members of the E-Car Club. Models such as the Renault Zoe, Nissan Leaf, and Renault Kangoo Maxi Van are available to hire (E-Car Club, 2015).

Autolib' is a car-sharing operation that exposes urban drivers to all-electric means of transport. This French network offers a cost-efficient and convenient way for people to get about by eliminating the barrier of high initial capital cost of owning an EV by joining a shared mobility scheme. Autolib' was launched in Dec. 2011 as a public–private partnership (PPP) between French holding company, Bolloré, the City of Paris, and the surrounding cities (Urban Foresight, 2014). The 2500 EVs are supported by a network of 4710 charging stations in Paris and environs. Primarily funded by Bolloré, the city contributed €35 million to build the appropriate infrastructure, parking spaces, and the development of a car-sharing programme. The public has easy access to EVs with a radius of less than a quarter of a mile to the next car-sharing hub. The programme has proved quite popular with 6.6 million trips, 178,000 individual subscribers, over 60 million kilometres, and the mitigation of approximately 7575 t of CO_2 from Parisian air. The car-sharing industry is ever-growing and with plans to implement a similar car-sharing network in London. The French company IER that operates Autolib' won the contract to run such a system in London that consists of 1400 charging stations across 700 sites in the city. The new proposed 'Source London' project will be managed and operated by the French company who will also provide the appropriate on-street parking spaces for EVs. In 2013, when the contract was awarded to IER, the Source London scheme had 1000 subscribers paying an annual fee of £10 to charge their vehicles for free at any of the city's charging points. The project will initially start with the introduction of 100 all-electric Bluecars, built by Bolloré. With a positive public up-take, the project will then ultimate expand to 3000 vehicles and by 2018 hope to have a network of 6000 public charging points around the city. Schemes such as Autolib' and Source London help build a positive relationship between the public and EV's moving towards sustainable transport.

10.5.12   Engaging the public to challenge misconceptions

The consumer will be the person that ultimately defines the success of the EV as an alternative to the conventional ICV. Without positive public relations between all stakeholders, the investments, policies, and efforts of moving towards a cleaner, safer, and healthier future are doomed to fail. It is therefore vital that the future consumer is engaged and educated. 'Evolution' is a Scottish-based event in partnership with the Scottish Enterprise, Transport Scotland, and the Energy Saving Trust. The aim of the event is to facilitate a public forum where the low emission technologies from the passenger car, vans, to electric bicycles are on display to the potential customer. The public are allowed the opportunity to test drive the latest EVs on the market. The event is attractive not only to the potential first-time electric buyers but also to EV owners. EV owners can share their experience with others and get up-to-date information on new technologies available.

Evolution is not only for the driver but is a social day for all the family making the EV event attractive to all ages. Evolution has been running since 2014 and the results have been very positive. A poll carried out at the first event showed that people were surprised at how advanced the technology was and that they were expecting to see technology at its very early stages; the event instilled confidence in the public that this is a well-established technology of the future and is here to stay. After the event, 83% of people who responded to a survey said that they were more likely to buy an electric car after attending the event. It is important that not only in Northern Europe but also globally that events such as Evolution are organized to engage the public and potentially attract the future consumer (Evolution, 2015).

There are a lot of misconceptions surrounding the EV including the misconceptions that they are slow, unreliable, and even unsafe. It has been recognized that the general public's perception will not change with presenting information alone but through tackling these issues head-on and demonstrate the advancements of this niche market. The issues that hinder the uptake of the EV are dated from old slow, unreliable, and unsafe EV models that existed in the past. There lies a certain responsibility on the media to support the market to encourage the move towards electrified transport and to report nonbias information to combat EV anxieties. The nonprofit association Fédération Internationale de l'Automobile (FIA) that governs motor sports worldwide aims to overcome any negative attitudes towards battery-powered vehicles with their Formula E fully electrified racing series (Connelly, 2012). This vehicle hopes to generate a positive rapport with the public and encourage the industry into a competitive market. The aesthetic image of a vehicle is an important factor to the customer and this Formula E series does just that. The company Formula E Holdings wants to shape perceptions through motorsports to make it an attractive and worthy alternative. Races will be held in London, Beijing, and Los Angeles. The races will purposely pass by iconic landmarks in the cities at speeds of up to 225 km/h (Urban Foresight, 2014). Events such as this innovative display of high-speed EVs will not address all the misconceptions but is a move towards making the public aware of the presence of EVs and slowly persuade people that they are desirable and a practical fit for their lifestyle. The FIA Foundation will continue to support research and development in alternative energy systems such as solar, biofuels, electric, and hybrids.

10.5.13   Fleet optimization, Route Monkey, UK

Route Monkey, a United Kingdom-based company, offers a range of resources for various fleets in terms of software to optimize the energy efficiency, utilization, and carbon footprint of their vehicles. Route Monkey works closely with companies who wish to go electric in order to reduce their fleets' carbon footprint. Route Monkey requires real-time data of the existing fleet requirements in order to run their complex algorithm and select the most suitable option for the fleet and optimize its benefits in moving towards EVs. This software solution provider helps to plan routes efficiently for over 400 clients across the United Kingdom, Europe, and America (Route Monkey, 2015). This has been a useful tool for financial directors, transport offices, and managing directors of various fleets. An electric logistic operator with a large fleet can use Route Monkey algorithms as a procurement process by creating a virtual trial of the electric fleet and apply real-life data to minimize the cost and optimize the efficiency of the fleet. The innovative technology will also calculate which vehicles would best suit particular journeys within the fleet and where they should be based to achieve the companies real-life task at minimum cost. Route Monkey easily shows how the EV can be more reliable and robust by effectively route scheduling and infrastructure planning. The unique software application helps drivers easily adapt to new behaviours or range awareness required when driving an EV. The software calculates the best possible route in terms of energy consumption and plans charging opportunities either overnight or rapid charging. The company offers free virtual trial to new potential customers where they take the drivers real-time data and runs it through the software, where on average a 10%–15% savings can be made should Route Monkey plan be implemented.

10.5.14   Visions for the future—The TEV project

So what will our future transport look like? Will it include the EV or is the EV just a bridge to where we are moving towards? The TEV Project (Tracked Electric Vehicle) is Will Jones. Jones, a well-respected inventor, battery expert, and founder of Philadelphia Scientific, asks all stakeholders to stop talking about issues of climate change and start acting on moving towards mitigating concerns that exist in the transport sector. TEV is a novel transport network that will potentially meet the needs of the 21st century. In 2014, the TEV Project was invited to present their findings to the US Department of Transport at a Federal Highway Administration's Exploratory Advanced Research (EAR) Program workshop (TEV Project, 2015b).

The TEV Project does not claim to have invented new technologies. The TEV Project is based on implementing appropriate existing technology and software in an environment where it will optimize energy and economic and environmental efficiency of this new revolutionary transport system. Once the infrastructure planning and policies keep pace with the EV industry, electrified transport will be a source of endless driving for people around the world. The EV will dominate the automobile industry but only with the support of efficient infrastructure such as TEV.

TEV is a unique system that allows existing EVs or vehicles to be retrofit so they can adapt to a tracked highway. The vehicle will have rubber tires so as to allow it to drive on the roads once alighting the TEV highway to allow the driver to continue their journey to their final destination. On approaching the TEV highway, the vehicle will be assessed to make sure it is suited to the track once it enters the network the vehicle is attached to track by a safety lock mechanism and from then on the vehicle will 'drive by wire'. The driver can avail of the luxury of a driverless car and enjoy the journey. The vehicle will accelerate to an appropriate speed before it joins others on the highway. The TEV system will make the electric car more viable and superseding the performance of the existing ICV engine.

This computer controlled system allows vehicles to travel at speeds of 120 miles per hour. This technology will allow vehicle distance to be at a minimum, which will increase efficiency and reduce aerodynamic drag in comparison with the conventional highway where this would not be safe to do so for manually driven vehicles. A single TEV track will provide the same capacity of 17 lanes of freeway. Should the TEV be at approximately 75% capacity, the network would cater for 29,700 cars per hour. The UK Highway Code has a rule of thumb of 2 s between every vehicle. To get an idea of how drastic a solution the TEV is, a study compared the aforementioned figures with the M6 near Manchester. Should vehicles on the motorway travel at 70 mph and with the appropriate spacing, a single lane will have the capacity of 1588 cars per hour. Not only was the TEV system compared with road transport but also with successful rail networks. The success of the French TGV (Train á Grande Vitesse) high-speed train as an innovative project of its time has proved very popular and efficient, cursing at speeds of over 180 mph. At peak hours where frequency is at its maximum, 16 trains an hour will transport 800 passengers each, giving a capacity of 12,800 people an hour. Given that a car can carry 4 passengers, this would equal to 3200 cars an hour, the capacity of almost two motorway lanes. London's Underground can transport 3000 people an hour with trains departing every 10 m, the equivalent of around 750 cars giving only half the capacity of a single lane motorway (TEV Project, 2015a).

The TEV Project is preferred to be privately funded and serve as a profit-making enterprise that will create jobs. The project is not exclusive to developed regions such as the United States and Europe but will target all countries that contribute to pollution through their transport network such as China, India, and other countries whose transport network relies heavily on oil and other fossil fuels. The scale of the project is considerable that stakeholders will have to take a risk that the uptake of the system will be successful. Automatic tolls will be the primary source of income to the TEV system. The question whether the ICV should be adapted to the TEV track is a controversial topic as it will jeopardize the 'clean' reputation of this innovative transport system. However, should ICV be included in the network, the payback period of the project will be much sooner than if the project was to discriminate against petrol and diesel vehicles.

The TEV is an ambitious vision but not only addresses issues concerning air pollution from road vehicles but also it will improve safety on the roads. The TEV track will not be affected by weather conditions, and the computerized system also eliminates the human error factor reducing human fatalities on the roads, saving thousands of lives every year. Congestion will be reduced because of the fast-moving flow of traffic, and the system also plans to provide underground parking for vehicles once it reaches its destination. The driver can disembark their car and the car will park itself while the driver continues their journey on foot. On arrival at the TEV station for the return trip, by using an application on the driver's smart phone, the vehicle will retrieve itself from the automatic parking zone to meet the driver. Robo-vans, robo-cabs, robo-minibuses, and robo-trains will be a driverless service and single-mode in nature or in other words will be fixed to the track. Goods and people may be transported long distances without the concern of driver's fatigue.

In order to have a truly sustainable system, TEV recognizes that the project will require investments in a robust electricity supply infrastructure. TEV will require a considerable amount of electricity to power the network and has recognized that wind power may be the solution although issues regarding the variable in frequency of wind as a source may not be the most reliable. TEV has stated that nuclear energy could have its place as a reliable power supply. Although the TEV Project is not in support of nuclear energy, it highlights that should nuclear power plants be managed correctly and monitored appropriately, nuclear energy may supply the energy required at times when wind energy may not meet its peak demand. The TEV Project is a truly ambitious, efficient, yet exciting vision of transport systems of the future (TEV Project, 2015b).

10.5.15   Renewable energy and its part to play in EV's success

A move towards alternative energy sources is inevitable as fossil fuel supply is rapidly depleting. Governments feed in tariffs resulted in considerable investments in renewable sources of energy. In the United Kingdom, the penetration of wind energy dominated the renewable energy sector especially in Scotland. These government incentives had a positive impact on the development of technology; however, the existing power network cannot cope with the excess power produced when the power produced supersedes demand. There are very few efficient ways of storing this excess wind energy that has resulted in considerable constraint payments paid to windfarms to curtail their energy production that subsequently has resulted in poor public perception and confidence in alternative energy sources. UK industrial figures show that in 2014, £53.1 million was paid to wind farms to 'switch off'. The lack of infrastructure or grid connection to transport this wind energy from Scotland, south of the border into England, is one of the reasons why a lot of this power has been lost. Scottish energy demands are met but transferring the energy to England has proved a challenge. Wind farmers receive compensation for not producing as they will lose out on feed in tariffs and energy production. It is reported that a third more earning is made when the facility is 'switched off' than if it were to produce energy. Britain's largest on-shore wind farm Whitelee, located outside Glasgow, received over £20 million constraint payments to constrain the power produced by its 215 wind turbines in 2014 (Scottish Power Renewables, 2015). The sum of these pay-outs has been justified by RenewableUK as there is a significant risk and opportunity cost that has to be considered when switching off (Mendick, 2015). In Apr. 2015, Tesla Energy announced a product to harness excess energy produced by renewable energy, incorporating solar energy and a storage system called the Tesla Powerwall. For £5900 a home can be powered with a solar recharged 14 kWh battery. With Tesla's Powerwall, future homes can become independent of the grid (Tesla, 2015a). To serve buildings with higher energy demand than a domestic household, Tesla introduced the Tesla Powerpack that will produce an infinite quantity of energy in the future.

The EV could harness this power that would alternatively be lost due to the failure of adequate energy storage systems. At times when wind energy supersedes peak demand, this energy could be used to recharge EVs, and similarly at times when wind production is low and needs support to supply peak demand, the EV can supply the energy stored in its battery to the grid. Therefore, the energy supply network frequency may be evened out and become a robust network free from the fear of any electric 'black outs'. EVs will help integrate more wind power onto the network and avoid switch-off scenarios in order to capture wind energy that would have been previously wasted.

An example of where EVs are proving popular is in the Orkney Isles off the coast of Scotland. In 2013, 103% of the locals energy demand was produced by renewable means predominately from on-shore wind turbines. Approximately 5 MW of power is lost as it cannot be stored as there is limited opportunity to export this power. Similar to other parts in Britain, the Orkney Isles network has to be curtailed. The uptake of EVs with locals has improved as the people use the technology as a form of energy storage to get optimum efficiency from their unique network management system and avoid any financial loses as compensation may not apply to curtail this system. (Urban Foresight, 2014).

10.5.16   Rapid charging network, UK and Ireland

To date, the major barriers that face the EV concern costs, driving range, and recharging facilities. While government incentives are in place to address the concerns relating to the initial capital cost of the EV, the recharging infrastructure is yet to be established in order to dissolve range anxiety issues. In 2013, the rapid charge network (RCN) project was established to support the EV industry by setting out a framework that will install 75 rapid chargers in the United Kingdom and Ireland connecting major cities, airports, and ports covering 1100 km of major roads (RCN, 2014). The project is co-financed by the European Union through their TEN-T programme. The system will be free to use on the existing Ecotricity's Electric Highway in the United Kingdom and ESB ecars in Ireland. There are various recharging technologies, and the non-uniform adaption of a universal charging system has been a source of concern for the potential consumer and investors and so the deployment of rapid charging points has been slow. First of its kind, this project is to establish a dialogue with the original equipment manufacturers (OEMs) in working towards building the world's first largest multi-standard EV RCN linking the United Kingdom and Ireland to continental Europe. RCN is led by Nissan with the help of other major EV manufacturers to this niche market such as Renault, BMW, and Volkswagen. The Irish electric utility and distribution network operator, the ESB, is also one of the main stakeholders in RCN.

Another important factor to be considered is that the system has to be appealing to the customer. RCN recognized that at the moment charging stations are being deployed without any development of the stations environment. It is important to invest in overhead shelters to make the plugging-in experience as comfortable as possible or somewhat comparable to the refuelling of a conventional vehicle at a gas station where there are canopies to protect the driver from the elements. Also, investors could take advantage of the roof space of these canopies and install PV modules to operate the charging points.

Along with installing 74 rapid charging points, as shown in Fig. 10.5.2, and operating and monitoring the network, the project aims to develop a comprehensive roadmap with appropriate recommendations and guidelines. Through sharing and exchanging of information, this will help other EU member states to take the step towards implementation of similar networks to roll out across Europe.

Fig. 10.5.2 The Rapid Charge Network's 74 charging stations in the United Kingdom and Ireland (RCN, 2014).

This €7,358,000 project is a practical and universal solution where manufactures, universities, major utilities, and member states join forces to contribute to the acceleration of EV uptake in the United Kingdom and Ireland (RCN, 2014).

10.5.17   The European Green Vehicle Initiative

The European Green Vehicle Initiative (EGVI) is a PPP set up to deliver green vehicle and their complementary technology solutions to many of the major issues that face communities, cities, industries, and society in Europe today. This project is a continuation on from the European Green Vehicles Initiative that ran from 2009 until 2013. The EGVI focuses on energy efficiency of vehicles and alternative sources of power through accelerating research, development, implementation, and demonstration of the resulting findings and allowing the growth of 'greener' road transport. The partnership works with the industry and research members to establish a collaboration that will identify research and innovative ideas. Horizon 2020, an EU framework programme for research and innovation, will financially support these proposed activities from 2014 to 2020. The EGVI roadmap is a document that outlines the vision of the initiative. This roadmap was written up with strong references to the three European Technology Platforms: ERTRAC, EPoSS, and SmartGrids (EVGI, 2015). All three of these platforms have their own goals where ERTRAC aims at building robust transport strategies and focusing on research and development in the transport sector and to support effective public-private investment; EPoSS looks at researching technologies of the future, which are industry driven to implement smart systems; and SmartGrids develop smart electricity networks of the future (EPoSS, 2015; ERTRAC, 2015; SmartGrids, 2015). The EVGI roadmap incorporates all three. This nonprofit making association set up in 2013 has 78 interested members from the industry and research sector and chaired by Jean-Luc di Paola-Galloni from Valeo who is also acting chair for ERTRAC. EGVI's recommendations will be delivered to European member states through the communication channel created by the initiative with the European Commission services.

10.5.18   The ultimate driver for EV's globally

Sustainability, carbon emissions, and climate change, to name a few, are topical 'buzz' words in the media today. One can say the industry has taken the opportunity to use these issues and exploited the situation for financial gains. Some are of the opinion that these are only 'rich' people's issues; however, the driver for change that does not discriminate ones economic status is health. By 2030, it is estimated that 60% of the world's population will be living in urban areas, and with high density of people in smaller areas, around 75% of the world's energy is consumed by cities and with energy consumption comes pollution. Many of the cities in the world do not meet the World Health Organization's (WHO) air quality standards. People are being exposed to harmful fumes (2.5 times higher than the recommended safety level) and are at higher risks of developing respiratory diseases and heart diseases and having a stroke, diabetes, cancer, and other long-term health problems contributed by air pollution. In April of 2014, WHO released a report estimating that 3.7 million people under the age of 60 had died from illnesses linked to outdoor air pollution in 2012 across the globe, which equates to approximately one out of every eight deaths (WHO, 2014). In the United Kingdom, more than 28,000 people died in 2010 from long-term exposure to poor quality air (BBC, 2014). Moving towards electrifying the transport sector not only creates good economic opportunities for stakeholders but also improves the air quality and therefore avoids health issues in communities as a whole and also savings for governments in regard to the health system as long-term benefits may be seen.

10.6 Conclusions

- The EV concept has been around for longer than one may think the first battery-powered vehicle been invented by Robert Anderson around 1834. The fall of the EV was contributed by its range constraints and to nonflexible technology to adapt to mass production.
- Global, European, and UK trends show an increase in population and vehicle registrations. Consequently, energy usage, CO_2 emissions, ambient temperatures, and ocean temperatures have all increased over time.
- Should global ocean temperatures increase by 0.1°C, 11 Mtonnes of CO_2 will be released into the atmosphere. This is equivalent to a further CO_2 environmental loading of 0.5%.
- The transport sector contributes to 23% of CO_2 emissions globally.
- The EV is seen as a solution of many major energy, environmental, and economic problems facing society today.
- It is essential that emissions are not shifted from the transport sector to the energy sector and technologies such as wind, solar, and hydro should be embraced to support the existing distribution network.
- A study by Unger et al. (2010a) discussed how the on-road sector produces more GHG than aerosols to counteract the warming effect to the atmosphere. The power sector produces a considerable amount of aerosol gases in comparison with the on-road emissions. Should electricity demand increase with EV penetration to the automobile market, aerosol production will increase in the power sector.
- Installation prices for renewable energy technologies have fallen dramatically in recent decades with electricity prices increasing.
- Projects to connect European countries with electric motorways are supported by the TEN-T project.
- How we will travel in the future is still to be defined, but one is confident that the EV will play an essential role to this young yet well-established industry.

References

Agenzia. (2015). *Agenzia per il controllo e la qualità dei servizi pubblici locali di Roma Capitale*. Available at: http://agenzia.roma.it/home.cfm?nomepagina=settore&id_settore=8. Accessed 06.02.15.

AMDEA. (2015). *Time to change; electricity prices*. Available at: http://www.t2c.org.uk/saving-energy/electricity-prices/. Accessed 20.03.15.

BBC. (2014). *BBC News—Air pollution link to 28,000 deaths*. Available at: http://www.bbc.co.uk/news/uk-26973783. Accessed 23.02.15.

Bloomberg New Energy Finance. (2014). *Renewable energy | Bloomberg new energy finance*. Available at: http://about.bnef.com/services/renewable-energy/. Accessed 20.03.15.

Bolton, P. (2014). Energy prices, House of Commons Library (UK Government). Available at: http://researchbriefings.parliament.uk/ResearchBriefing/Summary/SN04153, Accessed on 26.04.2017.

Connelly, G., 2012. MAKING CARS GREEN; The Cars of Tomorrow (March).

DECC. (2014). *Chapter 1 Energy*: (pp. 11–41).

DEFRA, 2013. 2013 Government GHG conversion factors for company reporting: Methodology paper for emission factors (July).

DEFRA. (2016). *Greenhouse gas reporting—Conversion factors 2016*. Available at: https://www.gov.uk/government/publications/greenhouse-gas-reporting-conversion-factors-2016. Accessed 16.08.16.

Department for Transport. (2016). *All licensed vehicles and new registrations (VEH01)—Statistical data sets - GOV.UK*. Available at: https://www.gov.uk/government/statistical-data-sets/veh01-vehicles-registered-for-the-first-time. Accessed 20.03.15.

Department for Transport. (2013). *Greenhouse gas emissions (ENV02)—Statistical data sets—GOV.UK*. Available at: https://www.gov.uk/government/statistical-data-sets/env02-greenhouse-gas-emissions. Accessed 20.03.15.

E-Car Club. (2015). *Electric vehicles E-Car*. Available at: http://www.e-carclub.org/what-is-e-car/electric-vehicles/. Accessed 06.02.15.

Eakins, B. W., & Sharman, G. (2010). *Volumes of the World's Oceans from ETOPO1*. Boulder, CO: NOAA National Geophysical Data Center. Available at: http://scholar/scholar?hl=en&btnG=Search&q=intitle:Volumes+of+the+World+'+s+Oceans+from+ETOPO1#0.

Ecotricity. (2015). *Our electric highway—For The road—Ecotricity*. Available at: http://www.ecotricity.co.uk/for-the-road/our-electric-highway. Accessed 05.02.15.

Elbilforening, N. (2012). *The Norwegian Charging Station Database for Electromobility—NOBIL*. Norway: The Norwegian Electric Vehicle Association. Available at: http://info.nobil.no/images/downloads/nobilbrosjyre.pdf. Accessed on: viewed 26.04.2017.

Eltis. (2015). *Electric mini buses | Eltis*. Available at: http://www.eltis.org/discover/case-studies/electric-mini-buses. Accessed 06.02.15.

Emmott, S. (2013). *10 billion*.

ENDS Europe. (2015). *ENDS Europe | Electric vehicle market still not fully charged*. Available at: http://www.endseurope.com/index.cfm?go=39163&rss=news. Accessed 09.02.15.

Energy Norway. (2015). *Welcome to Grønn Bil*. Available at:http://www.gronnbil.no/english/. Accessed 20.03.15.

Engineering Toolbox. (2015). *Solubility of gases in water*. Available at: http://www.engineeringtoolbox.com/gases-solubility-water-d_1148.html. Accessed 20.03.15.

EPoSS. (2015). *Objectives & mission*. Available at: http://www.smart-systems-integration.org/public/about/objectives-mission. Accessed 25.02.15.

ERTRAC. (2015). *About ERTRAC*. Available at: http://www.ertrac.org/index.php?page=what-is-ertrac. Accessed 25.02.15.

ESRL. (2015). *ESRL global monitoring division—Global greenhouse gas reference network*. Available at: http://www.esrl.noaa.gov/gmd/ccgg/trends/global.html. Accessed 23.03.15.

European Comission. (2015). *London*. Available at: http://urbanaccessregulations.eu/countries-mainmenu-147/united-kingdom-mainmenu-205/london. Accessed 20.03.15.

European Commission. (2012). *Road transport: A change of gear*.

EV Norway. (2015). *EV Norway—Projects*. Available at: http://www.evnorway.no/#/projects. Accessed 06.02.15.

EVGI, 2015. Presentation. Available at: http://www.egvi.eu/about-the-egvi-ppp/presentation. Accessed 25.02.15.

Evolution. (2015). *Evolution Show Scotland—Home*. Available at: http://scotland.evolutionshow.co.uk/. Accessed 06.02.15.

Gas2. (2015). *Tesla Model S sets all-time Norway car sales record—Gas 2.* Available at: http://gas2. org/2014/04/02/tesla-model-s-sets-all-time-norway-car-sales-record/. Accessed 20.03.15.

Hemingway, J., & Michaels, C. (2012). *Electricity generation and supply figures for Scotland, Wales, Northern Ireland and England, 2008 to 2011* (pp. 50–58).

IEA. (2014a). IA-HEV hybrid and electric vehicles; the electric drive accelerates.

IEA. (2013). *International Energy Agency Data Services.* Available at: http://data.iea.org/ IEASTORE/DEFAULT.ASP. Accessed 20.03.15.

IEA, (2014b). Railway handbook 2014 energy consumption and CO_2 emissions.

Intelligent Car Leasing. (2015). *Huge new study compares every UK Council's electric vehicle usage | UK Car Blog & News.* Available at: http://www.intelligentcarleasing.com/blog/ new-study-compares-every-uk-council-electric-vehicles/. Accessed 24.02.15.

IPCC, (2007). Contribution of working group I to the fourth assessment report of the intergovernmental panel on climate change.

Leibling, D., 2008. Car ownership in Great Britain (October).

Levitus, S., et al. (2012). World ocean heat content and thermosteric sea level change (0–2000 m), 1955–2010. *Geophysical Research Letters, 39*(10), 1–31. Available at: http://doi. wiley.com/10.1029/2012GL051106. Accessed 21.02.14.

Mendick, R. (2015). *Wind farms paid £1m a week to switch off.* Available at: http://www. telegraph.co.uk/news/earth/energy/windpower/11323685/Wind-farms-paid-1m-a-week-to-switch-off.html. Accessed 20.02.15.

Morice, C. P., et al. (2012). Quantifying uncertainties in global and regional temperature change using an ensemble of observational estimates: The HadCRUT4 data set. *Journal of Geophysical Research, 117*(D8), D08101. Available at: http://doi.wiley.com/10. 1029/2011JD017187. Accessed 21.02.14.

OECD. (2013). *Statistics/OECD Factbook/2013/Electricity generation.* Available at: http:// www.oecd-ilibrary.org/sites/factbook-2013-en/06/01/03/index.html?itemId=/content/ chapter/factbook-2013-43-en. Accessed 03.03.15.

RCN. (2014). *The development and study of a network of rapid charge points across the UK and Ireland.*

Route Monkey. (2015). *Route Monkey the algorithm of things.* Available at:http://www. routemonkey.com/. Accessed 06.02.15.

Scottish Power Renewables. (2015). *About the Windfarm | Whitelee Windfarm, Glasgow, Scotland.* Available at:http://www.whiteleewindfarm.co.uk/about_windfarm?nav. Accessed 19.02.15.

SmartGrids. (2015). *Background.* Available at: http://www.smartgrids.eu/Background. Accessed 25.02.15.

SMM, 2014. Ship of the Year 2014: All-electric car ferry ZeroCat 120, (3).

Tesla. (2015a). *Powerwall | The Tesla home battery.* Available at: https://www.tesla.com/en_ GB/powerwall. Accessed 18.08.16.

Tesla. (2015b). *Tesla Motors UK | Premium electric vehicles.* Available at: http://www. teslamotors.com/en_GB/. Accessed 20.03.15.

TEV Project. (2015a). *The TEV Project; Tracked Electric Vehicle System; Reference Booklet v1.14.*

TEV Project. (2015b). *The TEV Project to present to the US Department of Transportation on novel transport systems | The TEV Project.* Available at: http://www.tevproject.com/ 2014/10/20/the-tev-project-to-present-to-the-us-department-of-transportation-on-novel-transport-systems/. Accessed 17.02.15.

The Courier. (2014). *Supporters say Dundee electric taxis bid could make city an international "pioneer"* Available at: http://www.thecourier.co.uk/news/local/dundee/supporters-say-dundee-electric-taxis-bid-could-make-city-an-international-pioneer-1.400292. Accessed 24.02.15.

The Scotsman. (2015). *Edinburgh has highest electric vehicles usage in Scotland*. Available at: http://www.scotsman.com/news/transport/edinburgh-has-highest-electric-vehicles-usage-in-scotland-1-3690420. Accessed 25.02.15.

Transport Scotland. (2013). *Switched on Scotland: A roadmap to widespread adoption of plug-in vehicles* (pp. 1–82).

Trenberth, K. E., & Smith, L. (2005). The mass of the atmosphere: A constraint on global analyses. *Journal of Climate, 18*, 864–875.

U.S. EIA. (2015). *How much energy is consumed in the world by each sector?—FAQ—U.S. Energy Information Administration (EIA)*. Available at: http://www.eia.gov/tools/faqs/faq.cfm?id=447&t=1. Accessed 20.03.15.

UK Committee on Climate Change, 2010. The Fourth Carbon Budget Reducing emissions through the 2020s (December), pp. 1–188.

Unger, N., et al. (2010a). Attribution of climate forcing to economic sectors. *Proceedings of the National Academy of Sciences of the United States of America, 107*(8), 3382–3387. Available at: http://www.pnas.org/cgi/content/long/107/8/3382. Accessed 17.11.14.

Unger, N., et al. (2010b). Attribution of climate forcing to economic sectors. *Proceedings of the National Academy of Sciences of the United States of America, 107*(8), 3382–3387. Available at: http://www.pubmedcentral.nih.gov/articlerender.fcgi?artid=2816198&tool=pmcentrez&rendertype=abstract. Accessed 17.11.14.

Urban Foresight, 2014. Ev city casebook; 50 BIG IDEAS shaping the future of electric mobility.

Water Encyclopedia. (2015). *Carbon dioxide in the ocean and atmosphere—Sea, depth, oceans, important, system, plants, marine, oxygen, human, Pacific*. Available at: http://www.waterencyclopedia.com/Bi-Ca/Carbon-Dioxide-in-the-Ocean-and-Atmosphere.html. Accessed 20.03.15.

WHO. (2014). *Air quality deteriorating in many of the world's cities*. Available at: http://who.int/mediacentre/news/releases/2014/air-quality/en/. Accessed 23.02.15.

World Bank. (2014a). *Energy & mining | Data*. Available at: http://data.worldbank.org/topic/energy-and-mining. Accessed 20.03.15.

World Bank. (2014b). *Environment | Data*. Available at: http://data.worldbank.org/topic/environment. Accessed 20.03.15.

World Bank. (2014c). *Road sector energy consumption (kt of oil equivalent) | Data | Table*. Available at: http://data.worldbank.org/indicator/IS.ROD.ENGY.KT. Accessed 20.03.15.

World Bank. (2014d). *Urban development | Data*. Available at:http://data.worldbank.org/topic/urban-development. Accessed 20.03.15.

World Bank. (2015a). *Norway | Data*. Available at: http://data.worldbank.org/country/norway. Accessed 20.03.15.

World Bank. (2015b). *United Kingdom | Data*. Available at: http://data.worldbank.org/country/united-kingdom. Accessed 20.03.15.

ZERO. (2015). *What is Zero Rally?—Zero Rally 2012*. Available at: http://www.zerorally.com/aboute. Accessed 20.03.15.

Electric vehicles: Status and roadmap for India

Parimita Mohanty*, Yash Kotak[†]
*UN Environment, Asia Pacific office, [†]Heriot Watt University, Riccarton, Edinburgh, United Kingdom

11.1 Introduction

In India, transport sector is one of the fastest growing sectors. Within this sector, two-wheelers are the dominating sector, which accounts for almost 75% of total vehicle in country. These increasing numbers of vehicle raise the local challenges such as congestion on road and deterioration of air quality. Currently, transport sector is one of the major contributors of CO_2 emission and as per 2012–13 data; the transport sector is responsible for 14% of India's energy-related CO_2 emissions. An assessment carried out by Central Pollution Control Board surveyed that 75% cities are at very high risk of PM_{10} levels. Out of these, 50% of cities have critical level of NOx as well. Road transport, which is the main mode of transportation in India, has experienced increased activities in terms of increase in the number of vehicles. This is because as household income increases, the need to use an improved form of mobility also increases and households tend to move from nonmotorized to motorized form of mobility. The Indian auto industry is one of the largest in the world with an annual production of 23.37 million vehicles in financial year (FY) 2014–15 (as shown in Fig. 11.1.1), following a growth of 8.68% over the last year. The automobile production grew at compound annual growth rate (CAGR) of 10.5% over the last 10 years in India and is expected to grow at the same rate in near future.

Further, the growth in the number of automobiles in India leads to substantial growth in energy consumption. As per the study carried out by The Energy and Resources Institute (TERI), the projection of growth in energy consumption by the transport sector in India shows that there will be a seven times growth in energy use in transport sector between 2012 and 2047 (16 CAGR), out of which more than 90% of the share would be only from road transport (see Fig. 11.1.2; India Energy Security Scenario (IESS) 2047, 2016).

In India, currently, the transportation sector relies heavily on liquid fuel and accounts for about one-third of the total crude oil consumption; out of this, road transport accounts for more than 80% of the consumption (India Energy Security Scenario (IESS) 2047, 2016). Such heavy demand on liquid fuel draws an immediate attention on diversification of the use of conventional fuel.

Electric Vehicles: Prospects and Challenges. http://dx.doi.org/10.1016/B978-0-12-803021-9.00011-2

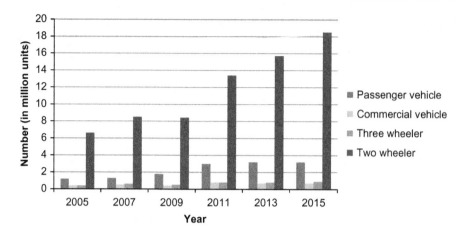

Fig. 11.1.1 Total production of automobiles in India.
Prepared by the authors with data from Society of Indian Automobile Manufacturers (SIAM).
(2016). [online] Available at: http://www.siamindia.com/statistics.aspx?mpgid=8&
pgidtrail=13 Accessed 05.01.17; National Electricity Mobility Mission Plan 2020. (2012).
National Electricity Mobility Mission Plan 2020 report [online]. Available at: http://dhi.nic.in/
writereaddata/Content/NEMMP2020.pdf Accessed 12.12.16.

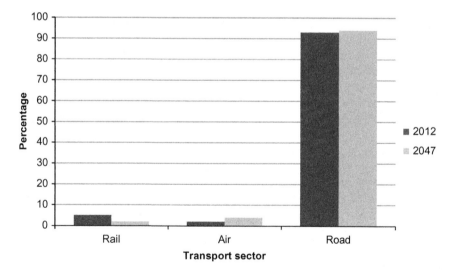

Fig. 11.1.2 Growth in energy consumption by the transport sector.
Prepared by the authors with data from India Energy Security Scenario (IESS) 2047. (2016).
[online] Available at: http://www.indiaenergy.gov.in/ Accessed 30.01.17.

Hence, there is a need to look into the potential role of clean fuels in powering the transport sector. In this regard, electric vehicles (EVs) offer a potential solution by offering a clean mode of mobility, provided the energy used to power electric vehicles is derived from non-polluting, renewable sources of energy.

11.2 Electric vehicle—An alternate mode of transport in India

'Electric vehicles' (EV) are defined as vehicles that use an electric motor for propulsion. The electricity used to run the motor could come either through 'transmission wires', as is the case with electric locomotives, metro trains, and trams or through a 'single or a series of connected batteries', as is the case in electric bikes and electric cars, or it could be generated on board using a fuel cell.

Again, battery-based EVs used in road transport include a large range of vehicles from electric two-wheelers, three-wheelers (rickshaws), cars, and electric buses. In this chapter, the term electric two-wheelers (E2Ws) is used for both electric bicycles and electric scooters while electric four-wheelers (E4Ws) is used for electric cars; E3W is used to refer to electric three-wheelers (including *E*-rickshaws) and E bus to refer to electric buses. In addition, plug-in electric vehicles are classified into two types: battery electric vehicles (BEVs) and plug-in hybrid electric vehicles (PHEVs).

A battery electric vehicle (BEV) runs entirely using an electric motor and battery, without the support of a traditional internal combustion engine, and must be plugged into an external source of electricity to recharge its battery. Like all electric vehicles, BEVs can also recharge their batteries through a process known as regenerative braking, which uses the vehicle's electric motor to assist in slowing the vehicle and to recover some of the energy normally converted to heat by the brakes. The modern PHEV is a hybrid vehicle that is powered by an internal combustion engine and an electric motor. Batteries power the electric motor for the initial 'all electric range' of the vehicle. When the batteries run low, the vehicle continues to travel on the engine using fuels such as gasoline, biofuels, natural gas, or hydrogen. Batteries can then be recharged whenever the vehicle is parked with access to any source of electricity (Andrew, 2016).

11.3 Characteristic of various models of electric vehicle available in India

There are several ways of classifying the EVs. Normally, it can be classified on the basis of their characteristic such as (i) charging time, (ii) driving range, and (iii) the maximum load-carrying capacity. Of these attributes, from the consumer perspective, charging time of batteries (i.e. the time required to charge the battery) and driving range (i.e. the maximum distance an EV can run when fully charged) are perhaps the two most important characteristics of an electric vehicle. Charging time depends on the input power characteristics (i.e. input voltage and current), battery type, and battery capacity, and the driving range depends upon the battery capacity and characteristic of the electric motor. The driving range of EVs could be as low as 20 km per charge to as high as 400 km per charge. Similarly, the top speed could go as high as 160 km/h in case of E4Ws and some E2Ws.

11.3.1   Characteristics of electric 4-wheeler

As mentioned in the earlier section, charging time and driving range are two important characteristic. In order to assess the characteristic of different E4Ws, few specification of commercially available model along with their charging time, driving range, and cost were collected. Fig. 11.3.1 plots the two characteristics of E4W, charging time, and driving range, with price for some popular E4Ws currently available in the Indian and global markets (Shukla, Dhar, Pathak, & Bhaskar, 2014).

Eighteen models of E4Ws by 14 automobile manufacturers are included here. To allow for comparability across models, the data for the latest (i.e. 2013 or 2014, whichever was the latest) editions of models were used. The price of vehicles ranged from USD 11,300–94,570. While most E4Ws have driving range around 150–200 km, there are some E4Ws with a range of over 300 km (Shukla et al., 2014). Similarly, the charging time is typically less than 8 h, but for some vehicles, the charging time is higher (greater than 12 h). In general, the driving range of the vehicle with the price range less than USD 60,000 is similar and does not vary much. Similarly, there is a very weak positive correlation between price of vehicles and charging time. This is due to two opposite trends related to battery quality and motor size: higher price vehicles have not only relatively larger battery sizes but also better battery designs.

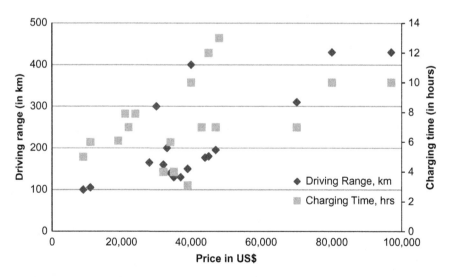

Fig. 11.3.1 Driving range, charging time, and price of E4Ws for Indian and global models. Prepared by the author with data from UNEP DTU Report; Shukla, P. R., Dhar, S., Pathak, M., & Bhaskar, K. (2014). Promoting low carbon transport in India [online]. Available at: https://www.researchgate.net/profile/Minal_Pathak/publication/271447336_Electric_Vehicle_Scenarios_and_a_Roadmap_for_India/links/54c867250cf238bb7d0de64f.pdf Accessed 20.07.16.

11.3.2 Characteristic of electric 2-wheeler

There are three types of E2Ws based on nature of technology: (a) electrical cycles and mopeds (ECM), (b) electrical scooters (ES), and (c) electrical motorcycles (EM). They vary with regard to top speed, price, and load-carrying capacity, as shown in Figs. 11.3.2–11.3.4.

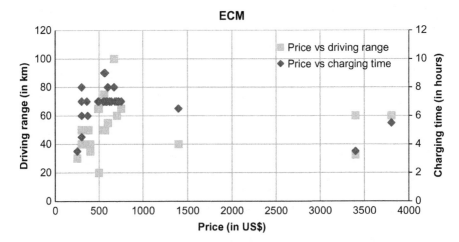

Fig. 11.3.2 Driving range, charging time, and price for electric cycles and mopeds (ECMs). Prepared by the author with data from UNEP DTU Report; Shukla, P. R., Dhar, S., Pathak, M., & Bhaskar, K. (2014). Promoting low carbon transport in India [online]. Available at: https://www.researchgate.net/profile/Minal_Pathak/publication/271447336_Electric_Vehicle_Scenarios_and_a_Roadmap_for_India/links/54c867250cf238bb7d0de64f.pdf Accessed 20.07.16.

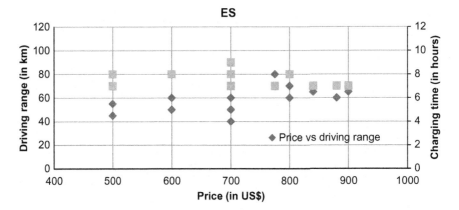

Fig. 11.3.3 Driving range, charging time, and price for electric scooters (ESs). Prepared by the author with data from UNEP DTU Report; Shukla, P. R., Dhar, S., Pathak, M., & Bhaskar, K. (2014). Promoting low carbon transport in India [online]. Available at: https://www.researchgate.net/profile/Minal_Pathak/publication/271447336_Electric_Vehicle_Scenarios_and_a_Roadmap_for_India/links/54c867250cf238bb7d0de64f.pdf Accessed 20.07.16.

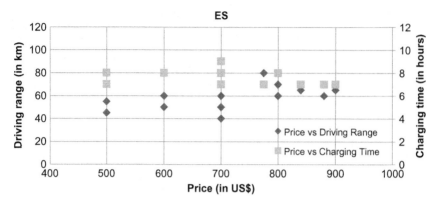

Fig. 11.3.4 Driving range, charging time, and price for electric motorcycles (EMs). Prepared by the author with data from UNEP DTU Report; Shukla, P. R., Dhar, S., Pathak, M., & Bhaskar, K. (2014). Promoting low carbon transport in India [online]. Available at: https://www.researchgate.net/profile/Minal_Pathak/publication/271447336_Electric_ Vehicle_Scenarios_and_a_Roadmap_for_India/links/54c867250cf238bb7d0de64f.pdf Accessed 20.07.16.

It can be observed from Figs. 11.3.2–11.3.4 that the charging time of many of the vehicles is in the same range while the driving range varies quite largely. It can also be noted that there is a wide driving range (20–100 km) with a price range of USD 400–600. In general, charging time and driving range differ with the type of E2Ws. Driving range does not increase much with price for EM and ES. Meanwhile, in the case of EM, there is a weak positive correlation, and driving range in general increases with price.

11.4 Past trend and current status of EV market in India

The current market for EVs is very small in India. Though there are different types of E2Ws (scooters and bikes), E4Ws (electric cars), and electric buses, the overall share of EVs is negligible (National Electricity Mobility Mission Plan 2020, 2012). There are few market players (companies) in the EV sector in India.

In the two-wheeler segment, Hero cycles, Electrotherm India, TVS Motor, and Hero Electric are manufacturing and selling electric two-wheelers. These electric two-wheelers are usually charged at the domestic supply voltage and, therefore, require no special adapter for charging the batteries. Normally, motors and other electrical kits for these vehicles are imported from China and other countries, whereas mechanical design and assembly of these bikes/two-wheelers are done here. In spite of a number of players selling these two-wheelers, the market share of these EVs is not large, primarily because of the high cost of the vehicle. However, recently, the offtake of these vehicles has picked up on account of subsidy scheme launched by the Government of India and concessions given by some state governments (National Electricity Mobility Mission Plan 2020, 2012).

In case of electric three-wheeler segment, there are few established automobile manufacturers. The first electric three-wheeler (Vikram SAFA) was developed by *Scooters India Ltd.*, Lucknow in 1996, and approximately 400 vehicles were made and sold. These vehicles ran on a 72 V lead-acid battery system. The model was discontinued due to very less market demand.

Mahindra & Mahindra Ltd. launched its first electric three-wheeler in 1999 including a new company named as MEML, based in Coimbatore, in 2001, to make and sell electric vehicles named Bijlee (National Electricity Mobility Mission Plan 2020, 2012). In 2004, MEML was closed down due to the lack of demand. However, Mahindra again started making electric vehicles at Haridwar plant in 2006 and continued to produce electric vehicles as per market demand.

Bajaj Auto Ltd., Pune, had also demonstrated their three-seater electric rickshaw in 2001. However, this product has not been commercially launched (Shukla et al., 2014).

Recently, many of the major cities in India including Delhi, Hyderabad, Jaipur, Lucknow, and Kolkata have witnessed locally assembled electric rickshaws that have emerged as a popular mode for last-mile connectivity in these cities. Powered by electric motors and batteries, these electric rickshaws fall somewhere in between auto-rickshaws and pedal-driven cycle rickshaws (see Fig. 11.4.1).

The different components (all key electrical components i.e. batteries, motors, and controllers) of these e-rickshaws are imported from China and then assembled locally. Since majority of the components are imported, there are no standard specifications for these components, and thus, the features and specifications of one rickshaw vary from that of the other. There are around 15–20 types or makes of such locally assembled e-rickshaws available in the market (as shown in Fig. 11.4.2). Although the safety standards of these vehicles are not at all ensured, it is becoming popular primarily because of its cost-effectiveness in terms of its initial cost of investment.

Fig. 11.4.1 Locally assembled electric rickshaws.
Prepared by the authors (Author's credit).

Fig. 11.4.2 Variety of locally assembled e-rickshaws available in Indian market.
Prepared by the authors (Authors credit).

In Delhi alone, there are more than 100,000 e-rickshaws, and therefore, to regulate
their use and address safety concerns, government has made amendments of the Motor
Vehicle Act in 2014. Due to these amendments, e-rickshaws must be registered, and
the drivers are required to have driving licenses. Around the year 2000, only a couple
of electric two-wheelers were available in the Indian market. However, the market has
expanded and over two dozen different two-wheelers are available in the market at
present. These include low-speed vehicles with a maximum speed of 25 km/h to
high-speed vehicles capable of achieving speeds up to 65 km/h. The driving range varies from 20 to 100 km (3 Wheeled Cheese, 2012).

In 2014, seven e-rickshaw dealers in New Delhi were interviewed by experts from
TERI, as part of the surveys to get an understanding of the nature of these e-rickshaw
dealers and their dealerships. As per the survey, dealers indicated that the dealerships
are of formal nature and employed about 3–6 employees. The dealers have claimed to
have registered themselves to sell e-rickshaws and dealt with more than one manufacturer. These dealers procure fully assembled e-rickshaws from manufacturers like
Yatri, Chetak, Sarthi, Mayuri, and Best and do not assemble e-rickshaws on their
own, though they provide repair facilities if any of the e-rickshaws is found to be
faulty. All of them sell autorickshaws along with e-rickshaws. According to these
dealers, the average price of an e-rickshaw has decreased from INR 91,860 to
INR 85,570.

The Reva Electric car company introduced the first electric car REVA in India in
the early 2000s that continues to sell a few units. Reva Electric focused on creating
affordable electric cars through advanced technology and launched its first model in
India in 2001 and in London in 2004. Total units of around 3200 cars have been sold
worldwide including approx 1500 cars that have been sold in India, mostly in Bangalore City. Another firm, named as Mahindra Group, which launched a revamped

version of the car in 2013, later acquired the firm. Powered with lithium-ion batteries, the new model allows for a top speed of 80 km/h and a driving range of 100 km with a single charge. With a charging time of 5 h, it is marketed to provide significant cost savings over a conventional car.

Motor Vehicle Act

According to the Motor Vehicles Act, 1988, '"motor vehicle" or "vehicle" means any mechanically propelled vehicle adapted for use upon roads whether the power of propulsion is transmitted thereto from an external or internal source and includes a chassis to which a body has not been attached and a trailer; but does not include a vehicle running upon fixed rails or a vehicle of a special type adapted for use only in a factory or in any other enclosed premises or a vehicle having less than four wheels fitted with engine capacity of not exceeding [twenty-five cubic centimetres]". The Central Motor Vehicles Rules (CMVR), 1989, notified under the Motor Vehicles Act, 1988, further provide a more specific definition of battery-operated vehicles. According to the rules, 'battery-operated vehicle' means a vehicle adapted for use upon roads and powered exclusively by an electric motor whose traction energy is supplied exclusively by traction battery installed in the vehicle. Provided that the following conditions are verified and authorized by any testing agency specified in rule 126, the battery-operated vehicle shall not be deemed to be a motor vehicle:

i. The 30 min power of the motor is less than 0.25 kW.
ii. The maximum speed of the vehicle is less than 25 km/h.
iii. Bicycles with pedal assistance which are (a) equipped with an auxiliary electric motor having a 30 min power less than 0.25 kW, whose output is progressively reduced and finally cut off as the vehicle reaches a speed of 25 km/h or sooner, if the cyclist stops pedalling and (b) fitted with suitable brakes and retroreflective devices, i.e. one white reflector in the front and one red reflector at the rear.

Mahindra has also come out with a completely electric vehicle e2o, manufactured in its green facility, which is offering innovative battery rental scheme. As of 2015, the on-road price of this vehicle is approximately USD 7000 (Mahindra Electric, 2016). So far, company had sold more than 1000 units and target to sell 500 units per month. The company has plans to expand to Europe and South Asian countries where EV sales are picking up and government incentives are available.

In 2000, BHEL developed an 18-seater electric bus. Its power pack consisted of an AC induction motor and a 96 V lead-acid battery pack. Around 200 electric vans were built and run in Delhi, with monetary support from the Ministry of New and Renewable Energy (MNRE). The major concern with these vehicles was their poor consistency, low life, and very high cost of the battery.

Further, in 2010, Toyota introduced the Prius Hybrid model in the Indian market. It was designed to run on electricity and petrol, to suit customer preference. It was built with twin-cam petrol engine that puts out 97 bhp, while on the other side it also has a sealed nickel-metal hydride (Ni-MH) battery, which increases the total power output to 134 bhp. Hence, in Indian market, it is considered as a car with three modes, namely, Eco, Power, and EV. The Prius Hybrid model has been followed up by introducing Camry Hybrid in 2013. The Camry Hybrid has a total of 202 bhp, out of which 158 bhp is from petrol and 44 bhp is from the electric motor (Naik, 2016).

Similar to Toyota, BMW also introduced its BMW i8 in India at Delhi Auto Expo in 2014. It has a regular lithium-ion battery of 7.1 kWh and three in-line petrol unit. I8 gives 228 bhp power output from gasoline engine and 129 bhp from electric motor, which adds up to a total of 357 bhp. Due to its high-power output, the electronically controlled top speed of i8 is 250 Km/h. BMW started its deliveries in August 2014 and managed to sell 1741 units by Dec. 2014 (Naik, 2016).

Recently, few Indian firms have announced plans to introduce electric cars in the short-to-medium term. The currently running hybrid and electric buses are being considered as pilot initiatives in a few cities. The Bangalore Municipal Transport Corporation has recently introduced an electric bus on a dense corridor in the city. Given the established two-wheeler manufacturing industry and the recent interest in EVs, India has the opportunity of strengthening domestic EV industry and emerging as a global leader in EV manufacturing market. Efforts are underway by electric vehicle manufacturers to provide options that can reduce charging time and increase awareness among consumers regarding lower fuel and maintenance costs of E4Ws compared with conventional cars.

11.5   Electric vehicle market forecast in India

It is estimated that the total potential demand for the full range of electric vehicles in India (mild hybrids to full electric vehicles) will be in the range of 5–7 million units in new vehicle sales by 2020. This will include 3.5–5 million pure electric two-wheelers (BEVs), 1.3–1.4 million HEV vehicles (4W, buses, and LCVs), and 0.2–0.4 million other pure electric vehicles (4W, buses, and LCVs) (Shukla et al., 2014). The level of penetration that is possible in 2020 is depicted in Table 11.5.1.

Table 11.5.1 **EV potential demands by 2020**

Electric vehicle/technology	Potential for xEVs (MUnits)
BEV 2W	3.5–5
HEV vehicles (4W, bus, and LCV)	1.3–1.4
Other BEV vehicles (3W, 4W, bus, and LCV)	0.2–0.4
Total	5–7

Prepared by the authors with data from NEMMP 2020 report; National Electricity Mobility Mission Plan 2020. (2012). National Electricity Mobility Mission Plan 2020 report [online]. Available at: http://dhi.nic.in/writereaddata/Content/NEMMP2020.pdf Accessed 12.12.16.

However, it is also clear that on their own these levels of penetrations are not expected to be achieved and will therefore require government support to spur industry investments. The next section briefly describes about various government support programmes that exist for promoting EV in India.

11.6   Electric vehicle-Programme and policies in India

Since the past few years, the Government of India has been supporting electric mobility efforts. In Mar. 2016, Piyush Goyal, the minister for Power, Coal, New and Renewable Energy stated his goal to make India 100% EV nation by 2030. He announced this at the conference of young Indian, organized by Confederation of Indian Industry. He added that government is making a scheme under which people will get electric cars for zero down payments and has to repay over a period of time from the savings made from fuel (PTI, 2016).

The few other existing supportive EV schemes are mentioned below.

11.6.1   National Electric Mobility Mission Plan (NEMMP)

The National Electric Mobility Mission Plan 2020 (NEMMP, 2020) was launched on 9 Jan. 2013 with a view to provide the future roadmap, establish common set of priorities, broad principles, and framework for promoting the adoption of the full range of electric mobility solutions for India. This plan can enhance national fuel security, provide affordable and environmentally friendly transportation, and enable the Indian automotive industry to achieve global manufacturing leadership.

NEMMP 2020 targets

- Target of deploying 5–7 million electric vehicles in the country by 2020, of which approx 4 millions are expected to be two-wheelers.
- Emphasizes importance of government incentives and coordination between industry and academia.
- Target of 400,000 passenger battery electric cars (BEVs) by 2020—avoiding 120 million barrels of oil and 4 million tons of CO_2.
- Lowering of vehicular emissions by 1.3%–1.5% by 2020.
- Total investment required—INR 20,000—23,000 Cr (approx 3 billion USD).
- There will be approx 180,000 new jobs created, which will lead to country's economic growth.

The NEMMP 2020 is developed with the basis for guiding all the future initiatives, schemes, policies, and other interventions of the government for electric mobility. It will also help shape the future priorities of the industry. While the 2020 levels of xEV penetration projected by NEMMP 2020 is being adopted as the national goals, the various specific formulations provided by the study including the level of demand incentives, their duration, and total investment required are being taken as an indicative estimate that will be fine-tuned during the process of formulating the various specific schemes and interventions. The various schemes and interventions for demand creation, xEV R&D, xEV infrastructure, etc. will be approved and rolled out separately as per govt rules. A mechanism for continued review, monitoring, and mid-course corrections is intended to be an integral part of the NEMMP 2020.

11.6.1.1 Current status of R&D in India and focus areas for xEV

At present, the Indian OEMs and component manufacturers have limited R&D capabilities across all key xEV components. OEMs like TVS (with 5–7 patents), Mahindra Reva (with 10 patents), Tata Motors, and Hero Electric (one patent each) have made some progress on xEV components like battery, power electronics, and electric motor. Indian component manufacturers have no patents yet, and foreign OEM sand manufacturers carry out most of the R&D in their home country.

An industry assessment indicates that high-priority R&D areas for India include battery cell technologies, which will drive affordability and adoption. As per this study, battery cells and battery management systems form the highest-priority areas, as they have a great impact on cost and performance of xEVs. It is expected that localized battery systems customized for Indian weather and traffic duty cycles will yield better performance. The next priority is accorded to xEV powertrain system integration, transmission systems (hybrid), electric motors, and power electronics (especially for HEVs/PHEVs). Localized power electronics can yield better performance at a lower cost, and low-cost motors by the use of nonrare earth magnets can be an important area of technology developed. To achieve the desired efficiency and performance from xEV vehicles, optimized xEV powertrain integration is an important R&D area as well. The technology priority areas for R&D are summarized in Table 11.6.1 as follows (NEMMP, 2020).

11.6.1.2 Existing infrastructure for charging of xEVs

The xEV rollout will require augmentation of existing infrastructure of power generation and transmission and also setting up of charging infrastructure for xEVs. The requirement of additional power generation on account of xEVs is not expected to be very high. However, significant charging infrastructure would be required to be set up particularly for buses. While, in case of buses, charging stations can be located

Table 11.6.1 **Priorities of technology for xEV vehicles development**

Technology	Priority
Battery cell	Priority 1
Battery management system	Priority 2
Power electronics (hybrids)	Priority 3
Electric motor	Priority 4
Transmission system (hybrids)	Priority 5

Prepared by the authors (Authors credit).

in the bus depots, for other vehicle segments, charging stations will need to be set up in apartment complexes, malls, parking garages, workplaces, public buildings, etc.

For charging infrastructure, the aim will be to develop this as a commercially viable business opportunity that attracts private investment. The government will need to facilitate this activity by taking a number of measures that include the following:

• Mandating charging infrastructure in public buildings
• Introducing standards for charging equipment
• Amending building laws to mandate provision for charging outlets
• Providing speedier access to land for setting up of charging infrastructure
• Allowing private retailing of power
• Uninterrupted electricity at reasonable costs for xEV recharge
• Speedy clearances
• Undertaking pilot projects for testing the efficacy of the various charging infrastructure models that can be adopted

11.6.1.3 Manufacturing strategies for xEVs in India

The Department of Heavy Industry has been supporting the growth of the Indian auto component sector through various programmes and is working on various new initiatives like the Automotive Skills Development Council, developing an Auto Component Technology Development fund, etc. These initiatives will need to be augmented, and efforts to improve the competitiveness of the Indian auto component supply chain will be the essential prerequisite for developing an efficient and competitive indigenous xEV component supply chain.

Meeting the domestic demand for xEVs will necessitate strengthening of the local manufacturing and supply base. The development of the local xEV supply chain is sought to be achieved through a phased approach for gradually building the domestic manufacturing capabilities for the various vehicle segments. In addition, this strategy will also involve the essential requirement of linking government support for demand creation to firm commitments by the industry to reach predefined levels of domestic value addition in terms of the percentage value of the local xEV components used in the xEV. This will help create, largely, the required impetus for the supply side investments and commitment from the industry for creating the xEV supply chain. The committed level for localization is also proposed to be gradually increased in a phased manner during the period wherein demand incentives continue to be provided.

In addition, in order to encourage manufacturers having higher level of localization content, the possibility of having distinguishing/gradation localization criterion for getting higher level of demand incentives (in terms of quantity of vehicles covered or the amount of incentive provided) can also be explored. Further, the possibility for the range of other enablers to fast track investments will also be seen.

As per the joint government-industry study, a three −/four-phased approach spread over the next 10 years, depending upon the vehicle segment, will be adopted to build the manufacturing capability for xEVs in India. In the case of the four-wheelers, the first stage of development (0–4 years) would involve the strengthening of the domestic assembly. During this phase, the industry will develop high capability in manufacturing with local assembly of xEVs using imported or local components. Simultaneously, the local sourcing can also be increased translating to moderate capabilities on this front. Further, the industry and government will also initiate investments in R&D, and product development centres as the current capabilities in these areas are quite low in the country. The second stage from the next 5–8 years will involve developing indigenized products. By this time, it is expected that the R&D and industry product development centres would have attained moderate capabilities for developing indigenized components (BMS, transmission system, electric motors, etc.). At this point, the country will target $\sim$25–30% of xEV components in terms of value to be sourced locally. In order to support the reasonably developed local supply chain, during this phase, the customs duty exemptions for xEV components will also be phased out completely.

11.6.2 Faster adoption and manufacturing of (hybrid) and electric vehicles (FAME)

Faster adoption and manufacturing of hybrid and electric vehicles in India is a part of the National Electric Mobility Mission Plan, which was launched on Apr. 2015. Under this scheme, there are four major subgroups: (a) vehicle system integration, (b) batteries and battery management system, (c) motors and controllers and power electronics, and (d) batteries and battery management system. Each of these subgroups will be having their own individual Centre of Excellence (COE) setup to obtain desired rapid results.

The scheme envisages Rs 795 crore supports in the first two fiscal years starting with 2015–16. Initially, the scheme is envisaged to implement in metropolitan and smart cities and gradually extended to all major cities in India. INR 14,000 crore is estimated as the total requirement to implement the scheme. Phase 1 of the scheme will be implemented over a 2-year period in 2015–16 and 2016–17 with an approved outlay of INR 795 crore, out of which INR 500 crore will be spent on demand incentives. Table 11.6.2 shows the break-up of the estimated outlay of the initial 2 years.

From Table 11.6.2, it can be noticed that there is significant amount invested on pilot projects. Few examples of current pilot projects are the following (Press Information Bureau, 2016):

- Seven-seater EV to be operated for last 2 km at Taj Mahal (one of the seven wonders of the world), Agra
- Use of two-wheelers EV for home delivery by Dominos and KFC
- Three- and four-wheelers for fruit and vegetable distribution and garbage collection
- EV for taxi, corporate hire, and rental scheme
- Last-mile connectivity from metro stations
- Hybrid and electric buses for public transport

Under this scheme, the customer can get the incentive in the form of lower cost of hybrid or electric vehicles at the time of its purchase. Manufacturers can claim the incentive from the government at the end of each month.

Table 11.6.2 Break-up of the estimated outlay of the initial 2 years of FAME

Areas	2015–16 INR Cr (million USD)	2016–17 INR Cr (million USD)
Technology platform (+ testing infra)	70 Cr (10.8)	120 Cr (18.6)
Demand infrastructure	155 Cr (24)	340 Cr (52)
Charging infrastructure	10 Cr (1.5)	20 Cr (3.1)
Pilot projects	20 Cr (3.1)	50 Cr (7.7)
IEC/operations	5 Cr (0.7)	5 Cr (0.7)
Total (INR)	260 Cr (40.3)	535 Cr (83.1)
Grand total (INR)		795 Cr (123 million USD)

Prepared by the authors with data from NEMMP 2020 report; National Electricity Mobility Mission Plan 2020. (2012). National Electricity Mobility Mission Plan 2020 report [online]. Available at: http://dhi.nic.in/writereaddata/Content/NEMMP2020.pdf Accessed 12.12.16.

As per the scheme, depending on technology, battery-operated scooters and motorcycles will be eligible to demand incentives as mentioned in Table 11.6.3. According to Indian government, FAME scheme is closely integrated with some of the leading other policies, schemes, and missions being operated (Press Information Bureau, 2016):

- Mission—Make in India
- National Transport policy
- National Action Plan for Climate Change (NAPCC)
  ○ National Solar Mission
  ○ National Mission for Sustainable Habitat (NMSH)
  ○ National Mission for Enhanced Energy Efficiency (NMEEE)

It is assumed that the FAME scheme will also help domestic manufacturers to develop and sell their vehicle in India and export it.

11.6.3   Other schemes

The Ministry of New and Renewable Energy (MNRE) ran one of the recent demand generation incentive programmes in India for xEVs, namely, the Alternate Fuels for Surface Transportation Programme (AFSTP). This programme provides demand incentives for all xEVs, amounting to a total of Rs 95 crore between 2010 and 2012. This was available to OEMs giving at least 1-year warranty and setting up 15 service stations across India.

Various states have also been providing incentives for electric vehicles. For instance, the Delhi government provides (i) a 15% subsidy on the base price, (ii) 12.5% exemption of VAT, and (iii) refund of road tax.

In addition, the Department of Science and Technology (DST) has a funding scheme for all industries, including automotive industry. The Council for Scientific and Industrial Research (CSIR) is also active, and R&D on lithium-battery technology

Table 11.6.3 **Incentives for electric vehicles in India under FAME Scheme**

Demand side incentives announced under FAME India		
Vehicle segment	**Minimum incentive (INR)**	**Maximum incentive (INR)**
Two-wheeler scooter	1800 (30 USD)	22,000
Motorcycle	3500	29,000
Three-wheeler autorickshaw	3300	61,000
Four-wheeler cars	11,000	1,38,000
LCVs	17,000	1,87,000
Bus	30,00,000 (47,000 USD)	66,00,000
Retro fitment category	15% or 30,000 if reduction in fuel consumption is 10%–30%	30% of kit price or 90,000 if reduction in fuel consumption is more than 30%
• Availed by buyers upfront at the point of purchase.		
• Manufacturers reimbursed by the Department of Heavy Industries.		

Prepared by the authors with data from NEMMP 2020 report; National Electricity Mobility Mission Plan 2020. (2012). National Electricity Mobility Mission Plan 2020 report [online]. Available at: http://dhi.nic.in/writereaddata/Content/NEMMP2020.pdf Accessed 12.12.16.

for xEVs is ongoing at the Central Electrochemical Research Institute, Karaikudi. Furthermore, certain state governments like the Delhi govt are also providing demand side subsidies in addition to VAT and road tax waiver. Owing to these various programmes, xEV two-wheelers have reportedly witnessed the growth of 20%, and Reva recorded a threefold rise in average monthly sales. However, these initiatives remain largely fragmented and short term.

11.7   Charging infrastructure for xEVs

The charging infrastructure broadly includes level 1 terminals, level 2 terminals (fast chargers), and level 3 terminals (rapid chargers). The typical time taken for charging by these chargers is 6–8 h, 3–4 h, and less than 30 min, respectively. These chargers also vary substantially in their costs; therefore, the charging terminals need to be created as per the requirement of the location. While level 1 charging terminals are appropriate at residences/offices, building/parking lots, etc., the level 2 charging terminals are more suited for areas where vehicle is likely to be parked for shorter but substantial duration of time like commercial areas (shopping malls), airport/railway stations, and also some at parking lots. The rapid chargers are best suited to be located at convenient locations like petrol pumps. A mix of these public charging points needs to be established on a pilot basis to evaluate consumption pattern and the optimal type and location mix for the different charging terminals.

The extent of linkage between the availability of public xEV charging infrastructure and adoption of xEV varies between different vehicle segments, for different usage patterns (average vehicle kilometres travelled). For instance, electric two-wheelers are less dependent on public recharging infrastructure for their off take as compared with electric four-wheelers. Further, within the two-wheeler segment, the consumers whose average vehicle mileage is lower are less impacted by the non-availability of public recharging infrastructure, as they can recharge their vehicles at home just like their mobile telephones. In addition, the fleet operators like the state-run intracity bus services have a totally different and unique recharging infrastructure requirement for their electric bus fleet. It is, therefore, necessary that the public charging infrastructure strategy and planning have to be unique to match the context in which it is being set.

Given the current status of xEV charging infrastructure in the country and the likely levels of infrastructure investment required, need to initially undertake pilot projects, lead time required for setting up infrastructure, and the likelihood for amendments/modifications in building byelaws and other laws by various governments (central, state, and local), there is likely to be a significant gestation time.

Besides ensuring availability of reliable, regular electricity supply, and adequate xEV charging infrastructure, the government will need to ensure standardization of batteries and recharging related components, as this is an essential ingredient for a successful rollout of recharging infrastructure.

The strategy of xEV infrastructure rollout involves a three-phased approach. The phase I (first year) or the initial preparatory phase will involve achieving immediate to short-term objectives. This will entail greater detailed and in-depth evaluation of various options, prioritization, and putting in place the required framework, enabling policies, charging infrastructure standards, laws, and undertaking detailed studies that will facilitate the rollout of the optimum xEV infrastructure. The various activities to be taken up immediately include the following:

(a) Introduction of standards for vehicle to grid interface, components, batteries, and other charging infrastructure.
(b) Assessment of the changes to the legal framework required for facilitating xEV infrastructure rollout for daytime recharging, public recharging infrastructure at petrol pumps, office locations, shopping centres, parking lots, mass transport terminals, streets, etc. will be required.
(c) Assessment of the various possible incentives for setting up of recharging infrastructure that may include tax breaks, grants, and soft loans.
(d) Other innovative electric charging business models, for example, smart metering and grid powering from batteries will also need to be examined and developed.
(e) Developing the infrastructure and business models for reuse and recycling of Li-ion batteries.

Table 11.7.1 Assessment of the segment-wise investments required for xEV infrastructure

Area	4W		2W	3W	
	HG/ HEV	HG/ HEV/B EV	HG/HEV	HG/HEV/BEV	
				HG/HEV	
Additional generation capacity (MW)	150,225		600	10–15	
Power infrastructure (Rs Crore)	700–800	1200–1300	3300–3400	40–50	75–85
Charging infrastructure (Rs Crore)	700–800	950–1000	–	40–50	70–80

Prepared by the authors (Authors credit).

(**f**) Developing strong linkages between the electric mobility initiative with the national renewable energy generation efforts: In case the energy generation is not clean and transmission losses remain high, the environmental benefits of shifting to electric mobility cannot be fully realized; therefore, in order to get the full environmental benefits of electric mobility, the quality of power generation as also the efficiency of power transmission and distribution needs to be addressed in a holistic manner.

In terms of the assessment made by the joint government-industry study, the amount of total investments needed for setting up the required infrastructure up to 2020 (both power and charging infrastructure), vehicle segment wise, is summarized in Table 11.7.1.

11.8   Existing business model of e-rickshaws

The business model for the most commonly used e-rickshaws in India is studied and is summarized below:

11.8.1   Upfront costs

The main upfront cost involved in the e-rickshaw business is the cost of the e-rickshaw. It was found from various surveys carried out by expert institutions that initial investment of about USD 1500 (inclusive of all taxes) is required to own and operate an e-rickshaw. Besides the e-rickshaw purchase cost, there are no other upfront costs for an e-rickshaw since these vehicles are not getting themselves registered or obtaining any kind of permits for operations. While there was no official registration being done, it was found that some sorts of route permits to ply the e-rickshaws in certain areas are being informally executed by the route union in many parts of the countries. The cost towards such route permits, and its validity varies from location to location.

11.8.2   Operational costs

The only daily operational cost involved is the cost of charging the batteries. In general, 4–8 electrical units (kWh) are consumed to charge a fully discharged battery to a full charge, and the battery runs for 8–10 h with one full charge. The batteries are generally charged at night, and typically, USD 0.7–1 is spent daily for charging the battery. Additionally, some of the e-rickshaw operator also charges the battery during the daytime to ensure more operations during the day and until late evenings. This additional day charging costs them about USD 0.3–0.4 (in addition to night charging). For those drivers, who are running rented e-rickshaws, the daily rentals include the cost of charging in many of the cases; e-rickshaw owners in this case charge the battery at night at their residence and recover the cost of charging in the rent that is charged.

11.8.3   Maintenance costs

The key maintenance cost involved in e-rickshaw is the battery replacement cost, which is around USD 200–400 including the guarantee period for 6 months. As per the e-rickshaw drivers, the life of the batteries is very unpredictable and do not exceed more than 7–8 months, and hence, replacing the batteries turns out to be a big recurring cost. In most of such cases, e-rickshaw drivers opt for a battery replacement scheme, in which the old batteries are handed over to seller, to fetch a discount on the price of the new batteries. However, this discounted price of batteries varies from place to place.

In addition to battery replacement, other general maintenance cost includes replacement of battery charger, motor, controller, and tyres. These costs are around USD 80 per annum.

Moreover, the minor maintenances such as repairing punctures, welding broken parts, and fixing the headlights/horn/indicators ranges between USD 2–3 over a period of 1 year. In cases where the e-rickshaws are given out on rent, it is the owner's responsibility to pay for the major maintenance cost such as battery replacement, motor replacement, and tyre replacement, while the e-rickshaw driver pays for the minor maintenances.

11.8.4   Revenue generation

The main source of revenue for the e-rickshaw drivers is the fares charged from the passengers. Although the fares vary from places to places, in general, it is a flat fare of USD 0.13–0.15 on a shared basis for a distance of 2–5 km. The fare increases if the passengers travel a longer distance. Other than fares, majority of the drivers do not have any secondary means of revenue. However, in few locations of Delhi and tier two cities, some drivers are paid USD 2–3 per month by business owners for advertising products/shops on the e-rickshaws.

11.8.5   Parking and charging infrastructure

There are no exclusive parking spaces for the e-rickshaws in the cities. The e-rickshaw drivers have managed themselves by parking on places such as roads near metro stations, markets, residential areas, and commercial centres. During night, majority of the e-rickshaws (generally around 70%–80%) are parked at the homes of the e-rickshaw drivers. Among drivers who have taken their e-rickshaws on rent, many of them park the vehicles at their homes or at e-rickshaw owner's house, and very few of them parks at parking facilities.

11.8.6   E-rickshaw driver perceptions

Few surveys have been carried out with the e-rickshaw drivers in Delhi, in the Sunderban in West Bengal, and few tier two cities in India to understand their perception regarding e-rickshaw. The findings and the observations are summarized below:

- Majority of the drivers felt that the passengers actually prefer e-rickshaws over cycle rickshaws and autorickshaws. The reason for this preference according to them is that their service is mostly cheaper besides being easier to board and deboard due to low clearance height. In few occasion (such as in case of the Sunderban), e-rickshaws have more steady ride without sudden jerk, which is preferred by older passengers, pregnant women, etc.
- They felt that the e-rickshaws are not rugged enough and they fear of causing accident. However on the other hand, few drivers felt that there is no danger of major accidents as the e-rickshaws runs at lower speed as compared with other vehicles.
- The e-rickshaw drivers had issues regarding the battery discharge. Therefore, they tend to constantly check the battery status and head back home when the battery charge goes below 30%.
- Few of the e-rickshaw drivers felt that the e-rickshaws were inadequately covered and therefore posed difficulties to passengers during winter and monsoon season.

Majority of the e-rickshaw drivers wanted the e-rickshaws to be registered because, by that way, it would give them a sense of ownership and prevent thefts. The problem of theft of e-rickshaws is very rampant in the city, and many people are forced to sleep in their e-rickshaws during the night to prevent these thefts. Once stolen, these drivers have no way of tracing their e-rickshaws and not even the option of registering a complaint, because they do not have documents proving the ownership or a unique vehicle number/identity of the vehicle. However, few of the e-rickshaw drivers were not in the favour of the registration of their e-rickshaws because they feared that, it would increase the overall costs, while some drivers feel that the registration process would not make a difference in their current situation.

The responses of the e-rickshaw drivers in terms of issuing a driving licenses/permit were positive, in majority of the cases. The reason for that is driving license would give them recognition for being able to drive the e-rickshaw and prevent underage and very old people from driving these vehicles. They also felt that issuing driving licenses would force drivers to drive safely and help in case of accidents.

11.8.7   E-rickshaw users' trip details

As per a survey carried out by TERI, half of the e-rickshaw users being surveyed were using the vehicle as a means of first- and last-mile connectivity, which highlights the fact that these e-rickshaws cater to the much-needed first- and last-mile connectivity options in the city. Another major chunk of the users (around 31%) were using the e-rickshaw for executing their main trips, thereby replacing other vehicles for the same trip purpose (see Fig. 11.8.1).

The average trip length of an individual passenger travelling in an e-rickshaw is 1.74 km. The estimated maximum and minimum trip lengths ranges between 0.5 and 4 km, which shows that the passengers prefer e-rickshaws for short lengths only, i.e. 80% of trips were less than 2 km.

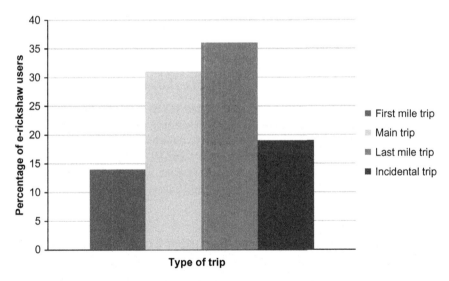

Fig. 11.8.1 Type of trips for which e-rickshaws were being used.
Prepared by the authors with data from Singh, S. (2014). A study of the battery operated
E-rickshaws in the State of Delhi [online]. Available at: https://ccsinternship.files.wordpress.com/
2014/06/323_study-of-the-battery-operated-erickshaws-in-the-state-of-delhi_shashank-singh.pdf
Accessed 02.01.17; TERI Report on Study of the battery-operated e-rickshaws in the State
of Delhi, 2014.

11.9   xEV Bus pilots in India

Besides, e-rickshaw, few initiatives have been taken in India to promote electric vehicles for public bus transportation. These are briefly mentioned as below:

11.9.1   Tata motors hybrid bus pilots in Delhi and Mumbai

During the 2010 Commonwealth Games in New Delhi, Tata Motors presented four CNG-electric hybrid low-floor starbuses to the Delhi Transport Corporation (DTC). Similar Tata hybrid buses are also being test run by Brihanmumbai Electric Supply Transport (BEST), Mumbai. A hybrid bus costs around USD 160,000 (according to year 2010). The price differential between an ICE bus and the Hybrid bus is almost USD 57500.

In Mumbai, these buses have helped BEST save more than 10% on fuel and lower emissions by 30%, as compared with conventional buses. The hybrid buses offer substantial improvement in fuel economy compared with a conventional bus. As a result, the usage of this technology leads to lower emissions, thereby contributing to cleaner air and a greener, more environment-friendly commercial passenger transportation application.

Tata Motors claims that its compressed natural gas (CNG) and electric hybrid bus, which comes fitted with start-stop technology, will offer 22% higher mileage and thus reduce emission by the same level.

11.9.2   Plug-in-hybrid bus by Ashok Leyland

Hinduja Group flagship Ashok Leyland unveiled HYBUS at the Auto Expo 2010. HYBUS was on roads for the first time in India during the 2010 Common Wealth Games in New Delhi. Ashok Leyland delivered two buses to the Organizing Committee of the Commonwealth Games to ferry VVIPs and VIPs during the inaugural ceremony in Delhi. The buses were also used later for the movement of media persons during the Games at Pragati Maidan, New Delhi (Sharma, 2014).

11.9.3   Bangalore metropolitan transport Corporation (BMTC) plans to introduce hybrid buses in its fleet

In a recent development, the Government of India has sanctioned 50 hybrid buses for the Bangalore Metropolitan Transport Corporation (BMTC) to be operated as a trial run on the city roads. This would place Bangalore among the first cities in India to run a hybrid bus fleet. This plan is expected to be implemented in the city by Jan. 2014 (Business Standard, 2010).

11.10 Existing challenges and strategies for faster adoption of EVs

The existing challenges that India is facing with respect to faster adoption of EVs include (i) consumer acceptability for EVs due to limited consumer awareness, (ii) current price-performance gap, (iii) limited manufacturing investments include the limited domestic manufacturing capabilities and non-existent supply chain, and (iv) the lack of xEV-related infrastructure.

In order to overcome these barriers, the creation of market for xEVs through greater consumer acceptability will have to be the starting point. This should be done along with the introduction of demand incentives and localization of manufacturing of EVs in order to rescue the price gap between the EVs and the normal ICE vehicles. Besides these, the efforts for addressing the infrastructure related issues need to be addressed.

The two most important drawbacks highlighted by the industry that need to be improved in the earlier and existing xEV demand incentive schemes are (i) the lack of continuity of the incentive policy/scheme for sufficiently long duration that is necessary to give confidence to the industry for committing large investments for setting up domestic manufacturing facilities. Currently, the existing schemes are extended on year-to-year basis without any assurance on their total duration and budget. As a result, the manufacturing/value addition in the country is currently restricted to assembly operations with critical components being imported. (ii) Simple and effective incentive delivery mechanism allows the incentive to reach the intended beneficiaries quickly and efficiently. Current schemes leave a lot of scope for the improvement for this. As such, the learning from efforts made so far highlight the importance of having upfront commitments on total size (number of vehicles covered and amount of incentive per vehicle), duration of a demand incentive policy/scheme, and an efficient incentive delivery mechanism for the scheme to be effective in faster adoption of xEVs, for spurring domestic investments and for the creation of domestic xEV manufacturing ecosystem.

In addition, the past efforts also did not have the desired level of synergy, continued top-level support, and ownership both in the government and industry. As such, most of the efforts undertaken fizzled out since they were isolated in nature, lacked collaborative approach and did not tackle all the issues holistically. In order to achieve the potential, a more systematic and collaborative approach is required with a clear long-term roadmap.

11.11 Overall analysis of EV

There are two major type of analyses that are carried out in this section: (a) strength, weakness, opportunity, and threats (SWOT) and (b) political, economical, social, technological, environmental, legal (PESTEL) (Bej, 2015).

11.11.1 PESTEL analysis

Political	Economical	Social	Technological	Environmental	Legal
Government is continuously supporting EV sector by launching policies like FAME, NEMMP Government wants to reduce national carbon emission through EVs	Stable economic policies Availability of finance with ease Economic incentives should be available easily Should have low duties and taxes Car ownership is increasing	Rural market could be beneficial/ profitable Increasing urbanization and income level Skilled labours available at cheapest cost Purchasing EV can symbolize the positive attitude towards saving nature/ environment	Thousands of engineers pass every year, which can add greater strength to the sector High-quality component suppliers are available Modern people have high acceptability rate of EVs	Indian government is trying to closely align the automotive emission regulations up to worldwide standards Moving towards EV can significantly help Indian government to achieve carbon emission targets	Tax deduction High import duties

11.11.2 SWOT analysis

Strength	Weakness	Opportunities	Threats
Low running cost Energy-efficient as compared with conventional cars Reduce dependence on foreign oil imports Low charging cost— less electricity price at charging time, i.e. night Increasing amount of research and development	High initial price Limited range of options available Driving range is almost five times lower than conventional cars Recharging time is significantly high Lack of spare parts, in case of repair or maintenance	More opportunity for R&D Increases job for local public Skilled labour available at cheaper rates Encourage consumer/ buyers with trained salespersons at dealer shops	Different manufacturers may have different charging plugs Existing charging infrastructure is not capable enough to support EVs growth The lack of compaction leads to the lack of designs Declining petrol and diesel prices can reduce EV sales

Manufacturing utilities of conventional cars can be used	Less sales decreases overall profit	Increase mass production to decrease initial cost
Can raise Indian economy, i.e. by revenue		

11.12   Recommendations

11.12.1   Charging infrastructure

As seen in Section 11.6, there are some initiatives taken to grow the charging infrastructure. However, it is a fact that India still does not have enough infrastructures that is required to deploy the EVs into market. Hence, it is creating an environment of inconvenience in the people for adoption on EVs.

11.12.2   Public awareness

Though government has announced many an inactive to have India as 100% on EVs by 2030, it is vital that public have enough of awareness about the advantage of EVs. According to Gill (CEO of Hero Electric), 'awareness continues to be abysmally low. The government should conduct large-scale awareness programmes. Individual OEMs will concentrate only on limited markets where they are present'. According to Gill, only 30% of Indians are aware about EVs, which is the key reason for slow deployment of EVs. She suggested that EV awareness should be increased along with the increasing policies and subsidies for EVs.

One such example for raising public awareness and promoting green vehicles is of Hero electric company. This company is providing 100 electric scooter free rides to Delhi people. According to the CEO of the company, the outcomes of this initiative were significant as 6000 free rides are completed so far.

11.12.3   Lack of manufacturers

One of the reasons for slow progression of EV into Indian market is the lack of manufacturers, specifically four-wheelers. In addition, FAME scheme is only supportive to Indian manufacturers, as they are only eligible for incentives.

11.12.4   Emission standards

Some Indian cities had made a drastic change in vehicular technology such as replacing petrol and diesels three-wheelers to CNG to attain the emission goals. However, Indian government can propose a plan towards stricter emission standards rather than replacing the complete category of vehicles.

11.12.5 EV as public transport (bus system)

Many cities of India are missing with public transport, and hence, they are developing their public transport infrastructure and vehicles (buses). Hence, it could be a good idea that those public transport buses could be as electric buses. However, some cities of India have an existing infrastructure for bus rapid transit system (BRTS), where it would be relatively easy to modify the existing infrastructure and run electric buses. In addition, the structure could easily add rapid chargers, demonstrated by ABB company, which can charge an electric bus within 15 s.

11.12.6 Integration of electric mobility within urban policies and plans

Integration of electric mobility into urban policies and plans such as national solar mission and smart grid would encourage consumers to use renewable electricity and use it to charge their electric vehicles. This could be done by installing a smart metre onto consumers house, i.e. owner of EVs. The consumer can charge the car at lower rate of electricity generated by renewable sources, rather than paying high rates of electricity generated by traditional sources. The other advantage is that the car owners can use EVs as a source of storing electricity, which could be used for critical household appliances in case of power shortage.

11.12.7 Financial support

Considering the current status of EV into Indian market, it seems to be essential that EVs should gain substantial amount of financial support by government. It has been noticed that, initially, there were 20 companies involved in Society of Manufacturers of Electric Vehicles (SMEV), but after the drop out of funding AFSTP, nearly eight manufacturers shut down their business. Moreover, there was 40% decline in a number of units manufactured by Mahindra Reva, after the termination of subsidy in the year 2012–13. However, after the launch of FAME, manufacturers are returning to Indian market.

Furthermore, Indian government can provide (a) concessions in sales tax, excise and custom duty and (b) providing significant grants for research and development of areas such as vehicle technology, battery technology, and battery-charging infrastructures.

11.13 Steps to promote EVs

11.13.1 Charge while parked facilities

Finding a parking space in most of the cities in India is difficult and expensive task. If some parking space were pre-booked with charging facility and incentivised rates for parking then, it would encourage people to move towards EVs.

11.13.2   Arrangements for EV in traffic

In many cities of India, there are cycle tracks that are being constructed along with bus lanes. Hence, it could be a good idea that slow-moving (speed less than 25 km/h) electric two-wheelers are allowed on cycle tracks. This would increase the safety and avoid congestion. Similar sort of prioritized arrangements can be made for four-wheeler electric cars along with public taxis and buses.

11.13.3   Incentives for EVs

For faster adoption of EVs, government could implement some of the incentives to local people, such as the following,

- Specific lanes for EVs on national highways
- Designated toll booths for EVs
- Permitting to EVs to drive on BRT lanes
- Concessional charging tariffs of EVs during nonpeak hours
- More incentives for solar PV-based and storage-integrated charging stations

11.13.4   Motivating the performance oriented market

One of the ideas to promote EVs is to arrange a competition between different companies and publicize it as much as possible to local people. It has been noticed from the United States (US) EV market increased since Tesla Motors Inc. has shown that EVs can perform better than conventional power-driven vehicles. Hence, Indian government can arrange such campaign, and market/companies can participate and see what other competitors are performing and what they can do to increase themselves.

11.14   Conclusion

A growth of vehicles in India is inevitable. There are various initiatives taken in the country to promote electric mobility that could not yield the desired results mainly due to the higher cost of EVs, challenges in battery technology, limited range of EVs, and the lack of infrastructure including consumer mindset. However, with the advent of these new mission plans such as 100% EV nation by 2030 and schemes such as FAME to provide subsidies on electric and hybrid cars will encourage the manufacturers to introduce more EVs in Indian market.

References

Andrew, A. F. (2016). *Professor Emeritus.* Davis, CA: Department of Mechanical and Aeronautical Engineering, University of California.

Bej, S. (2015). Feasibility analysis of electric vehicles in India [online]. Available at: http://www.slideshare.net/SushovanBej/feasibility-analysis-of-electric-vehicles-in-india Accessed 05.07.16.

Business Standard, (2010). Business standard official website [online]. Available at: http://www.business-standard.com/article/press-releases/ashok-leyland-s-hybus-india-s-first-plug-in-cng-hybrid-bus-110100700136_1.html Accessed 29.01.17.

3 Wheeled Cheese, (2012). Launch of new report on auto-rickshaws in India [online]. Available at: https://3wheeledcheese.wordpress.com/2012/02/10/launch-of-new-report-on-auto-rickshaws-in-india/ Accessed 01.07.16.

India Energy Security Scenario (IESS) 2047, (2016). [online] Available at: http://www.indiaenergy.gov.in/ Accessed 30.01.17.

Mahindra Electric, (2016). Mahindra Electric official website [online] Available at: http://mahindrareva.com/product/explore-the-e2o.aspx Accessed 25.06.14.

Naik, A. (2016). Top 5 hybrid/electric cars in India [online]. Available at: http://auto.ndtv.com/news/top-5-hybrid-electric-cars-in-india-757164 Accessed 02.08.16.

National Electricity Mobility Mission Plan 2020, (2012). National Electricity Mobility Mission Plan 2020 report [online]. Available at: http://dhi.nic.in/writereaddata/Content/NEMMP2020.pdf Accessed 12.12.16.

Press Information Bureau, (2016). Anant Geete launch's the scheme for faster adoption and manufacturing of (hybrid &) electric vehicles in India—Fame India [online]. Available at: http://pib.nic.in/newsite/PrintRelease.aspx?relid=118088 Accessed 10.07.16.

PTI, (2016). The Economic Times. India aims to become 100% e-vehicle nation by 2030 [online]. Available at: http://economictimes.indiatimes.com/industry/auto/news/industry/india-aims-to-become-100-e-vehicle-nation-by-2030-piyush-goyal/articleshow/51551706.cms Accessed 15.07.16.

Sharma S. (2014). BMTC floats tenders for hybrid buses [online]. Available at: http://www.deccanherald.com/content/389418/bmtc-floats-tenders-hybrid-buses.html   Accessed 28.01.17.

Shukla, P. R., Dhar S., Pathak, M., Bhaskar, K., (2014). Promoting low carbon transport in India [online]. Available at: https://www.researchgate.net/profile/Minal_Pathak/publication/271447336_Electric_Vehicle_Scenarios_and_a_Roadmap_for_India/links/54c867250cf238bb7d0de64f.pdf Accessed 20.07.16.

Further Reading

Singh S. (2014). A study of the battery operated E-rickshaws in the State of Delhi [online]. Available at: https://ccsinternship.files.wordpress.com/2014/06/323_study-of-the-battery-operated-erickshaws-in-the-state-of-delhi_shashank-singh.pdf Accessed 02.01.17.

Society of Indian Automobile Manufacturers (SIAM), (2016). [online]. Available at: http://www.siamindia.com/statistics.aspx?mpgid=8&pgidtrail=13 Accessed 05.01.17.

Recharging of electric cars by solar photovoltaics

Irene Illescas García, Michael Jeffrey
Edinburgh Napier University, Edinburgh, Scotland, United Kingdom

12.1 The solar meadow farm at Edinburgh College

The solar plant at Edinburgh College, Midlothian Campus, South East Scotland (Fig. 12.1.1), is a £1.2 M, equal-partnership project between Edinburgh College and Scottish and Southern Electricity (SSE) Energy Solutions, who installed and maintain the site. Located in Dalkeith, to the south-east of Edinburgh, the plant was commissioned and built for a number of different purposes: to produce carbon-free, renewable energy; to generate an extra revenue stream for the college; and to make good use of a brown field site.

For the college, there is also the opportunity to use it as a research and learning environment, where students can get hands-on experience with modern renewable technology. The site has now been made dual-purpose by planting wild meadow grasses and encouraging the growth of local wildlife, making this plant Scotland's first solar meadow. An area of wetland and a bank of beehives on the site will further expand the potential ecological benefit to the area.

The 627.5 kWp installation, as shown in Fig. 12.1.2, comprises 2560 solar panels each rated at 230 W, 32 P Power-One Aurora Trio-20.0-TL inverters, and 20 kW inverters and attendant cabling, framing, and housing. In fact, over 11 km of cable

Fig. 12.1.1 Midlothian solar meadow farm.

Electric Vehicles: Prospects and Challenges. http://dx.doi.org/10.1016/B978-0-12-803021-9.00012-4

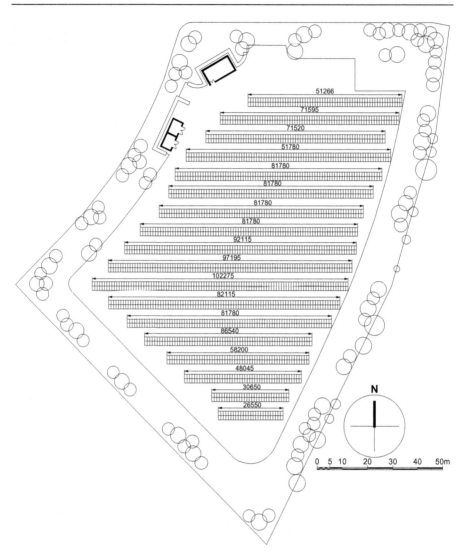

Fig. 12.1.2 Site plan of the meadow.

was laid out to connect the AC and DC sides of the system, while almost 1000 ground screws were planted to support the module mountings.

In 1 year, the site is predicted to produce 568 MWh of electrical energy, enough energy to power 170 homes. Feeding directly into the national grid, this energy will offset the conventional electricity production to save approximately 293,000 kg of CO_2 per year.

The photovoltaic (PV) modules used for the plant are Astronergy crystalline PV module CHSM6610P series. The polycrystalline modules have a nominal power output of 230 W, with an overall module efficiency of 14%. Modules are connected in series strings of 20 modules each, with 4 parallel strings per inverter. Power-One

Aurora Trio inverters are used, with a maximum power input of 20 kW each a nominal efficiency of 98%. The modules are mounted on framing orientated due south, with an inclination of 25.

While the site has now been completed and in fact started producing on the 28 March 2013, contractual details have only recently been finalized. This proved problematic for the completion of this project and for the college in general, as the site and data access were not always available.

There were a number of issues to overcome in the first year of operation. Initially 6 out of the 32 inverters were not available for the parallel connection, thus preventing maximum yields for the first 2 months of operation. This limiting effect culminated in the inverters only being able to operate at 12 kW instead of the 20 kW expected, i.-e. these inverters were only operating at 60% of their maximum input power.

A problem with transformer balancing and variable phasing caused the substation to trip out on the solar meadow side. This has meant that a total of 4–5 weeks generation have been lost. Both of these factors have contributed to the shortfall in production during year 1 of operation. By May 2014, 440 MWh of power had been generated against a projected standard assessment procedure (SAP) projection of 431 MWh that included reduction due to shading. Subsequently, further works have been carried out to install fans and heat sinks to all the inverters to increase heat dissipation. In addition, the transformer taps have been changed on the national grid side to lower the operating level to a point where phase shifts and grid fluctuations will have less impact at peak times.

It is also worth mentioning that from Sep. 2015 to Mar. 2016 a total capacity of 35.1 MW electric has been installed in solar farms equal or larger than 1 MW in Scotland (see Fig. 12.1.3), the total installed capacity of PV in Scotland being approximately 248 MW in Q1 2016, including the solar farms with an installed capacity below

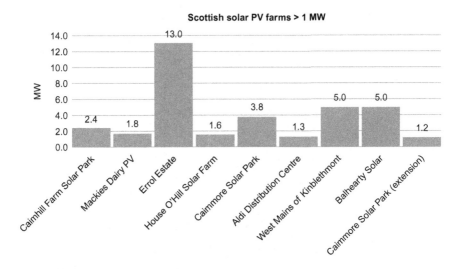

Fig. 12.1.3 Scottish solar PV farm >1 MW.
Data from Federschmidt (2016).

1 MW electric and private solar rooftop arrays. In 2015, all installed PV plants in Scotland generated a total electricity output of 187 GWh, a share of 0.86% of the overall renewable energy electricity generation (Scottish Renewables, 2016).

12.1.1 Shading analysis of the solar meadow farm

In the first stage of the project, a site survey was undertaken. This was done in order to ascertain what measures were required to undertake the later experimental work, to ensure the reference documents such as site plans and electrical connections were accurate and also to determine what shading, if any, was present on the site.

Shading is an important factor in the operation of any solar installation, as it not only cuts down on the solar energy available to the system to be converted to electrical output but also interferes with system operation: one shaded module can affect the performance of an entire string.

A basic shading analysis was undertaken at five points (Fig. 12.1.4) towards the south end of the site where shading effects are more severe. The main obstacles to be considered were the tree lines to the south-east and south-west and the earth wall enclosing the site.

Elevations and positions were derived from both the site plan and measurements taken by theodolite. These elevations were then used to construct horizon lines for each point. Examples are shown in Figs. 12.1.5 and 12.1.6.

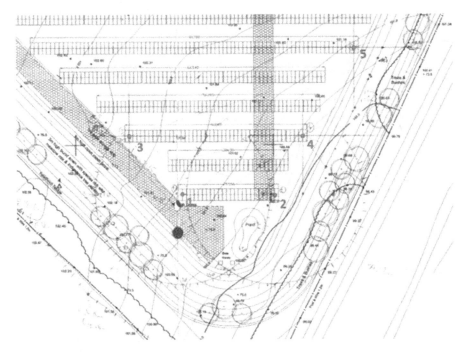

Fig. 12.1.4 Shading analysis locations (based on Fig. 12.1.2).

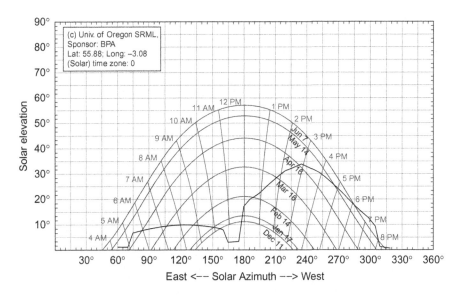

Fig. 12.1.5 Shading at point 2.

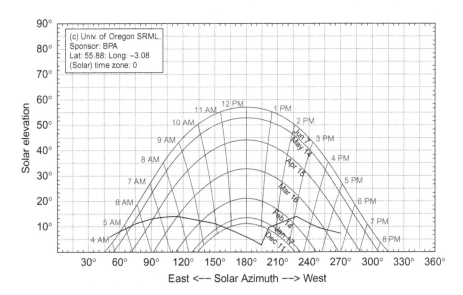

Fig. 12.1.6 Shading at point 5.

The shading analysis indicates that a significant portion of incident radiation will be lost to shading effects for the southernmost rows of modules. This shading will be more serious in the afternoon for the southerly points and in the morning for the easterly points.

Table 12.1.1 **Green energy supply**

Month	Monthly energy (kWh/m²)	Farm monthly energy (kWh)	Shade delivery percentage (%)	Output energy with shading (kWh)
January	5.27	19,659	70.00	13,762
February	13.93	52,013	80.00	41,610
March	12.10	45,152	91.00	41,088
April	17.24	64,364	99.00	63,720
May	27.71	103,440	100.00	103,439
June	36.47	136,113	100.00	136,113
July	25.45	95,003	100.00	95,003
August	21.90	81,749	99.00	80,931
September	10.05	37,515	96.00	36,140
October	13.42	50,092	88.00	44,081
November	5.17	19,299	69.00	13,316
December	3.41	12,713	54.00	6865
Total	192.12	717,111.28		643,543.12

Finally, Table 12.1.1 below summarizes the shade delivery percentage for each month and the output energy that the solar farm can produce due to the shadings. As shown, in months like December where the output is lower than in other months, the shade percentage is higher as the sun's path in winter months is generally lower in the sky, thus producing more shadows between the PV modules.

Fig. 12.1.7 shows that from April to August the output energy is virtually unaffected by shadows, the month of December being the most affected, with a reduction in the output energy of 46%. During a whole year, the facility will have a loss of around 10.2% in its output.

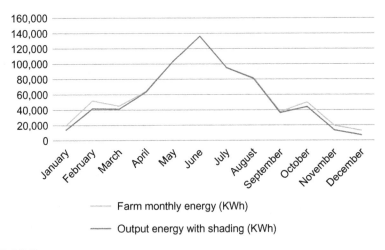

Fig. 12.1.7 Energy output.

12.1.2   Experimental measurements at solar meadow farm

Experimental data were collected at the site to fulfil two goals. First, to obtain detailed, specific data on the day-to-day operation of the plant, and second, to assess the accuracy of solar models used to predict such values as in-plane (or slope) irradiation, PV module cell temperature, and cell efficiency.

The experimental setup consisted of the following:

- Heat flux sensors mounted directly on the back panel of one of the PV modules. These were used to determine the rate of heat transfer between the module and its surrounding air.
- K-type thermocouples. Two of these were mounted in a similar way to the heat flux sensors to determine cell temperature, while another two were used to record air temperature under shade.
- Solar pyranometers. These were mounted on tripods: one above the modules facing directly upwards to record the horizontal solar irradiation and two in the plane of the modules (with equal inclination and orientation) to record the slope irradiation.
- Two data loggers, which the sensors were wired into and recorded 5 min periodic averaged values of the sensor outputs.

The sensors were positioned around the solar modules as shown in Figs. 12.1.8 and 12.1.9.

Data from each of the temperature, flux, and irradiation sensors were logged using two Grant 2020 series 'squirrel' data loggers, taking measurements every 5 min throughout the day. These datapoints constituted the average value (temperature, irradiation, or heat flux) registered over the 5 min period, forming the basis for the calculation process.

Fig. 12.1.8 Position of pyranometers.

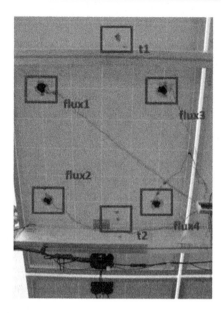

Fig. 12.1.9 Position of flux sensors.

The final data values required were the power output of the solar module being measured. Unfortunately, no method of automated logging of module output was available, and a compromise had to be made. Manual readings from the string inverter corresponding to the chosen PV module were taken, 9 a.m., 12 noon, and 4 p.m., over the course of 3 days. The module output was simply estimated as a fraction (1/80) of the input power to the inverter, as per technical specifications.

12.1.3 Calculation process: Slope irradiation, cell temperature and cell efficiency

The comparison of measured and calculated variables was completed according to Fig. 12.1.10.

This being a linear calculation process, each stage relying on the previous, any errors or inaccuracies picked up would be propagated through the calculations, making an accurate, independent assessment of each stage difficult. For this reason, calculations were based on measured data along with previous-stage calculated values.

12.1.3.1 Slope irradiation

As indicated in Fig. 12.1.11, the principal input variables to determine the slope irradiation (once the position of the slope has been decided) are the horizontal irradiation and the time of day (and date), which allow the position of the sun in the sky to be calculated. As data readings were taken over 5 min time intervals, not at an exact time,

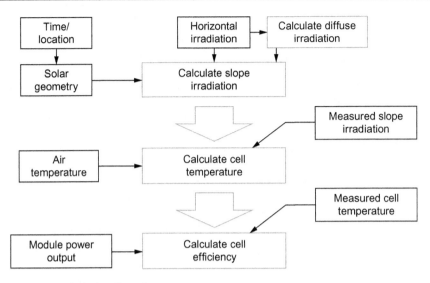

Fig. 12.1.10 Calculation flow chart.

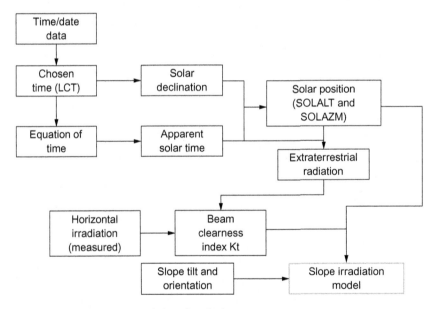

Fig. 12.1.11 Solar geometry and slope irradiation.

it was decided to follow the approach used by Clarke et al. (2007) and take the mid-point of the time period as the datapoint, i.e. 2.5 min before the logged time.

As used by Muneer et al. (2004), Fig. 12.1.11 demonstrates how time/date and irradiation data were used with solar geometry equations to calculate slope irradiation.

The final important factor to be determined was the diffuse irradiation component, i.e. the solar energy emitted across the whole hemisphere of the sky rather than directly from the sun. This was achieved through use of separate models for relating the relationship between the global clearness index $k_t = I_G/I_E$ and the horizontal diffuse-to-global ratio $k = I_D/I_G$ (where I_G is the global irradiation, I_E is the extraterrestrial irradiation, and I_D is the diffuse).

This relationship allows us to calculate the diffuse irradiation, rather than directly measure it, which is a critical step towards determining the slope irradiation.

Three different equations (12.1.1), (12.1.2), and (12.1.3) were selected to generate a range of results; Eq. (12.1.3) was taken from software programme Calc4–08, Muneer et al. (2000), and the others from Clarke et al. (2007). They are shown below with coefficients selected for Edinburgh:

$$k = 0.8721 + 1.7619k_t - 6.2135k_t^2 + 3.9467k_t^3, \ 0.25 \le k_t \le 0.8 \text{ for } k_t > 0.8 \quad (12.1.1)$$

$$k = 0.8798 + 1.7195k_t - 6.1193k_t^2 + 3.8769k_t^3, \ 0.2 \le k_t \le 0.85 \quad (12.1.2)$$

$$k = 1.006 - 0.317k_t + 3.1241k_t^2 - 12.7616k_t^3 + 9.7166k_t^4 \quad (12.1.3)$$

Once both the diffuse and global (total) horizontal irradiation are known, it becomes possible, for a given collector inclination and orientation, to calculate the global slope irradiation (generally referred to as the slope irradiation). This method is adapted directly from the windows in buildings (Muneer et al., 2000) software. Three quantities are calculated separately for the plane of the collector:

- Beam irradiation
- Diffuse irradiation
- Ground-reflected irradiation

Beam irradiation depends on the global and diffuse horizontal irradiation, the sun's angle of incidence on the centre of a solar panel (SOLINC) and the sun's altitude as seen in the sky above a solar panel (SOLALT). It is set to zero if either SOLALT is $<7°$ or SOLINC is $>90°$, in other words, if the sun is not in 'view' of the collector. Otherwise, the beam component of slope irradiation is given by Eq. (12.1.4) from the software programme Calc4–08 (Muneer et al., 2000):

$$Beam_{slope} = Beam_{horizontal} \frac{\cos SOLINC}{\sin SOLALT} \quad (12.1.4)$$

Diffuse slope irradiation is more complex to determine, as in the model proposed by Muneer et al. (2004), the sky is not considered isotropic. This method has been shown to give better results for diffuse irradiation (Muneer et al., 2004). Complete algorithmic details for the calculation of the above-mentioned slope irradiation and computations are provided in the latter reference. The final value used for the slope irradiation during the given period is simply the sum of the beam, diffuse, and ground-reflected components.

12.1.3.2 Cell temperature

The cell temperature, similarly to the slope irradiation, was estimated using three different models. The NOCT model is based on the behaviour of a solar module under certain test conditions and utilizes a simple calculation relating the solar irradiation to the temperature (Eq. 12.1.5):

$$T_c = T_a + \frac{G_{slope}}{G_{noct}} \left(T_{c,noct} - T_{a,noct} \right) \left(1 - \frac{\eta_{stc}}{\tau\alpha} \right) \tag{12.1.5}$$

where T_c is the cell temperature, T_a is the air temperature, G_{slope} is the global slope irradiation, G_{noct} equals 800 W/m2, $T_{c,noct}$ and $T_{a,noct}$ are the cell and air temperatures at NOCT, η_{stc} indicates the cell efficiency at standard test conditions (STC), and $\tau\alpha$ is related to the transmissivity-absorptivity of the module to solar irradiation.

The HOMER software model alters Eq. (12.1.5) to include a linearly variable rather than static cell efficiency. The HOMER cell temperature is shown in Eq. (12.1.6) (HOMER Energy, 2013):

$$T_c = \frac{T_a + \left(T_{c,noct} - T_{a,noct} \right) \dfrac{G_{slope}}{G_{noct}} \left[1 - \dfrac{n_{stc} \left(1 - \alpha_p T_{c,stc} \right)}{\tau\alpha} \right]}{1 + \left(T_{c,noct} - T_{a,noct} \right) \dfrac{G_{slope}}{G_{noct}} \dfrac{\alpha_p n_{stc}}{\tau\alpha}} \tag{12.1.6}$$

In the third case, a thermal model was implemented based on the method proposed in Aldali et al. (2013). This avoids the assumption of the previous methods that the module's thermal parameters will not change under different circumstances such as air temperature or irradiation.

Fig. 12.1.12 shows that there are three main mechanisms for thermal energy transfer from the cell (or module) to its surroundings. These correspond to convective losses to the air on both sides of the cell and radiative losses to the sky (radiative losses to the ground are much smaller and are neglected).

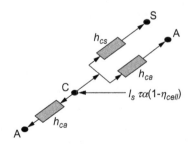

Fig. 12.1.12 Heat transmission from a solar module.
From Aldali, Y., Celik, A., & Muneer, T. (2013). Modeling and experimental verification of solar radiation on a sloped surface, photovoltaic cell temperature, and photovoltaic efficiency. *Journal of Energy Engineering, 139*(1), 8–11.

The convective losses are a function of the temperature difference between cell and air (Eq. 12.1.7):

$$\text{Convective loss} = 2h_{ca}(T_c - T_a) \tag{12.1.7}$$

The radiative losses are the same but to the sky (Eq. 12.1.8):

$$\text{Radiative loss} = h_{cs}\left(T_c - T_{sky}\right) \tag{12.1.8}$$

in which the effective sky temperature, T_{sky}; the air heat transfer coefficient, h_{ca}; and the sky heat transfer coefficient, h_{cs}, can be determined from Eqs. (12.1.9)–(12.1.11) (Aldali et al., 2013):

$$T_{sky} = 0.0552T_a^{1.5} \tag{12.1.9}$$

$$h_{ca} = 5.67 + 3.8v \tag{12.1.10}$$

$$h_{cs} = \frac{\sigma\varepsilon_c\left(T_c^4 - T_{sky}^4\right)}{T_c - T_{sky}} \tag{12.1.11}$$

The thermal model cell temperature value is found by combining the energy losses with the heating effect from the sun (Eq. 12.1.12):

$$T_c = \frac{I_{slope}\tau\alpha(1 - \eta_{cell}) + h_{cs}T_{sky} + 2h_{ca}T_a}{h_{cs} + 2h_{ca}} \tag{12.1.12}$$

where I_{slope} is the incident irradiation. The value for T_c is derived iteratively as T_c affects the heat transfer coefficients h_{cs} and h_{ca}.

12.1.3.3 Cell efficiency

While the other variables could be measured fairly directly, the cell efficiency had to be estimated from the measured data (using Eq. 12.1.13):

$$\text{Cell efficiency} = \frac{P_{out}}{P_{in}} = \frac{P_{module} / \text{No.of cells}}{\text{Solar irradiation} \times \text{cell area}} \tag{12.1.13}$$

P_{module} was derived from the manually recorded inverter power readings at each time interval. This was done in two ways to get the best possible result: firstly, as an average of a set of instant power readings taken around the sample time, and, secondly, from the change in total inverter energy reading over the time period, divided by the sample time. Each reading had its own advantages. In general, the instant readings were highly variable, whereas the averaged readings were adversely affected by the low resolution of the energy counter (nearest 0.1 kWh):

$$\eta_{cell} = \eta_{stc}\left[1 + \alpha_p(T_c - T_{c,stc})\right] \tag{12.1.14}$$

12.1.4 PV module analysis: Slope irradiation, cell temperature and cell efficiency

12.1.4.1 Slope irradiation

The measurement of slope irradiation was relatively simple and reliable, as the use of a professionally calibrated pyranometer aligned to match the slope and orientation of the solar module in question gave accurate readings with which to check the calculated values against. More complex was the method used for calculating the slope irradiation from only the horizontal global irradiation, requiring a number of different steps and intermediate quantities (as covered in Section 12.1.3); regardless it was expected that the calculation accuracy would be relatively high, as has been demonstrated in the papers presenting the methods used (Aldali et al., 2013; Clarke et al., 2007).

The three models for determining the clearness index and hence the slope irradiation gave similar results, the best of which is shown in Fig. 12.1.13, derived from the Clarke summer model (Clarke et al., 2007).

It can be seen that Fig. 12.1.13 shows an excellent, almost entirely linear, correlation for the site studied.

While all three methods show a high degree of accuracy, as expected, the calculations optimized for the given location give slightly better results. Table 12.1.2 makes a direct comparison in terms of the same quantities highlighted in Fig. 12.1.13: gradient, y-axis intercept, and regression coefficient, R^2, between the computed and measured quantities.

The difference between using the seasonal and monthly models follows the assessment in Clarke et al. (2007), namely, that the increased complexity of using monthly

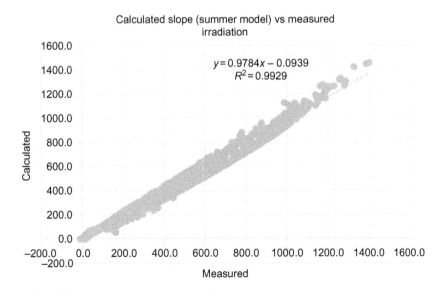

Fig. 12.1.13 Slope irradiation results.

Table 12.1.2 Slope irradiation model results

Model	Gradient	Offset	R^2
Summer	0.9784	0.0939	0.9929
June	0.9761	0.3164	0.9929
Muneer	0.9738	0.405	0.9897

coefficients (as in the June model) in a project gives a low return in increased accuracy (in this case little to none).

12.1.4.2 Cell temperature

As detailed in Section 12.1.3, the cell temperature was modelled in three ways: the simple but widely used NOCT or nominal operating cell temperature model, the slightly more complex one used in the HOMER software, and finally the full thermal model. It was expected that each would give successively better results when compared with the cell temperature directly measured from the back of the PV module, with the thermal model being significantly more accurate than the other two due to its consideration of a greater number of factors. Fig. 12.1.14 shows the output of the thermal model only across the full-time range of the experiment, while Table 12.1.3 compares the outputs of the three models.

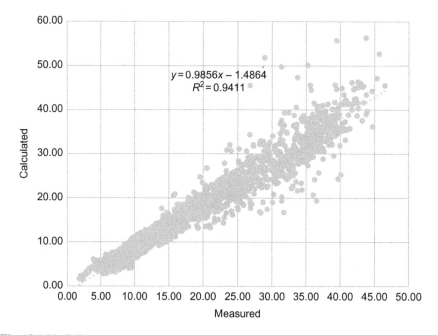

Fig. 12.1.14 Cell temperature results.

Table 12.1.3 **Cell temperature model comparison**

Model	Gradient	Offset	R^2
NOCT	1.0662	0.7332	0.9401
Homer	0.8876	2.303	0.9452
Thermal	0.9856	−1.4864	0.9411

These results clearly show the relatively high reliability of the simple NOCT method. Comparing the three methods, we can see the reliability (or R^2 value) of every method is around 94%. A larger difference can be seen in the gradient, which indicates the average percentage error if we utilize the given method of calculation. Here, the thermal model gives the best results, corresponding to only a 1.5% degree of error compared with 6.6% for the NOCT model and 11.3% for the HOMER model.

It is proposed that with the inclusion of reliable and high-resolution wind data, the thermal model could be optimized to give even better results than it has done, particularly at higher temperatures when the measured value starts to vary further from that calculated.

12.1.4.3 Cell efficiency

The cell efficiency differed from the other two quantities since it is not a directly measured quantity. Efficiency is derived from the power in to useful power out from a system, in this case solar irradiation and electrical power. Since automatic logging of electrical power was not possible as part of the experiment and manual measurements had to suffice, two problems arose: which method of determining the average power value for the time period to use (as described in Section 12.1.3) and assuring accuracy in the timing of measurements. Thus, only a very limited set of data, recorded manually, could be used for evaluating the efficiency model and results for Eq. 12.1.14 presented in Fig. 12.1.15.

As can be seen in Fig. 12.1.15, a fairly good correlation is seen over the course of a clear day.

The range over which the measured cell efficiency varied was approximately correct and had a close concordance to the computed values, as shown in Fig. 12.1.16.

An average efficiency value of close to 16% is observed.

12.1.4.4 Overall system

As covered previously, the estimated generation of the solar meadow farm is 568,611 kWh over a full year's operation. This is a significant amount of energy, avoiding the production of 293,000 kg of CO_2. The actual energy produced in year 1 was 439,276 kWh.

The solar meadow at Edinburgh Midlothian campus is more than just a valuable financial investment; it is a firm indication of the viability and importance of large-scale solar in Scotland. It supports the strong likelihood of the future uptake of the projects of this kind as part of a greener, more sustainable energy solution.

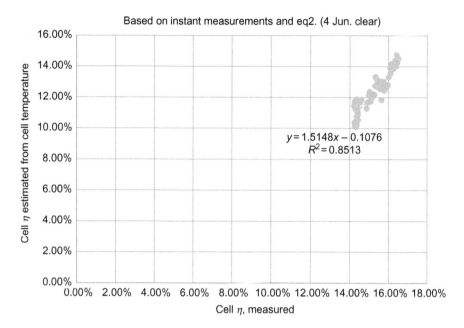

Fig. 12.1.15 Clear-day efficiency.

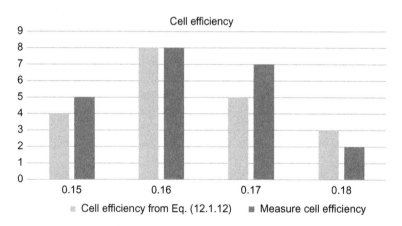

Fig. 12.1.16 Cell efficiency: measured and computed values (note y-axis is the frequency count).

12.2   Future plans: Implementation of a solar charging station for e-cars at Edinburgh college

In the past few years, while in the United Kingdom, there has been an increase in the demand of electric cars; only a small amount of solar charging stations has been installed. This report aims to carry out a study that consists on the determination of the optimal orientation and inclination of the solar panels according to different designs and performing a technical and financial analysis and an environmental impact assessment of the facility in order to establish the feasibility of such a project.

12.2.1   Calculation process: Slope irradiation, cell temperature and cell efficiency

The process to calculate the slope irradiation, cell temperature, and cell efficiency was the same carry out in the calculations done for the solar meadow farm; see Section 12.1.3.

12.2.1.1   Slope irradiation

The methodology followed in this section was the same as that used in Section 12.1.3. The different steps to reach the slope irradiation were carried out using a software programme adapted directly from Windows in Buildings (Muneer et al., 2000). Some modifications in the software programme related to the process of calculating the diffuse irradiation were done, and the slope irradiation was calculated according to the different tilt angles in order to find the optimal orientation and inclination of the solar panels.

To calculate the diffuse irradiation, a new diffuse ratio (k) was used according to the article 'Monthly averaged-hourly solar diffuse radiation model for the UK' (Muneer, Etxebarria, and Gago, 2014). The figure below (Fig. 12.2.1) shows the relation between the diffuse ratio (k) and the clearness index (k_t), showing a single regression curve for United Kingdom.

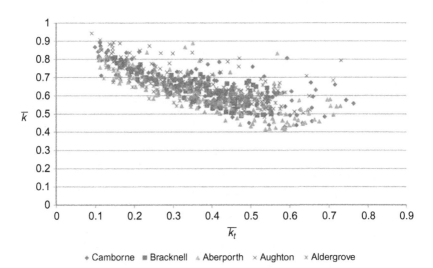

Fig. 12.2.1 Monthly-average hourly k-k_t plot for UK (locations arranged in an increasing order of latitude).
From Muneer, T., Etxebarria, S., & Gago, E. (2014). Monthly averaged-hourly solar diffuse radiation model for the UK. *Building Services Engineering Research and Technology, 35*(6), 573–584.

Table 12.2.1 List of angles and orientations to be studied

Orientation	Angles								
South	0°	10°	20°	30°	35°	40°	45°	50°	55°
East-west	0°	10°	20°	30°	35°	40°	45°	50°	55°

The resulting regression model equation was the following, Eq. 12.2.1, (Gago et al., 2010):

$$k = 0.89k_t^2 - 1.185k_t + 0.95 \qquad (12.2.1)$$

To analyse the optimum inclination, the annual slope irradiation for tilt angles between 0 degree and 55 degrees facing south and east-west was evaluated.

Table 12.2.1 shows a list of the angles and orientations that were considered for the present study.

12.2.1.2 Cell temperature

Cell temperature was calculated according to Eq. (12.1.6), and the PV module selected to carry out the calculations was an ASP-400GSM monocrystalline. Some of its technical specifications are shown in Table 12.2.2.

The cell temperature was calculated according to different scenarios (different tilt and azimuth angles) in order to know if the variation of temperature from one tilt and azimuth position to another would be significant or not.

The angles studied to obtain the cell temperature for every month and every hour were as follows:

- South orientation, $\alpha = 30$ degrees, $\alpha = 40$ degrees, and $\alpha = 50$ degrees
- West-east orientation, $\alpha = 0$ degrees, $\alpha = 30$ degrees, $\alpha = 40$ degrees, and $\alpha = 50$ degrees

Table 12.2.3 shows the cell temperature for the PV module facing south at 40 degrees. The cell temperature will be different from the hourly temperature only in the hours when radiation is emitted (coloured cells in the tables).

Table 12.2.2 Technical specifications PV module

Parameter	Value
Model	ASP-400GSM
Rated power (Pmax)	400 W
Maximum power voltage (Vmp)	49.25 V
Maximum power current (Imp)	7.92 A
Open-circuit voltage (Ioc)	59.62 V
Short-circuit current (Voc)	8.42 A
Dimensions	(1977 × 1315 × 34 mm)
Module efficiency	15.2%

Table 12.2.3 Cell temperature—PV module facing south at 40 degrees

Month/ time					Hourly cell temperature, 40 degrees							
	Jan	Feb	Mar	Apr	May	Jun	Jul	Aug	Sep	Oct	Nov	Dec
1	1.87	1.94	2.98	4.14	6.18	9.08	10.52	11.09	8.48	6.63	5.33	4.93
2	1.65	1.70	2.68	3.72	5.86	8.72	10.18	10.73	8.08	6.29	5.09	4.69
3	1.48	1.51	2.47	3.42	5.61	8.45	9.93	10.46	7.79	6.03	4.91	4.51
4	1.31	1.32	2.25	3.11	5.36	8.18	9.67	10.18	7.50	5.77	4.72	4.32
5	1.20	1.20	2.10	2.90	5.20	8.00	9.50	10.00	7.30	5.60	4.60	4.20
6	1.31	1.32	2.25	3.52	6.70	9.96	11.23	11.00	7.50	5.77	4.72	4.32
7	1.70	1.76	2.98	6.01	9.55	12.75	13.90	13.55	9.33	6.37	5.15	4.75
8	2.66	3.00	7.03	10.42	13.89	16.79	17.83	17.64	13.32	8.41	6.46	5.79
9	4.70	8.34	11.29	15.22	18.51	21.02	21.91	22.05	17.78	13.38	10.11	7.03
10	5.68	12.02	15.35	19.65	22.70	24.79	25.60	26.06	22.01	17.19	14.92	8.69
11	9.78	15.16	18.71	23.27	26.03	27.81	28.57	29.31	25.51	20.40	18.48	14.39
12	10.31	17.04	20.67	25.42	28.00	29.61	30.32	31.26	27.58	22.32	20.46	15.85
13	10.73	17.53	21.25	26.24	28.65	30.33	31.00	31.99	28.37	23.01	20.95	16.36
14	11.04	16.59	20.36	25.64	27.92	29.88	30.52	31.43	27.76	22.40	19.86	15.79
15	7.81	14.34	18.12	23.56	25.81	28.21	28.85	29.52	25.74	20.46	17.21	11.01
16	7.45	11.35	14.87	20.26	22.53	25.43	26.07	26.51	22.61	17.59	13.07	10.02
17	6.02	6.72	11.39	16.57	18.81	22.19	22.93	23.10	19.23	13.57	10.09	9.45
18	5.46	5.91	7.87	12.92	15.04	18.78	19.60	19.65	15.90	12.14	9.24	8.84
19	4.62	4.98	6.55	9.59	11.54	15.27	16.24	16.37	13.28	10.85	8.32	7.92
20	4.00	4.30	5.75	8.05	9.30	12.50	13.75	14.55	12.20	9.90	7.65	7.25
21	3.50	3.74	5.09	7.12	8.56	11.69	12.99	13.73	11.32	9.13	7.10	6.70
22	2.99	3.18	4.44	6.20	7.82	10.88	12.22	12.91	10.44	8.35	6.55	6.15
23	2.60	2.75	3.93	5.48	7.25	10.25	11.63	12.28	9.75	7.75	6.13	5.73
24	2.21	2.32	3.41	4.75	6.68	9.62	11.03	11.64	9.06	7.15	5.70	5.30

12.2.1.3 Cell efficiency

The equation used to calculate the cell efficiency was Eq. (12.1.4). Finally, the average cell efficiency according to different tilt and azimuth angles had a value of 16.2%.

12.2.2 Design of the solar charging station: First phase

In this first phase of the design, three different designs with different shapes were analysed (see Figs. 12.2.2–12.2.4) with the purpose of knowing, for each case, the final output generated, the number of panels, geometry, etc. All this information will be essential to determine which design would be the most suitable for the solar charging station in terms of annual solar energy provided and feasibility. To start with this first phase, an area of 400 m², 20 × 20 m was considered to build the solar charging station.

12.2.2.1 Design 1: South orientation

The tilt angles studied for this design were 30 degrees, 40 degrees, and 50 degrees. All dimensions, the number of PV modules that may be installed on the roof, the annual energy generated, and the slope irradiation of this facility (Fig. 12.2.2) for different tilt angles are presented in Table 12.2.4.

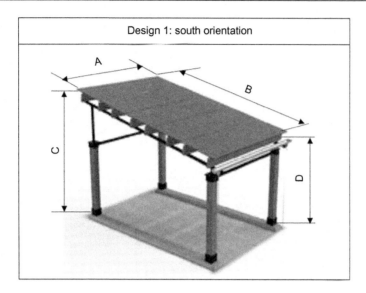

Fig. 12.2.2 South facing design.
Based on Formfonts, 2016.

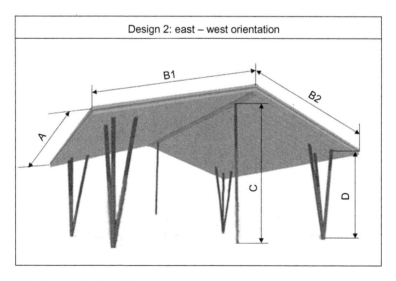

Fig. 12.2.3 West-east facing design.
Based on Formfonts, 2016.

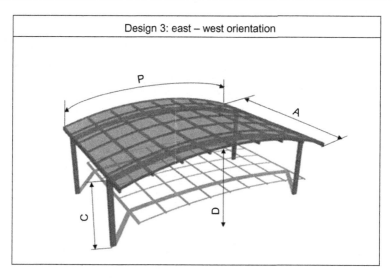

Design 3: east – west orientation

Fig. 12.2.4 West-east facing curved design.
Based on Formfonts, 2016.

Table 12.2.4 **Main characteristics of Design 1**

Design	Tilt angle	Roof length	Roof width	Height, max	Height, min	N° PV module	Output	Slope irradiation
Unit	α	A (m)	B (m)	C (m)	D (m)	–	(MWh)	(kWh/m^2)
1A	30°	20	23.1	14	2.5	180	78.6	1037
1B	40°	20	26.1	19.3	2.5	195	85.7	1044
1C	50°	20	31.1	26.3	2.5	240	103.7	1026

12.2.2.2 Design 2: East-west orientation

In this type of design (Fig. 12.2.3), half of the panels will be facing west, and the other half will be facing east. The tilt angles studied for this design were 0 degree (facing any orientation), 30 degrees, 40 degrees, and 50 degrees. All the relevant data for these designs are shown in Table 12.2.5.

Design 2A yields the maximum slope irradiation for this orientation (west-east) but with the drawback that the number of panels to be installed is minimum. As a general observation, it could be said that this V-shape design is less efficient than that of the south-facing designs but it allows for half the height with the same number of panels.

This is the worst case presented (design 2D); in comparison with design 1C, the energy generated is 24% lower. In design 1C, the total energy obtained was 103.7 MWh/yr compared with 78.6 MWh in the present design. On the other hand, if the

Table 12.2.5 Main characteristics of Design 2

Design	Tilt angle	Roof length	Roof width	Height, max	Height, min	N° PV module	Output	Slope irradiation
Unit	α	A (m)	B1+B2 (m)	C (m)	D (m)	–	(MWh)	(kWh/m^2)
2A	0°	20	20	—^{a}	2.5	150	56.5	893.8
2B	30°	20	23.1	8.3	2.5	180	64.1	846.5
2C	40°	20	26.1	10.9	2.5	200	68.8	817.2
2D	50°	20	31.1	14.4	2.5	240	78.6	778.3

^{a}The height of the solar roof could be any with an area of 400 m^2.

present design is compared with the previous designs in terms of height, this design with the same height as design 1A generates the same amount of energy, but the latter uses more PV modules, 240 versus 180. Therefore, this design would be less cost-effective.

12.2.2.3 Design 3: East-west orientation

For this curved design, Fig. 12.2.4, heights of 10 and 5 m were used. As the roof is curved, each string of PV modules will be placed at different angles; in order to determine the orientation of each string of PV modules and calculate the number of strings, some sketches were done in AutoCAD (see Figs. 12.2.5 and 12.2.6). Table 12.2.6 presents the main characteristics of the two designs presented above.

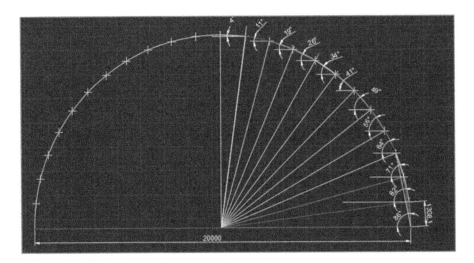

Fig. 12.2.5 Design 3A—Orientation of PV modules on the roof.

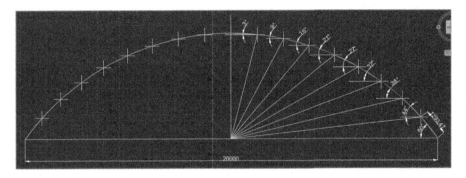

Fig. 12.2.6 Design 3B—Orientation of PV modules on the roof.

Table 12.2.6 **Main characteristics of Design 2**

Design	Tilt angle	Roof length (m)	Roof perimeter (m)	High max (m)	High min (m)	N° PV module	Output (MWh)	Slope irradiation (kWh/m^2)
Unit	α	A (m)	P (m)	C (m)	D (m)	–	(MWh)	(kWh/m^2)
3A	Various	20	31.4	12.5	2.5	240	77.8	769.5
3B	Various	20	23.2	7.5	2.5	180	64.4	849.3

12.2.2.4 Design summary

All the results are summarized in the table below (Table 12.2.7). The table is arranged according to the shape of the design, tilt angle, number of panels used, slope irradiation, generated energy per year, and height of the solar roof. To carry out the second phase of the design, some specifications were defined according to the results

Table 12.2.7 **Data recollected from all designs—first design phase**

	Design	Shape	Tilt angle	PV No.	Slope irradiation (kWh/m^2)	Output (MWh)	Height, max (m)
First design South	1A		$\alpha = 30°$	180	1037	78.6	14
	1B		$\alpha = 40°$	195	1043	85.6	19.3
	1C		$\alpha = 50°$	240	1026	103.7	26.3
Second design W-E	2A		$\alpha = 0°$	150	893.82	56.47	–
	2B		$\alpha = 30°$	180	846.48	64.14	8.3
	2C		$\alpha = 40°$	200	817.23	68.8	10.9
	2D		$\alpha = 50°$	240	778.28	78.6	14.4
Third design W-E	3A		$\alpha = ‡$	240	769.48	77.78	12.5
	3B		$\alpha = ‡$	180	849.31	64.39	7.5

shown in Table 12.2.7; for example, the maximum height for the facility would be 8.5 m; therefore, in this first phase, designs 1A, 1B, 1C, 2C, 2D, and 3A were rejected.

The next chapter covers a second phase of design, where both designs chosen will be further studied and new designs will be created.

12.2.3   Design of the solar carport: Second phase

In this second phase, it was desirable to take a look at the Edinburgh College complex located in the Dalkeith area and study how new carports could be designed according to the available space. Fig. 12.2.7 below shows the area where the facility could be installed. If solar carport was placed in this area, the overall costs could be reduced, as the site preparation costs would be avoided.

For this second phase, two things were taken into account: the area of the car park and its accessibility. Generally, the dimensions of a standard parking space are 2.4 m width and 4.8 m depth (Jackson, 2016).

12.2.3.1   Design 4: South orientation

In the first design phase, all the south-facing designs were rejected because their heights exceeded the maximum allowed height; therefore, the aim of this second phase of the design will be to reduce this height with new models:

- By decreasing the inclination of the roof
- By modifying the width and length of the area, maintaining always the area of 400 m^2

Fig. 12.2.7 Edinburgh college—Dalkeith area.
From Google maps.

For the south orientation, three kinds of designs were created. The first one with an area of 80 × 5 m, the second one with an area of 40 × 10 m (see Figs. 12.2.8 and 12.2.9), and the third one with the dimensions established in the first phase, 20 × 20 m, solving the previous problem by decreasing the height of the solar carport.

Based on this, the geometrical dimensions of the solar roof and energy output (MWh/yr), slope irradiation (kWh/m^2), and number of PV modules will be calculated.

Design 4A

This design consists on a string of parking spaces with an area of 80 × 5 m. The tilt angles studied were 30 degrees, 40 degrees, and 50 degrees. The geometry of this design and further calculations for the following angles are shown in Table 12.2.8.

Design 4B

This design will consist on a double string of parking spaces with an area of 40 × 10 m; see Fig. 12.2.10.

For this area, an inclination of the roof >30 degrees won't be possible because the height exceeds the maximum height allowed; therefore, the analysed angles were 20 degrees and 30 degrees (see Table 12.2.9).

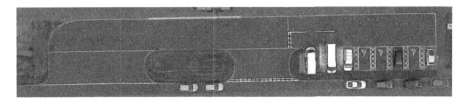

Fig. 12.2.8 Area to place the designs number 4.
From Google maps.

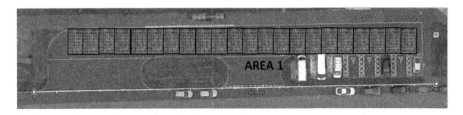

Fig. 12.2.9 Area to place the designs 4A.
Modified from Google maps.

Table 12.2.8 Relevant data for the Design 4A

Shape	Name	Tilt angle	Width	Length	Height[a]	Inclined surface	No. PV mod.	Output	Slope I
		α	x (m)	l (m)	h (m)	c (m)		(MWh)	(kWh/m²)
	4AA	30°	5	80	2.88	5.77	180	78.62	1037
	4AB	40°	5	80	4.2	6.52	200	87.96	1044
	4AC	50°	5	80	5.96	7.78	240	103.72	1026

[a]The maximum height of the facility is determined by the roof height (h) and the 2.5 m minimum height requirement.

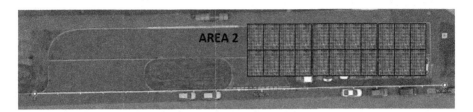

Fig. 12.2.10 Area to place the design number 4B.
Modified from Google maps.

Design 4C
The area chosen to install this solar carport is shown in Fig. 12.2.11.

As shown in Table 12.2.10, tilt angles >15 degrees will not be permissible due to the roof height being >6 m.

12.2.3.2 Design 5: East-west orientation

This solar carport will be facing east-west. In this case, a new design was prepared, in which the height of the carport was smaller while the energy production was the most effective possible; the best tilt angle for this kind of orientation was 0 degrees, and the design would look like this (Fig. 12.2.12).

The area chosen to place this design is shown in Fig. 12.2.13.

This design was done for angles of 30 degrees, 40 degrees, and 50 degrees.

Table 12.2.11 below summarizes the most relevant data including the dimensions of the facility, total number of PV modules, and total output by all PV modules.

12.2.3.3 Design 6: East-west orientation

Fig. 12.2.14 shows the different areas chosen to build the carport, area A with the dimensions of 40 × 10 m and area B with dimensions of 67 × 6 m. Heights of 5 and 3 m were chosen for the designs.

Design 6A
For a height of 5 m, the tilt angle of each string of PV modules was calculated with AutoCAD (see Fig. 12.2.15).

Fig. 12.2.16 shows a sketch in order to see how the life-size solar carport would look.

In this case, the roof can accommodate 12 strings of PV modules and 20 PV modules along its 40 m of length; therefore, the total number of PV modules will be 240.

For a height of 3 m, the arc roof will be divided into nine parts; the inclination angles for the PV modules are shown in Fig. 12.2.17 below.

Table 12.2.9 Relevant data for the Design 4B

Shape	Name	Tilt angle	Width	Length	Height	Inclined surface	No. PV mod.	Output	Slope I.
		α	x (m)	l (m)	h (m)	c (m)		(MWh)	(kWh/m^2)
	4BA	20°	10	40	3.64	10.64	160	68.04	1010
	4BC	30°	10	40	5.67	10.64	180	78.62	1037

Fig. 12.2.11 Area to place the design number 4C.
Modified from Google maps.

Table 12.2.12 includes the most relevant data for the two designs, 6AA and 6AB.

Design 6B
For a height of 5 m, Fig. 12.2.18 shows how the design would finally look. As this solar carport design did not seem very appropriate in terms of shape, it was rejected.

For a height of 3 m and a roof length of 9.43, the roof arc will be divided into seven parts. The inclination angles for the PV modules are shown in Fig. 12.2.19 below.

Table 12.2.13 includes the most relevant data for this design 6BB.

12.2.3.4   Summary of the results by design shape, first and second phase

Table 12.2.14 summarizes all the designs developed in the first and second phase of the design, disregarding the designs with a maximum height over 8.5 m. This table shows all the relevant data in order to choose the best design for the project.

Table 12.2.10 Relevant data for the Design 4C

Shape	Name	Tilt angle	Width	Length	Height	Inclined surface	No. PV mod.	Output	Slope I.
		α	x (m)	l (m)	h (m)	c (m)		(MWh)	(kWh/m²)
	4CA	10°	20	20	3.52	20.3	156	63.17	961.4
	4CB	15°	20	20	5.35	20.7	159	66.17	988.0

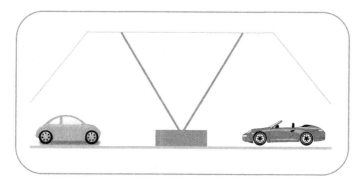

Fig. 12.2.12 Sketch design number 5.

Fig. 12.2.13 Area to place the design 5.
Modified from Google maps.

12.2.3.5 Design critique

Among all the designs proposed in the table above, the best designs were those facing south, in particular designs 4AB and 4 AC; these designs feature the best final energy output. These designs consisted of only a row of cars, which allows for an increase of the tilt angle of the roof in order to obtain more slope irradiation. From

Table 12.2.11 Relevant data for the Design 5

Shape	Name	Tilt angle	a = c	b	Length	Height	Inclined surface	No. PV mod.	Output	Slope L.
		α	(m)	(m)	l (m)	h (m)	r (m)		(MWh)	(kWh/m²)
	5A	30°/0°	5	10	20	2.9	5.77	160	58.6	870.15
	5B	40°/0°	5	10	20	4.2	6.53	180	64.5	851.26
	5C	50°/0°	5	10	20	5.96	7.78	200	69.5	824.5

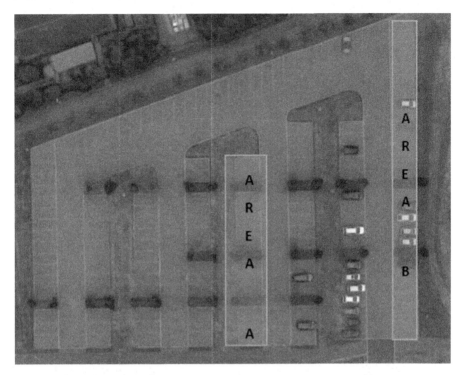

Fig. 12.2.14 Area to place the designs 6.
Modified from Google maps.

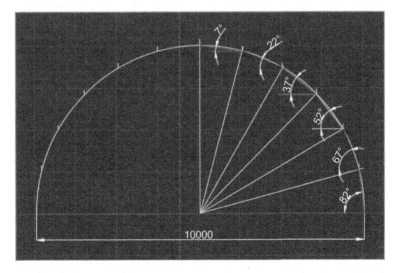

Fig. 12.2.15 Design 6AA—Orientation of PV modules on the roof.

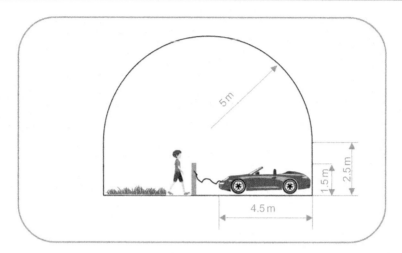

Fig. 12.2.16 Sketch of design 6A.

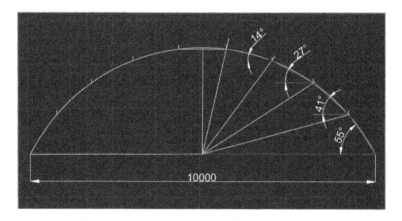

Fig. 12.2.17 Design 6AB—Orientation of PV modules on the roof.

these two designs, design 4AB had better features in terms of height, around 1.8 m lower than design 4 AC, and because this south orientation at 40 degrees was the most efficient one, which means more benefits in terms of the amount of energy produced. Therefore, the optimal design was found to be the design facing south at 40 degrees.

It is also worth mentioning that south orientation may not always be the best option for solar carports; it will depend on the existing solar carport layout or the shadows cast by adjacent buildings or foliage. For example, according to Table 12.2.14, for east-west orientation, the best design for the solar carport would be design 5B, which would generate 64.5 MWh/yr.

Table 12.2.12 Relevant data for Design 6A

Shape	Name	Tilt angle	Width	Length	Height	Roof perimeter	No. PV mod.	Output	Slope I
		α	x (m)	l (m)	h (m)	p (m)		(MWh)	(kWh/m^2)
	6AA	Various	10	40	5	15.7	240	78.18	773.37
	6AB	Various	10	40	3	12.25	180	63.34	835.45

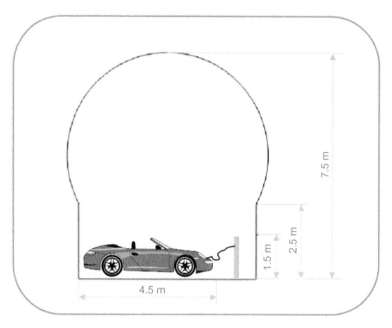

Fig. 12.2.18 Sketch of design 6B.

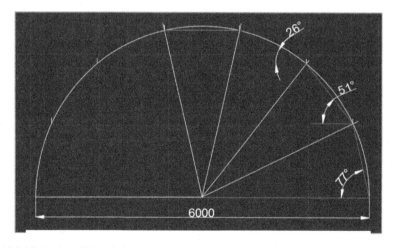

Fig. 12.2.19 Design 6BA—Orientation of PV modules on the roof.

12.2.3.6 Characteristics of the chosen design

The best position to place the modules according to the chosen design was horizontally, as shown in Fig. 12.2.20, because it allows for more PV modules to be installed, thus generating more output where the dimensions of the carport area will be of 403 m^2 and the roof area of 526.5 m^2.

Table 12.2.13 Relevant data for the Design 6B

Shape	Name	Tilt angle	Width	Length	Height	Roof perimeter	No. PV mod.	Output	Slope I
	6BB	α Various	x (m) 6	l (m) 67	h (m) 3	p (m) 9.43	238	(MWh) 79.67	(kWh/m²) 794.76

Table 12.2.14 Summarized data table of all possible designs

Orientation	Design	Shape	Tilt angle	No. PV Modules	Slope I (kWh/m2)	Output (MWh/yr)	Height, max[a] (m)
W-E	2B		$\alpha = 30°$	180	846.48	64.14	8.3
	3B		$\alpha = ‡^b$	180	849.31	64.34	7.5
W-E	6AA		$\alpha = ‡$	240	773.37	78.18	7.5
	6AB		$\alpha = ‡$	180	835.45	63.34	5.5
	6BB		$\alpha = ‡$	238	794.76	79.67	5.5
	5A		$\alpha = 30°$	160	870.15	58.6	5.4
W-E	5B		$\alpha = 40°$	180	851.26	64.5	6.7
	5C		$\alpha = 50°$	200	824.5	69.5	8.46
	4AA		$\alpha = 30°$	180	1037	78,62	5.38
	4AB		$\alpha = 40°$	200	1044	87.96	6.7
	4AC		$\alpha = 50°$	240	1026	103,72	8.46
South	4BA		$\alpha = 20°$	160	1009.6	68.04	6.14
	4C2		$\alpha = 30°$	180	1037	78.62	8.17
	4CA		$\alpha = 10°$	156	961.4	63.17	6.02
	4C2		$\alpha = 15°$	159	988.0	66.17	7.85

[a]This height includes the minimum height of 2.5 m.
[b]Strings of PV modules oriented at different tilt angles.

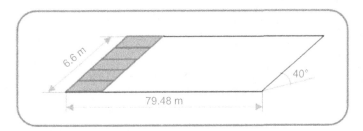

Fig. 12.2.20 Horizontal positions of the PV modules.

Therefore, the chosen design with south orientation at 40 degrees is made up of 200 PV modules rated at 400 Wp placed horizontally in a roof area of 526.5 m². This design can generate 87.96 MWh/yr as it can receive a slope irradiation of 1044 kWh/m². The installed capacity is 80 kWp.

Taking into account that the standard car parking area for a single car is 4.8 × 2.4 m, as the proposed facility has a ground area of 79.5 × 5.07 m, this solar charging station will be able to provide space for 33 vehicles. The solar carport has a minimum height of 2.5 m and a maximum height of 6.7 m.

The pictures below show how the facility would look with real dimensions (see Figs. 12.2.21–12.2.23).

12.2.4   Design of the PV system

This chapter deals with the selection and the needed amount of inverters, PV modules, charger stations, and its connection to the grid.

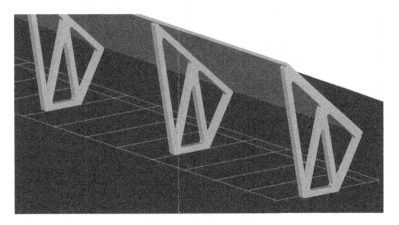

Fig. 12.2.21 Solar carport, bottom view.

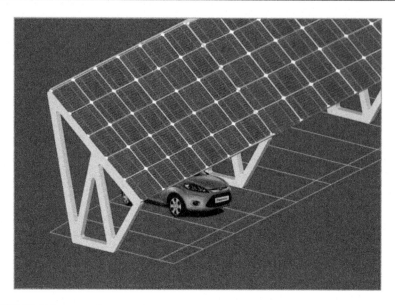

Fig. 12.2.22 Solar carport, upper view.

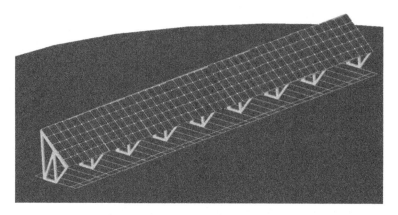

Fig. 12.2.23 Full view of the solar carport.

12.2.4.1   Selection of the inverter

The chosen inverter was a TRIO-20.0-Tl with a maximum power input of 20 kW and a peak efficiency rating of 98.3% (Power-One, 2016). In order to calculate the number of inverters needed by the facility, an example provided by SMA called 'Example Design of a PV Array' (SMA, 2016) was used.

Table 12.2.15 **Temperature conditions**

Temperature data	Edinburgh
Tmax	19.1°C
Tmin	1.2°C
Tmax cell	32°C
Tmin cell	1.2°C
Tstc	25°C

PV modules per string and maximum number of strings

The temperature data of Edinburgh are necessary to perform the following calculations; see Table 12.2.15.

Eq. (12.2.2) was used to calculate the maximum open-circuit voltage of the PV module, which had a value of 65.3 V, the open-circuit voltage of the PV module being 59.62 V, the voltage temperature coefficient being $-0.4\%/°C$, and the temperature at STC and minimum ambient temperature being $-23.8°C$:

$$V_{DCmaxMOD} = V_{OC}{}^* \left(1 + \frac{T_{DCUocMOD}{}^* \Delta T_{LOW}}{100\%} \right) \qquad (12.2.2)$$

where

- $V_{DCmaxMOD}$: maximum PV module voltage
- V_{oc}: open-circuit voltage of the PV module
- $T_{DUCocMOD}$: voltage temperature coefficient
- ΔT_{LOW}: temperature at STC and minimum ambient temperature (Tcellmin-Tstc)

The open-circuit voltage decreases as temperatures rise. The minimum PV module open-circuit voltage is calculated with the following Eq. (12.2.3) where the value of the voltage of the PV module at maximum power is 49.25 V and the temperature at STC and maximum cell temperature is 7°C, the result of the minimum PV open-circuit voltage being 47.87 V:

$$V_{DCminMOD} = V_{mpp}{}^* \left(1 + \frac{T_{DCUocMOD}{}^* \Delta T_{max}}{100\%} \right) \qquad (12.2.3)$$

where

- $V_{DCminMOD}$: minimum PV module voltage
- V_{mpp}: voltage of the PV module at maximum power
- ΔT_{max}: temperature at STC and maximum cell temperature (TmaxMOD-Tstc)

The PV modules within a string will have the same current as the string because they are placed in series. Eq. (12.2.4) was used to calculate the maximum PV module current, which had a value of 8.39 A, the short-circuit current of the PV module being 8.42 A, and the current temperature coefficient being 0.05%/°C:

$$I_{DCmAXstr} = I_{SC} * \left(1 + \frac{T_{DClocMOD} * \Delta T_{max}}{100\%} \right)$$

(12.2.4)

where

- $I_{DCmaxSTR}$: maximum string current
- I_{SC}: short-circuit current of the PV module
- $T_{DClocMOD}$: current temperature coefficient

The maximum string voltage must not exceed the maximum permitted system voltage of the photovoltaic modules. The maximum number of modules per string turned out to be of 15; see Eq. (12.2.5), the maximum input voltage of the inverter being 1000 V:

$$n_{maxMODSTR} \leq \frac{V_{DCmaxWR}}{V_{DCmaxMOD}}$$

(12.2.5)

where

- $n_{maxMODSTR}$: maximum number of PV modules per string
- $V_{DCmaxINV}$: maximum input voltage of the inverter

The minimum number of PV modules per string that should be installed would be of nine, the minimum MPP voltage of inverter being 450 V; see Eq. (12.2.6):

$$n_{minMODSTR} \leq \frac{V_{DCmppminWR}}{V_{DCminMOD}}$$

(12.2.6)

where

- $n_{minMODSTR}$: minimum number of PV modules per string
- $V_{DCmppminINV}$: minimum MPP voltage of inverter

The optimum number of strings per array must not be less than the minimum number of strings and must not exceed the maximum number. In order to avoid system damage such as high occurring current flows, it would be better not to choose the maximum number of strings. Therefore, the number of PV modules per string is calculated as follows:

$$n_{minMODSTR} \leq n_{MODSTR} \leq n_{maxMODSTR}$$

where n_{MODSTR} is the number of PV modules per string.
Therefore,

$$9 \leq 15 \leq 15$$

Knowing the values of the power of the inverter, 20,000 W, and the maximum power of the PV modules, 400 W, the minimum and maximum number of strings needed to achieve the total power can be calculated. The minimum and maximum number of strings needed according to Eqs. (12.2.7) and (12.2.8) was three in each equation:

$$n_{\min STR} = \frac{P_{DCGEN}}{P_{\max MOD} * n_{MODSTR}} \tag{12.2.7}$$

$$n_{\max STR} = \frac{I_{DCmaxINV}}{I_{DCmaxSTR}} \tag{12.2.8}$$

where

- P_{DCGEN}: power of the inverter
- P_{maxMOD}: maximum power of the PV modules

The optimum number of strings per array must not be less than the minimum number of strings and must not exceed the maximum number:

$$n_{minSTR} \le n_{STRCHOOSE} \le n_{maxSTR}$$

Therefore,

$$3 \le 3 \le 3$$

Number of PV modules handled by one inverter and total number of inverters

The maximum voltage and current values of the string shouldn't exceed the maximum voltage and current values of the inverter. This can be determined using Eqs. (12.2.9) and (12.2.10):

$$V_{Array} = V_{DCmaxSTR} = n_{MODSTR} * V_{DCmaxMOD} \tag{12.2.9}$$

$$I_{DCmaxSTRrray} = n_{STR} * I_{DCmppMOD} \tag{12.2.10}$$

Once it has been verified, the total number of PV modules that can be handled by one inverter can be obtained.

Knowing that the maximum number of strings is 13 and that each string consists of 3 modules, the total number of PV modules handled by one inverter will be 39.

The design chosen had 200 PV modules, so if an inverter can be connected to 39 PV modules, the total number of inverters that the facility would need would be

$$n_{TOTAL INV} = \frac{200}{39} = 5.1 \sim 5 \; Inverters$$

12.2.4.2   Selection of the PV module

See Section 12.2.1.

12.2.4.3 Selection of the charging station

There arc three main types of EV chargers (Jackson, 2016):

- Slow charging (up to 3 kW), suitable for charging during 6–8 h overnight.
- Fast charging (7–22 kW) that can fully recharge some models in 3–4 h.
- Rapid charging (43–50 kW), able to provide an 80% charge over 30 min. They can be used in AC or DC.

The charging station chosen for the facility was a twin rapid charger CHAdeMO, an AC rapid charger. This charger installed at Napier University (see Fig. 12.2.24) features a charge time of around 1 h and can provide electricity to the electric vehicle to drive around 80 mi. Table 12.2.16 below shows the charging times for Renault Zoe. For the proposed facility, three twin rapid charger stations were installed.

Fig. 12.2.24 Rapid charger CHAdeMO at Napier University.
From Plugshare.com. (2016). PlugShare. [online] Available at: http://www.plugshare.com/?location=54084 Accessed 09.08.16.

Table 12.2.16 Charging times for Renault Zoe

Charging times					
Charger type	Phases	Current (A)	Voltage (V)	Power (kW)	Charge time
Very slow	1	10	230	2.3	9.5 h
Slow	1	16	230	3.7	6.0 h
Fast	1	32	230	7.4	3.0 h
AC-Rapid	3	32	230	22	1.0 h
DC-Rapid	3	63	230	43	0.5 h
Battery Swap	–	–	–	–	90 s

Data from Renault 2014, Tesla 2014, cited in Muneer, T., Milligan, R., Smith, I., Doyle, A., Pozuelo, M., & Knez, M. (2015). Energetic, environmental and economic performance of electric vehicles: Experimental evaluation. *Transportation Research Part D: Transport and Environment*, 35, 40–61, p. 52.

12.2.4.4   Layout

The PV systems will be connected to the grid; no storage was taken into account as the electric vehicle batteries themselves will act as storage.

Fig. 12.2.25 below shows how the connection of the PV system to the grid would be.

Some of the advantages of not using battery storage in the connection to the grid are that there are no additional costs involved.

Fig. 12.2.26 below shows the circuit diagram of the proposed 80 kW solar carport facility. It has been divided into five channels (arrays), each made up of 39 PV

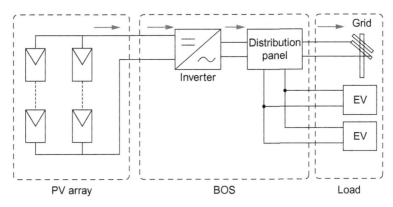

Fig. 12.2.25 Typical grid-connected PV system without battery storage.
From Narayan, N. (2013). Solar charging station for light electric vehicles. A design and feasibility study. Master of Science Thesis. Delft University of Technology.

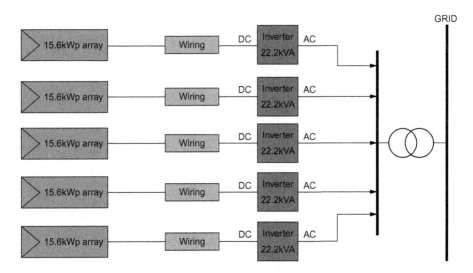

Fig. 12.2.26 Schematic circuit diagram of the 80kWp facility.

modules rated at 400 Wp connected to the inverter that converts DC power from the PV modules into AC power. The facility will be connected to the national grid through a transformer.

12.2.5 Driving behaviour

This chapter will aim to calculate the average driving distance by a single vehicle in a day and the average energy consumption of the vehicle per day. To this effect, a report called 'Energetic, environmental and economic performance of electric vehicles: Experimental evaluation' (Muneer et al., 2015) was used.

12.2.5.1 Average driving distance

According to the Department for Transport, in 2014, each person in England travelled around 6500 mi annually (covering all means of transport), where cars accounted for 78%, i.e. the number of miles travelled by car was around 5067 (8107 km) per year (Department for Transport, 2015). Therefore, the average trip length per car and per person would be of around 22 km, while in Scotland, the average car journey per person, also in 2014, was 20.8 km (Transport Scotland, 2015).

Trips in progress by time and day in the UK

Fig. 12.2.27 below shows the number of journeys with a vehicle along the day during weekdays and weekends. As shown, peak journeys take place around 8 a.m. and 4 p.m.

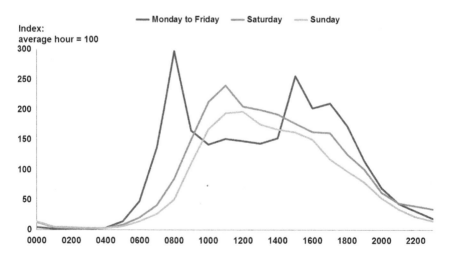

Fig. 12.2.27 Trips in progress by time of day and day of week.
From Department for Transport (2015). National travel survey: England 2014. [online] England, p.19. Available at: https:/www.gov.uk/government/uploads/system/uploads/attachment_data/file/457752/nts2014-01.pdf Accessed 20.05.16.

According to the National Travel Survey, education and work have a big impact on travel patterns, because it is the time when people go to work and school and come back home; Fig. 12.2.28 shows the different kinds of trips, such as trips by leisure, most of that at weekends; shopping, where the most shopping trips are made between 9 a.m. and 3 p.m. and one-fifth on Saturdays or commuting; and business where 68% of the trips start between 6 a.m. to 9 a.m. and 4 p.m. to 7 p.m.

12.2.5.2   Estimation of energy consumption by electric vehicles

One way to obtain this information would be to analyse and calculate the amount of best-selling electric cars in the United Kingdom and, based on the manufacturer's datasheet, to calculate the average energy consumption of all these cars. However, sometimes, the energy consumption specified by the manufacturer doesn't match reality; therefore, based on the report mentioned before (Muneer et al., 2015), it was decided to take the data from this study instead of any manufacturer's data.

Car model: Renault Zoe e-car

The French manufacturer Renault introduced their e-car Zoe in the year 2013. Edinburgh Napier University obtained the very first model that was made available. In this study, the Renault Zoe e-car was used for an experimental evaluation, where the speed and energy were recorded in a journey from Morningside to Leith. To calculate the consumption per kilometre (kWh/km) of this car, information about time, speed, and altitude of this journey was required, which was being logged at specific times during the test drives.

With these data collected, the next step was to calculate the driven distance in metres, the gradient in radians (angle of inclination), and the acceleration in metre per second squared of the car. Once these unknown quantities are known and some

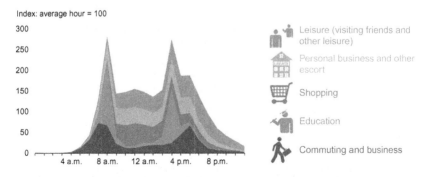

Fig. 12.2.28 Trips in progress by start time and purpose, Monday to Friday.
From Department for Transport (2015). National travel survey: England 2014. [online] England, p.19. Available at: https://www.gov.uk/government/uploads/system/uploads/attachment_data/file/457752/nts2014-01.pdf Accessed 20.05.16.

Table 12.2.17 Data Renault Zoe e-car

A, m^2	g, m/s^2	μ	ρ	m, kg	Cd
2.75	9.81	0.030	1.23	1693	0.28

data from the Renault Zoe car (see Table 12.2.17), Eq. (12.2.11) below needs to be used in order to obtain the energy consumption by a single car:

$$E = \left[\mu mg\cos\theta + mg\sin\theta + \frac{1}{4}CdAp\left(v_f^2 + v_i^2\right) \right]\Delta d + \frac{1}{2}m\left(v_f^2 - v_i^2\right) \qquad (12.2.11)$$

This equation takes the following into account:

- *Tyre friction*: $[\mu mg\cos\theta]\Delta d$
- *Hill climbing*: $[mg\sin\theta]\Delta d$
- *Wind drag*: $\left[\frac{1}{4}CdAp\left(v_f^2 + v_i^2\right)\right]\Delta d$
- *Change in kinetic energy*: $\frac{1}{2}m\left(v_f^2 - v_i^2\right)$

The most relevant features of the Renault Zoe for all calculations are shown below where

- A: front area of the car
- g: gravity
- μ: friction coefficient
- ρ: air density
- m: weight
- Cd: aerodynamics coefficient

By applying the time, speed, and altitude data and the manufacturer's data (Table 12.2.17), the amount of energy that would be obtained during the different types of driving such as acceleration, cruise, and deceleration was determined.

The breakdown of events during a driving cycle is illustrated in Fig. 12.2.29. The chart illustrates that 47% of the driving cycle is used to decelerate; in this drive mode, a gain of energy is generated, where 27% of that deceleration is due to the descending gradient that results in a greater potential for energy recovery (Walsh, Muneer, and Celik, 2011).

According to the study conducted by Muneer et al., the Renault Zoe model was reported to have a power usage of 12% higher than the manufacturer's values. The analyses show that the average energy consumption by a single car, the Renault Zoe, was 0.164 kWh/km compared with 0.146 kWh/km (Muneer et al., 2015).

The purpose of studying the driving behaviour has been to estimate how much energy an electric car could use in a day. As the facility is going to be placed in Edinburgh, the average of kilometres that a car drives per day is of 20.8 km, as it was mentioned before.

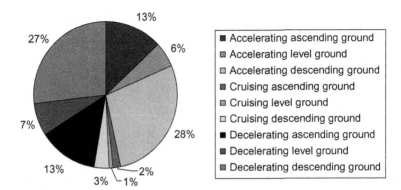

13%

27%

6%

7%

28%

13%

2%

3% 1%

- ■ Accelerating ascending ground
- ▣ Accelerating level ground
- ▢ Accelerating descending ground
- ■ Cruising ascending ground
- ▣ Cruising level ground
- ▢ Cruising descending ground
- ■ Decelerating ascending ground
- ■ Decelerating level ground
- ▣ Decelerating descending ground

Fig. 12.2.29 Driving cycle in Edinburgh City.
From Walsh, J., Muneer, T., & Celik, A. (2011). Design and analysis of kinetic energy recovery system for automobiles: Case study for commuters in Edinburgh. *Journal of Renewable and Sustainable Energy*, 3(1), 013105.

Therefore, the energy consumption per day of this specific car could be calculated with Eq. (12.2.12):

$$E = \text{km travelled by car per day}^* \text{ Energy consumption}$$
$$= 3.41 \text{ kWh/day} \tag{12.2.12}$$

12.2.6 Number of vehicles to be charged during a day by the solar carport

The average daily slope irradiation per year in Edinburgh being around 2.85 kWh/m^2, the 200 PV modules of the facility could generate a daily output of 240.08 kWh/daily; see Eq. (12.2.13):

$$E_{daily} = A_m \, \eta_{sys} \, I_{tilt} \, No._{PV} = 240.08 \text{ kWh/daily} \tag{12.2.13}$$

According to Section 12.2.5, the consumption of a car was estimated to be 0.164 kWh/km; therefore, with 1 kWh, a car could drive 6.1 km or 3.8 mi. Thus, with an amount of energy of 240.08 kWh generated by the facility daily, 912.3 mi could be driven in a day. If the average trip per passenger by car was 20.8 km or 13 mi in a day (see Section 11.2), a total of 70 cars could be powered.

Therefore, the energy consumption for these 70 cars per day would be 238.7 kWh, the energy that a car consumes per day being 3.41 kWh.

12.2.6.1 Energy production and energy consumption by the facility

The monthly energy production and energy consumption for the facility can be seen in the picture below along the year.

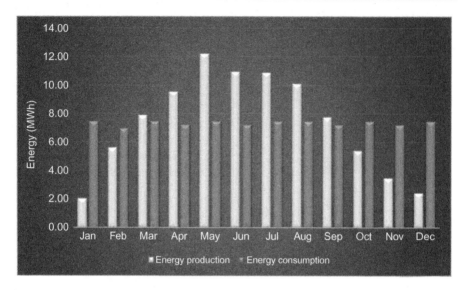

Fig. 12.2.30 Energy production vs. Energy consumption.

Fig. 12.2.30 shows that in some months, the first and the last ones, the energy produced by the solar facility does not meet the demand. However, this problem could be solved by using the national grid and in the months where the facility produces energy in excess; this energy will be sold to the grid.

12.2.6.2 Load profile

In order to model the monthly energy consumption of electric cars, it was necessary to make a load profile that simulated the load behaviour. The number of electric vehicles that would be charged was defined in the previous section. This load behaviour could be simulated based on trips made during a day according to the journey and peak times, as mentioned in Section 12.2.5. However, as this solar charging station is located at the college, it is difficult to estimate how the load would change over the months and throughout the year, because students usually do not come and go at the same times.

It was thus decided to define the range of hours when the vehicles could be charged, estimating that this energy supply would be constant in that period, that way the load profile would be simplified.

The possible hours chosen for charging the vehicles were from 8:00 to 21:00, as these are the opening hours of the college. The charts below show the energy consumption of the 70 cars and the energy generated by the facility in some months (see Figs. 12.2.31 and 12.2.32).

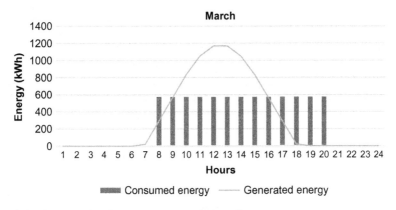

Fig. 12.2.31 Consumed and generated energy in March.

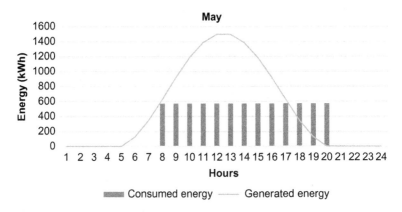

Fig. 12.2.32 Consumed and generated energy in May.

12.2.6.3   Financial analysis

One of the aims of this report was to determine the economic viability of the solar carport facility. To conduct a study on the market situation, a financial plan will be developed for three different scenarios: the first scenario for the present year 2016; the second scenario for 2020, as our client Edinburgh College could carry out the project around this date; and the third scenario based on unsubsidised generation.

Various methods such as payback period time, net present value (NPV), internal rate of return (IRR), and the debt-service coverage ratio (DSCR) were studied to determine the feasibility of the project.

12.2.6.4   Scenario 1

This scenario will be developed according to the present year, 2016.

Expenses: OPEX and investment costs

Operating Expenditure (OPEX) The main components of a carport are PV systems, roofs, frames, and foundations. The budget needed for the investment (80 kWp of installed capacity) is presented below, with a breakdown of all components of the PV facility. All the data used for this analysis were taken from a study published by Fraunhofer Institute for Solar Energy Systems entitled 'Current and Future Cost of Photovoltaics' (ISE, 2015).

The total operating expenditures were found to be 15.4 £/kWp. These costs were divided as shown in Table 12.2.18 below.

Therefore, the operational expenses for this 80 kWp solar carport facility will be £1232 per year.

Investment costs The investment costs represent the measurable technical factors in money involved in the production. For the proposed facility, it will be necessary to take into account the costs for the following:

- PV modules
- Inverters
- Balance of system
- Charging station
- Frames and foundations

The balance of system includes items such as the cost of the installation, mounting system, infrastructure, transformers, grid connection, wiring, planning and documentation (ISE, 2015), mounting structure, and grid connection accounting for the most expensive cost.

The costs distribution in percentage for the proposed facility and the breakdown of the balance of costs are shown in Figs. 12.2.33 and 12.2.34.

PV modules, inverters, and balance of system costs are based on a ground-mounted PV plant; therefore, charging stations, frames, and foundations costs are not included in this document. Presently, the charging stations are manufactured by the staff of Edinburgh College at an approximate cost of £600. As aforementioned, the number of chargers to be installed was three. On the other hand, the frame and foundation costs were estimated to be around £200/kWp (Jackson, 2016).

Table 12.2.19 below shows the breakdown of all the total distributed costs.

Therefore, the total expenses of the facility reach an amount of £90,832, £1232 from operational expenses annually and £89,600 from the initial investment.

Table 12.2.18 Operating expenditures—current year, 2016

OPEX			Total costs
Maintenance and operating cost	25%	3.85 £/kWp	£308
Replacing and cleaning cost	35%	5.40 £/kWp	£432
Insurance and taxes	40%	6.15 £/kWp	£492
Total	100%	15.4 £/kWp	£1232

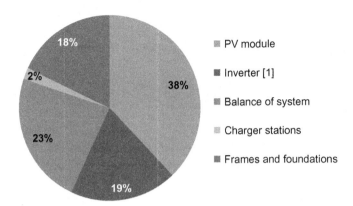

Fig. 12.2.33 Main costs distribution.

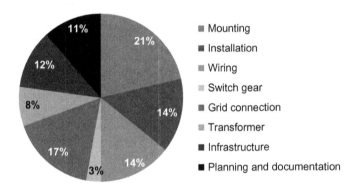

Fig. 12.2.34 Balance of System costs distribution.

Revenues

The revenues will be obtained by two different mechanisms:

- Feed-in tariffs
- Selling the electricity generated to the electric car fleet

Feed-in tariff (FIT) earnings The feed-in tariff applicable to the proposed facility was 2.70 p/kWh, and the export tariff was 4.85 p/kWh (Ofgem.gov.uk, 2016).

It is important to note that if the facility capacity of the project is below 50 kW, the payments from FIT will be much better, 4.59 p/kWh instead of 2.70 p/kWh (Ofgem. gov.uk, 2016). According to Ofgem, the FIT payments will last for 20 years for PV system (Recc.org.uk, 2016). The total energy generated for the facility in the first year was 87.96 MWh; thereafter, the PV modules will suffer a degradation of 0.4% per year (Stu, 2014), and therefore, the energy generated will be affected year by year.

Table 12.2.19 Detailed cost distribution

Investment cost			Total costs
PV module	55%	425 £^{a}/kWp	£34,000
Inverter^b	11%	85 £/kWp	£17,000
Balance of system	34%	260 £/kWp	£20,800
Mounting	8%	61 £/kWp	£4500
Installation	5%	38 £/kWp	£3000
Wiring	5%	38 £/kWp	£3000
Switch gear	1%	2.5 £/kWp	£600
Grid connection	6%	15.5 £/kWp	£3500
Transformer	2%	5 £/kWp	£1600
Infrastructure	4%	10.5 £/kWp	£2400
Planning and documentation	4%	10.5 £/kWp	£2400
Total^c	100%		£71,800
Charging stations	–	600 £/unit	£1800
Frames and foundations		200 £/kWp	£16,000
Total			£89,600

^{a}Currency exchange approximately 1 € = 0.77 £.
^{b}Taking into account the replacement of the inverter every 10 years.
^{c}Frames and foundations not included.

Selling the electricity generated to the electric vehicles fleet In order to sell the electricity to the electric vehicles, consideration was given to the possibility of setting two different prices.

As aforementioned, the estimated hours during which the vehicles are going to be charged are 8:00–21:00. In this interval, from 8:00 to 9:00 and from 18:00 to 21:00, the cost of electricity was decided to be 18 p/kWh, because in that period most of the time the energy should be bought from the grid, thus generating an additional cost. For therest of the hours, the price of the electricity would remain at its present value, 12 p/kWh, which means that 70% of the total time the price of electricity would be lower, thus benefitting the consumers.

12.2.6.5   Scenario 2

This scenario will be developed for the year 2020, where the feed-in tariff may have a lower rate and also the PV system costs will decrease.

Expenses: OPEX and investment costs
Operating Expenditure (OPEX) According to the same study published by Fraunhofer Institute for Solar Energy Systems, by 2050, the operating expenditure will have been reduced to 7.7 £/kWp (ISE, 2015). The estimation for 2020 is 14.1 £/kWp (see Table 12.2.20).

Therefore, the operating expenditure for this scenario will be £1128 per year.

Table 12.2.20 **Operating expenditures—year 2020**

OPEX			Total costs
Maintenance and operating cost	25%	3.53 £/kWp	£282
Replacing and cleaning cost	35%	4.94 £/kWp	£395
Insurance and taxes	40%	5.64 £/kWp	£451
Total	100%	14.1 £/kWp	£1128

Investment costs Fig. 12.2.35 above shows an approach including the assumptions for 2050 where the costs of the PV system will decrease between 610 €/kWp in the worst-case scenario and 280 €/kWp in the best-case scenario. It is estimated that by 2020, these costs will be 935 €/kWp in the worst-case scenario and 880 €/kWp for the best-case scenario. For the purposes of this project, the average of these two costs was used. The estimated distributed costs for 2020 in pounds are shown in Table 12.2.21 below, where the charging station costs were also reduced.

The total expenses will decrease by 8.8% by 2020 compared with 2016.

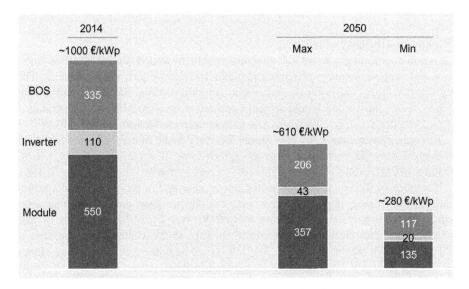

Fig. 12.2.35 PV system costs in 2015 combining minimum and maximum assumptions. From Fraunhofer-Institute for Solar Energy Systems (ISE) (2015). Current and future cost of photovoltaics. [online] Berlin: Mara Marthe Kleiner. Available at: https://www.agora-energiewende.de/en/topics/-agothem-/Produkt/produkt/88/Current+and+Future+Cost+of+Photovoltaics/ Accessed 18.07.16.

Table 12.2.21 **Detailed cost distribution**

Investment cost			Total costs
PV module	55%	384 £/kWp	£30,720
Inverter[a]	11%	75 £/kWp	£15,000
Balance of system	34%	235 £/kWp	£18,800
Charging stations	–	400 £/unit	£1200
Frames and foundations	–	200 £/kWp	£16,000
Total	100%		£81,720

[a]Taking into account the replacement of the inverter every 10 years.

Revenues
Feed-in tariff (FIT) earnings According to Ofgem, feed-in tariffs for 2019 will have a rate of 1.76 p/kWh, and export tariffs will have a rate of 4.85 p/kWh (Ofgem.gov.uk, 2016).

12.2.6.6 Scenario 3

Feed-in tariffs are decreasing as each year passes; also the costs for PV technology are falling quickly; therefore, subsidies are no longer guaranteed in the future. In this scenario, the same assumptions as in Scenario 2 have been taken into account, where FIT was disregarded.

CRC Energy efficiency scheme
It is worth mentioning that this UK government scheme aimed at cutting carbon emissions and improving energy efficiency in public and private sector organizations. The organizations that qualify for this scheme are organizations in the public and private sectors across the United Kingdom that consume over 6000 MWh/yr and that as a whole, account for over 10% of the UK CO_2 emissions (Gov.uk, 2014a, 2014b).

The organizations must buy allowances for every tonne of carbon they emit; these allowances may be purchased from the government at a current price of £16.90 per tonne of CO_2, or they may be bought in the secondary market (Gov.uk, 2016a, 2016b).

If smaller facilities such as the solar carport presented in this report could participate in this scheme, they could save around £370 per year, considering that solar energy has a carbon footprint of 50 g CO_2e/kWh (see Fig. 12.2.36, invalid source specified) and that electricity generates as average in the United Kingdom around 300 g CO_2e/kWh (Ecotricity.co.uk, 2016). It should be noted that this revenue cannot be claimed if the PV system is receiving a feed-in tariff.

12.2.6.7 Financial assumptions

First, it was assumed that the bank would lend 100% of the investment costs over a term of 15 years at a fixed rate of 3%. The project life is 25 years, and the duration of feed-in tariff is 20 years. The inflation rate (RPI) for FIT payments over 20 years was considered to be 2% and the electricity price inflation over the 25 years' lifetime of the project 3%.

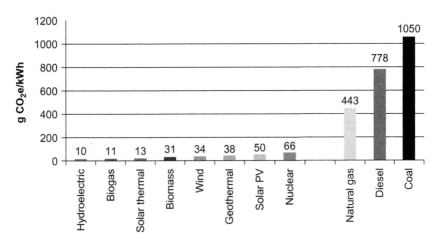

Fig. 12.2.36 Carbon footprint of various energy sources.
From Nugent D., & Sovacool, B. K. (2014). Assessing the lifecycle greenhouse gas emissions from solar PV and wind energy: A critical meta-survey, *Energy Policy, 65*, 229–244.

Apart from that, the lifetime for the PV modules, the mounting system, and the wiring was determined to be of 25 years (Narayan, 2013), whereas the lifetime of the inverter was 10 years; therefore, it will have to be replaced once.

Payback period, NPV, IRR and DSCR
All these equations (Eqs. 12.2.14–12.2.17) were applied in the financial analysis in order to determine the feasibility of the facility. The payback period time, the net present value (NPV), the internal rate of return (IRR), and the debt-service coverage ratio (DSCR) were calculated for the different scenarios (Prentice, 2015):

$$Payback = \frac{Investment\ costs}{Net\ annual\ cash\ in\ flow} \tag{12.2.14}$$

$$NPV = -C_O \sum_{t=0}^{n} \frac{C_t}{(1+r)^t} \tag{12.2.15}$$

where t is the number of years, n the project time, r the discount rate in %, Ct the cash flow in year t, and Co the initial investment:

$$NPV = -C_O \sum_{t=0}^{n} \frac{C_t}{(1+r)^t} = 0 \tag{12.2.16}$$

$$DSCR = \frac{Net\ operating\ income}{Debt\ service} \tag{12.2.17}$$

Table 12.2.22 Financial results for the solar charging station

	Scenario 1	Scenario 2	Scenario 3
Payback	8.8 years	8.4 years	9.3 years
NPV (25 years)	£175,033	£170,676	£144,374
IRR (25 years)	15%	15%	13%
DSCR (15 years)	1.9	1.96	1.71

Financials the three different scenarios

As shown in Table 12.2.22, the financial results for the facility were very profitable. The payback period between the scenarios ranged from 8.8 to 9.3 years; the NPV that determined the profitability of the project was found to be higher in Scenario 1, with a benefit of £170,676, and the percentage of IRR showed the financially robust project.

According to the results, even in Scenario 3, which represents the worst-case scenario—assuming that the facility will not benefit from subsidies—the economic benefits are still positive.

This project has broadly proved the feasibility of the solar charging stations.

12.2.7 Environmental analysis

To conclude, a life-cycle assessment was carried out in order to determine the environmental impact of this project. As solar charging stations are a relatively novel form of deployment and limited information was found to conduct the LCA for the whole installation, the life-cycle assessment was carried out for the main components: PV modules, balance of system, inverters, and mounting system.

The amount of CO_2 saved when using solar energy as a source to generate electricity instead of the electricity from the grid will also be determined.

12.2.7.1 Life cycle assessment (LCA) of the project

The LCA is important to determine the life of products, materials, system, process, and impact on the environment (Asif and Muneer, 2006).

Eco-audit of the PV module selected

As the PV module is the most important element in the facility, the LCA was carried out using the software CES EduPack, and the eco-audit tool was used to evaluate the environmental impact of the PV module by focusing on two well-understood environmental stressors, CO_2 footprint and energy usage identifying which of the main life phases (material, manufacture, transport, use, and end of life) is the most demanding of all (see Fig. 12.2.37).

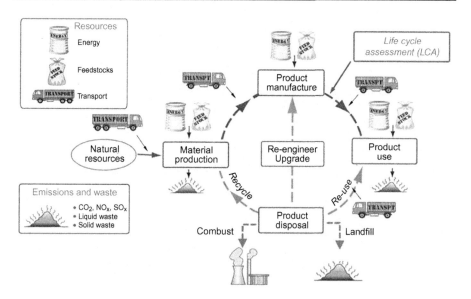

Fig. 12.2.37 The product life-cycle.
From Granta. (2016). Granta's eco audit methodology [online]. Available at: https:/www.
grantadesign.com/eco/audit.htm Accessed 25.07.16.

Materials

A typical structure of a monocrystalline silicon PV module can be seen in Fig. 12.2.38
(Mohammad Bagher, 2015).

   All the data entered in the software programme such as the materials that make up
the PV module, the kind of process used to manufacture them, and their masses were
based in a study carried out by G. J. M. Phylipsen and E. A. Alsema. These data were
adapted to the characteristics of the PV module used for this project.

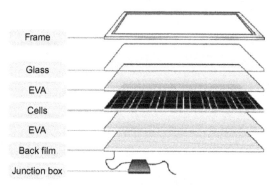

Fig. 12.2.38 Monocrystalline solar
cell structure.
From Mohammad Bagher (2015).

Transport assumptions

The PV modules used, ASP-400GSM, are produced in the United States. The energy and CO_2 released in the use of transport must be taken into account to carry out the LCA. The distance from the United States to Edinburgh is approximately 6500 km; see Fig. 12.2.39. This data will be considered to carry out the LCA.

Use

The last step to conclude the LCA was to define the use of the PV module during its lifespan, which was 25 years. As its mode of operation is static, the power rating, duty cycle, and the product efficiency were defined.

With all the data collected, the programme could provide detailed information about the breakdown of energy usage and CO_2 footprint of a single PV module or for the 200 PV modules used for the facility.

Fig. 12.2.40 below shows that the largest energy demand and the largest release of CO_2 occur during the production phase of the materials where the most polluting material was the silicon, with a percentage of 78% (see Table 12.2.23).

The results extracted from CES EduPack report for the 200 PV modules used are shown in Table 12.2.24; the table shows all the energy consumption and CO_2 footprint of each individual phase. A total of 70,300 kg of CO_2 would be emitted during all the life phases of the product, and 998,000 MJ would be required during the lifespan of the PV module after the end of its life.

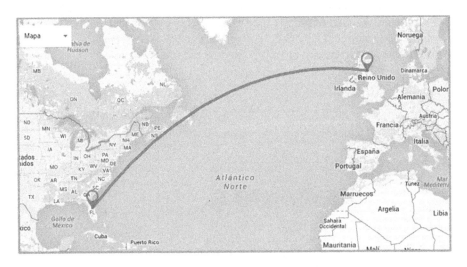

Fig. 12.2.39 Distance from the United Kingdom to the United States.
Modified from Freemaptools (2016).

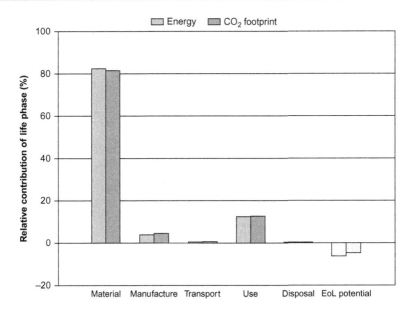

Fig. 12.2.40 Energy and CO2 footprint details of a single module.
Data from CES Edupack, 2016.

12.2.7.2 Life cycle assessment of the balance of system (BOS) and system mounting

For these elements, the LCA was based in a study entitled 'Life cycle assessment of a medium-sized photovoltaic facility at a high latitude location', at the Napier University's Merchiston Campus (Muneer et al., 2006).

Balance of system

The components to be taken into account in BOS are the inverters, cables, fuses, and transformers. Due to the lack of data, this report only considers the inverter and cables. According to Peyser (2010), the contribution of greenhouse gases from BOS is relatively small compared with other components of PV systems.

The inverter used for the installation at Napier University was a Fronius (IG60) with a nominal power of 4.6 kW, the amount of CO_2 released per inverter being 850 kg, and the embodied energy 1 MWh (Muneer et al., 2006). Based on this information, the CO_2 released by the inverter selected for the solar carport, an Aurora Trio of 20 kW of nominal power, would be 3700 kg of CO_2 per inverter and 4.35 MWh of embodied energy. As the facility has five inverters, the total amount of CO_2 released would be 18,500 kg and the energy used would be 22 MWh.

For the facility, the use of around 450 kg of copper for cabling was estimated; the amount of CO_2 released was estimated being 2280 kg and 8.4 MWh of embodied energy for the facility of 400 m^2.

Table 12.2.23 Detailed breakdown of individual material phase for a single module

Component	Material	Recycled content (%)	Part mass (kg)	Qty.	Total mass (kg)	Energy (MJ)	%
	Single crystalline silicon, photovoltaics	Virgin (0%)	2,2	1	2,2	3,7e+03	78,0
	Alkaline-earth lead glass	Virgin (0%)	28	1	28	7e+02	14,7
	EVA (Shore A95/D50, 12% vinyl acetate)	Virgin (0%)	1,9	2	3,8	3e+02	6,3
	PVC (flexible, Shore A60)	Virgin (0%)	0,7	1	0,7	42	0,9
Total				5	35	4,7e+03	100

Data from CES EduPack (2016).

Table 12.2.24 **Individual life-phase CO$_2$ footprint details**

Phase	Energy (MJ)	Energy (%)	CO$_2$ footprint (kg)	CO$_2$ footprint (%)
Material	9,39e+05	94,0	6,56e+04	93,4
Manufacture	4,69e+04	4,7	3,75e+03	5,3
Transport	7,11e+03	0,7	505	0,7
Use	720	0,1	51,1	0,1
Disposal	4,7e+03	0,5	329	0,5
Total (for first life)	9.98e+0.5	100	7.03e+04	100
End-of-life potential	−6.5e+04		−3.66e+03	

Data from ES EduPack (2016).

System mounting

With regard to the mounting of the system, for the 520 m^2 photovoltaic roof installation where the 200 PV modules are going to be placed, the use of around 9750 kg including spigots, vertical rails, tie brace, and schuco rails was estimated (Muneer et al., 2006).

The amount of CO$_2$ released was estimated to be 21,000 kg and 159 MWh of embodied energy.

12.2.7.3   *Energy payback time (EPBT) and **global warming potential (GWP)** summary*

Based on the LCA, it was estimated that the facility (PV modules, BOS, and system mounting) would release a total amount of 130 tonnes of CO$_2$ and it would have about 488 MWh of embodied energy. Table 12.2.25 below shows a summary of the most relevant data.

Once the total energy used to manufacture the PV system components has been calculated, it would be interesting to determine if the PV facility is a net positive producer of energy in its 25 years of lifetime.

The energy payback time estimates the energy reimbursement of a product; for the proposed facility, the energy payback time obtained was 5.55 years (see Eq. 12.2.18) (McEvoy, Markvart, and Castañer, 2012):

$$EPBT = \frac{\text{Total energy input during the system life cycle}}{\text{Yearly energy generation by the system}} \qquad (12.2.18)$$

The EPTB for this facility in the present study was estimated to be 5.55 years.

The global warming potential (GWP) is a measurement to quantify the effects on global warming; therefore, knowing that the facility will release 130,580 kg of CO$_2$ and that the lifetime of the installation is 25 years, the ratio of the lifetime emissions of CO$_2$ for the solar carport was found to be 0.059 kg CO$_2$/kWhe (see Eq. 12.2.19) (Muneer et al., 2006):

Table 12.2.25 Summary embodied energy and CO$_2$ released by the facility

Element	Number	Material	Mass (Kg)	Embodied energy (MWh)	Embodied energy (MJ)	CO$_2$ released (kg)
PV module (400 Wp)	200	Mixed	6940	277	998,000	70,300
BOS						
Inverter (20 kW)	10[a]	Mixed	355 × 2	22 × 2	79,200 × 2	18,500 × 2
Cables	—	Copper	450	8.4	30,240	2280
Mounting	—	Steel	9750	159	572,400	21,000
Total				488.4	1,759,040	130,580

[a]Assuming the replacement of the inverter every 10 years.

$$\mathrm{GWP} = \frac{\text{Total } CO_2 \text{ released during the system life cycle}}{\text{Energy generation along the lifetime of the system}} \qquad (12.2.19)$$

12.2.7.4 CO_2 emissions saved

In order to determine the amount of CO_2 saved by this facility, it was necessary to determine the carbon content of the UK grid. To this effect, measurements were taken during 10 days at specific times (10 a.m., 17 p.m., and 22 p.m.) in August through a website that provides live data on how much electricity is being made nationally and the percent from fossil fuels, nuclear, or renewable energy, giving information about the grams of CO_2 emitted for every unit of electricity generated in the United Kingdom.

As the average carbon content of the UK grid was 274 g CO_2/kWh during these 10 days of measurements and, according to Fig. 12.2.36, the carbon footprint that solar energy has is 50 g CO_2/kWh, the amount of CO_2 saved by this facility yearly generating an output of 87.96 MWh would be

$$CO_2 \text{ saved} = (0.274 - 0.05) \times 87,960 = 19.7 \text{ tonnes of } CO_2$$

12.3   Conclusions

12.3.1   Solar meadow farm at Edinburgh College

The performance of solar irradiation, PV cell temperature, and efficiency models have been assessed through a study done on a commercial solar plant in the Edinburgh area. An experiment to measure a range of data quantities was designed and implemented, and a survey of the site to obtain an estimate of the degree of shading was performed. To overcome limitations on data available to the researcher, 3 days of manually recorded measurements were made to support the experimental data.

The main findings were the following:

- The shading characteristic of the site was inaccurate, mostly in the winter months.
- Shading was most severe at the south end of the site.
- Shading on the solar modules will adversely affect the performance of the plant.
- The seasonal model proposed by Clarke et al. (2007) to convert solar irradiation data from horizontal intensity to intensity received on a slope has been proved to be very accurate.
- The prediction of a solar photovoltaic module's cell temperature from environmental data such as air temperature and solar irradiation was shown to be fairly accurate and reliable across three different calculation methods (NOCT, HOMER software, and thermal method).
- The calculation of cell efficiency over the 3 days of manual data recordings was performed, the average measured value being 16%.

12.3.2   Solar charging station at Edinburgh College

The proposed design of the solar charging station has met all expectations, proving to be a cost-effective and environmentally friendly design and technically and financially viable. The main findings of this research are listed below:

- The average solar energy available in Edinburgh was found to be 2.47 Kwh/m^2 daily, and due to the low temperatures in the city, using a temperature model, it was determined that the cell efficiency of the PV module chosen rose up from 15.2 to 16.2%.
- It was found that the best tilt and azimuth angle to install the PV modules was with south orientation at 40 degrees, where the design arrangement could generate 87.96 MWh of energy per year.
- The car park area of approximately 400 m^2 provides parking spaces for 33 vehicles. The roof area will have enough space to accommodate 200 monocrystalline PV modules rated at 400 Wp; also five inverters Aurora Trio rated at 20 kW, and three twin AC rapid charger CHAdeMO charging stations were required. This facility will be connected to the national grid.
- According to the average car journey per person in the United Kingdom, 20.8 km, and the average energy consumption by an electric car, 0.164 kWh/km, the total energy demand for a single electric car in a day is 3.41 kWh. According to these results, the facility would be able to feed 70 EVs during a day.
- A financial analysis was carried out for three different scenarios. The result of the financial analysis for Scenario 2 during a lifetime of 25 years is shown in Table 12.3.1 below:

The payback, i.e. the recovery of the investment for Scenario 2 starts in year 8; the net present value (NPV) analysis shows the best-case NPV of £176,676 and an internal rate of return (IRR) of 15%.

- In the environmental analysis, the energy payback time (EPBT) was found to be recovered within 5 years and 5 months, and the global warming potential was 0.059 kg CO_2/kWh$_e$. On the other hand, the CO_2 saved for this facility as compared with the electricity produced by the UK grid was 19.7 t of CO_2.

Table 12.3.1 Financial analysis

Expenditures		Revenues	
Operating expenditures (annually)	£1128	Feed-in tariff and export tariff	£53,532
PV modules	£30,720	Sale of electricity	£430,006
Inverters	£15,000		
Balance of system costs	£18,800		
Charging stations	£1200		
Frames and foundations	£16,000		
Buying electricity from the grid (25 years)	£81,063		
Interest to pay 3%	£20,960		
Total in 25 years	£211,943	Total in 25 years	£483,538

Acknowledgements

In the preparation of this chapter, the authors have sourced data from a number of references a list of which is provided below. The authors would like to express their thanks to the authors/ publishers of those references.

Fig. 12.1.1–12.1.2 Fig. 12.1.3	Federschmidt C., 2016. MSc report on 'Charging solar station design for electric vehicles', Edinburgh Napier University, Scotland
Figs. 12.1.4–12.1.7 Figs. 12.1.8 and 12.1.9 Figs. 12.1.10 and 12.1.11 Fig. 12.1.12	https://www.researchgate.net/publication/ 274079251_Modeling_and_Experimental_ Verification_of_Solar_Radiation_on_a_ Sloped_Surface_Photovoltaic_Cell_ Temperature_and_Photovoltaic_Efficiency
Figs. 12.1.13–12.1.16 Fig. 12.2.1	http://futurecitiesenviro.springeropen.com/ articles/10.1186/s40984-015-0008-5
Figs. 12.2.2–12.2.4 Figs. 12.2.5 and 12.2.6 Figs. 12.2.7 and 12.2.8 Figs. 12.2.9 and 12.2.11 Fig. 12.2.12 Figs. 12.2.13 and 12.2.14 Figs. 12.2.15–12.2.23 Fig. 12.2.24	https://formfonts.com https://maps.google.com/ Based on https://maps.google.com/ Based on https://maps.google.com/ http://www.plugshare.com/? location=54084
Fig. 12.2.25	http://repository.tudelft.nl/islandora/object/ uuid:743ccf85-3d61-4ae5-bfc1- e414865fe1ca?collection=education
Fig. 12.2.26 Figs. 12.2.27 and 12.2.28	https://www.gov.uk/government/uploads/ system/uploads/attachment_data/file/ 433994/vls-2015-q1-release.pdf
Fig. 12.2.29	http://www.engineersedge.com/material_ science/free_body_diagram_13073.htm
Fig. 12.2.30	https://www.researchgate.net/publication/ 241620714_Design_and_analysis_of_ kinetic_energy_recovery_system_for_ automobiles_Case_study_for_commuters_ in_Edinburgh

Continued

Continued

Figs. 12.2.31 and 12.2.32	
Figs. 12.2.33 and 12.2.34	
Fig. 12.2.35	https://www.agora-energiewende.de/en/topics/-agothem-/Produkt/produkt/88/Current+and+Future+Cost+of+Photovoltaics
Fig. 12.2.36	http://www.sciencedirect.com/science/article/pii/S0301421513010719
Fig. 12.2.37	https://www.grantadesign.com/eco/audit.ht
Fig. 12.2.38	http://article.sciencepublishinggroup.com/pdf/10.11648.j.ajop.20150305.17.pdf
Fig. 12.2.39	https://www.freemaptools.com/how-far-is-it-between-edinburgh_scotland-and-new-york_-usa.htm
Fig. 12.2.40	CES EduPack, 2016
Tables 12.1.1, 12.1.2, 12.1.3, 12.2.1–12.2.3, 12.2.4–12.2.7, 12.2.8–12.2.14, and 12.2.15	
Table 12.2.16	http://www.sciencedirect.com/science/article/pii/S1361920914001783
Table 12.2.17	
Tables 12.2.18–12.2.22	
Tables 12.2.23 and 12.2.24	CES EduPack, 2016
Table 12.2.25	
Table 12.3.1	

References

Aldali, Y., Celik, A. N., & Muneer, T. (2013). Modeling and experimental verification of solar radiation on a sloped surface, photovoltaic cell temperature, and photovoltaic efficiency. *Journal of Energy Engineering, 139*(1), 8–11.

Asif, M., & Muneer, T. (2006). Life cycle assessment of built-in-storage solar water heaters in Pakistan. *Building Services Engineering Research and Technology, 27*(1), 63–69.

Clarke, P., et al. (2007). Technical note: An investigation of possible improvements in accuracy of regressions between diffuse and global solar irradiation. *Building Services Engineering Research and Technology, 28*(2), 189–197.

Department for Transport (2015). National travel survey: England 2014. [online] England, p. 19. Available at: https://www.gov.uk/government/uploads/system/uploads/attachment_data/file/457752/nts2014-01.pdf Accessed 20.05.16.

Ecotricity.co.uk. (2016). UK grid live-our green energy-ecotricity. [online] Available at: https://www.ecotricity.co.uk/our-green-energy/energy-independence/uk-grid-live Accessed 09.08.16.

FraunhoferInstitute for Solar Energy Systems (ISE). (2015). *Current and future cost of photo-voltaics. [online]*. Berlin: Mara Marthe Kleiner. Available at: https://www.agora-energiewende.de/en/topics/-agothem-/Produkt/produkt/88/Current+and+Future+Cost+of+Photovoltaics/ Accessed 18.07.16.

Gago, E., Etxebarria, S., Tham, Y., Aldali, Y., & Muneer, T. (2010). Inter-relationship between mean-daily irradiation and temperature, and decomposition models for hourly irradiation and temperature. *International Journal of Low Carbon Technologies*, 6(1), 22–37.

Gov.uk. (2014a). CRC energy efficiency scheme: Qualification and registration—Detailed guidance—GOV.UK. [online] Available at: https://www.gov.uk/guidance/crc-energy-efficiency-scheme-qualification-and-registration Accessed 28.07.16.

Gov.uk. (2014b). CRC energy efficiency scheme: Qualification and registration—Detailed guidance—GOV.UK. [online] Available at: https://www.gov.uk/guidance/crc-energy-efficiency-scheme-qualification-and-registration Accessed 28.07.16.

Gov.uk. (2016a). 2010 to 2015 Government policy: Energy demand reduction in industry, business and the public sector—GOV.UK. [online] Available at: https://www.gov.uk/government/publications/2010-to-2015-government-policy-energy-demand-reduction-in-industry-business-and-the-public-sector. Accessed 30.07.16.

Gov.uk. (2016b). All vehicles (VEH01)—Statistical data sets—GOV.UK. [online] Available at: https://www.gov.uk/government/statistical-data-sets/all-vehicles-veh01#table-veh0105 Accessed 31.05.16.

HOMER Energy, 2013. HOMER software. [online] Available at: http://homerenergy.com/software.html Accessed 19.08.13.

Jackson, C. (2016). *Solar car parks: A guide for owners and developers.* BRE National Solar Centre. [online] (No. 1092193). Available at: http://www.bre.co.uk/filelibrary/nsc/Documents%20Library/NSC%20Publications/BRE_solar-carpark-guide.pdf Accessed 20.05.16.

McEvoy, A., Markvart, T., & Castañer, L. (2012). *Practical handbook of photovoltaics.* Waltham, MA: Academic Press.

Mohammad Bagher, A. (2015). Types of solar cells and application. *American Journal of Optics and Photonics*, 3(5), 94. Available at: http://article.sciencepublishinggroup.com/pdf/10.11648.j.ajop.20150305.17.pdf.

Muneer, T., Abodahab, N., Weir, G., & Kubie, J. (2000). *Windows in buildings: Thermal, acoustic, visual and solar performance* (1st ed). Oxford: Butterworth-Heinemann.

Muneer, T., Gueymard, C., & Kambezidis, H. (2004). *Solar radiation and daylight models* (1st ed.). Amsterdam: Elsevier Butterworth Heinemann.

Muneer, T., Etxebarria, S., & Gago, E. (2014). Monthly averaged-hourly solar diffuse radiation model for the UK. *Building Services Engineering Research and Technology*, 35(6), 573–584.

Muneer, T., Milligan, R., Smith, I., Doyle, A., Pozuelo, M., & Knez, M. (2015). Energetic, environmental and economic performance of electric vehicles: Experimental evaluation. *Transportation Research Part D: Transport and Environment*, 35, 40–61.

Muneer, T., Younes, S., Lambert, N., & Kubie, J. (2006). Life cycle assessment of a medium-sized photovoltaic facility at a high latitude location. *Proceedings of the Institution of Mechanical Engineers, Part A: Journal of Power and Energy*, 220(6), 517–524.

Narayan, N. (2013). Solar charging station for light electric vehicles. A design and feasibility study. Master of Science Thesis. Delft University of Technology.

Ofgem. (2016). Feed-in Tariff (FIT) generation & export payment rate Table 1 April 2016–31 March 201. [online] Available at: https://www.ofgem.gov.uk/system/files/docs/2016/04/01_april_2016_tariff_table.pdf Accessed 20.07.16.

Peyser, J. (2010). *Tracing a path forward: A study of the challenges of the supply chain or target metals used in electronics.* Washington, DC: RESOLVE Inc.

Power-one. (2016). Aurora Trio. [online] Available at: http://www.thepowerstore.com/datasheets/PowerOne/trio-20.0_27.6-tl-us.pdf Accessed 09.06.16.

Prentice, D. (2015). Finance notes and lecture materials: Payback ARR NPV file. http://moodle. napier.ac.uk/pluginfile.php/986436/mod_resource/content/1/2.2%20Payback%20ARR% 20NPV.pdf Accessed 08.08.16.

Recc.org.uk. (2016). Feed in tariffs-consumers-renewable energy consumer code (RECC). [online] Available at: https://www.recc.org.uk/consumers/feed-in-tariffs Accessed 06.08.16.

Scottish Renewables (2016). *Renewables in numbers.* Available at: https://www. scottishrenewables.com/sectors/renewables-in-numbers/ [Accessed 30 Nov. 2016].

Stu, S. (2014). A small issue of PV solar panel degradation. [online] PV solar panels UK. Available at: https://www.heatmyhome.co.uk/solar-panels/small-issue-pv-solar-panel-degradation#. V2LtFPnhDIU Accessed 16.06.16.

Transport Scotland (2015). Transport and travel in Scotland 2014. [online] Edinburgh. Available at: http://www.transport.gov.scot/sites/default/files/documents/rrd_reports/ uploaded_reports/j389989/j389989.pdf Accessed 05.06.16.

Walsh, J., Muneer, T., & Celik, A. (2011). Design and analysis of kinetic energy recovery system for automobiles: Case study for commuters in Edinburgh. *Journal of Renewable and Sustainable Energy, 3*(1), 013105.

Further Reading

Ackermann, T., Andersso, G., & Soder, L. (2000). Distributed generation: A definition. *Elsevier Science. Electric Power Systems Research*, 195–204.

Aldali, Y. (2015). *Solar thermal and photovoltaic electrical generation in Libya.* MSc Mechanical Engineering ThesisEdinburgh: Napier University.

ASP (2016). ENF Ltd. [online] Es.enfsolar.com. Available at: http://es.enfsolar.com/pv/panel-datasheet/Monocrystalline/3487 Accessed 01.06.16.

Azadfar, E., Sreeram, V., & Harries, D. (2015). The investigation of the major factors influencing plug-in electric vehicle driving patterns and charging behaviour. *Renewable and Sustainable Energy Reviews, 42*, 1065–1076.

Botsaris, P., & Filippidou, F. (2009). Estimation of the energy payback time (EPR) for a PV module installed in north eastern Greece. *Applied Solar Energy, 45*(3), 166–175.

Bre National Solar Centre (2014). Planning guidance for the development of large scale ground mounted solar PV systems. [online] Bre. Available at: https://www.bre.co.uk/filelibrary/ pdf/other_pdfs/KN5524_Planning_Guidance_reduced.pdf Accessed 16.05.16.

Butcher, L. (2016). Electric vehicles and infrastructure. Number CBP07480. [online]. *House of Commons Library*, 1–3. Available at: http://researchbriefings.parliament.uk/ ResearchBriefing/Summary/CBP-7480 Accessed 31.05.16.

California Energy Commission (2007). High-performance photovoltaics carports: Design through deployment. [online] Richmond. Available at: http://www.energy.ca.gov/ 2013publications/CEC-500-2013-090/CEC-500-2013-090.pdf Accessed 05.08.16.

Cherrington, R., Goodship, V., Longfield, A., & Kirwan, K. (2013). The feed-in tariff in the UK: A case study focus on domestic photovoltaic systems. *Renewable Energy, 50*, 421–426.

Christensen, C. and Barker, G. (2001). Effects of tilt and azimuth on annual incident solar radiation for United States locations. [online] Washington, pp. 1–8. Available at: http://www. builditsolar.com/References/EFFECTS_OF_TILT_AND_AZIMUTH_ON_ANNUAL_ INCIDENT_SOLAR_RADIATION.pdf Accessed 06.06.16.

Climate.nasa.gov. (2016). Scientific consensus: Earth's climate is warming. [online] Available at: http://climate.nasa.gov/scientific-consensus/ Accessed 11.06.16.

Department for Business, Energy & Industrial Strat, 2016. Renewable energy planning database monthly extract. [Online] Available at: https://www.gov.uk/government/publications/renewable-energy-planning-database-monthly-extract Accessed 20.09.16.

Department for Business Energy and Industrial Strategy (2016). Energy consumption in the UK [online] London, p. 7. Available at: https://www.gov.uk/government/uploads/system/uploads/attachment_data/file/541163/ECUK_2016.pdf Accessed 16.07.16.

Department of Energy and Climate Change (2015). 2014 UK greenhouse gas emissions, provisional figures. Annex: 1990–2013 UK greenhouse gas emissions, final figures by end-user sector including uncertainties estimates. [online] London, p. 17. Available at: https://www.gov.uk/government/uploads/system/uploads/attachment_data/file/416810/2014_stats_release.pdf Accessed 19.07.16.

Department of Energy and Climate Change (2016). Updated energy and emissions projections 2015. [online] London, p. 12. Available at: https://www.gov.uk/government/uploads/system/uploads/attachment_data/file/501292/eepReport2015_160205.pdf Accessed 16.08.16.

Engholm, A., Johansson, G. and Åhl Persson, A. (2013). Life cycle assessment of Solelia Greentech's photovoltaic based charging station for electric vehicles. [online] Uppsala Universitet. Available at: http://www.diva-portal.org/smash/get/diva2:626019/FULLTEXT01.pdf Accessed 09.07.16.

Esrl.noaa.gov. (2016). ESRL Global Monitoring Division-Global Radiation Group. [online] Available at: http://www.esrl.noaa.gov/gmd/grad/solcalc/glossary.html#E Accessed 13.05.16.

Etxebarria Berrizbeitia, S. (2013). Inter-relationship between mean-daily irradiation and temperature, and decomposition models for hourly radiation and temperature. MSc Energy and Environmental Engineering Edinburgh: Napier University.

Goos, D. (2015). Feasibility study of a solar charging facility for electric vehicles in Munich. Master thesis Edinburgh: Napier University.

Gov.scot. (2015). International low carbon. [online] Available at: http://www.gov.scot/Topics/Environment/climatechange/international Accessed 04.06.16.

Gov.uk. (2013). Cars (VEH02)–Statistical data sets—GOV.UK. [online] Available at: https://www.gov.uk/government/statistical-data-sets/veh02-licensed-cars#history Accessed 31.05.16.

Granta. (2016). Granta's eco audit methodology. [online] Available at: https://www.grantadesign.com/eco/audit.htm Accessed 25.07.16].

Grove, J. (2015). Vehicle licensing statistics: Quarter 1 (Jan-Mar) 2015. [online]. Department of Transport. Available at: https://www.gov.uk/government/uploads/system/uploads/attachment_data/file/433994/vls-2015-q1-release.pdf Accessed 31.05.16.

HM Government (n.d.). Carbon Plan. [online] p. 5. Available at: https://www.gov.uk/government/uploads/system/uploads/attachment_data/file/47621/1358-the-carbon-plan.pdf Accessed 31.06.16.

Höök, M., & Tang, X. (2013). Depletion of fossil fuels and anthropogenic climate change—A review. Energy Policy, Elsevier, 52, 797–809.

House of Commons, 2012. The renewables obligation. [online] Available at: www.parliament.uk/briefing-papers/sn05870.pdf Accessed 19.08.13.

IEA.org. (2016). September: How solar energy could be the largest source of electricity by mid-century. [online] Available at: http://www.iea.org/newsroomandevents/pressreleases/2014/september/how-solar-energy-could-be-the-largest-source-of-electricity-by-mid-century.html Accessed 04.07.16.

International Energy Agency Photovoltaic Power Systems Programme (2011). Life cycle inventories and life cycle assessments of photovoltaic systems. [online] Available at: http://www.seas.columbia.edu/clca/Task12_LCI_LCA_10_21_Final_Report.pdf Accessed 19.07.16.

Jadraque Gago, E. (2011). *The use of photovoltaic solar energy as an energy source in the residential housing sector*. PhD Granada: University of Granada.

Jordan, D., & Kurtz, S. (2012). *Photovoltaic degradation rates—An analytical review. [online]*. NREL, National Renewable Energy Laboratory. Available at: http://www.nrel.gov/docs/fy12osti/51664.pdf Accessed 16.06.16.

Kelly, I. (2013). Analysis of solar modelling techniques through experiment on a 620kWp solar power plant at Dalkeith, Scotland, Edinburgh: Master thesis. Napier University.

Lane, B. (2016). Electric vehicle market statistics 2016—How many electric cars in UK &#63. [online] Nextgreencar.com. Available at: http://www.nextgreencar.com/electric-cars/statistics/ Accessed 31.05.16.

Lavado, M. (2015). *Large scale solar power plant flat roof and carport installation in Finland*. Degree Programme in Energy Engineering: Bachelor thesis Lappeenranta: University of Technology.

Martins Nunes, P. (2015). *Enabling solar electricity with electric vehicles in future energy systems*. Thesis doctoral in Sistemas Sustentáveis de Energia Universidad de Lisboa.

Muneer, T., Gueymard, C., & Kambezidis, H. (2004). *Solar radiation and daylight models*. Oxford: Elsevier Butterworth Heinemann.

Myrenaultzoe.com. (2016). Charging | My Renault ZOE electric car. [online] Available at: http://myrenaultzoe.com/index.php/zoe-description/charging/ Accessed 07.06.16.

Nansai, K., Tohno, S., Kono, M., Kasahara, M., & Moriguchi, Y. (2001). Life-cycle analysis of charging infrastructure for electric vehicles. *Elsevier. Applied Energy, 70*(3), 251–265.

Nugent, D., & Sovacool, B. K. (2014). Assessing the lifecycle greenhouse gas emissions from solar PV and wind energy: A critical meta-survey. *Energy Policy, 65*, 229–244.

Ofgem (2013). Applying for the Feed-in Tariff (FIT) scheme. [online] Ofgem.gov.uk. Available at: https://www.ofgem.gov.uk/environmental-programmes/feed-tariff-fit-scheme/applying-feed-tariff-fit-scheme Accessed 15.07.16.

Phylipsen, G., & Alsema, E. (1995). *Environmental life-cycle assessment of multicrystalline silicon solar cell modules*. The Netherlands: The Netherlands Agency for Energy and the Environment, NOVEM. p. 65.

Plugshare.com. (2016). PlugShare. [online] Available at: http://www.plugshare.com/?location=54084 Accessed 09.08.16.

Solar Power Portal, 2012. Malcolm Allen Housebuilders invest in one of Scotland's largest solar arrays. [Online] Available at: http://www.solarpowerportal.co.uk/case_studies/malcolm_allen_housebuilders_invest_in_one_of_scotlands_largest_solar_2356

Rahmani, V. (2015). *Analysis of energy delivery of the Edinburgh college solar PV farm: Effect of shading*. MSC Renewable Energy, thesis Edinburgh: Napier University.

Renewable Energy Policy Network for the 21st Century (REN21) (2016). Renewables 2016 global status report. [online] Paris, pp. 27–29. Available at: http://www.ren21.net/wp-content/uploads/2016/06/GSR_2016_Full_Report1.pdf Accessed 03.07.16.

Rugg, P., 2012. Renewable energy planning guidance note 2: The development of large scale (>50kW) solar PV arrays. [online] Available at: http://www.solartrade.org.uk/media/2%20Large%20Scale%20Solar%20PV%20%20August%202012.pdf Accessed 19.08.13.

SMA (2013). *Central inverter planning of a PV generator planning guidelines. [online]*. SMA Solar Technology AG Available at: http://files.sma.de/dl/1354/DC-PL-en-11.pdf Accessed 16.06.16.

SMA. (2016). Sunny central 800CP XT/850CP XT/900CP XT. [online] Sma.de. Available at: http://www.sma.de/en/products/solarinverters/sunny-central-800cp-xt-850cp-xt-900cp-xt.html#Downloads-14440 Accessed 18.06.16.

The Florida Solar Energy Center (2014). Cells, modules, and arrays. [online] Fsec.ucf.edu. Available at: http://www.fsec.ucf.edu/en/consumer/solar_electricity/basics/cells_mod ules_arrays.htm Accessed 20.06.16.

The International Energy Agency (2010). World energy outlook, 2010. [online] France, pp. 95–96. Available at: http://www.worldenergyoutlook.org/media/weo2010.pdf Accessed 30.06.16.

The International Energy Agency (2013). World energy outlook 2013. [online] France, p. 71. Available at: https://www.iea.org/publications/freepublications/publication/WEO2013. pdf Accessed 30.06.16.

The International Energy Agency (2014). Technology roadmap solar photovoltaic energy. energy technology prospective. [online] Paris, p. 5. Available at: https://www.iea.org/pub lications/freepublications/publication/TechnologyRoadmapSolarPhotovol taicEnergy_2014edition.pdf Accessed 29.06.16.

Tools, F. (2016). How far is it between Edinburgh, Scotland and New York, USA. [online] Freemaptools.Com. Available at: https://www.freemaptools.com/how-far-is-it-between-edinburgh_-scotland-and-new-york_-usa.htm Accessed 10.07.16.

Tutiempo Network, S. (2016). Clima en El Reino Unido Página 2. [online] www.tutiempo.net. Available at: http://www.tutiempo.net/clima/El_Reino_Unido/GB_2.html Accessed 30.05.16.

Tverberg, G. V. (2012). World energy consumption since 1820 in charts. [online] Our Finite World. Available at: https://ourfiniteworld.com/2012/03/12/world-energy-consumption-since-1820-in-charts/ Accessed 29.06.16.

Drive cycles for battery electric vehicles and their fleet management

13

Ross Milligan
Edinburgh College, Dalkeith, United Kingdom

13.1 Introduction

This presented research work is focused on analysis of real operational data from an electric vehicle fleet both in proprietary data logging and reading and investigating the data coming from the vehicles own electronic control unit. In this work, >50 electric vehicle data have been monitored for 4 years test period. The key characteristics of the electric vehicles operations, e.g. journeys, speeds, distances, routes, and vehicle energy consumption, have been investigated over the evaluation period. It has been noticed that driving cycle patterns have significant impact on vehicle's energy intensity. Also, different operational modes, e.g. acceleration, deceleration, cruise, ascent, and decent, have been evaluated and validated using real operational data for determining the regenerative braking efficiency of the electric vehicle fleet. This research has attempted to estimate vehicular driving patterns in the Edinburgh region and to offer an option of battery electric vehicles for sustainable mobility.

Nissan LEAF

Leading Environmentally friendly *Affordable Family* car

One of the vehicles used for this experiment work was the compact five-door Nissan LEAF (Fig. 13.1.1) electric vehicle; this vehicle has kinetic energy recovery systems (KERS). Different KERS may be built purely electric, purely mechanical, or hybrid mechanical/electric differing for vehicle applications. As previously stated, this vehicle has an all-electric energy recovery system to replace energy into the traction battery when braking or under overrun conditions.

Table 13.1.1 gives key vehicle performance figures over a 37-year period used as comparison across makes and models; the author has compared old diesel engine vehicles with modern battery electric vehicles indicating a marked improvement across performance figures.

This research aims to offer a control strategy and an alternative to the conventional fossil-fuel vehicle the effectiveness of any control strategy depends on accurate data representation to given routes and drive cycles (Table 13.1.2).

Battery electric vehicles have been used for this task with the 'car-chasing' technique employed to measure the college-specific driving cycle. 'Car chasing' is the action of following the vehicle in front whilst maintaining speed with the traffic flow

Electric Vehicles: Prospects and Challenges. http://dx.doi.org/10.1016/B978-0-12-803021-9.00013-6

Fig. 13.1.1 Nissan LEAF automobile.

Nissan LEAF (second generation)

Product information

Small car with a length of 4.45 m × 1.77 m wide × 1.55 m height
80 kW (109 hp) peak
280 Nm maximum torque
Max speed of 145 km/h (90 mph)
175 km range (108 mi)
17.3 kWh/100 km consumption

Battery and charging

Lithium ion, 24 kWh, standard charge in 6–8 h (100%)
Quick charge in 30 min (80%)
Charging connector front

Price and market entry

£23,500 VAT included
Available in the United Kingdom since 2012 (Nissan UK, 2016).

Table 13.1.1 Battery electric vehicles compared with previous economical diesel vehicles (Milligan, 2016)

	1977 2.1D Peugeot 504	19841.6D Ford Fiesta	2012 Mitsubushi i-MiEV	2014 Nissan LEAF Acenta
Best MPG	33	74	112 equivalent	169 equivalent
CO$_2$ (g/km)	222.6	127	0	0
Horsepower	59	54	66	109
Torque	82	70	145	254
Weight	1210	835	1070	1493
0–60	23 s	19 s	11 s	9.7 s
Top speed	83	<90	81 (Limited)	93 (Limited)
Price	n/a	£15,000	£23,499	£18,490

Table 13.1.2 SWOT analysis for electric vehicles (Knez, Celik, & Muneer, 2014)

Strengths	Weaknesses
• Eco-friendly • Silent • Low cost of ownership • Cheaper to run • Energy savings—achievable from regenerative braking system • Simpler mechanism	• Needs time to recharge • The lack of recharging infrastructure • Changing batteries is expensive
Opportunities	*Threats*
• Government subsidies for ownership • No congestion charge • Lower taxes • Increasing fossil-fuel costs	• Competition in form of electric hybrids, alternative fuels, hydrogen-powered cars • Rise in cost of electricity

and conditions. The speed and energy use were recorded for the vehicle that was driven along the principal arteries of the city of Edinburgh, Scotland. In both places, urban and suburban routes were covered for different times of the day.

Results are presented to quantify the energetic, environmental, and economic performance indices for the driven vehicle. A discussion is also provided on the potential for reduction of carbon emissions from the transport sector by provision of environmentally friendly means of generating electricity.

Inevitably from a climate change perspective, the vehicular release of such large amounts of CO_2 will need to be examined. In this respect, the possible link between human population growth, automobile population growth, global CO_2 concentration, and temperature is presently (European Association for Battery Electric Vehicles, 2009) explored. Furthermore, a critical review of the present road transport needs relating to energy demand for UK needs to be examined. This will determine if alternative sustainable mobility modes could be introduced to reduce current and future vehicular emissions.

With legislation and changing government policy, Fig. 13.1.2 indicates the actual reduction and proposed further reduction of grid carbon intensity to the 2030 level of 50 g CO_2e/kWh; this reduction will play a vital role in the commitment to air quality issues and will necessitate renewable energies to support the grid demand.

The Scottish government has committed to almost complete decarbonization of the road transport sector by 2050. As such, a major element of this transformation will be a shift towards the electrification of road transport. A sustainable fleet of electric vehicles aligns with Scottish investment in a renewable energy sector. After all, a quarter of Europe's tidal and offshore wind potential lies in Scotland (Scotland, 2011). Scotland has set itself a most ambitious target to acquire the equivalent of all of Scotland's electricity needs to come from renewable sources by 2020 (Committee on Climate Change, 2015).

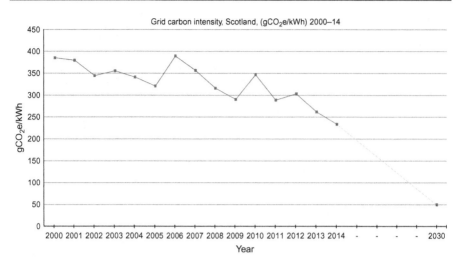

Fig. 13.1.2 Grid carbon intensity (Committee on Climate Change, 2015).

The main drivers for the above actions have been identified as the following:

- Climate change
- Energy security through exploitation of renewable energy resource
- Air quality and noise pollution
- Public health
- Economic opportunities and job creation

The UK road transport emitted 33 million tonnes of CO_2, accounting for approximately 22% of the total UK CO_2 emissions. Following the Kyoto Protocol agreement of Dec. 1997, the United Kingdom voluntarily agreed to reduce the 1990 level of total 'greenhouse gas' emissions by 12.5% by the year 2010. Within this context, the UK government is therefore committed to a 20% reduction in CO_2 emissions over the same time period. One of the means by which the government aims to achieve this target is by reducing emissions from road traffic (Department of the Environment, 2000).

Within this research paper, levels of NO_x pollutants were considered and accounted for as studies have shown that NO_x can cause lung irritation and lowering people's resistance to pneumonia, bronchitis, and other respiratory infections. In the presence of sunlight, NO_x can react to produce a photochemical smog.

If hydrocarbons are also present, ozone and VOC can be produced, which has a similar health effect to NO_x. All are harmful pollutants in their own right.

Although higher concentrations of NO_x are found in city areas, the NO_x emissions must be accounted for when undertaking this study as it is detrimental to health; toxin whether emitted locally from the vehicle or from the power generation source will have the same end effect.

Figs. 13.1.3 and 13.1.4 indicate the concentrations levels are higher in England. The greenhouse gas effect of nitrous oxide itself is hundreds of times greater than carbon dioxide; it is the fourth largest contributor to greenhouse gas global warming (RAC, 2012).

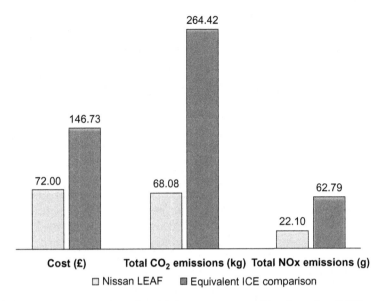

Fig. 13.1.3 Pollutant comparison Scottish figures—generated by the author (Milligan & Muneer, 2015).

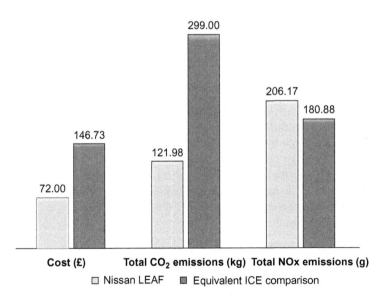

Fig. 13.1.4 Pollutant comparison English figures—generated by the author (Milligan & Muneer, 2015).

This higher level of NO_x must be addressed by both the correct utilization of 'clean' emission vehicles and supporting legislation to generation companies in an attempt to 'clean up' the generation process as the mobility relies on clean energy generation to support the new BEV technology as there is a significant difference between Scottish and UK figures.

Fig. 13.1.4 indicates the calculated results from an experimental journey from central Scotland to southern England and indicates the differences in pollutants dependant on the energy mix at source in the electricity generation process.

Vehicle end of life (VEOL)—this phase is also very dependent on the actual practice in real life, and only a limited amount of data with large variability exists in the open literature. It is unclear what the state-of-the-art disposal and recycling technologies will be for 2015 vehicles and beyond; the VEOL treatment of which will not likely be carried out until 2030. There is an assumption the vehicle and battery expected life span is in the region of 150,000 km, and this is reported by Erikson (2014) although life spans have been reported as between 150,000 and 300,000 km (Hawkins, Gausen, & Strømman, 2012). End-of-life treatment will consist of common procedures as applicable to ICE vehicles but will have the added treatment impact associated with dismantling and material disposal process of the battery pack. VEOL can open further studies for future discussion.

13.2 Edinburgh College BEV—Early adopters

The college has 14 electric vehicles; there is also an electric minibus and 4 electric vans available for staff use. There are a total of 12 charging points, 3 located at each campus. The electric vehicles are for staff use only and for college business, and Fig. 13.2.3 indicates the user plots captured over a 5-year period. The college has leased the electric vehicles since 2011 with the first year operating as a trial period, following full rollout of vehicles to the four main campuses. Trials are still frequently undertaken to understand the efficiency of the vehicles in serving the operational needs of the staff at the college to maximize the integration to the changing business requirements.

In the early trials, Edinburgh College manufactured six charging posts for its own needs, and the use does differ from the government demand for fast charging as the vehicles are used through the day and are always returned at night for charging; this was the nature of college business for the initial trials.

Fig. 13.2.1 illustrates the experimental charge point. This was a 3 kW slow charger with simple operation that had two LED functions, one for on-charge and the other for off-charge. A locked-off compartment ensured that the supply to the vehicle could not be interrupted without authorization.

13.2.1 Charger evaluation

Edinburgh College designed charge post is a Scotland first and production has been increased to 12 posts by the end of 2012 with CE approval requested in line with the IET code of practice for electric vehicle charging equipment installations standards.

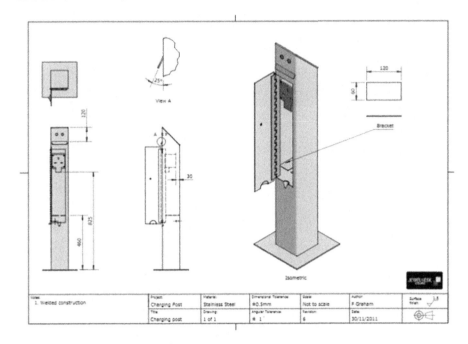

Fig. 13.2.1 Authors own design—college concept post (Milligan, 2016).

The design of the concept charging post was determined as necessary to be of absolute minimum maintenance and ease of use for the user.

The charging post was designed in-house using welded stainless steel box section that will be an easier section to manufacture for quick fabrication.

The post height was determined to be sufficiently tall, so it could be seen from all positions within the average vehicle, thus reducing the risk of damage due to vehicle impact. These posts will be bolted to the ground at college strategic locations (Fig. 13.2.2).

These slow charger posts were the forerunner to the fast charge posts that were recently installed as part of a major Edinburgh College project offered to all colleges across Scotland. In total, 24 fast charge points were installed across Scotland, the United Kingdom, as part of this Edinburgh College BEV project.

The uptake of college electric mobility will ultimately necessitate improved control over 'dumb charging' that is to monitor the activity and electricity supplied to the vehicle (Fig. 13.2.3).

One of the main concerns of the national BEV user is charging availability and will the BEV infrastructure remain robust enough to support it. The infrastructure network as of Aug. 2016 was becoming far reaching across Scotland; the map below indicated the state and locations of the rapid charger network in Scotland, the United Kingdom.

As of Aug. 2016 in Scotland, there are currently 164 rapid chargers (see Fig. 13.2.4), and these are not including Tesla Supercharger network. A further 3 are awaiting replacement/resiting in Moffat, Hawick, and Brodick, and there are another 11 awaiting installation and grid connection.

Fig. 13.2.2 Authors own design installed—college concept post (Milligan, 2016).

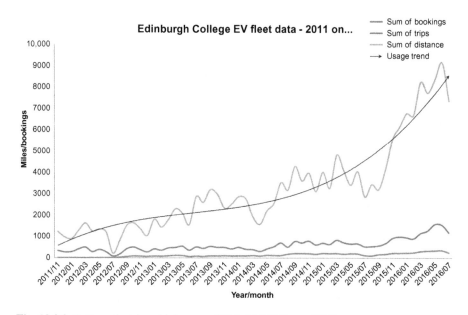

Fig. 13.2.3 Battery electric vehicle usage 2011–16 (Milligan, 2016).

Due to the high cost of these charge facilities, their location must be strategic, publically accessible, and continually under review to the changing requirements of the BEV.

All charge points will be commissioned with 'back-office' data capture and are continually monitored to ensure functionality; these 'live' data are transmitted to administration for control and update of location map systems.

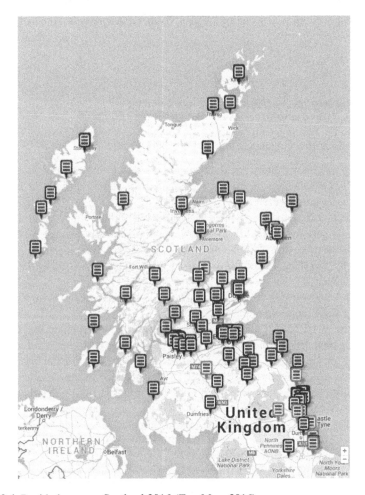

Fig. 13.2.4 Rapid chargers—Scotland 2016 (Zap-Map, 2016).

All-electric journeys in EVs and PHEVs offer the greatest emission reduction and therefore should be facilitated as much as possible. Transport Scotland plans to deploy a network of rapid chargers at intervals of least every 50 mi on Scotland's primary road network.

The majority of journeys undertaken in Scotland are well within the driveable range of an EV.

The Transport Scotland literature suggests that 94% of journeys in Scotland are under 40 km, with the average trip length in a car being only 12.1 km (Transport Scotland, 2016).

For many users, ownership of an EV is unlikely to be a constraint on their ability to make longer distance journeys. Findings have indicated from the 2010 National Travel Survey that 37% of households in Scotland with regular access to a car also had access to a second vehicle (Transport Scotland, 2011), which would allow the use of a fossil-fuelled vehicle for longer journeys.

From a trial conducted by one of the test Nissan LEAF vehicles, the following key points were ascertained:

- *576 kWh* energy used
- *2181* mi travelled
- *248* trips
- *82* charging sessions (average duration 3 h and 39 min)
- Mobile for *13%* of month
- Plugged in for *34%* of month
- Idle/parked for *52%* of month

The 'street post' energy metre accuracy standard stated in this thesis is in accordance with BS EN 50470-3 (class b ±1%); this standard is stated in the APT installation manual with document reference 'EV-202-USE'.

This Nissan LEAF trial indicated 3.77 mi/kWh but the unacceptable mobility period of only just over 13%. It does however indicate the 'typical' 87% idle/parked/plugged-in time that is indicative of a fossil-fuel vehicle 'parked time' as researched by the RAC Foundation in 2013 that reports this is 90%–95% (Barter, 2013) (Fig. 13.2.5).

Table 13.2.1 illustrates the annual impact in both frequency and financial of the strategically sited rapid charger.

Taking the 2014/2015 financial year as an example, we revealed that 85.47% of trips are less than 10 mi long, and the most popular time of travel is midday—with 15.75% of trips carried out between 12:00 and 13:00 h. This information is used to inform policy and procedures relating to EVs, for instance, when vehicles are likely to be on-charge and the likely impact of Grid CO_2 intensity during that demand time.

Total cost of ownership (TCO) of a vehicle is the sum of the annualized fixed (purchasing) costs of the vehicle, variable costs composed of maintenance, repair and

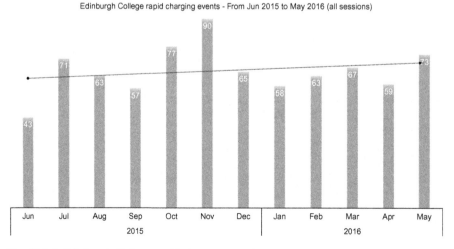

Edinburgh College rapid charging events - From Jun 2015 to May 2016 (all sessions)

Fig. 13.2.5 Edinburgh College rapid charger usage frequency (Milligan, 2016).

Table 13.2.1 Edinburgh College rapid charger (Milligan, 2016)

Edinburgh College rapid charger—usage report (from Jun. 2015 to May 2016)	
Total sessions (all users)	786
Total energy supplied	5242 kWh
Total cost of energy supplied	£378
Edinburgh College only sessions	178
Edinburgh College only energy supplied	1490 kWh
Cost of Edinburgh College only energy supplied	£105
Connector used	*Frequency*
DC (CHAdeMO)	692
DC (CCS)	55
AC	39
Vehicle	*Sessions*
Aixam Mega City Electric	1
BMW i3 (BEV)	0
BMW i3 Rex	56
Mitsubishi i-MiEV	29
Mitsubishi Outlander	213
Nissan e-NV200	63
Nissan LEAF	332
Peugeot iOn	1
Renault ZOE	18
Smart ForTwo ED	2
Tesla Model S	4
Volkswagen e-Golf	1
Volkswagen Golf GTE	13
Unspecified	53
2015 Electricity use and costs	
'Day' rate charge sessions (07:00–00:00)	618
'Day' rate energy supplied	4042 kWh
'Day' rate cost (@ £0.073401/kWh)	£297
'Night' rate charge sessions (00:00–07:00)	36
'Night' rate energy supplied	259 kWh
'Night' rate cost (@ £0.063384/kWh)	£16
2016 Electricity use and costs	
'Day' rate charge sessions (07:00–00:00)	122
'Day' rate energy supplied	878 kWh
'Day' rate cost (@ £0.069268/kWh)	£61
'Night' rate charge sessions (00:00–07:00)	10
'Night' rate energy supplied	62 kWh
'Night' rate cost (@ £0.057455/kWh)	£4

tyres, and fuel or electricity costs, for a standard distance driven per year. The purchasing costs of the vehicle consist of the platform and any applicable combination of ICE, transmission, battery, and electric generator and motors.

Wu, Inderbitzin, and Bening (2015) state that although the TCO of electric vehicles may become close to or even lower than that of conventional vehicles by 2025; our own findings add evidence to past studies showing that the TCO does not reflect how consumers make their purchase decision today, and the ownership is based on preference or tailpipe emissions.

13.2.2 Staff mobility

Due to the nature of the college transport requirements, over 85% of journeys are under 10 mi long due to the travel distance between campuses and local journeys that the vehicles are mainly used for. Similar figures have been reported by Weiss et al. (2014) in a previous study. Fig. 13.2.6 confirms the frequency of travel over 19000 recorded trips for BEV short-range mobility.

The field data used in this study were collected as part of a nationwide BEV project in Midlothian, Scotland, the United Kingdom. Nissan LEAFs, e-NV200s, and Mitsubishi i-MiEVs were trialled for a 6-month period. The participants were selected from different genders and geographical areas in order to achieve a representative sample of the drivers in the area.

The data loggers installed in the vehicles were configured to read information from vehicle sensors available on the vehicle's control area network (CAN) bus and to store

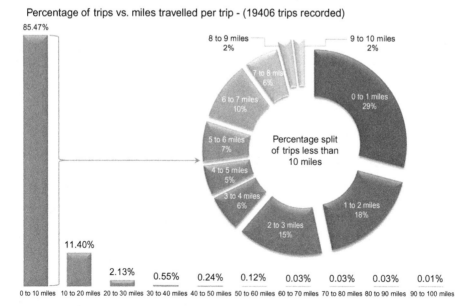

Fig. 13.2.6 Percentages of travel distance (Milligan, 2016).

these data in the logger's internal memory along with the vehicle's GPS position. GPS data and CAN bus messages were logged every 5 s and every 1 s, respectively, when the vehicle ignition was on. Specifically, the vehicle's velocity was logged every second from the CAN bus.

Transient real-world driving cycles are essential for EV powertrain design, battery management systems, battery range estimation, and provision of better information to EV users. The developed driving cycle would aid in the design of EVs that are operating in urban, rural areas, and medium-sized cities. In addition, the developed driving cycle would allow electricity grid analysis, economic, and lifecycle studies to be conducted with a higher degree of confidence for the government and the vehicle buyer.

The route is depicted below in Fig. 13.2.7, travelling by the most direct route and utilizing the A7 primary trunk road at all times on both the outward and return journey.

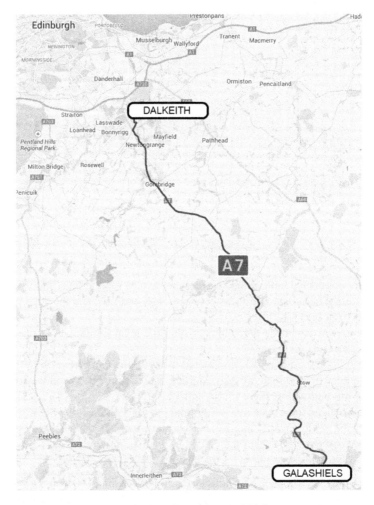

Fig. 13.2.7 Experimental commute route (Google Maps, 2014).

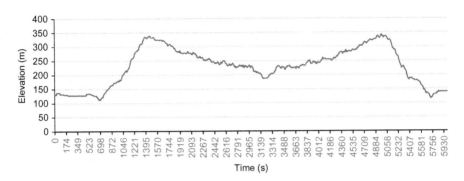

Fig. 13.2.8 50 mi monitoring cycle—elevation—from Midlothian to Galashiels (Milligan, 2016).

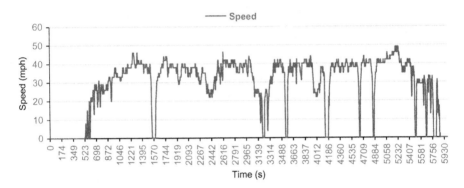

Fig. 13.2.9 50 mi monitoring cycle—speed— from Midlothian to Galashiels (Milligan, 2016).

The test was conducted both in the summer and winter weather conditions where the ambient external air temperature was between 16°C and 2°C, and the vehicle internal temperature was set to 20°C.

The driver was operating the vehicle within the legal limitations of the road and surrounding traffic conditions.

This mobility experiment was conducted utilizing live data when the vehicle was driven on a journey, and then, a return journey was concluded. The elevation and the drive cycle are outlined in Figs. 13.2.8 and 13.2.9.

The journey was from Dalkeith, Midlothian, to Galashiels in the Scottish Borders region, and the vehicle was being driven by a staff member. This is in regards to the method used to record the data as indicated by Chaari and Ballott (2012) distinguished by Andre (1996) that stated that

- the car-chasing technique is used to measure the driving conditions presenting a lower risk of influencing the driver's behaviour,
- the vehicle is being driven by its usual 'daily' driver.

13.2.3 50 Mile route

13.2.3.1 Midlothian to Galashiels, south east of Edinburgh city, Scotland UK

It is also necessary to obtain a full working knowledge of the traffic situations and road conditions to accurately determine whether there is a real benefit from the BEV under real-world driving conditions. The test cycles will be conducted on a primary route into the city of Edinburgh for realistic urban and rural road conditions. The test cycles will be obtained from quantitative measured data, and the tests will represent the actual traffic/vehicle journey at differing times of the day and night (Esteves-Booth et al., 2001).

The primary trunk road A7—highlighted on map below—was used for the 50 mi monitoring experimental trials as the topographical location encompasses mixed driving styles (Fig. 13.2.10).

The A7 is one of the primary routes from Scotland to England in the United Kingdom and is a main thoroughfare for passenger cars and commercial vehicles accessing the Borders region. This review utilizes a section of road between Dalkeith, Midlothian region, and Galashiels in the Borders region.

This controlled comparative drive and recharge utilized the same road section for three different manufacturers of BEV currently available in the European market:

- Nissan LEAF
- Renault Kangoo ZE
- Nissan e-NV200

The objective was to compare the energy usage to cover the same distance over the same road section. The route was driven by all vehicles, and data were collected to

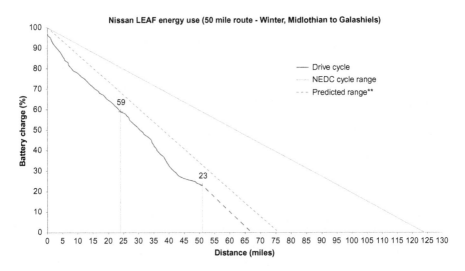

Fig. 13.2.10 50 mi experimental journey—energy actual (Milligan, 2016).

reflect the energy usage. The Nissan e-NV200 van was driven twice, unladen, and repeated again laden with 500 kg as a payload.

The initial section of the journey has an incline to a maximum elevation in excess of 300 m above sea level followed by a decent to the destination; this data have a mirror image for the return journey as the same route was used.

In the winter test, the weather was dry and clear with external air temperature of 2° C; the vehicle required defrosting prior to driving. When interrogated at the data port, the electronic control unit memory indicated that the battery temperature was 0°C.

The Nissan LEAF like other electric vehicle manufacturers incorporates a high-voltage battery heater so as to avoid frost damage to this unit; this too will require its own vehicle battery energy as a power source. The evaluation of thermal comfort inside is of importance as this will reduce driver stress and fatigue and has been studied in great detail by Alahmer et al. (2011). The authors present study understands that whilst maximizing range is important so is maintaining a comfortable cabin temperature.

The auxiliary energy values on Fig. 13.2.11 are the heating and air-conditioning energy demands required to defrost the Nissan LEAF vehicle; this was in excess of 3.75 kW for approximately 6 min, and the energy peaked at 4.25 kW on initial demand all of which had an adverse effect on the vehicle range when compared with the summer trials.

The lower trace value on Fig. 13.2.11 indicates the consumption of auxiliary energy required to provide energy for the headlights and entertainment systems excluding heater system.

Table 13.2.16 shows the NEDC; anticipated and actual distances travelled for the chosen vehicle and all journeys were single occupancy, so it is predicted that this will

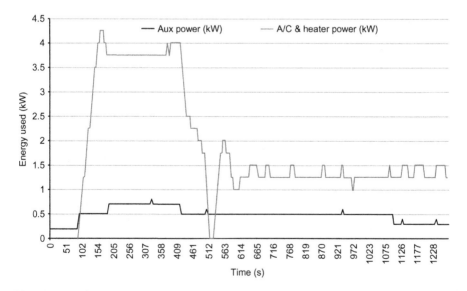

Fig. 13.2.11 A/C, heater, and Aux demand—initial 30 min of the journey (Milligan, 2016).

be the 'best'-case monitoring analogy; the Nissan e-NV200 was under test in both the unladen and the laden condition.

From the 50 mi journey comparisons as shown in Table 13.2.2, it was found that there was a significant difference between the manufacturer's given distance data and the distance actually achievable. The actual distance travelled was less than the dashboard displayed value that when combined gave the user concerns with the achievable range and the possibility of failing to reach the destination. Furthermore, due to the reduced temperature, the achievable distance was reduced further when operated in the winter months and gave rise to potential worry should there be the need to detour or other unforeseen circumstances.

The most demanding situation was when the vehicle was utilized in winter and was in the laden condition; in this situation, the vehicle had the greatest percentage difference between actual and estimated range.

This journey was successful through the experimental period, and this information was shared with the vehicle drivers so that all users had the knowledge that under all conditions the journey was achievable, but added range restrictions were introduced with the reduction in temperature and when laden.

Table 13.2.2 does reflect a decreased range capability in winter, and also, it was found that winter conditions seem to result in an unjustified decrease in use, and a substantial share of battery capacity is redundant. This mobility research found that this was not due to the technical constraints of the vehicles but concerns of the drivers using the EVs in those conditions. A study by Morrissey, Weldon, and O'Mahony (2016) showed that the charging behaviours of EV users vary depending on the location of the charging infrastructure and the known reliance on them; this behaviour will require to be conditioned to maximize the efforts of the local authorities.

The theory of planned behaviour (TPB) is extended with emotional reactions towards the electric vehicle and vehicle driving in general. Emotions and attitude towards the electric vehicle are the strongest determinants of usage intention. Reflective emotions towards vehicle driving and perceived behavioural control factors also play a significant role. Differences in the relative importance of the determinants of usage intention based on environmental concern, behaviour, and social values are also considered within this decision. In general, people who are more inclined to use an electric vehicle are less driven by emotions towards the electric vehicle and more by reflective emotions towards vehicle driving and take more perceived behavioural concerns into account.

Smith (2008) focuses on the factors affecting the usage intention of electric vehicles, a newly developed product that can lead to a fundamental change in sustainable mobility behaviour.

What will make or break the successful introduction of electric mobility in the real world is consumer acceptance (Verhoef et al., 2008).

Therefore, insights into the motivations and barriers of this acceptance, especially by early adopter market segments, are important for a successful introduction of the electric vehicle.

Additionally, for the electric vehicle adoption, *the place where one lives* may be important. Living in an urban area can be an opportunity for electric vehicle driving,

Table 13.2.2 50 mi journey comparisons (Milligan, 2016)

Vehicle type	Summer						Winter					
	Max range in miles (NEDC cycle)[a] acclaimed mileage given by manufacturer	Estimated range in miles as indicated while driving the BEV (displayed on the vehicle dash board)	Actual trip miles to depletion	Energy efficiency miles per kWh	% Difference		Max range in miles (NEDC cycle) acclaimed mileage given by manufacturer	Estimated range in miles as indicated while driving the BEV (displayed on the vehicle dash board)	Actual trip miles to depletion	Energy efficiency miles per kWh	% Difference	
					Actual vs. estimated (%)	Actual vs. NEDC (%)					Actual vs. estimated (%)	Actual vs. NEDC (%)
Nissan LEAF	124.00	93.0	81.1	3.68	−12.80	−34.60	124.00	75.0	76.5	3.66	+2.00	−38.31
Renault Kangoo ZE	106.00	70.0	68.3	3.2	−2.43	−35.57	106.00	70.0	59.3	3.0	−15.29	−44.06
Nissan eNV200 Unladen	105.00	n/a	n/a	n/a	n/a	n/a	105.00	70.0	68.0	3.57	−2.86	−35.24
Nissan eNV200 Laden (500 kg)	105.00	n/a	n/a	n/a	n/a	n/a	105.00	86.0	68.0	3.72	−17.14	−35.24

All routes carried out in drive mode 'D' and without enabling 'ECO' mode.
Summer trips undertaken without heating, air conditioning, or other auxiliary systems being active.
Winter trips carried out with heating and air conditioning on, set to 20°C, plus all other available auxiliary systems (such as lights and entertainment system) switched on to replicate the actual drive of a typical user during their commute.
[a]Manufacturer's quoted maximum available range on NEDC cycle details (Nissan UK, 2016; Renault UK, 2015; Barlow et al., 2009).

as driving in the city often implies short trips, which reduces the problem of the range and anxiety of an electric car. But on the other hand, users may have restricted charging facilities at home due to location and infrastructure.

If the daily driving requirement distance is less than 60 mi in most applications, this will be within the capacity of a BEV; however, drivers become 'uncomfortable' if the remaining driving range becomes less than 30 mi. Public accessible rapid chargers will relieve the drivers' range anxiety even if the driver can rely on it without having to use it.

This was found in a case study by the Tokyo Electric Power Company when they introduced a trial EV with a capacity of 80 km in 2007. The vehicle's daily driving distance is in the region of 40 km allowing for the EV to easily cover the local area before recharge was required.

Although the driver had been informed that the vehicle battery capacity was more than sufficient to cover 80 km, due to worries that he would run out of power, the driver was hesitant to take advantage of this information (CHAdeMO, 2007).

To relieve this anxiety, the company installed a quick charger; car usage dramatically increased with the monthly driving distance exceeding 1400 km, which was more distance than normally covered by conventional vehicles in the same area. Of notable mention is that the driver hardly used the quick charger. If so, why did the driving distance suddenly increase to the extent that it did? It was all psychological. In other words, the driver, knowing that he could charge up the car's battery at any time, gave him peace of mind resulting in longer (and probably more relaxed) drives.

Hence, the author discovered that quick chargers contribute to both the charging efficiency and increasing driving distances. However, even if the chargers are actually not used, the nearby installation of quick chargers provides drivers with a feeling of comfort that induces users to maximize EV usage.

It is imperative that charging systems in the early stages minimize the total cost in the required infrastructure, and as a charging system develops, then, it is a natural progression to fit quick chargers to replace the normal AC outlets.

Not all electric cars are the same, and there is an *accepted* range of around 60–90 mi. Up to this range, battery packs albeit costly are compact and manageable in size and weight. Vehicles with this range are already produced today at *affordable* prices and do not need a dramatic technology break through. Nevertheless, in the public and popular media, there is a sense that such vehicles are not an adequate replacement for traditional internal combustion engine (ICE)-powered car's long range, fast refuelling, and high performance.

Such a view ignores the potential of small BEVs to deliver adequately and economically on the actual mobility requirements of the majority of the population.

This economic and sustainable transport however does come with the expectation of route planning, and the user must partially focus on an accurate charging strategy that will suit the individuals' needs. This approach must be acceptable to each user and should not become an onerous task.

13.2.4 Experimental and simulation results

The vehicle dynamics and energy consumption (VEDEC) simulation software has been developed, which is capable of calculating power and energy requirements for any vehicle during driving. The software also computes energy savings that are achievable from regenerative braking system when compared directly with the energy requirements of the same vehicle without the system. Simulations take detailed account of energy consumed during level cruise, acceleration, and gradient-climbing modes. For the purpose of auditing the latter driving mode, topography data may be keyed in using topography maps or directly logged using an onboard altimeter. The energy analysis work is largely based on the work of Rubin and Davidson (2001).

For the simulation study, motor efficiency was accounted for as a prerequisite value. The current market models utilize induction motors, and their efficiency can be as high as 96% at full load, though 94% is more common. Efficiency for a lightly load or no-loaded induction motor is poor because most of the current is involved with maintaining magnetizing flux. As the torque load is increased, more current is consumed in generating torque, whilst current associated with magnetizing remains fixed.

For the test procedures, the software was set at 95% for motor efficiency and 60% for regeneration efficiency.

The principle theory is the conservation of energy and several analytical and experimental data such as rolling resistance, aerodynamic drag, and mechanical efficiency of power transmission for the general driving performance of the BEV on this route, which will be conducted by the daily driver.

Regenerative charging of the BEV will be monitored with respect to the change of operating conditions of a model vehicle such as inclined angle of road and vehicle speed; the chosen route will encompass all of this and rural and urban driving.

The test BEV was driven on four routes that involved different driving styles. The vehicle was driven from a known charge point, and on each departure, the high-voltage battery energy level was at 100% state of charge. The vehicle was driven to its destination point, and the energy reduction was recorded, and the vehicle was put on charge. Distance, time, time on charge, and when plugged in to the charge point, the energy was recorded along with the time taken to recharge the vehicle battery pack.

This was conducted for the following data monitoring routes:

- Urban
- Extraurban
- Rural

The results were compared with the simulation software, and conclusions were drawn from the accuracy of the software and which route gave a higher degree of accuracy than the others in this trial.

The value of the rolling friction coefficient between the tyre and the road did have a significant impact on the measured simulation results. Variations of this coefficient were altered to determine the simulation changes and the effect that it had on energy modelling. Tyre and vehicle manufacturers will be required to ensure that their BEV tyre will comply with an ultra-low rolling friction factor coupled with significant grip (traction) for vehicle and occupant safety under all road conditions.

Eq. (13.2.1) was used to determine the friction coefficient for the simulation experimental work.

Friction coefficient (Muneer, Clarke, & Cullinane, 2008).

Friction coefficient equation stated as

$$\mu = \left[\frac{E}{\Delta d} - \frac{1}{4}CdA\rho\left(V_f^2 + V_i^2\right)\right] \cdot \frac{1}{mg}$$

$$\mu = \left[\frac{720,000}{1000} - \frac{1}{4} \cdot 0.272 \cdot 2.754 \cdot 1.2\left(22.35^2 + 22.35^2\right)\right] \cdot \frac{1}{1875 \cdot 9.81} = 0.0269$$

$$(13.2.1)$$

where

Cd is the coefficient of drag
A is the frontal area
ρ is the air density
m is the mass
g is the gravity
V_f is the velocity final
V_i is the velocity initial
E is the energy
Δd is the distance

Tyre rolling friction will influence the overall experimental results as a high-coefficient friction tyre will require greater energy to rotate and therefore give a reduction in overall distance travelled due to the increased effort.

The rolling friction factors used in the experiments were between 0.025 and 0.03 dependant on tyre pressure and road surface material. Through calculation as in Eq. (13.2.1), the value for this vehicle type with occupants was 0.0269.

Due to the BEV's limited range, they are sensitive to the effects of high rolling resistance, so all manufacturers will specify a tyre with the lowest friction coefficient whilst commanding vehicle safety and low noise.

Fig. 13.2.12 illustrates a typical BEV tyre rating (point 1, 2, and 3):

1. Premium tyre and lowest rolling resistance for greatest economy
2. Safe wet road holding and will allow maximum breaking on a wet surface
3. Tyre rolling noise indication (76 dB is the set industry standard)

13.2.5 Drive cycle

To measure the driving patterns experienced in the city of Edinburgh, the author carried out a study with the use of data logging equipment into a test vehicle, in this case a Nissan LEAF model. Measurements of elements such as distance travelled, speed, rate of acceleration, rate of deceleration, and difference in altitude to determine incline rates and time for journey were recorded.

These data were then transferred to computer for analysis and development of driving patterns along the main commuter routes into the city centre.

For the kinetic energy recovery system, the most important phase to focus upon is the deceleration phase, as this determines the amount of potential energy that can be extracted.

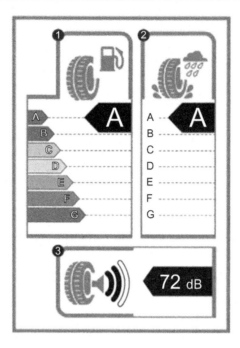

Fig. 13.2.12 Tyre performance rating standards (Automobile Association, 2013).

Three routes were identified as experimental test routes, and the scope was to determine factors that were unique to that route from known parameters. A route was planned, and the BEV was driven to the point of destination. The scope of intention is therefore a representative plot of driving behaviour within a given city or a region and is characterized by speed and acceleration.

There is an increasing need to simulate driving cycles for a particular area of study. Within this research, the author utilized a software tool using actual experimental inputs that accurately represent real driving behaviour within that location.

The mobility activity was determined for this particular journey and is illustrated in Figs. 13.2.13 and 13.2.14. These findings will have a significant effect on the range capabilities of the vehicle as acceleration periods are energy depleting and short deceleration periods with greater acceleration periods will reduce the possibility of any regenerative energy recovery.

The software was developed using fundamental dynamics equations and written within the VBA environment of Microsoft Excel software. The difference between computed and measured net traction energy ranged between +7.9% and −4.5%. In this respect, Figs. 13.2.15–13.2.17 may be referred.

This route utilized one of the primary arteries into the city of Edinburgh from the south, and due to the nature of this route, there was multiple stop-start situations at junctions, and all speeds were determined by the traffic flow.

Fig. 13.2.18 indicates the mobility characteristics when the vehicle was utilized within restricted urban conditions. In initial experiments, the author attempted modelling with 30 s intervals, and subsequently, the speed was recorded also at 30 s

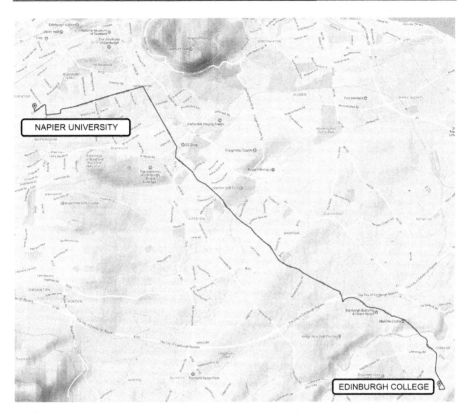

Fig. 13.2.13 From Midlothian to Napier—town route (Milligan, 2016).

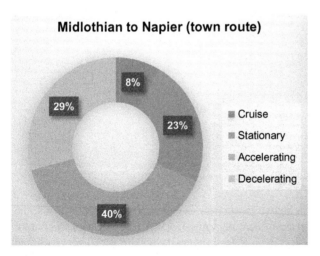

Fig. 13.2.14 From Midlothian to Napier mobility activity (Milligan, 2016).

Fig. 13.2.15 Urban
mobility energy, distance,
and speed—overview
(Milligan, 2016).

	Simulation		Actual	
E used	2.672	kWh		kWh
E regen	0.342	kWh		kWh
E tot	2.331	kWh	2.160	kWh
Avge speed	22.75	mph		
Total dist	12,161.70	m		

E Accuracy	107.90	diff.	0.171 kWh

Actual time (s)	1196
Actual dist (m)	11,909
Actual av. speed (mph)	22.27

Sim time (s)	1196
Sim dist (m)	12,162
Sim av. speed (mph)	22.75

Sim dist accuracy (%)	102.1
Sim speed accuracy (%)	102.1

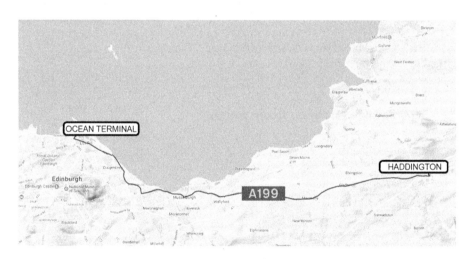

Fig. 13.2.16 Extraurban route (Milligan, 2016).

intervals, but this gave a high degree of inaccuracy as with this type of mobility and the
30 s time frames; there can be significant data missing that would render the output
inaccurate. For the finalized experimental testing, the dashboard speed was recorded
at 1 s interval for the entire journey, and these data were the input into the software.
The resultant output was accurate as no data were lost when modelled using this

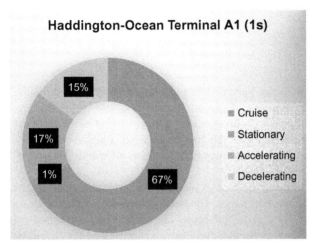

Haddington-Ocean Terminal A1 (1s)

- Cruise
- Stationary
- Accelerating
- Decelerating

15%

17%

1%

67%

Fig. 13.2.17 From Haddington to Ocean Terminal mobility activity—primary route (Milligan, 2016).

	Simulation		Actual	
E used	6.732	kWh		kWh
E regen	0.315	kWh		kWh
E tot	6.416	kWh	6.72	kWh
Avge speed	37.53	mph		
Total dist	29,008.39	m		

E Accuracy	95.48	diff.	−0.304 kWh

Actual time (s)	1729
Actual dist (m)	30,095
Actual av. speed (mph)	38.94

Sim time (s)	1729
Sim dist (m)	29,008
Sim av. speed (mph)	37.53

Sim dist accuracy (%)	96.4
Sim speed accuracy (%)	96.4

Fig. 13.2.18 Extraurban A1 route—overview (Milligan, 2016).

reduced time frame. Fig. 13.2.15 gives an overview of the energy used in simulation and the comparison with the actual value for this 12.1 km journey.

As can be seen in Fig. 13.2.15, the modelled energy usage comparison is within 8% of the actual energy used, and the modelled speed is within 3% of the actual speed as displayed on the vehicle dashboard. It is understood that the energy metre standards for APT street posts are active (kWh) = *BS EN 50470-3* (class B ±1%). This gives confidence that the simulation software and derived values are the basis of a reliable data set:

$$100\% = 1 : 1 \text{ relationship}$$

The comparison of the simulated speed and distance indicates a 2.1% overestimation between the estimated and the measured data.

This route utilized one of the primary arteries into the city of Edinburgh from the east, and due to the partially rural nature of this route evidenced, a higher percentage of 'cruise' until the experiment reached built-up areas, and then, all speeds were determined by the traffic flow.

Figs. 13.2.19 and 13.2.20 indicate the type of mobility activity when utilizing the primary and alternative routes into Edinburgh. The results obtained from the 1 s tests illustrated longer periods of cruise and on both routes chosen gave similar periods of acceleration and deceleration. Both of these routes were chosen because they are standard routes that will be taken by the commuter on a daily basis. The software simulation gave between 95% and 97% accuracy in determining the total energy usage for the journey.

Due to the dual-carriageway nature of this primary route, higher more consistent speeds were achieved, and this resulted in a significant 67% cruise section.

The prolonged section of higher speed activity resulted in greater energy use although the journey time was approximately 25% reduced when compared with the alternative route. Modelled energy use for this route was within 5% of the actual energy depletion.

The alternative route for this analysis can be seen in Fig. 13.2.19. This route is slightly shorter in distance, but due to the urban sections, this test encountered traffic congestion and slower road speeds.

This alternative route gave the greatest consistency when modelled. The simulated energy was within 3% of the actual energy consumed to conduct this route.

Fig. 13.2.19 From Haddington to Ocean Terminal mobility activity—alternative route (Milligan, 2016).

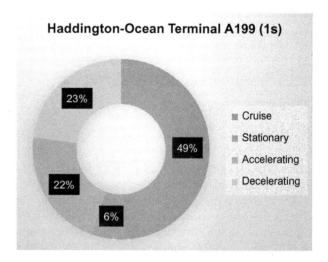

	Simulation		Actual	
E used	5.638	kWh		kWh
E regen	0.490	kWh		kWh
E tot	5.148	kWh	5.28	kWh
Avge speed	25.40	mph		
Total dist	27,040.70	m		

E Accuracy	97.50	diff.	−0.132 kWh

Actual time (s)	2381
Actual dist (m)	27,835.7
Actual av. speed (mph)	26.15

Sim time (s)	2381
Sim dist (m)	27,041
Sim av. speed (mph)	25.40

Sim dist accuracy (%)	97.1
Sim speed accuracy (%)	97.1

Fig. 13.2.20 Extraurban A199 route—overview (Milligan, 2016).

Table 13.2.3 Distance travelled comparison and error (Milligan, 2016)

Test number	Vehicle odometer reading (converted to m)	GBmapometer		Google maps		Simulation	
		Distance (m)	Conformity (%)	Distance (m)	Conformity (%)	Distance (m)	Conformity (%)
1	27842.0	27440	98.6	27358.8	98.3	28031.74	99.3
2	29766.5	29590	99.4	29611.9	99.5	30872.73	96.4
3	11909.0	12350	96.4	11748.2	98.6	12161.70	97.9
4	12231.0	12340	99.1	11748.2	96.1	12147.66	99.3
5	30095.0	29680	98.6	29611.9	98.4	29008.39	96.4
6	27835.7	27390	98.4	27358.8	98.3	27040.70	97.1
Total			98.4		98.2		97.7

As can be seen, all of these routes did bring into question the distance error that the author was aware of and gives detailed analysis in the thesis "Critical evaluation of the battery electric vehicle for sustainable mobility."

Table 13.2.3 illustrates the known accuracy of the route distances taken from proprietary sources when compared with the vehicle own dashboard. The dashboard accuracy can also be questioned; however, the author can state that for UK legislative purposes, any speedometer/odometer must be accurate to within −0% to >+10% of the displayed value (Hansard & Whitty, 2001).

This is an EU directive that states that speedometers must 'over'-read, and they are not allowed to 'under'-read the value; for example, if the cars velocity is actually

30 mph, then the speedometer must read 30 mph or up to 10%. The speedometer will become less accurate as the road tyres wear giving them a smaller rolling circumference.

Under the test conditions, both gbmapometer and Google Maps gave similar results to each other with less than 2% discrepancy when compared with the vehicle manufacturers' results taken from the dashboard.

The tolerance threshold from Google Maps and gbmapometer varies dependant on the map scale in line with the terrestrial reference system (TRS) as laid down by Ordnance Survey (Crown Copyright, 2016). The software model results were deemed confident as they were within 1% of the two proprietary route mapping tools when tests were taken across six independently devised test routes.

The Excel visual basic for application (VBA) programme uses the speed, altitude, and time sample data, as well as the vehicle manufacturer's data and data pertaining to the energy recovery system. From that, it determines the amount of energy available during differing combinations of events such as acceleration, cruise, deceleration, and stoppages, as well as the road conditions: ascending, level, or descending (Table 13.2.4).

From Sighthill/Midlothian campuses to Moto Services, Pirnhall, Stirling, Central Scotland (Fig. 13.2.21):

Summer testing done, no heating or auxiliary systems were used. Start point was Sighthill campus, and end point was Midlothian campus.

Winter testing done on same route, start point was Midlothian campus. End point should have been the Sighthill campus.

The vehicle was driven in gear selector mode D, and the driver did not operate the 'economy' feature (ECO).

Comparative results were indicated as 73.6 mi completed in summer and 71.6 mi in winter conditions.

Table 13.2.4 Overview of experimental parameters (Milligan, 2016)

Route	Distance (m) Simulated	Distance (m) GPS actual.	Distance (m) Accuracy(%).	Av. Speed (mph) Simulated	Av. Speed (mph) GPS actual.	Av. Speed (mph) Accuracy(%).	E used (Kwh) Simulated	E used (Kwh) Measured Charge.	E used (Kwh) Accuracy(%).	Notes
Hadd to OT A199	28431.7	27842.0	102.1	24.94	24.42	102.1	4.759	4.560	104.4	
Hadd to ot a1	30872.7	29766.5	103.7	42.63	40.23	106.0	5.925	5.280	112.2	
Mid to nap town	12161.7	11909.0	102.1	22.75	22.27	102.1	2.209	2.160	102.3	
Nap to mid town	11877.7	12231.0	97.1	22.61	23.29	97.1	1.832	1.920	95.4	
Ot to Hadd A1	29008.4	30095.0	96.4	37.53	38.94	96.4	6.110	6.720	90.9	
Ot to Hadd A199	27040.7	27835.7	97.1	25.40	26.15	97.1	4.892	5.280	92.6	
Total			99.8			100.1			99.6	

Quick adjust			
Rolling friction factor	0.026		
Motor efficiency	95%		
Regenerative efficiency	60%		

Motor eff.	Regen. eff	RFF
▲	▲	▲
▼	▼	▼

Energy total

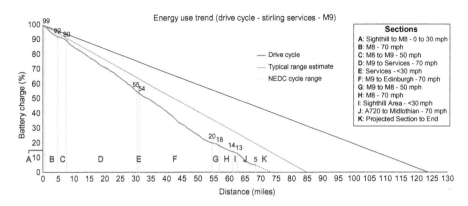

Fig. 13.2.21 From Midlothian to Stirling services—energy discharge (Milligan, 2016).

The range prediction calculation is based on past-driving data, collected from the vehicle, the user, and the environment. Aftermarket tracking equipment is fitted to the vehicle and will give data in relation to the driver's requirements and storing times, GPS coordinates, and driver behaviour such as excessive braking, acceleration, and cornering forces. From the GPS coordinates, it is easy to calculate travel distances, and from the hardware, interrogating the vehicle electronic control unit and CANbus data via the diagnostic port such as state-of-charge levels and temperature of the high-voltage battery can be determined. Driver profile is based on recording these data at time of departure and will be aligned with past data analysis and historical evidence of driver behaviour; all of these will be inputs to act as a 16-point 'data set' approach to estimate the electric vehicle range. The data set approach will model the regression activities to find the best fitting estimation based on current state-of-charge level and past driver behaviour (previous SOC level, weather information and temperature, average speed, and traffic information).

This drive cycle experiment will target certain routes and road networks to give the electric vehicle user confidence that the vehicle will be able to achieve the distance expected under real driving conditions and advise of any discrepancy in the data being relayed back to the driver.

This discrepancy will be measured so as to give the driver a real-world data set that is realistic when taking into account the inputs as indicated in Fig. 13.2.22. It is essential to record all dependency parameters as indicated in Fig. 13.2.22 as changes here will have an effect on the potential distance that can be travelled.

The experimental work conducted within this research agrees with Boretti (2013) where the discussion of analysis of the energy flows to and from the battery of a latest Nissan LEAF covering the urban dynamometer driving schedule (UDDS). This analysis provides a state-of-the-art benchmark of the propulsion and regenerative braking efficiencies of electric vehicles with off-the-shelve technologies.

Whilst the propulsion efficiency approaches 90%, the round trip regenerative braking efficiency reaches 70%, values previously achieved only with purely mechanical systems, and gives the Nissan LEAF car a very efficient regeneration system without having a negative effect on the driving experience for the car user and its passengers.

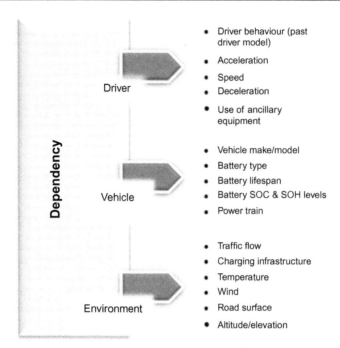

Fig. 13.2.22 The required 16-point data set for the experimental range prediction (Milligan & Muneer, 2015).

This regeneration system will have a bearing on the experimental results obtained and allow the altitude change to have an effect on the energy present to propel the vehicle further for the given route; however, this cannot be done in isolation alone as the behaviour of the driver will have a significant effect to utilizing the vehicles' potential range, and there are no two drivers alike, therefore making this an ever-changing variable.

To ensure that all driving conditions were met, it was decided to conduct the real-world tests on routes that vehicle users will encounter such as rural, urban, extraurban, and intercity journeys. And to obtain greater accuracy, the tests were conducted across different seasons. Experimental findings were determined for the summer and winter months; this allowed data analysis of the air conditioning and heater consumption and the auxiliary energy usage so as to allow thorough vehicle defrosting prior to the test drive and maintaining an inside vehicle temperature of 20°C for the comfort of the driver conducting the test.

This section critically analyses two objectives. Information and data were collected to determine comparisons between dashboard discrepancies and to determine reliability of the manufacturer's figures. One was to evaluate the accuracy of the computer algorithm in car display when compared with the accuracy of the NEDC figures that all car manufacturers must comply with. The other was to accurately measure the discrepancy in the dashboard displayed value when the vehicle is used over differing weather conditions.

For the car purchaser and user, the NEDC value will be the value that the manufacturer claims, and this will go towards the decision of the buyer at time of vehicle purchase. This information will also determine the strategic positioning of the charging system infrastructure throughout the nations' road network, as this positioning must be based on the 16-point data set to calculate the actual range achievable by the vehicles.

Tests were conducted and data collected in the summer months when the air temperature was between 16°C and 20°C and again in the winter months when the air temperature was between −2°C and 2°C; this information proved to be invaluable as the distance the vehicle could be driven was reduced as the temperature decreased. Interior cabin temperatures were set to 20°C in the summer trials, and cooling ventilation conditions were utilized in the winter months.

In accordance with the journey, taken auxiliary energy was used for equipment such as heated seats and in-car entertainment as the driver deemed to be acceptable for the journey as there is nothing to gain by trying to conserve all auxiliary energies as this is not likely to happen when the vehicle is being used in conditions such as the daily commute. Under the test conditions, the radio, heater (set at 20°C), and heated seats were used if applicable as this will be the normal condition when used by the commuter.

The in-car dashboard display is all the driver can get as an indication of what distance is available, as its information will be used similar to a fuel gauge in an internal combustion engine conventional car; this display accuracy will be essential to the vehicles' range in journey planning (see Fig. 13.2.23). Prior to any journey, the user must check that the previous user has put the returned vehicle on-charge on termination of that journey. Under ideal conditions, the battery electric vehicle will have a full charge on commencement of the journey, thus giving it maximum range capabilities. The electric vehicles that were under test here have got in the region of <100 mi range when fully charged, and this is approximately one-fourth the range of a conventional fuel car when full of fossil fuel, so this requires a change of human behaviour so to remove the risk of range issues and battery depletion.

Fig. 13.2.23 In-car display—suggested distance the vehicle can travel shown on the right-hand side instrument (Milligan, 2016).

The Nissan LEAF test car had full charge as indicated by the dashboard display prior to the start of each experiment; this was 'real-time' analysis, and as the cars were driven, the energy reduction was recorded as the vehicle was moving towards its destination.

Recording stopped when the car had been driven to near-critical or actual battery depletion that was beyond when the car was warning the driver of imminent electrical system failure on the dashboard display. Prior to the start of each journey, the vehicle was fully charged as indicated by 100% useable energy displayed on the dashboard and checking that the charge supply had stopped from the charging post indicating that further charging was not possible. This was all verified by interrogating the vehicle electronic control unit and CANbus via the diagnostic port using a hardware plug-in to determine the charge of the high-voltage traction battery state; this experiment gave a lower value of 97% for the state of charge (SOC) when fully charged and not 100% as expected. The hardware plug-in values are shown in Fig. 13.2.24, and this indicates a maximum available energy capacity of 22 kWh.

GIDs = value representing state of charge of the battery. One GID is equivalent to 75–78 Wh of available energy and in-keeping with the traction battery energy capabilities. This was correctly calculated at the experimental stage as repeated tests indicated a fully charged HV battery indicated 22 kWh and 284 GIDs.

Fig. 13.2.24 screenshot was indicative of the values stored within the car electronic control unit (ECU). This was a fully charged car, and the available energy was 2 kWh less than the manufacturer's stated capacity of 24 kWh. This would indicate that a fully charged battery is displaying 100% to the driver; however, the ECU is recording a lower value.

This test was repeated over many Nissan LEAF cars and over many time frames, and the values of Fig. 13.2.24 were constantly repeated, and the maximum value that could be attained was 22 kWh, giving 2 kWh margin for safe battery control and removing the risk of overcharging irrespective of the charging source that the car is plugged into.

The drive cycle data are gathered live for the experiment using the vehicles own electronic control unit (ECU) memory and the CANbus diagnostic port using a LUJII ELM327 V1.5 miniwireless OBD2 diagnostic scan tool that has been retrofitted. This equipment will record off-line, store, and display battery management system

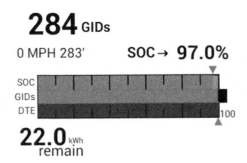

Fig. 13.2.24 High-voltage traction battery information from CANbus port (Milligan, 2016).

information pertinent to the state of health (SOH), temperature, state of charge (SOC), number of fast charges, rapid charges, and auxiliary power usage to operate systems.

The purpose of this analysis is to determine the drive cycle information that is gathered from journeys taken at random, and the vehicle is not being driven to any specific drive patterns nor route. These are journeys that have been encountered within the daily usage of the battery electric vehicle; a sample has been taken to illicit four different drive cycles, and with all journeys, a typical commuter driver was driving not a test driver. All of these cycles were employing the 'car-chasing' technique and following the traffic conditions, so the car speed was governed by road condition legislative constraints and the speed of the vehicle in front. It was not necessary to plan the journey to certain times of the day as this would not be expected when using a conventional vehicle nor was the route taken all the information given was that the car had to get from point A to point B without detailed route planning.

The journeys driven utilized the following road types:

(a) Urban cycle
(b) Rural cycle
(c) Mixed combined (extraurban) cycle
(d) Intercity cycle

For the four given drive cycles, data were collected from the summer months for the routes and again for the winter months, and comparisons were examined in relation to the energy discharge rate and the route and distance travelled.

Table 13.2.5 indicates the discrepancy in actual distance travelled as ratified by Google Maps when compared with the computational dashboard value as displayed to the driver and the range inaccuracies when compared with the NEDC value. To accurately determine if there was a difference in results due to climatic conditions, this experimental test approach was taken over two seasons and the variance recorded by live data recording in the vehicle.

The percentage difference was recorded and compared across all routes and both seasons. It can be seen from Table 13.2.5 that the vehicle was less energy-efficient and had the greatest percentage difference when used in the winter months.

The extraurban route chosen was from Dalkeith, Midlothian, Scotland driving east of the city of Edinburgh to the main A1 dual carriageway. This route encompasses mixed driving styles and speeds, and the test was conducted over sections of 30–40 mph urban roads and national speed limit 50–60 mph roads and 60–70 mph dual-carriageway roads.

The route analysis in Fig. 13.2.25 was taken from a section of A1 road in an easterly direction from the city of Edinburgh, from Midlothian, Dalkeith, to Cockburnspath, southeast of Dunbar in the Scottish Borders region, where the vehicle was then driven back along the same route for the return journey.

The extraurban drive cycle route with the drive characteristics indicated comparing speed over time. It can be clearly seen that there are significant periods when the car is maintaining a speed in the region of 50–60 mph at the dual-carriageway section and there will be negligible energy recovery over this time period (Fig. 13.2.26). Fig. 13.2.27 above encompasses the total (return) drive cycle, and the only periods where regeneration is applicable are at the start and at the end of the test.

Table 13.2.5 Nissan LEAF range comparisons (Milligan, 2016)

Route	Summer						Winter					
	Max range in miles (NEDC cycle) acclaimed mileage given by manufacturer	Estimated range in miles as indicated while driving the BEV (displayed on the vehicle dash board)	Actual trip miles	Energy efficiency miles per kWh	% Difference Actual vs. estimated (%)	Actual vs. NEDC (%)	Max range in miles (NEDC cycle) acclaimed mileage given by manufacturer	Estimated range in miles as indicated while driving the BEV (displayed on the vehicle dash board)	Actual trip miles	Energy efficiency miles per kWh	% Difference Actual vs. estimated (%)	Actual vs. NEDC (%)
1—Rural	124.00	93.0	81.1	3.68	−12.80	−34.60	124.00	75.0	76.5	3.66	+2.00	−38.31
2—Extra urban	124.00	93.0	84.0	3.50	−9.68	−25.00	124.00	83.0	69.2	3.47	−16.63	−44.19
3—Intercity	124.00	85.0	73.6	3.34	−13.41	−31.45	124.00	92.0	71.6	3.30	−22.17	−42.26
4—Urban	124.00	93.0	89.0	4.04	−4.30	−28.23	124.00	86.0	71.0	3.70	−17.44	−42.75

Manufacturer's quoted maximum available range is 124 mi on NEDC cycle as expressed by Nissan (2015) and Barlow et al. (2009).
All routes carried out in drive gear mode 'D' and without 'ECO' mode.
Winter trips carried out with heating and air conditioning plus all other available auxiliary systems (such as lights and entertainment system) switched on.
Summer trips carried out with no heating, air conditioning, or other auxiliary systems active.
Winter trips carried out with heating, air conditioning on, and interior temperature set to 20°C.

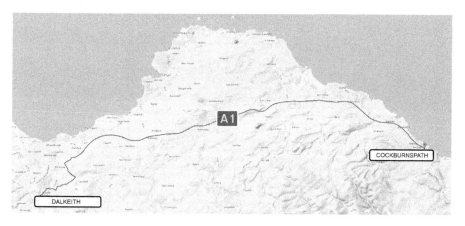

Fig. 13.2.25 Extraurban drive cycle route (Google Maps, 2014).

Fig. 13.2.26 Extraurban—elevation (Milligan, 2016).

Indication from the hardware plug-in, Table 13.2.6 below, gives values pertaining to the battery condition both before and after the experiment; Table 13.2.7 indicates fully charged with maximum battery energy available, and the 33 Wh is the energy that has been used out of the available 22 kWh available as can be seen after the vehicle has completed the experiment; 19,792 Wh has been used, thus only leaving 0.6 kWh remaining.

The battery after the test is very close to depletion and will require to be recharged fully prior to the next drive cycle operation.

The extraurban drive and discharge cycle are shown below in Fig. 13.2.28.

Fig. 13.2.28 indicates the experimental variance in the suggested and available distance that can be travelled, as can be seen the dashboard display algorithm is indicating 94 mi available to the user.

In the summer months, the vehicle has available energy to allow 84 mi, but in the winter months, only 69 mi are available. This is an 18% average reduction in range that may have an adverse effect on the driver and vehicle reaching their destination. The climatic conditions (wind, rain, and temperature) are the only variables here as the vehicle was driven by the same driver and all controls were set for the tests.

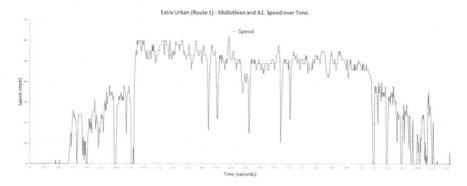

Fig. 13.2.27 Extraurban—speed (Milligan, 2016).

Table 13.2.6 Battery analysis—start (Milligan, 2016)

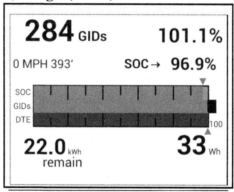

Battery condition at test start.

Table 13.2.7 Battery analysis—end (Milligan, 2016)

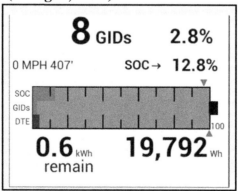

Battery condition at test end

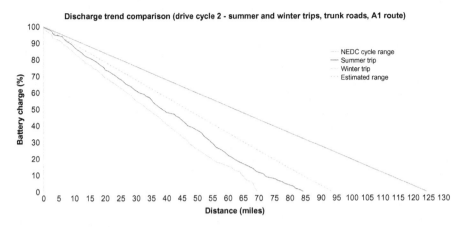

Fig. 13.2.28 Summer and winter—discharge actual (Milligan, 2016).

The greater the demands of the auxiliary power required for interior heating, the more of an adverse effect this will have on the vehicles' 'distance to empty'.

These results have come from the vehicle being used in day-to-day commute conditions, and the results show a marked decrease in potential range due to the reduced temperature conditions and associated demands on the energy from the high-voltage battery pack. This range deficiency was also recognized in the work of Farrington and Rugh (2000) where the impact of auxiliary loads was discussed.

Additional energy is being used to supply the auxiliary heating and air conditioning that is required to maintain the inside car temperature set at 20°C (see Fig. 13.2.29); this along with inaccuracies in the dashboard display algorithm gives the driver a

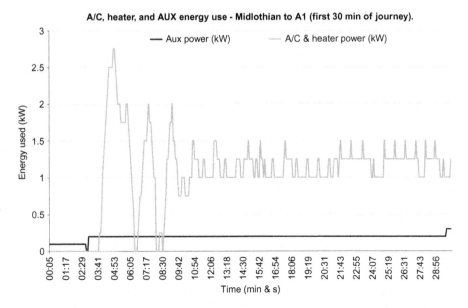

Fig. 13.2.29 A/C, heater, and Aux demand (Milligan, 2016).

different (lower) range achievable to what is being displayed and to the NEDC manufacturer values. For user confidence, the dashboard display algorithm must be reworked to give a higher degree of accuracy to keep the distance in line with what can actually be achieved when being driven by the user.

It should not be omitted to state in the findings that the BEV has got a 12-volt battery that will power the lights and auxiliary devices and a 350+ volt high-voltage traction battery to give traction to the road wheels and give auxiliary heating. This high-voltage battery must maintain the 12-volt battery charge at the same time as supplying all of the other vehicle needs so accurate recording and display values are of paramount importance as are drain and balancing.

Chart of the auxiliary energy consumption:

Fig. 13.2.29 indicates the energy consumed by the auxiliary battery to maintain the inside cabin temperature at an acceptable 18–20°C when the exterior temperature on the winter test was between 4°C and 6°C. It has to be remembered that this vehicle does not have an alternator to allow battery recharging so any recharging of the auxiliary battery will be through energy transfer from the HV traction battery.

13.2.6   City of Edinburgh commute

Edinburgh College has four sites located within Edinburgh and the Lothian region, Scotland, as shown in Fig. 13.2.30. Due to the campus locations, the BEV is an effective low emission mode of transport that is suitable for the across city commute.

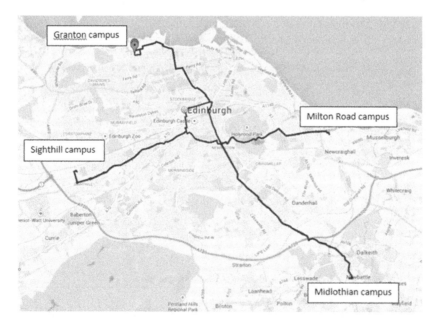

Fig. 13.2.30 Edinburgh College campus locations (Google Maps, 2014).

The intercampus commute has to be achieved at differing times of the day encountering different driving conditions and throughout the complete academic session. Experimental testing was conducted to determine if the BEV can fully service all campuses as a means of intercampus travel at any time in the year as a requirement to the staff throughout the day and the night.

In this drive cycle method, it is not expected that the driver should adapt his or her drive patterns to the limited range of the vehicle but it may require the use of quick charging to compensate for the restricted range available.

Chargers are located within each of the campus grounds, and there are other publically accessible chargers available within easy reach of the commuter undertaking this journey.

A real-world drive cycle was conducted in both summer and winter seasons to see if all campuses could be within the reach of the BEV without the need for plug-in time. Often, the cars are just participating in multi-drop use—stopping for less than 2 min, and therefore, the plug-in time would have a negligible effect on the battery chemistry state of charge, and therefore, this analysis would be the worst case situation for the BEV and driver.

The route was split into four experimental sections:

Section A: from Granton campus to Midlothian campus   (travel distance 13.1 mi)
Section B: from Midlothian campus to Milton Road campus   (travel distance 6.6 mi)
Section C: from Milton Road campus to Sighthill campus   (travel distance 9.3 mi)
Section D: from Sighthill campus to Granton campus   (travel distance 5.9 mi)

The commute was across the city of Edinburgh to each Edinburgh College campus location and then repeated, which is the ultimate ideal range that can be achieved. This will give any user the confidence required to complete this journey under any road and weather condition.

The campus locations are not equidistant, but the route taken will be the most direct. Each journey will have differing road and traffic conditions, and each section will require different driving styles that will change to suit the surrounding requirements.

The drive cycle was conducted within two seasons in the year, taking into account the warmer ambient summer temperature and the colder winter temperature. This is an important aspect as it will evidence any difference in BEV performance across the same route accounting for the external differences such as temperature change, wind, and rain. An important factor will be the need for extra interior cabin temperature for the driver to be comfortable and safe for the journey.

This extra heating will ultimately come from the transfer of energy from the high-voltage (HV) traction battery to the heater system, thus removing energy that would otherwise drive the vehicle.

The continual demand on heating and auxiliary systems has indicated a measurable loss of the available energy. This has been calculated a 2% energy reduction for every 15 min to maintain the interior cabin temperature at 21°C. This result was recorded with the vehicle stationary, and the exterior temperature was 5°C. It can be predicted

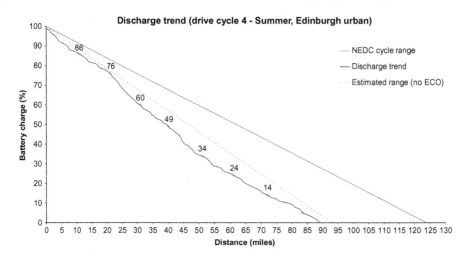

Fig. 13.2.31 Summer—experimental drive cycle—discharge actual (Milligan, 2016).

that greater energy will be used to support these systems when the vehicle is moving due to wind chill and the increase auxiliary load.

Summer drive cycle:

Fig. 13.2.31 depicts the discharge curve across all of the journey sections; the energy available in the summer is sufficient to complete return journeys between campuses, and as shown in discharge Table 13.2.8, the reduction in energy is indicated for the route.

The actual discharge curve for the summer drive cycle gave a reduced range by 8% from the dashboard prediction and a significant 30% reduction from the manufacturer's given figures as taken from the New European Drive Cycle (NEDC).

The ambient and battery temperatures were taken both at the start and at the end of the drive cycle in Table 13.2.9. The available energy was measured both at the start and at the end confirming that at the end of the four intercampus commute the battery was close to depletion is shown in Table 13.2.10.

Table 13.2.8 Discharge summer split per section (Milligan, 2016)

100%–86% charge section A—from Granton Campus to Midlothian Campus (first visit)
86%–76% charge section B—from Midlothian Campus to Milton Road Campus (first visit)
76%–60% charge section C—from Milton Road Campus to Sighthill Campus (first visit)
60%–49% charge section D—from Sighthill Campus to Granton Campus (first visit)
49%–34% charge section E—from Granton Campus to Midlothian Campus (second visit)
34%–24% charge section F—from Midlothian Campus to Milton Road Campus (second visit)
24%–14% charge section G—from Milton Road Campus to Sighthill Campus (second visit)
14%–5% charge section H—from Sighthill Campus to Granton Campus (second visit)
5%–0% charge section I—projected figures to depleted battery

Table 13.2.9 Summer temperatures (Milligan, 2016)

External temps (°C):
Start of trip = 10.0°C
End of trip = 12.0°C
Battery temps:
Start of trip = 9.5°C
End of trip = 12.1°C

Table 13.2.10 Battery capacity (Milligan, 2016)

Available kWh at start = 21.9
Available kWh at end = 1
Energy used = 20.011 kWh
Energy efficiency = 3.70 mi/kWh

This energy consumption was confirmed by hardware installed interrogating the BEV's electronic control unit at the start of the experiment and at the end Figs. 13.2.32 and 13.2.33 for summer and Figs. 13.2.34 and 13.2.35 for the winter season. The end parameters indicate close to full depletion of the usable available energy.

Winter drive cycle:

The aim was to accept the drivers behaviour, and 'normal conditions' were set to get a typical approach to the vehicle user. This drive cycle was conducted in the colder winter months when it was necessary to heat the passenger compartment to between 20°C and 22°C to make the journey safe and comfortable.

This drive cycle experiment as shown in Fig. 13.2.36 indicated a reduction of the available range to an actual 65 mi that was approximately 22% less than the dashboard display and nearly 50% less than the manufacturers NEDC stated figures of 125 mi.

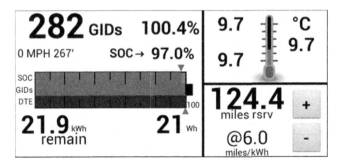

Fig. 13.2.32 Initial energy available (Milligan, 2016).

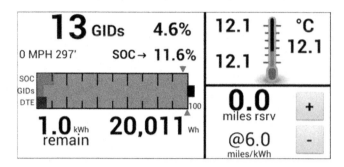

Fig. 13.2.33 Final energy available (Milligan, 2016).

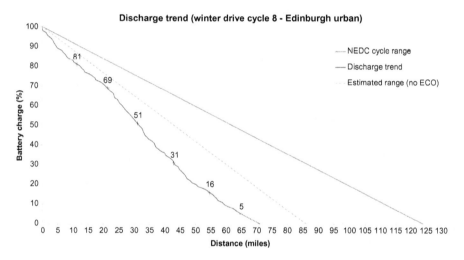

Fig. 13.2.34 Winter—experimental drive cycle—discharge actual (Milligan, 2016).

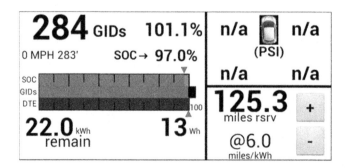

Fig. 13.2.35 Initial energy available (Milligan, 2016).

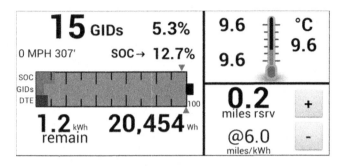

Fig. 13.2.36 Final energy available (Milligan, 2016).

Table 13.2.11 indicates that there were return journeys that could not be achieved, and this would necessitate BEV charging to be made available when an 'all campus' return journey was required.

The winter drive cycle could not be fully completed due to depleted battery prior to the end of the experiment (Table 13.2.12).

The results in Table 13.2.13 were recorded, and as a result of the extra energy being consumed to heat the passenger compartment, full completion of the intercampus journeys was not achieved. This heater power peaked at 3 kW and was automatically controlled by the interior vehicle temperature sensor; the temperature was set to 20°C as the set point and the system feedback cycled the variable until the desired temperature was achieved. This is shown in Fig. 13.2.37.

Table 13.2.11 Discharge winter split per section (Milligan, 2016)

100%–81% charge section A—from Granton Campus to Midlothian Campus (first visit)
81%–69% charge section B—from Midlothian Campus to Milton Road Campus (first visit)
69%–51% charge section C—from Milton Road Campus to Sighthill Campus (first visit)
51%–31% charge section D—from Sighthill Campus to Granton Campus (first visit)
31%–16% charge section E—from Granton campus to Midlothian Campus (second visit)
16%–5% charge section F—from Midlothian Campus to Milton Road Campus (second visit)
5%–0% charge section I—projected figures to depleted battery
From Milton Road Campus to Sighthill Campus (second visit)—not achieved
From Sighthill Campus to Granton Campus (second visit)—not achieved

Table 13.2.12 Winter temperatures (Milligan, 2016)

External temps (°C):
Start of trip = 2.0°C
End of trip = −1.0°C
Battery temps:
Start of trip = 3.8°C
End of trip = 9.6°C

Table 13.2.13 Battery capacity (Milligan, 2016)

Available kWh at start = 22.0
Available kWh at end = 1.2
Energy used = 20.454 kWh
Energy efficiency = 3.47 mi/kWh

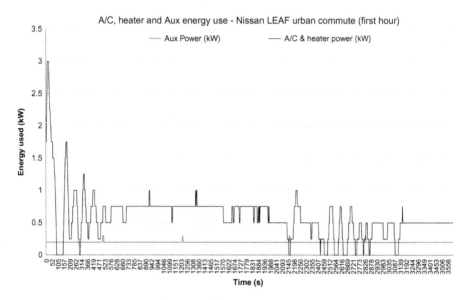

Fig. 13.2.37 A/C and auxiliary energy use (Milligan, 2016).

The auxiliary systems such as lights, radio, battery management, and vehicle control systems consumed a constant 200 W and the A/C and heater system consumed between 0 and 3000 W of power to maintain the constant passenger compartment temperature due to the vehicle interior heat losses sustained in this commute.

The following actual measured results were calculated to make a comparison of the losses accrued in heating and auxiliary systems consumed against the total energy available:

Auxiliary energy used in the first hour of commute was 723 Wh.

A/C and heating energy used in the first hour on commute was 2146 Wh.

This gave the combined additional energy used in the first hour of commute as 2.869 kWh, which is 13% of the overall total available energy.

As indicated from both Fig. 13.2.31 summer and Fig. 13.2.36 winter experiments, the BEV range has been affected by the change in ambient temperature; more energy will be utilized in cabin heating this compounded with wind, rain, and additional auxiliary equipment being used in the winter that had the effect of reducing the distance that can be travelled.

All sections of the experiment were conducted with the same driver; however, it is also noted that the slope of each discharge curve is not consistent from each of the sections driven.

All sections driven are different depicting the varying road conditions and the traffic encountered in the city of Edinburgh. The greatest differing sections are A and C, where section A is not a steep as section C; these are displaying differing driving styles required to achieve the journey that is having a greater depletion effect on the battery range; this supports the studies conducted by Neubauer, Brooker, and Wood (2012).

Figs. 13.2.38 and 13.2.39 illustrate section 'A' speed and elevation charts; these routes are across southeast Edinburgh, Granton campus to Midlothian campus with an average speed less than 20 mph and 14 stop/start occurrences. The acceleration is progressive, which is indicative with this drive cycle route and the vehicle motion, whilst decelerating will induce energy regeneration back to the traction battery.

Section C Milton Road to Sighthill campus drive cycle shown in Figs. 13.2.40 and 13.2.41 has a significant steeper slope to the discharge curve. This indicates much

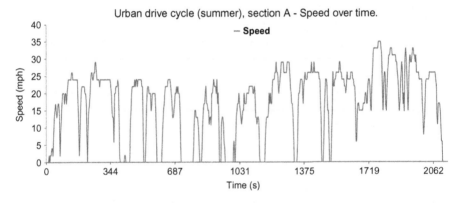

Fig. 13.2.38 Section A drive cycle—speed (Milligan, 2016).

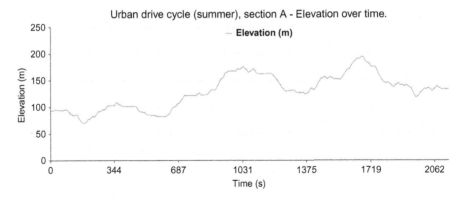

Fig. 13.2.39 Section A drive cycle—elevation (Milligan, 2016).

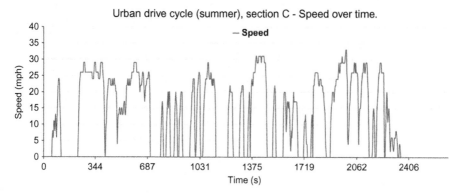

Fig. 13.2.40 Section C drive cycle—speed (Milligan, 2016).

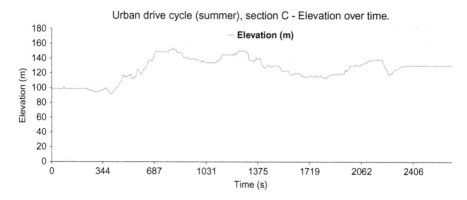

Fig. 13.2.41 Section C drive cycle—elevation (Milligan, 2016).

faster acceleration to gain road speed in order to maintain flow with the surrounding city traffic; it is still an average of 20 mph but with 24 stop/starts. The frequency of this experiment was conducted 10 times to give the results as indicated to calculate the coefficient of variation (see Table 13.2.14).

High discharge current required to accelerate the BEV due to the drive cycle conditions is indicated in the experiment, and this is evidenced by standard deviation calculations to determine the variance as results show in Table 13.2.14. The regeneration efficiency will be reduced as there is a high number of events where the speed drops below the 'regeneration cut-off', so here, there will not be any regenerative charging possible.

Coefficient of variation was calculated to illustrate the rate of change that was present in the section drive cycles. Section C gave the highest rate of change for both acceleration and significantly higher when in the deceleration mode.

Table 13.2.14 Intercampus journeys standard deviation (Milligan, 2016)

	Average acceleration	Standard deviation	Coefficient of variation	Average deceleration	Standard deviation	Coefficient of variation
Section A	0.50	0.33	0.66	−0.53	0.35	0.66
Section B	0.46	0.29	0.63	−0.65	0.40	0.61
Section C	0.55	0.41	0.74	−0.52	0.49	0.94
Section D	0.48	0.26	0.54	−0.55	0.37	0.67

Standard deviation equation (MTSU, 2008)

$$\sigma = \sqrt{\dfrac{\displaystyle\sum_{i=1}^{n}(x_i - \bar{x})^2}{n-1}} \qquad (13.2.2)$$

where

n is the number of data points
x_i is each of the values of the data
Σ is the sum of

The battery pack voltage discharge is shown in Fig. 13.2.42 for the urban commute. This is taken over the entire journey until the HV battery pack was close to depletion. The voltage drop was measured at 63.88 V between 100 and 0% SOC.

Nissan LEAF indicated values on Table 13.2.15 are those provided by the vehicle driver display. Actual values are those derived from hardware installed through the vehicle CANBUS. Nissan LEAF battery capacity is quoted by Nissan as 24 kWh. Actual useable capacity is 22 kWh at 100% max charge as indicated by control unit interrogation.

At 100% charge as indicated by the vehicle display, control unit interrogation indicates that the pack is 97.1% SOC.

At complete discharge as indicated by the vehicle, interrogation indicates that the pack still has approximately 12% capacity in reserve.

The vehicle display stops showing range available at around 5% (indicated) battery charge and also stops displaying the remaining battery percentage. This is designed to

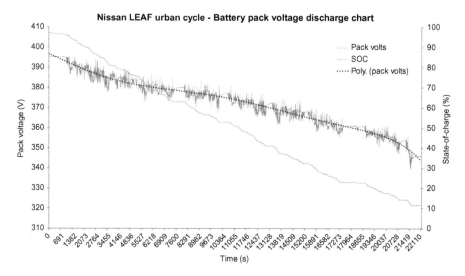

Fig. 13.2.42 Battery pack voltage discharge curve for the urban drive cycle (Milligan, 2016).

Table 13.2.15 State-of-charge equivalent—Nissan (Milligan, 2016)

Nissan LEAF				
SOC		Pack volts	kWh	
Indicated (%)	Actual (%)	Actual	Indicated	Actual
100	97.1	395.42	24	22
75	75	380.93	18	16.5
50	50	372.77	12	11
25	25	358.56	6	5.5
0	11.6	346.08	0	2.552

(NB. Actual figures derived from ECU interrogation)

encourage drivers to seek recharge during normal use and allow a 'reserve' for them to be able to do so, but this is approaching depletion.

Mitsubishi i-MiEV full discharge voltage is based on reading of 320 V for 90% discharged battery pack according to installed hardware. The vehicle indicators displayed 0% charge and zero range remaining (see Table 13.2.16).

13.2.7 Long range mobility

As part of the BEV research, a vehicle was driven on a 1000 mi return journey that would remove the vehicle from the local known charge point infrastructure and make the driver become fully reliant on the installed infrastructure. A significant problem is finding the shortest path for a BEV when it has a limited battery capacity; therefore, it must stop and recharge at certain locations (Table 13.2.17).

The driving range limit and the lack of charging infrastructure are the two main characteristics of EVs at the current adoption stage.

If the number of chargers is less than what we need, it means that the BEV drivers may have to wait for hours to get a fully charged battery that will discourage the user and eventually influence the market penetration.

Table 13.2.16 State-of-charge equivalent—Mitsubishi (Milligan, 2016)

Mitsubishi i-MiEV		
SOC (indicated) (%)	Pack volts	kWh
100	360	16
75	349	12
50	338	8
25	327	4
0	316	0

Table 13.2.17 Seasonal comparisons—overview (Milligan, 2016)

Urban route: Edinburgh	Summer				Winter			
	% of Battery used	Average speed (mph)	Estimated energy use[a] (kWh)	Notes	% of Battery used	Average speed (mph)	Estimated energy use[a] (kWh)	Notes
Section A: Granton Campus to Midlothian	14	20.4	3.08	Mostly uphill. Active stop/go traffic	19	19.3	4.18	Uphill section. Active stop/go traffic
Section B: Midlothian Campus to Milton Road	10	20.3	2.20	Mostly downhill. Less stopping	12	22.0	2.64	Downhill. Less stop/go. Higher average speed
Section C: Milton Road Campus to Sighthill	16	19.6	3.52	Active stop/go and sharp acceleration	18	20.0	3.96	Sharp acceleration and deceleration
Section D: Sighthill Campus to Granton	11	17.9	2.42	40+ mph roads, then stop/go and downhill	20	20.3	4.40	Mainly downhill, but extra heating increased kWh use[b]

Additional notes:
All routes carried out in a 2014, second generation Nissan LEAF 'Acenta', using drive mode 'D' and without 'ECO' mode.
Winter trips carried out with heating and air conditioning plus all other available auxiliary systems (such as lights and entertainment system) switched on and set to 20°C.
Summer trips carried out with no heating, air conditioning, or other auxiliary systems activated.
[a]Based on percentage of battery power used, assuming 100% charge representing 22 kWh available from Nissan LEAF battery pack.
[b]Section D (winter) also included a brief stop, so vehicle consumed extra energy in reheating when journey commenced again.

Therefore, it is of considerable importance to perform in-depth research on the multimodal BEV use because the battery technology is a main competitive factor in the EV market.

To identify the energy consumed under practical applications, the BEV was driven from Edinburgh, Scotland to Bristol, England. This was done over multiple days, in multiple sections, and it was made possible utilizing the 'Ecotricity electric highway' rapid charger network that is an aligned series of rapid chargers running along the main A701-M74-M6-M5 route from Scotland to South West England illustrated in Fig. 13.2.43. The Ecotricity electric intercity route was a UK government initiative that started in Jul. 2011 with a single charger, but now, the network encompasses almost the entire motorway network (Ecotricity, 2016).

The positioning of these rapid chargers is of paramount importance to the success; too close together and the cost of the network will be prohibitive, and too far apart, the journey may not be achieved. As discussed by Taylor et al. (2010), it is stated that with current technology vehicles, the saving could impact significantly on journeys less than (100 km) 62 mi charge range that is in line with the UK governments strategic aim to site rapid chargers at least 50 mi intervals on the primary road network (Alba, 2015).

And this work is verified with the previous given parameters as stated within this paper.

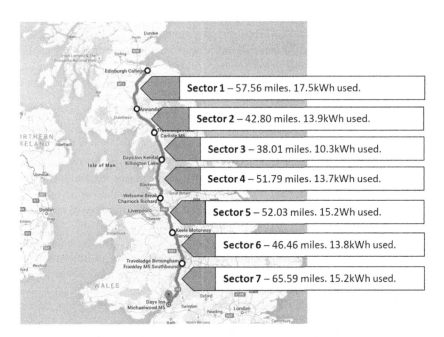

Fig. 13.2.43 Intercity long-range route—from Scotland to England (Google Maps, 2014).

The Nissan LEAF BEV was fitted with data recording equipment, and this was used to record the 1000 mi return journey, and this experiment will be reliant on robust infrastructure support with strategically positioned 50 kW rapid chargers available and maximize the available energy and distance from the car capabilities.

The vehicle was driven in accordance to the traffic laws, weather conditions, and drivers' capabilities.

From GPS location: Midlothian town centre, Edinburgh, Scotland, the United Kingdom

Latitude	55.883987
Longitude	−3.079659

To GPS location: Bristol town centre, Bristol, England, the United Kingdom (Google, 2016)

Latitude	51.454513
Longitude	−2.58791

The route taken was the quickest and most direct; this was 380 mi from start to finish that is given by the Automobile Association route planner as a journey of 6 h and 30 min (Automobile Association, 2015).

The route planner time taken is assuming an ICE vehicle. In this BEV experiment, the driving time was 8 h and 30 min, but the total time including all charging and rest stops was 14 h. This 5 h and 30 min difference was the time used in vehicle charging on either a 'fast' or 'rapid' charger and also for allowing the charge post to be cleared by the previous user so making it available. This situation occurred on more than one occasion, and the post was not available on demand.

The journey for detailed analysis was split into seven sectors; this gave an accurate representation of the sections, and due to the terrain and elevation, the cost of the trip, fuel costs, emissions, and energy consumed could be directly compared with ICE vehicles.

The time factor for the user was also taken into consideration as this type of journey is out with the expected user requirements. The BEV user must consider this factor as an acceptable duration of journey time and compared with an ICE vehicle.

The output of these will have an effect on the practicality of this type of mobility for a long-range journey.

The electric vehicle was driven from Edinburgh to Bristol, and the distance travelled is shown in Fig. 13.2.43 between the charge stops. This just-in-time strategy for the charge interval was of paramount importance for a successful journey as the BEV did not have sufficient range in reserve to make the next stop and charge point due to the extra demand placed either on the traction battery on the higher speed sections or on the sections with a greater and constant incline.

The charge sections driven gave actual measured data of the following (Table 13.2.18):

Table 13.2.18 **Charge section data (Milligan, 2016)**

Section 1	3.23 mi/kWh	5.17 km/kWh (poor weather, trunk road plus inclination)
Section 2	3.08 mi/kWh	4.93 km/kWh (poor weather, 60 mph section)
Section 3	3.70 mi/kWh	5.92 km/kWh (poor weather, 50 mph section, regeneration)
Section 4	3.78 mi/kWh	6.05 km/kWh (55 mph, motorway section)
Section 5	3.42 mi/kWh	5.47 km/kWh (55 mph, motorway section)
Section 6	3.37 mi/kWh	5.39 km/kWh (55 mph, motorway section)
Section 7	4.23 mi/kWh	6.77 km/kWh (slipstream behind HGV, motorway section)

It was noted that the tractive efficiency varied constantly dependant on road conditions, but instantaneous values on road sections gave data between 1.38 and 4.91 mi/kWh.

The BEV was driven at all times within the legislative speed limits and according to the prevailing weather conditions, but it was determined within the course of the experimentation drive that the available tractive energy was consumed at a greater rate when the vehicle was driven at a higher speed.

13.2.7.1   1000 Mile route comparison (CO_2 impact)

This long-range intercity journey was carried out in a 2014 second-generation Nissan LEAF 'Acenta'. All charging was free to the user, and this is current practice at this time; however, it is under review by the government, and it is expected that future studies in this field may show the local authorities requesting a payment for the charger use. The rapid charge complexities and different charge (monetary) methods with no apparent consistency across multiple authorities vary from the following:

- Rapid charge from £2.00 per session then 9 pence/kWh but with a monthly subscription of £8.00 (Chargemasterplc, 2016).
- Rapid charges from 15 pence/kWh—flat rate (Transport Evolved, 2016).
- Rapid charges from £5.50 for the first 45 min then 15 pence/min after that with a maximum cost of £12.00 (Transport Evolved, 2016).
- Rapid charges from £3.80 for the first hour then £5.00/h thereafter (Zap-Map, 2016).
- Rapid charge nominal rate of £4.50 per session regardless of time—flat rate; this figure was utilized in the comparison tables this nominal £4.50 rate adopted for this study (Chargemasterplc, 2016).

These are publically accessible rapid chargers; costs borne by the user will finance installation, support, infrastructure, and the supplying company's profit. As indicated by Energy Savings Trust (2015), the rapid charger will charge to 80% in approximately 1 h, whereas according to UK Power Limited (2015), the domestic user will experience an electricity 'plug-in-tariff' costs between £0.094 and £0.104/kWh for their electricity, but the charge time to 80% will be in the region of 6–7 h.

Table 13.2.19 CO₂e impact and ICE vehicle comparison per section (Milligan, 2016)

Day 1	Distance (miles)	Energy use (kWh)	CO$_2$e impact—UK figures (kg)	CO$_2$e impact— Scottish figures (kg)
Section 1				
Edinburgh to Roadchef Annandale Water (M74)	57.56	17.5	8.09	4.11
Section 2				
Roadchef Annandale Water (M74) to Moto Carlisle/Southwaite (M6 Southbound)	42.80	13.9	6.42	3.27
Section 3				
Moto Carlisle/Southwaite (M6 southbound) to Welcome Break Killington Lake (M6 southbound)	38.01	10.3	4.76	2.42
Section 4				
Welcome Break Killington Lake (M6 southbound) to Welcome Break Charnock Richard	51.79	13.7	6.33	3.22
Section 5				
Welcome Break Charnock Richard to Keele Services	52.03	15.2	7.02	3.57
Section 6				
Keele Services to Moto Frankley	46.46	13.8	6.3	3.24
Section 7				
Moto Frankley to Michaelwood (M5 northbound)	65.59	15.2	7.02	3.57
Total	354.24	99.6	46.03	23.41

	Distance (miles)	Fuel used and cost	CO$_2$ emissions	Notes
Equivalent ICE comparison	354.24	46.00 l £47.84	77.53 kg	2014 Ford Focus 1.6 (manual)

- Note 1—Table 13.2.19 indicates the CO_2e impact; these UK figures are based on UK grid carbon intensity of 462.19 g CO_2e/kWh for 2015 as given by the government's Department for Environment, Food and Rural Affairs (DEFRA, 2015).
- Note 2—the Scottish grid carbon intensity of 235 g CO_2e/kWh for 2013 is also given in Table 13.2.19; this is reported in the paper by the Committee on Climate Change (2015).
- Note 3—the CO_2e impact figures in Table 13.2.19 are calculated at nearly a 51% difference between the UK and Scottish values that indicate a clear difference in the carbon intensity between England and Scotland power generation methods that can be evidenced by the greater percentage reliance on renewable technologies in Scotland over than fossil-fuel-fired power stations primarily using polluting oil, gas, and coal.
- Note 4—this journey can be directly compared with the fuel and environmental cost of a current similar-sized ICE passenger vehicle, the 2014 Ford Focus 1.6 emissions at 136 g/km as shown in Table 13.2.19. According to Lane (2015), this similar-sized vehicle will give an expected 35 mpg under this drive cycle, and the financial cost will be nearly £100 for the return journey directly borne by the user and of which 72% is taken by the government as tax.
- Note 5—this petrol cost is variable, and market fluctuations will determine the final garage forecourt cost to the user. In Table 13.2.19, it has been stated (Automobile Association, 2015) that the fuel cost had been determined by calculating the Dec. 2015 average UK fuel price of £1.04/l (95 octane unleaded). This cost has fallen by 9% in the last 12-month period.

It should be noted that in the latest fuel price report as indicated by the Automobile Association (2016), the average UK price of 95 octane unleaded has increased to £1.11/l.

At the time of experiment journey as illustrated in Table 13.2.20, the following costs and data were utilized to determine the comparative pollution emissions from an internal combustion vehicle.

Therefore, 131.00 l of unleaded used divided by 4.546 gives the number of gallons. That result, multiplied by 6 kWh, determines the amount of energy used to create 131.00 l of fuel.

The calculated figure is *172.9 kWh*—the grid CO_2 intensity of which is *79912.7 g (79.9 kg)*.

Add that to the emissions caused by burning the fuel, and the CO_2 impact of an ICE for 1000 mi is *300 g/km*.

Tables 13.2.21 and 13.2.22 demonstrate the difference in total environmental costs and pollutants when the BEV is used primarily in Scotland or used in England.

As illustrated, when the BEV is driven in England, the total pollutants emitted are overall higher due to the power generation method. The Nissan LEAF CO_2 emissions for the test journey were almost twice the level due to the greater reliance on fossil fuel for current power generation. These findings agree with the National Statistics given from Department of Energy and Climate Change in Table 13.2.23.

Table 13.2.20 Overview between ICE vehicle and BEV CO₂ and cost comparison (Milligan, 2016)

	Distance (miles)	Energy used	Total cost of charges[a]	Cost per mile	Cost per kWh	CO₂ at tailpipe	CO₂e impact of power gen. UK figures[b]	CO₂e impact of power gen. Scottish figures[c]
Nissan LEAF	1008.6	269.5 kWh	£72.00 Actual £0	£0.07	£3.74 Actual £0	nil	124.56 kg	68.08 kg

	Distance (miles)	Amount of fuel used	Total cost of fuel[e]	Cost per mile	Cost per litre	CO₂ at tailpipe	CO₂e impact of fuel production UK figures[f]	CO₂e impact of fuel production Scottish figures[c]
Equivalent ICE comparison[d]	1008.6	131.00 l	£136.24	£0.15	£1.04	220.75 kg	79.90 kg	43.67 kg

Total carbon impact for 1000 mile trip (UK figures)

Nissan LEAF	Equivalent ICE Vehicle
124.56 kg CO₂e	300.65 kg CO₂e

Total carbon impact for 1000 mile trip (Scottish figures)

Nissan LEAF	Equivalent ICE vehicle
68.08 kg CO₂e	264.42 kg CO₂e

Trip carried out in a 2014, second generation Nissan LEAF 'Acenta'.
[a]Ecotricity 'electric highway' charge posts used. All are currently free to use, so cost assumes 16 logged charges during trip at nominal £4.50 (flat rate per session).
[b]Based on UK grid carbon intensity of 462.19 g CO₂e/kWh for 2015 DEFRA (2015).
[c]Based on Scottish grid carbon intensity of 252.6 g CO₂e/kWh for 2013 Committee on Climate Change (2015).
[d]2014 Ford Focus 1.6 emissions at 136 g/km and 35 mpg Lane (2016).
[e]Fuel cost calculated using December average UK fuel price of £1.04/l (95 octane unleaded) Experian Catalist (2015).
[f]Approximately 6 kWh of energy is used to produce 4.546 l (1 gal) of unleaded fuel The Long Tail Pipe (2015).

Table 13.2.21 Overview between ICE vehicle and BEV CO_2, NO_x, and cost comparison (Milligan, 2016)

	Distance (miles)	Energy used	Total cost of charges[a]	Cost per mile	Cost per kWh	CO_2 at tailpipe	NO_x at tailpipe	CO_2e impact of power gen. Scottish figures[b]	CO_2e impact of power gen. English figures[b]	NO_x impact of power gen. Scottish figures[c]	NO_x impact of power gen. English figures[c]
Nissan LEAF	1008.6	269.5 kWh	£72.00 Actual £0	£0.07	£3.74 Actual £0	nil	nil	68.1 kg	121.9 kg	22 g	206 g

	Distance (miles)	Amount of fuel and energy used	Total cost of fuel[d]	Cost per mile	Cost per litre	CO_2 at tailpipe[e]	NO_x at tailpipe[f]	CO_2e impact of power gen. Scottish figures[b]	CO_2e impact of power gen. English figures[c]	NO_x impact of power gen. Scottish figures[d]	NO_x impact of power gen. English figures[f]
Equivalent ICE comparison	1008.6	131.00 l 172.9 kWh	£136.24	£0.15	£1.04	220.7 kg	48.6 kg	43.7 kg	78.2 kg	14 g	132 g

[a]Ecotricity 'Electric Highway' charge posts used. All currently free to use so cost assumes 16 logged charges during trip at nominal £4.50 (flat rate per session).
[b]Based on UK Grid Carbon Intensity of 462.19g CO_2e/kWh for 2015. Source: DEFRA (2015).
[c]Based on Scottish Grid Carbon Intensity of 252.6g CO_2e/kWh for 2013. Source: Committee on Climate Change (2015).
[d]2014 Ford Focus 1.6 emissions at 136 g/km. 35 mpg. Source: Lane (2016).
[e]Fuel Cost calculated using December average UK fuel price of £1.04 per litre (95 Octane Unleaded). Source: Experian Catalist (2015).
[f]Approximately 6 kWh of energy is used to produce 4.546 litres (1 gallon) of unleaded fuel. Source: The Long Tail Pipe (2015).

Table 13.2.22 **Scottish and English environmental impact (Milligan, 2016)**

	Nissan LEAF	Equivalent ICE vehicle
Total impacts for trip (Scottish figures)		
CO_2	68.1 kg CO_2e	264.4 kg CO_2e
NO_x	22 g NO_x	62 g NO_x
Total impacts for trip (English figures)		
CO_2	121.9 kg CO_2e	299.0 kg CO_2e
NO_x	206 g NO_x	180 g NO_x

Table 13.2.23 **Energy generation trends—2015 data (Department of Energy and Change, 2015)**

	Scotland	Wales	Northern Ireland	England	UK total
2013					
Coal (%)	20.4	44.4	34.0	38.8	36.4
Gas (%)	10.3	17.3	45.8	30.3	26.7
Nuclear (%)	34.9	16.7	–	17.5	19.7
Renewables (%)	32.0	10.3	19.5	11.8	14.8
Oil and others (%)	2.4	11.3	0.7	1.6	2.4
2014					
Coal (%)	20.3	35.7	28.3	31.1	29.7
Gas (%)	5.4	24.1	49.1	34.3	29.8
Nuclear (%)	33.3	9.3	–	17.3	18.8
Renewables (%)	38.0	16.3	22.2	15.6	19.1
Oil and others (%)	2.9	14.6	0.3	1.7	2.6

13.3 System control management and monitoring

For an efficient management system, the control systems of business BEV allocation booking will be analysed. This system is designed and implemented to ensure that the vehicles are fully utilized and that their use can be a 'planned' structured event. This will be of great benefit as a vehicle management tool that, with bespoke design and execution, will ensure that the vehicles are suitable for the intended journey. This system will complement GPS tracking for fleet control. A part understanding of the technology was required to offer clarity and to reinforce the vehicle capabilities.

All of this became available on the Edinburgh College BEV booking system. A simple-to-use self-service system would give the user the ability to book a vehicle, thus removing it from general availability at this time.

The system performance and control was required to be an easy-to-use, easy-to-manage, and cloud-based reservation and scheduling system. Developed in-house and based on open-source software from http:/www.bookedscheduler.com/, this open software allows bespoke booking systems to be developed and is an ideal interface for staff to manage and adapt their own bookings.

On-screen field greys represent times the cars are available. Red fields show that a vehicle cannot be booked. To start a booking, the user will click on a start time. This generates a pop-up screen where a start time can be set, a recurrence, details added on the nature of the trip, plus any participants.

Clicking 'save' generates an email that can be printed out so that it can be handed over to facilities staff at each campus in exchange for the eCar keys. The booking can also be printed straight from the system. Bookings are editable and can be transferred to other staff members by administration if necessary.

The system allows reports to be generated for visibility of multiple statistics such as monthly number of bookings and most bookings per user/group.

As all users must have a college computer systems account, they will access the system, and in doing so, they are identified with a unique user number and their email name, and address is stored against the requested activity. This personal identifier is required as the username to access the booking management system. The user states the time and date that the vehicle is required for and submits this information. An automated response is sent to the applicants email confirming the request has been successful.

All authorized staff can see other bookings, but they can only modify their own by accessing the system and following the on-screen instructions.

The system was required for resource allocation to ensure that all staff had access to a user identifiable and controlled management system. In the early developing trials of the project, any existing users migrated from previous database.

All new users have to register using their email address. Permissions are set by assigning each user a 'group' based on their main home campus. This determines which eCar they are able to take out. Global announcements can be made through the front page (dashboard) to remind users of holidays, etc.

Driving licence checks are in place, and vehicle use and charging training is conducted; importantly, the user cannot access the system until these competencies are complete. The system is intuitive and self-policing. When people see on-screen that a vehicle is already out, they cannot book it. If necessary, bookings can be controlled by having each one set to ask for authorization when it is submitted (feedback).

Table 13.3.1 displays the data in table format. The actual monthly data are displayed graphically in Figs. 13.3.1 and 13.3.2, which also give an indication of the salient points that occur within the business such as peak travel times and frequency of low activity due to academic student/staff holidays.

The booking system went live in 2012 Tables 13.3.1 and 13.3.2 give a snapshot figures from May for the last 4 years followed by the annual figures.

The month of May was chosen due to the high frequency of mobility activity generated by activities in the conclusion to the end of the academic term. May has a higher activity of workplace assessors using vehicles that any other month.

Table 13.3.1 Individual month illustration (Milligan, 2016)

May 2012—3 bookings
May 2013—101 bookings (annual 97% increase)
May 2014—114 bookings (annual 12% increase and number of bookable vehicles doubled due to project expansion)
May 2015—190 bookings (annual 40% increase plus an additional four bookable vehicles)
May 2016—329 bookings (annual 43% increase)

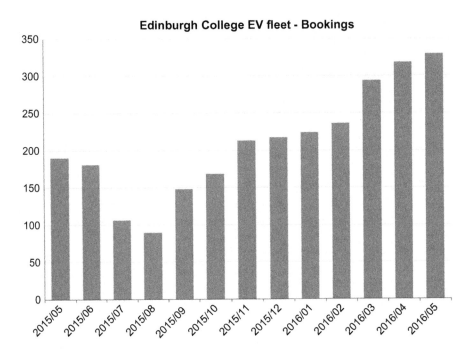

Fig. 13.3.1 Annual illustration of the management system—May 2015–16 (Milligan, 2016).

Table 13.3.2 illustrates the increased user frequency over a 4-year period, and this increase also accounts for all campus locations increasing their available vehicles by either a factor of two or three as the initiative became established.

Fig. 13.3.3 indicates the whole fleet usage trend. This usage pattern is supported by the works of Pearre et al. (2011) that takes into account all activity and illustrates the increase in frequency as an increased number of users travel further as drivers become more confident in the BEV.

Currently, the Milton Road Campus has the highest demand for electric car use, with the Granton Campus having the lowest demand. Notwithstanding this, given the nature of activity, the usage of the electric cars at each of the college campuses fluctuates throughout the academic year.

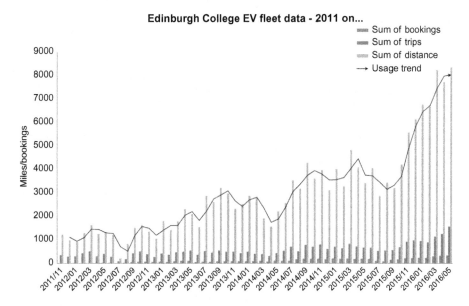

Fig. 13.3.2 Annual usage trend 2011–16 (Milligan, 2016).

Table 13.3.2 Annual illustration (Milligan, 2016)

2012—285 bookings per annum
2013—963 bookings per annum
2014—1601 bookings per annum
2015—2083 bookings per annum
Jan. to May 2016—1401 bookings to date (projected to be in excess of 3360 per annum)

Whilst there is a booking system in place that records journey lengths, their origins, and destinations, it was difficult to identify if there have been many occurrences of staff trying to book a car and being unsuccessful due to a lack of availability of cars.

In Apr. 2015, a 'SurveyMonkey' (Survey Monkey, 2015) was used to identify the acceptance within the college and to determine the users' perception of key drivers within the project.

The anonymous 'SurveyMonkey' was offered to all authorized eCar users and consisted of the same eight questions over the 2-year focus period. The questions were multiple choice and rated the activity out of 10 with 1 being unacceptable and 10 being exceptional. The final question was 'free text', and this gave the participants an opportunity to offer an expression or state their opinion.

Figs. 13.3.4 and 13.3.5 illustrate the findings from this in the form of a 'word cloud'. It can be seen in the early questionnaire that the concerns portrayed in the free text were with the lack of vehicles to satisfy the user requirements, and in the later study, the complaints were with staff block booking and preventing the staff group access.

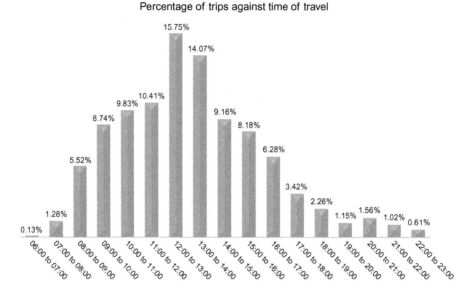

Fig. 13.3.3 Typical travel times and frequency (Milligan, 2016).

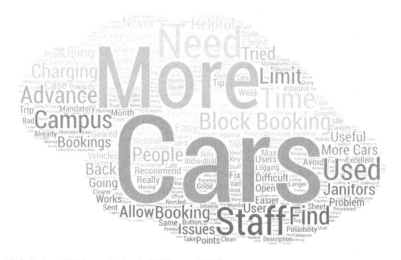

Fig. 13.3.4 Apr. 2015 word cloud (Milligan, 2016).

A study by Thompson et al. (2011) reviewed a demand-responsive transport system (DRTS) and the growing pressure of higher levels of service in transport systems. This present DRTS and the associated financial constraints, increasing the number of vehicles at a reasonable cost, were implemented.

In response to this earlier survey, extra vehicles were introduced to the initiative in an attempt to increase availability across all campuses. This proved to be on the whole

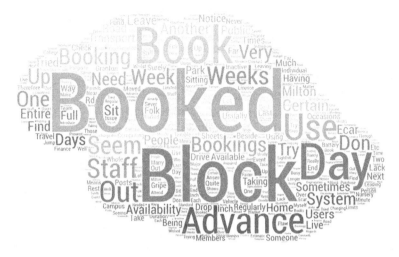

Fig. 13.3.5 Apr. 2016 word cloud (Milligan, 2016).

successful, but with the increased number of vehicles introduced, an increased number of users followed that put a greater focus on the charging infrastructure.

In an attempt to improve on the activity for all users, additional vehicles have been incorporated at each campus, and the maximum time a vehicle can be booked is now restricted by permissions being set on their user profile.

The majority of staff has only single campus user access, and this will be their primary campus. The rationale behind this was to remove any confusion and the requirement to identify vehicle registration numbers as it was a concern that cars may not be returned to the 'correct' campus and the possibility of a campus being without its allocation of vehicles.

The volume of vehicles is currently marginal to the user's needs, and any downtime such as infrastructure failure or vehicle of road for servicing can cause disappointment.

The current study realized that the demand for flexibility through additional vehicles was being requested by personnel reducing their dependency on the 'grey fleet'.

In the public sector, evidence indicates that grey fleet makes up around 57% of the total road mileage. According to the figures by the Office of Government Commerce, across the whole of the sector, this could add up to as much as 1.4 billion miles every year!

Whilst managing the duty of care to employees driving for work is a legal requirement, this includes employees driving their own vehicles for work (Act, 1974).

The Health and Safety at Work Act, 1974 states that

It shall be the duty of every employer to ensure, as far as is reasonably practicable, the health, safety and welfare at work of all employees.

Giving an alternative to ICE vehicles can be a sustainable option as in excess of 400,000 tons of CO_2 are emitted, on average, from grey fleet cars over 1.4 billion public sector miles. This is an annual carbon profile that would take 550,000 UK trees their whole lifetimes to offset (Office of Government Commerce, 2008).

A method of embedding the vehicles into the business infrastructure was determined. It was decided that the battery electric vehicles (BEVs) should be correctly controlled so as to allow maximum staff usage evenly distributed across the four campuses. The newly combined colleges did not have any existing pool cars to allow staff to travel between sites, so the system had to be designed from the ground up.

It is important for businesses to have a successful induction of new products and to realize the key factors to maintain their existence and their future success as indicated by Bayus, Erickson, and Jacobson (2003), and it is agreed that a fundamental important part is that the key factors are realized and the initial period of existence is given the support from the staff that will experience the BEV and its acceptance as a means of corporate transport.

References

Act (1974). Health and safety at work Act 1974, 30(3). Available at: http://www.legislation.gov.uk/ukpga/1974/37/contents.

Alahmer, A., et al. (2011). Vehicular thermal comfort models: A comprehensive review. *Applied Thermal Engineering, 31*(6–7), 995–1002.

Alba, M. (2015). Switched on Scotland : A roadmap to widespread adoption of plug-in vehicles switched on Scotland : A roadmap to widespread adoption of plug-in vehicles.

Andre, M. (1996). *Driving cycle development: Characterisation of the methods.* Warrendale, Pennsylvania: SAE International.

Automobile Association. (2013). Tyre labelling. *Service and Repair.* Available at: http://www.theaa.com/motoring_advice/safety/tyre-label-fuel-noise-grip.html Accessed 30.07.16.

Automobile Association. (2015). Route planning. *The AA.* Available at: http://www.theaa.com/route-planner/index.jsp Accessed 02.10.15.

Automobile Association. (2016). Fuel price report (September 2016). *Experian Catalyst,* (September), 9–10. Available at: http://www.theaa.com/resources/Documents/pdf/motoring-advice/fuel-reports/september2016.pdf Accessed 25.10.16.

Barlow, T., et al. (2009). *A reference book of driving cycles for use in the measurement of road vehicle emissions.* p. 280. Available at: http://www.trl.co.uk/online_store/reports_publications/trl_reports/cat_traffic_and_the_environment/report_a_reference_book_of_driving_cycles_for_use_in_the_measurement_of_road_vehicle_emissions.htm\nhttps://www.gov.uk/government/uploads/system/uploads/att.

Barter, P. (2013). Cars parked 95% of time. *Reinventing Parking.* Available at: http://www.reinventingparking.org/2013/02/cars-are-parked-95-of-time-lets-check.html Accessed 28.03.14.

Bayus, B. L. B., Erickson, G., & Jacobson, R. (2003). The financial rewards of new product introductions in the personal computer industry. *Management Science, 49*(2), 197–210. Available at: http://pubsonline.informs.org/doi/abs/10.1287/mnsc.49.2.197.12741.

Boretti, A. (2013). A fun-to-drive, economical and environmentally-friendly mobility solution. *Journal of Power Technologies, 93*(4), 194–201.

Chaari, H., & Ballot, E. (2012). Fuel consumption assessment in delivery tours to develop eco driving behaviour. pp. 1–12. Accessed 18 October 2016.

CHAdeMO (2007). Characteristics of CHAdeMO quick charging system. 4(October), pp. 818–822.

Chargemasterplc (2016). Chargemaster powering the future. Available at: https://chargemasterplc.com/ Accessed 22.12.15.

Committee on Climate Change (2015). Reducing emissions in Scotland: 2015 progress report. (March), p. 60.

Crown Copyright (2016). Ordnance survey. V3.0. Available at: https://www.ordnancesurvey.co.uk/docs/support/guide-coordinate-systems-great-britain.pdf Accessed 10.01.17.

DEFRA. (2015). Road transport emissions. *Transport Statistics Bulletin*. Available at: https://uk-air.defra.gov.uk/assets/documents/reports/empire/naei/annreport/annrep99/app1_29.html Accessed 28.09.14.

Department of Energy and Climate Change (2015). Energy trends (government publication).

Department of the Environment (2000). New directions in speed management—A review of policy. Transport of the Environment Regions, London.

Ecotricity. (2016). Ecotricity. *Britain's Green Energy*. Available at: https://www.ecotricity.co.uk/for-the-road/our-electric-highway Accessed 20.08.16.

Energy Savings Trust. (2015). ChargePlace Scotland. *Transport Scotland*. Available at: http://www.energysavingtrust.org.uk/scotland/businesses-organisations/transport/electric-vehicles-chargeplace-scotland Accessed 30.05.15.

Erikson, E. H. (2014). Daimler AG—Life cycle. *International Encyclopedia of the Social Sciences*, 9, 286–292.

Esteves-Booth, A., et al. (2001). The measurement of vehicular driving cycle within the city of Edinburgh. *Transportation Research Part D: Transport and Environment*, 6(3), 209–220. Available at: http://linkinghub.elsevier.com/retrieve/pii/S1361920900000249.

European Association for Battery Electric Vehicles. (2009). Energy consumption, CO2 emissions and other considerations related to Battery Electric Vehicles. *Energy*, (April), 1–21.

Experian Catalist, A. U. L. (2015). Fuel price report. (July), p. 1. Available at: http://www.theaa.com/motoring_advice/fuel/.

Farrington, R., & Rugh, J. (2000). Impact of vehicle air-conditioning on fuel economy, tailpipe emissions, and electric vehicle range. *Earth Technologies Forum*, (September). http://www.nrel.gov/docs/fy00osti/28960.pdf. Available at: http://www.smesfair.com/pdf/airconditioning/28960.pdf.

Google (2016). Google Maps. Available at: https://www.google.co.uk/maps Accessed 02.01.16

Hansard & Whitty. (2001). Speedometer accuracy. *Parliament Business*. Available at: http://www.publications.parliament.uk/pa/ld200001/ldhansrd/vo010312/text/10312w01.htm Accessed 01.07.16.

Hawkins, T. R., Gausen, O. M., & Strømman, A. H. (2012). Environmental impacts of hybrid and electric vehicles—A review. *International Journal of Life Cycle Assessment*, 17(8), 997–1014.

Knez, M., Celik, A. N., & Muneer, T. (2014). A sustainable transport solution for a Slovenia town. *International Journal of Low-Carbon Technologies*, 1–7 Available at: http://ijlct.oxfordjournals.org/cgi/doi/10.1093/ijlct/ctu007.

Lane, B. (2015). Next green car. *Next Green Car Limited*. Available at: http://www.nextgreencar.com/emissions/ngc-rating/ Accessed 15.04.15.

Lane, B. (2016). Ford focus emissions. *Next Green Car*. Available at: http://www.nextgreencar.com/view-car/53716/ford-focus-1.6-style-105ps-s6-petrol-manual-5-speed/Accessed 25.10.16.

Milligan, R. (2016). Critical evaluation of the battery electric vehicle for sustainable mobility (PhD thesis).

Milligan, R., & Muneer, T. (2015). A comparative range approach using the Real World Drive Cycles and the Battery Electric Vehicle. *SAE International, 28*, 1–5.

Morrissey, P., Weldon, P., & O'Mahony, M. (2016). Future standard and fast charging infrastructure planning: An analysis of electric vehicle charging behaviour. *Energy Policy, 89*(February), 257–270. Available at: http://www.sciencedirect.com/science/article/pii/s0301421515302159.

MTSU. (2008). Standard deviation. *Finance*, 1–14.

Muneer, T., Clarke, P., & Cullinane, K. (2008). *The electric scooter as a means of green transport*. Edinburgh Napier University.

Neubauer, J., Brooker, A., & Wood, E. (2012). Sensitivity of battery electric vehicle economics to drive patterns, vehicle range, and charge strategies. *Journal of Power Sources, 209*, 269–277. Available at: http://dx.doi.org/10.1016/j.jpowsour.2012.02.107.

Nissan, 2015. *Nissan LEAF Technical*. Training manual NMTN9217AE. Accessed 24 August 2015.

Nissan UK (2016). Nissan LEAF. Available at: https://www.nissan.co.uk/?&cid=psmM9W SXFpD_dc|D Accessed 09.02.13.

Office of Government Commerce (2008). Grey fleet best practice (June), Grey Fleet Stakeholder Forum. Accessed 11 October 2016.

Pearre, N. S., et al. (2011). Electric vehicles: How much range is required for a day's driving? *Transportation Research Part C: Emerging Technologies, 19*(6), 1171–1184. Available at: http://dx.doi.org/10.1016/j.trc.2010.12.010.

RAC. (2012). What is a Euro 6 Diesel. *RAC Motoring Services*. Available at: http://www.rac.co.uk/drive/news/motoring-news/euro-6-and-diesel-vehicles/ Accessed 27.05.14.

Renault UK. (2015). Renault. *Renault Sales Literature*. Available at: https://www.renault.co.uk/vehicles/new-vehicles/zoe.html Accessed 09.06.15.

Rubin, E., & Davidson, C. (2001). *Introduction to engineering and the environment*. McGraw Hill Available at: https://books.google.co.uk/books/about/Introduction_to_Engineering_and_the_Envi.html?id=rrvlAAAACAAJ&redir_esc=y.

Scotland, A. (2011). Reducing Scottish greenhouse gas emissions. *Environment*, (December).

Smith, R. A. (2008). Enabling technologies for demand management: Transport. *Energy Policy, 36*(12), 4444–4448.

Survey Monkey (2015). Survey Monkey. Available at: https://www.surveymonkey.com/mp/lp/sem-lp-5b/?&utm_campaign=UK_Search_Alpha_Brand_1&utm_medium=ppc&cmpid=brand&mobile=0&cvosrc=ppc.google.survey+monkey&adposition=1t1&creative=151735303098&network=g&cvo_adgroup=survey+monkey&cvo_campaign=UK_Search_Alpha_B Accessed 01.01.15.

Taylor, M. A. P. et al. (2010). Planning for electric vehicles—Can we match environmental requirements, technology and travel demand? 12th WCTR, Lisbon, pp. 1–18.

The Long Tail Pipe (2015). Evaluating the full transportation and energy life-cycle. Available at: https://longtailpipe.com/ebooks/green-transportation-guide-buying-owning-charging-plug-in-vehicles-of-all-kinds/gasoline-electricity-and-the-energy-to-move-transportation-systems/the-6-kwh-electricity-to-refine-gasoline-would-drive-an-electric-car-the-sam Accessed 29.11.15.

Thompson, R. G., et al. (2011). Determining the viability of a demand responsive transport system. *Methods*, 5.

Transport Evolved (2016). Transport evolved. Available at: https://transportevolved.com/ Accessed 05.05.16.

Transport Scotland. (2011). Travel and transport in Scotland. *Government Publications.* Available at: http://www.gov.scot/Publications/2011/08/31092528/3 Accessed 15.05.14.

Transport Scotland. (2016). Scottish household survey. *Travel Diary Results.* Available at: http://www.transport.gov.scot/statistics/scottish-household-survey-travel-diary-results-all-editions Accessed 03.04.16.

UK Power Limited. (2015). UK Power. *Power Comparison Publication.* Available at: https://www.ukpower.co.uk/?r=googlectbrand&utm_source=google&utm_medium=cpc&gclid=CLvu5IXT588CFUQz0wod-3IJHw Accessed 28.10.15.

Verhoef, E., et al. (2008). *Pricing in road transport.* Cheltenham: Edward Elgar Available at: https://www.e-elgar.com/shop/pricing-in-road-transport?___website=uk_warehouse.

Weiss, C., et al. (2014). Capturing the usage of the German car fleet for a one year period to evaluate the suitability of battery electric vehicles—A model based approach. *Transportation Research Procedia, 1*(1), 133–141. Available at: http://dx.doi.org/10.1016/j.trpro.2014.07.014.

Wu, G., Inderbitzin, A., & Bening, C. (2015). Total cost of ownership of electric vehicles compared to conventional vehicles: A probabilistic analysis and projection across market segments. *Energy Policy, 80,* 196–214.

Zap-Map (2016). Zap-Map statistics. Available at: https://www.zap-map.com/statistics/#region Accessed 02.02.16.

Index

Note: Page numbers followed by *f* indicate figures, *t* indicate tables, and *b* indicate boxes.

Printed in the United States
By Bookmasters